Communities for Clean Energy Justice and Equity in Grid Modernization

IEEE Press
445 Hoes Lane
Piscataway, NJ 08854

Communities for Clean Energy Justice and Equity in Grid Modernization

Edited by

Mohammadreza Daneshvar
University of Tabriz
Iran

Behnam Mohammadi-Ivatloo
LUT University
Finland

Amjad Anvari-Moghaddam
Aalborg University
Denmark

IEEE Press Series on Power and Energy Systems
Ganesh Kumar Venayagamoorthy, Series Editor

IEEE Press
WILEY

Published by John Wiley & Sons, Inc., Hoboken, New Jersey.
Published simultaneously in Canada.

Library of Congress Cataloging-in-Publication Data:

Names: Daneshvar, Mohammadreza, editor. | Mohammadi-Ivatloo, Behnam, editor. | Anvari-Moghaddam, Amjad, editor.
Title: Communities for clean energy justice and equity in grid modernization / edited by Mohammadreza Daneshvar, University of Tabriz, Iran, Behnam Mohammadi-Ivatloo, LUT University, Finland, Amjad Anvari-Moghaddam, Aalborg University, Denmark.
Other titles: IEEE Press series on power and energy systems.
Description: Hoboken, NJ : Wiley, [2025] | Series: IEEE Press series on power and energy systems | Includes bibliographical references and index.
Identifiers: LCCN 2025000013 (print) | LCCN 2025000014 (ebook) | ISBN 9781394265718 (hardback) | ISBN 9781394265732 (adobe pdf) | ISBN 9781394265725 (epub)
Subjects: LCSH: Renewable energy sources. | Sustainability. | Environmental justice. | Energy facilities.
Classification: LCC TJ808 .C67 2025 (print) | LCC TJ808 (ebook) | DDC 304.2/8–dc23/eng/20250304
LC record available at https://lccn.loc.gov/2025000013
LC ebook record available at https://lccn.loc.gov/2025000014

Cover Design: Wiley
Cover Image: © Justin Paget/Getty Images

Set in 9.5/12.5pt STIXTwoText by Straive, Chennai, India

Contents

About the Editors

Mohammadreza Daneshvar, PhD, is an Assistant Professor, Founder, and Head of the Laboratory of Multi-Carrier Energy Networks Modernization at the Faculty of Electrical and Computer Engineering, University of Tabriz, Tabriz, Iran. Prior to that, he was a postdoctoral research fellow in the field of modern multi-energy networks at the Smart Energy Systems Laboratory of the University of Tabriz for two years. He obtained his MSc and PhD in Electrical Power Engineering from the University of Tabriz, Tabriz, Iran, all with honors. He has (co)authored more than 50 technical journal and conference articles, 9 books, 28 book chapters, and 10 national and international research projects in the field. Dr. Daneshvar is a member of the Editorial Board of the *Energy and Built Environment* journal and the Early Career Editorial Board of the *Sustainable Cities and Society* journal. He also served as the Guest Editor for the *Sustainable Cities and Society* and *Sustainable Energy Technologies and Assessments* journals. Moreover, he serves as an active reviewer for more than 90 top journals of the IEEE, Elsevier, Springer Nature, Wiley, Taylor & Francis, and IOS Press and was ranked among the top 1% of reviewers in Engineering and Cross-Field disciplines based on Publons global reviewer database. His research interests include smart grids, transactive energy, energy management, renewable energy sources, integrated multi-energy systems, grid modernization, electrical energy storage systems, sustainable cities and society, microgrids, energy hubs, machine learning and deep learning, digital twin, and optimization techniques and AI.

Dr. Behnam Mohammadi-Ivatloo, PhD, is a Professor of Sector Coupling in Energy Systems at LUT University, Lappeenranta, Finland. He has a mix of high-level experience in research, teaching, administration, and voluntary jobs at the national and international levels. He was a Principal Investigator (PI) or C-PI for more than 20 externally funded research projects, including grants from EU Horizon and Business Finland. He is a Senior Member of IEEE since 2017 and a Member of the Governing Board of the Iran Energy Association since 2013, where he was elected as President in 2019. He is also an Editor for *IEEE Transactions on Power Systems* and *IEEE Transactions of Transportation Electrifications*. His research interests include integrated energy systems, sector coupling, renewable energies, energy storage systems, microgrids, and smart grids.

Amjad Anvari-Moghaddam (S'10–M'14–SM'17) is a Full Professor and Leader of the Intelligent Energy Systems and Flexible Markets (iGRIDS) Research Group at the Department of Energy (AAU Energy), Aalborg University, where he is also serving as the Vice-Head of Department (Research) and the Coordinator of Integrated Energy Systems Laboratory (IES-Lab). His research interests include planning, control and operation management of microgrids, renewable/hybrid power systems and integrated energy systems with appropriate market mechanisms. He has (co-)authored more than 300 technical articles, 7 books, and 17 book chapters in the field. Dr. Anvari-Moghaddam serves as a Member of the Council for Energy Efficient Transition at the Danish Ministry of Climate, Energy and Utilities, a Member of Reference Group 5 "Climate, Energy and Mobility" for Horizon Europe (HE) at the Danish Agency for Education and Research (UFS), a Board Member of Energy Mission at Aalborg University, and the Chair of the IEEE Denmark Section. Additionally, he holds the position of Editor-in-Chief for *Academia Green Energy* journal and has served as Associate Editor for several leading journals such as the *IEEE Transactions on Power Systems*, *IEEE Systems Journal*, and *e-Prime* Elsevier. He is an active member of IEC Standardization Subcommittees 8B-WG3, WG6, and WG8 as well as of the European Committee for Electrotechnical Standardization (CENELEC/SR-73, TC-8X, TC-95X, and SR-120). Furthermore, he has contributed as a technical committee member to various IEEE PES/IES/PELS and CIGRE working groups. He was the recipient of 2020 and 2023 DUO–India and SPARC Fellowship Awards, DANIDA Research Fellowship grant from the Ministry of Foreign Affairs of Denmark in 2018 and 2021, IEEE-CS Outstanding Leadership Award 2018 (Halifax, Nova Scotia, Canada), and the 2017 IEEE-CS Outstanding Service Award (Exeter-UK).

List of Contributors

Peyman Afzali
Department of Energy (AAU Energy)
Aalborg University
Aalborg
Denmark

Muzaffar Ali
Islamabad Electric Supply Company
Islamabad
Pakistan

Saqib Ali
National Fertilizer Company
Multan
Pakistan

and

Postdoc Fellow
Zagreb University
Croatia

Banafsheh Alipour
Department of Electrical Engineering
Shahid Bahonar University of Kerman
Kerman
Iran

Amjad Anvari-Moghaddam
Department of Energy (AAU Energy)
Aalborg University
Aalborg
Denmark

Hamidreza Arasteh
Power Systems Operation and
Planning Research Department
Niroo Research Institute
Tehran
Iran

and

Center for Renewable Energy and
Microgrid
Huanjiang Laboratory
Zhejiang University
Zhejiang
China

Arash Asrari
Department of Electrical and
Computer Engineering
Purdue University Northwest
Hammond
IN
USA

Salah Bahramara
Department of Electrical Engineering
Sanandaj Branch
Islamic Azad University
Sanandaj
Iran

Mahdi Bedrood
Department of Electrical Engineering
Shahid Bahonar University of Kerman
Kerman
Iran

Monica Bhatia
School of Business Management
NMIMS (Deemed to be University)
Indore
India

and

School of Technology, Management and Engineering
NMIMS (Deemed to be University)
Indore
India

Milad Biazar
Tavanir Company
Deputy of Power Distribution
Tehran
Iran

Emre Çakır
R&D Department
Karsan Automotive
Bursa
Turkey

Pradyumn Chaturvedi
EED
VNIT
Nagpur
India

Mohammadreza Daneshvar
Laboratory of Multi-Carrier Energy Networks Modernization
Faculty of Electrical and Computer Engineering
University of Tabriz
Tabriz
Iran

Rahman Dashti
Clinical-Laboratory Center of Power System & Protection
Faculty of Intelligent Systems Engineering and Data Science
Persian Gulf University
Bushehr
Iran

Sobhan Dorahaki
Department of Electrical Engineering
Qatar University
Doha
Qatar

Asghar Akbari Foroud
Electrical and Computer Engineering Faculty
Semnan University
Semnan
Iran

Reza Gharibi
Clinical-Laboratory Center of Power System & Protection
Faculty of Intelligent Systems Engineering and Data Science
Persian Gulf University
Bushehr
Iran

Homayoun Ghasemnejad
Department of Electrical Engineering
Shahid Bahonar University of Kerman
Kerman
Iran

Mohammad Hasan Ghodusinejad
School of Renewable Energies and Environmental Engineering
College of Interdisciplinary Science and Technologies
University of Tehran
Tehran
Iran

Mahyar Gholizadeh
Tavanir Company
Deputy of Power Distribution
Tehran
Iran

Nastaran Gholizadeh
Electrical and Computer Engineering
University of Alberta
Edmonton
Alberta
Canada

Furkan Hanilçi
R&D Department
Karsan Automotive
Bursa
Turkey

Sogand Hosseinalipour
Department of Electrical Engineering
Shahid Bahonar University of Kerman
Kerman
Iran

Akhtar Hussain
Department of Electrical and Computer Engineering
Laval University
Quebec City
Quebec
Canada

Noon Hussein
Electrical and Computer Engineering
University of Alberta
Edmonton
Alberta
Canada

Zohreh Kaheh
School of Technology and Innovations
University of Vaasa
Vaasa
Finland

Vikas Khare
School of Technology, Management and Engineering
NMIMS (Deemed to be University)
Indore
India

and

School of Business Management
NMIMS (Deemed to be University)
Indore
India

Yiğit Çağatay Kuyu
R&D Department
Karsan Automotive
Bursa
Turkey

Sheraz Azam Mirza
National Grid
Birmingham
UK

Sahar Mobasheri
Department of Electrical Engineering
Shahid Bahonar University of Kerman
Kerman
Iran

Niki Moslemi
Power Systems Operation and Planning Research Department
Niroo Research Institute
Tehran
Iran

Petr Musilek
Department of Electrical and Computer Engineering
University of Alberta
Edmonton
Alberta
Canada

Ehsan Naderi
Department of Electrical Engineering
College of Engineering and Computer Science
Arkansas State University
Jonesboro
AR
USA

Amir Naseri
School of Energy Engineering and Sustainable Resources
College of Interdisciplinary Science and Technology
University of Tehran
Tehran
Iran

Younes Noorollahi
School of Energy Engineering and Sustainable Resources
College of Interdisciplinary Science and Technology
University of Tehran
Tehran
Iran

Setareh Peirov
School of Renewable Energies and Environmental Engineering
College of Interdisciplinary Science and Technologies
University of Tehran
Tehran
Iran

Iman Rahmati
Electricity Market Monitoring & Studies
Iran Grid Management Company
Tehran
Iran

Arash Rajaei
Department of Electrical Engineering
Shahid Bahonar University of Kerman
Kerman
Iran

Masoud Rashidinejad
Department of Electrical Engineering
Shahid Bahonar University of Kerman
Kerman
Iran

Aamer Raza
Islamabad Electric Supply Company
Islamabad
Pakistan

Marziyeh Razeghi
School of Energy Engineering and Sustainable Resources
College of Interdisciplinary Science and Technology
University of Tehran
Tehran
Iran

Ashkan Safari
Faculty of Electrical and Computer Engineering
University of Tabriz
Tabriz
Iran

Miadreza Shafie-khah
School of Technology and Innovations
University of Vaasa
Vaasa
Finland

Behrooz Vahidi
Department of Electrical Engineering
Amirkabir University of Technology (Tehran Polytechnic)
Tehran
Iran

Hossein Yousefi
School of Renewable Energies and Environmental Engineering
College of Interdisciplinary Science and Technologies
University of Tehran
Tehran
Iran

and

School of Energy Engineering and Sustainable Resources
College of Interdisciplinary Science and Technology
University of Tehran
Tehran
Iran

Preface

Nowadays, the significant development of energy systems along with the changes in people's lifestyles has led to the day-by-day growth of energy consumption in its diverse types. Indeed, energy has become an essential tool for completing different processes in people's daily work and is considered as one of the prominent welfare commodities for humanity around the world. In such a circumstance, providing continuous and fair access to multi-carrier energy for consumers across various geographical places will be of great importance in realizing an appropriate degree of convenience for them. In this regard, energy justice and equity have emerged as crucial analytical tools for energy developers and policymakers aiming to drive energy plans and decisions toward a modern energy community with fair and equal access to multi-carrier energy. Energy justice and equity are developed to bridge informed energy choices for consumers by advancing decision-making tools that assist energy planners in achieving grid modernization goals. As endeavors in modernizing multi-energy networks are going to make the energy infrastructure clean with zero-carbon, creating a suitable structure has become a critical step in realizing modern energy grids. In this regard, multi-energy communities can play a holistic energy structure role in grid modernization due to their capabilities in procuring tremendous advantages and user-friendly services to end-users. Such features have made multi-energy communities susceptible for implementing energy justice and equity schemes in modern energy grids. However, how multi-energy communities can simultaneously promote carbon-free energy production and distribution, as well as welcome energy justice and equity programs, is a key question that needs to be addressed in deep detail. As the realization of energy justice and equity is an utmost requirement for public satisfaction and welfare, there is a great need for coordinating and implementing the relevant schemes in the clean multi-energy community structure in line with grid modernization objectives. As a pioneering book in the field, it covers energy justice and equity plans for grid modernization, as well as presents innovative ways for coordinating and implementing them in sustainable and clean energy communities with a high/full share of renewable energy sources. Moreover, this book states fundamental theories, technologies, and solutions for real-world problems in the operation and planning of clean energy communities for energy justice and equity. It also introduces advanced methods for the deployment of state-of-the-art technologies to realize energy justice and equity goals in the modern structure of energy grids under the cleaner multi-energy mix architectures.

The current book consists of 19 chapters. Chapter 1 analyzes energy justice and equity for the sustainable development of modern energy grids. It clarifies the real place of energy justice and equity in the suitable development of future modern multi-carrier energy networks. Chapter 2 emphasizes the symbiotic relationship between energy justice and grid modernization. This chapter highlights a comprehensive approach that combines technological advancements with social equality, laying the foundation for a more inclusive and sustainable energy future. Chapter 3 inspects the development and application of decision-making tools designed to enhance the integration of principles of justice and equity into the strategic planning and execution of clean energy projects. It also underscores the significance of leveraging innovative tools to navigate the complexities of decision-making, fostering a holistic approach that prioritizes social, economic, and environmental considerations. Chapter 4 delves into the critical role of multi-energy communities in shaping a holistic energy infrastructure for grid modernization. It also proposes innovative approaches to provide practical insights into the coordination and implementation of principles within sustainable and clean energy communities considering a high share of renewable energy sources. Chapter 5 focuses on equity-oriented energy allocation in mixed and underprivileged communities. It mostly addresses energy equity issues that arise when allocating energy among community members in different cases under the high penetration of renewables. Chapter 6 investigates the promotion of energy equity and justice among the participants in developing a fair multi-energy community and environment. Moreover, it discusses the structure, characteristics, and schemes of a multi-energy system that serves all the key players on a fair basis to promote justice among them. Chapter 7 reveals various meanings and definitions of the energy community, particularly when integrating social science aspects with technical, environmental, and economic factors. It also presents an optimal model for energy communities, where justice in peer-to-peer exchanges is evaluated to ensure performance accuracy. Chapter 8 explores the role of energy storage systems for energy justice and equity. It conducts this assessment by considering the systematic use of big data methods to ensure that energy storage remains on equal and fairgrounds. Chapter 9 examines energy trading mechanisms for energy justice and equity as well as reviews the tradability of energy carriers and their relation to energy justice. Furthermore, it presents the interaction between energy justice and energy-related trading mechanisms as a multi-faceted nexus that covers inter alia, energy access, energy affordability, energy subsidy, and energy efficiency. Chapter 10 discusses the position of peer-to-peer energy communities and their importance in expanding social welfare and escalating energy justice. It also covers the key concepts and required theory and validates them by implementing an analytical study. Chapter 11 delves into the multifaceted dimensions of sustainable development within the realm of energy communities, with a specific focus on fostering energy justice and equity. It aims to provide insights into how sustainable energy development can be harnessed as a catalyst for broader societal equity by examining case studies, policy frameworks, and technological interventions. Chapter 12 scrutinizes the nuanced discourse on energy justice and the electrification of both urban and rural areas. Moreover, it assesses the concept of energy justice and its associated challenges, emphasizing its importance for promoting social and economic development in these regions. Chapter 13 concentrates on the imperative research topic of enhancing social justice in electrical energy distribution by proposing practical solutions within distribution companies. It also underscores the pivotal role of justice in decision-making processes within the electricity industry, with a specific emphasis on its impacts on customers. Chapter 14 investigates the integration of artificial intelligence (AI) in the energy sector, with a focus on its role in promoting justice and equity. It also clarifies the impact of AI on the energy transition and highlights opportunities and challenges in revolutionizing planning, operation, control, and maintenance. Chapter 15 covers AI-based marketing analysis for realizing renewable energy justice and equity. It provides a clear vision for the transformative potential of AI in marketing analysis as a catalyst for realizing renewable energy justice and equity. Chapter 16 states the impact of machine learning (ML) and AI on energy equity investment prioritization. It also clarifies the potential and challenges associated with integrating ML and AI in equitable and sustainable energy investment prioritization techniques. Chapter 17 intends the vulnerability of communication platforms to malicious cyberattacks as a critical challenge in the operation of modern-day electricity networks. It develops a remedial action scheme that implements AI to mitigate different combinations of unserved energy and voltage violations caused by a coordinated false data injection attack. Chapter 18 focuses on the policy dimensions of energy justice and equity. Moreover, it undertakes a comprehensive examination of the existing policy landscape in this domain to illuminate the current state of affairs. Ultimately, Chapter 19 examines the current status of various global regions concerning energy justice and equity. It reviews different regions for coordinated collective actions in terms of experience sharing and regional cooperation based on geographical criteria.

As any research achievement may not be free of gaps, the Editors kindly welcome any suggestions and comments from respectful readers to improve the quality of this work. Interested readers can share their valuable comments with the Editors via mohammadrezadaneshvar95@gmail.com.

Editors

Mohammadreza Daneshvar
Laboratory of Multi-Carrier Energy Networks Modernization, Faculty of Electrical and Computer Engineering, University of Tabriz, Tabriz, Iran

Behnam Mohammadi-Ivatloo
Department of Electrical Engineering, School of Energy Systems, LUT University, Lappeenranta, Finland

Amjad Anvari-Moghaddam
Department of Energy (AAU Energy), Aalborg University, Aalborg, Denmark

1

Energy Justice and Equity for the Sustainable Development of Modern Energy Grids

Mohammadreza Daneshvar[1] and Amjad Anvari-Moghaddam[2]*

[1] *Laboratory of Multi-Carrier Energy Networks Modernization, Faculty of Electrical and Computer Engineering, University of Tabriz, Tabriz, Iran*
[2] *Department of Energy (AAU Energy), Aalborg University, Aalborg, Denmark*

1.1 Introduction

Nowadays, global development has found energy to be a vital requirement for realizing scheduled plans in different areas. This trend highlights a necessary alarm of the critical need for purposeful upgrading of energy networks due to the day-by-day growth of energy demand in its various carriers [1]. How this change needs to be conducted that not only caters to the multi-energy load but also delivers carbon-free and affordable energy to the consumer side is a key question in the expansion pass of modern energy grids. To address it, diverse endeavors are formed around the world to assess practical plans for the successful switching from the current centralized traditional systems to the modern zero-emission and cost-effective energy units [2]. In this respect, the universal trend toward clean energy is solidified by unveiling the importance of reliable access to environmentally friendly energy. The prominence of this issue was mostly identified when 193 members of the United Nations signed the agreement for sustainable development in 2015 [3]. In the seventh goal of sustainable development, the main motivation is "to ensure access to affordable, reliable, sustainable and modern energy for all" by 2030 [3]. However, the realization of this commitment is back to the availability of the financial budgets and accessibility to the required technologies [4]. The absence of the mentioned two key factors has deprived some countries of this development. Hence, how sustainable development in the energy world can be feasible for all countries, particularly for those that just have limited access to the required budgets and technologies, is an uncertain issue in the process of implementation of the United Nations pledge. Herein, another

*Corresponding author: m.r.daneshvar@tabrizu.ac.ir.

Communities for Clean Energy Justice and Equity in Grid Modernization, First Edition.
Edited by Mohammadreza Daneshvar, Behnam Mohammadi-Ivatloo, and Amjad Anvari-Moghaddam.

issue is properly distinguishing sustainable development efforts and endeavors for reinforcing energy privilege [5]. What can be done here is to explore the different dimensions of energy justice and equity in various layers of sustainable development. In this respect, recent literature has various perspectives on the effects of energy justice and equity in the transition to renewable energy. For instance, energy justice and equity are crucial for human development in [6] given its necessity for establishing social justice by procuring efficient and clean energy services for consumers. In another work [7], the impact of community acceptance of the planning of offshore grid connections on energy justice is examined. It is analyzed to find a practical way to rapidly implement renewable energy systems for clean energy supply. How artificial intelligence can improve grid flexibility and energy efficiency and how this improvement can positively affect energy justice and equity were the main questions in [8] that are conducted by exerting different methods to remove the related barriers for automated smart homes. Herein, the key question can be dedicated to the accessibility of various types of energy, which is why some consumers don't have access to reliable and affordable energy. Yet, the technology of multifarious decentralized systems has been mature enough to be deployed in different conditions for supplying energy. This can be known as the prominent dimension of energy justice and equity because it has a direct connection with the convenience of end users. Therefore, the development of modern energy networks is expected to be conducted in line with the realization of energy justice and equity plans in their different dimensions. The current chapter aims to scrutinize the interconnections between the sustainable development of modern multi-carrier energy networks and the feasibility of energy justice and equity schemes with the aim of providing an appropriate view for the purposeful development of future multi-energy networks.

1.2 Energy Justice and Equity for Sustainable Development

Recently, most contemporary debates in the energy world have revolved around specific topics in the domain of energy poverty alleviation, climate change decline, security, and reliability in energy access [9]. The practical plans for energy justice and equity can improve the accessibility to different dimensions of energy for all end users, reducing energy poverty and declining climate change by promoting the high usage of carbon-free energy systems. As the related plans need to be designed to provide better access to energy, it is expected that energy justice and equity enhance the security and reliability of energy delivery in energy grids. Energy has become the undeniable principle of realizing different goals of sustainable development leading to opening new questions in the context of linking sustainable development criteria and energy justice and equity. Herein, the seventh goal of sustainable development is formed to clarify and raise public awareness that access to affordable and clean energy is an important universal right [9]. In this respect, embedded rights refer to those actions that protect communities and individuals from energy-oriented harms and burdens in equitable access to energy [10]. Social, environmental, and economic rights are taken into account as the three most important domains of such actions, and their features are illustrated in Figure 1.1 [10].

1.2.1 Economic Rights

According to Figure 1.1, from the dimension of economic rights, the protection from energy-oriented threats is targeted at communities and individuals to provide economic livelihood and well-being for them. Two criteria can be considered for the assessment of economic rights. The first one refers to the development of

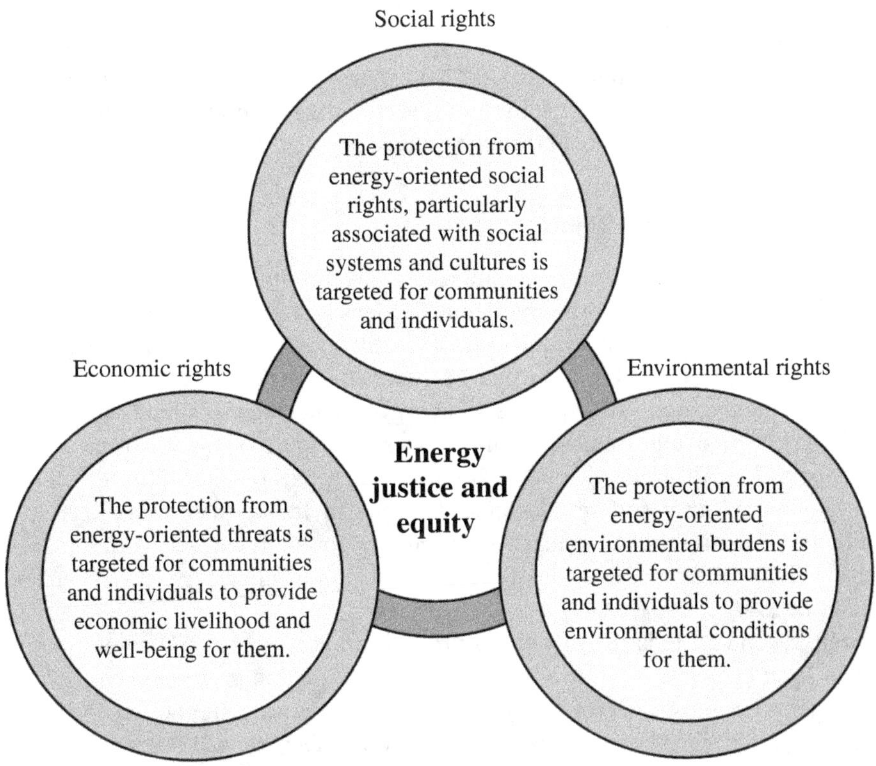

Figure 1.1 Three main domains of energy justice and equity as embedded rights. Source: Ciplet [10]/with permission of Elsevier.

energy projects that are not created to intend debt, economic exploitation, and predatory lending schemes as the main objective. This issue is more important for energy justice and equity as there are many energy systems that are basically developed in a way that relies on different payments from consumers [11]. This can be addressed by concentrating on the usage of various technologies of renewable energy systems as decentralized systems in geographical areas to meet the related end users [12]. Such actions will not only be cost-effective but also will have no harmful effects on the environment. Indeed, decentralized energy systems offer a variety of energy production options that enable the system to act cost-effectively in managing energy harvesting and serving that increase the system's flexibility in the control and management of energy sources. The distributed nature of decentralized energy systems in generating and supplying energy distinguishes them from centralized systems in terms of clean, affordable, reliable, and flexible energy serving.

The second important thing in the economic right is the protection of communities and individuals from the displacement of livelihoods through finance and energy projects. The displacement of livelihoods can negatively affect the achievement of some opportunities for special groups, such as women in finding appropriate work in different sectors [13]. Hence, proper distribution of energy systems and purposeful exploitation of diverse energy sources can help protect livelihoods from displacement through finance and energy projects.

1.2.2 Environmental Rights

As Figure 1.1 indicates, from the dimension of environmental rights, the protection from energy-oriented environmental burdens is targeted at communities and individuals to provide environmentally friendly conditions for them. Similar to economic rights, two criteria can also be intended for scrutinizing environmental rights. The avoidance of exposure to energy-oriented hazards, emissions, and risks is the first one that especially implies the rights of marginalized social groups as they disproportionately experience such hazards [10]. Indeed, energy development can result in ecological burdens in the form of emissions and different negative health impacts, particularly for marginalized communities that may be affected by various factors, such as indigeneity, age, class, gender, and race [14, 15].

The second issue is the involuntary resettlement of people and communities from local ecologies and land that should not occur due to development plans for energy systems. This concern also encompasses privatization or enclosure of common or public resources that may negatively affect the well-being and rights of individuals. Relying on decentralized clean energy systems in the sustainable development of modern energy networks can significantly alleviate the need for involuntary resettlement and displacement in the implementation process of energy systems resulting in not violating the environmental rights of communities and realizing energy justice and equity.

1.2.3 Social Rights

From the dimension of social rights, the protection from energy-oriented social rights is targeted for communities and individuals to provide energy-associated rights. These energy-associated social rights can be considered for different parts of the social layer, such as structures of resistance, social systems, cultures, and local contexts. Herein, three criteria can be assessed for realizing social rights. The first one necessitates the full recognition of livelihood practices, resource management, and local land tenure in designing the framework of energy-related social rights [10]. On the other hand, as different harms and displacement of energy projects back to the lack of sufficient recognition from local land tenure systems in most cases, explicit attention should be spent on heightened vulnerability related to various dynamics of inequality such as gender. The second issue implies protection from the desecration of social structures, cultural and social values, and landscapes [10]. In this context, energy projects can play a beneficial role in severing communities and people from their social structures, cultures, values, and landscapes. Eventually, the third issue is the assurance of the protection of communities and individuals who resist energy projects from violence, criminalization, coercion, manipulation, and tactics of repression in the process of energy project development [10]. Herein, powerful interests have applied several tactics to secure quell and land resistance to different energy projects.

1.3 Energy-Poverty Nexus in Energy Justice and Equity

One of the important aspects of energy justice and equity schemes is to address energy poverty and inequality in supplying and accessing diverse carriers of energy for different groups of consumers. Thus, the energy-poverty nexus is to delve into various aspects of the relationships between energy and poverty. From the viewpoint of socio-energy systems, it is defined as negative feedback and the ensemble of complexes that exacerbate and reinforce poverty in accessing multi-carrier energy, creating inconvenient conditions for communities, households, and individuals [16, 17]. Energy poverty can refer to diverse domains of energy accessibility, but its main reason is the lack of sufficient access to energy and related sources that are essential for the convenience of consumers [18, 19]. There can be multifarious reasons for energy poverty. For instance, it can come

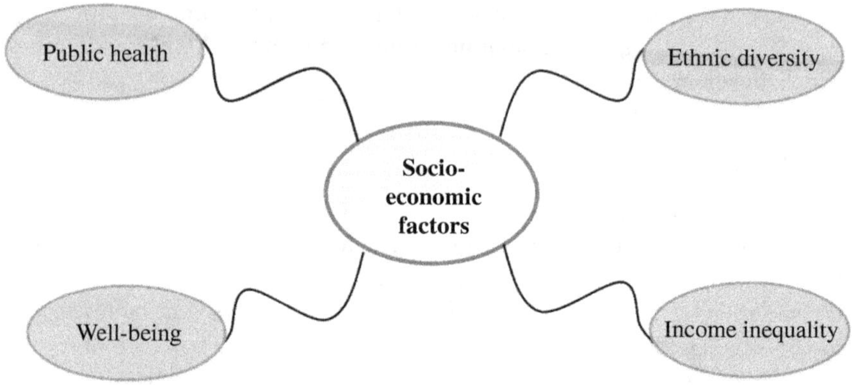

Figure 1.2 Socio-economic factors in dealing with energy poverty. Source: Adapted from [23–25].

from technological limitations and, as a result, being deprived of decent energy services [20]. In wealthy countries, energy poverty may indicate the inability to efficiently pay for energy, particularly for low-income end users due to the high energy burden [21, 22]. Therefore, socio-economic reasons can affect the nexus of energy and poverty and can be used in sustainable development schemes for modern energy networks. In this respect, Figure 1.2 illustrates socio-economic factors in dealing with energy poverty [23–25]. According to this figure, public health is one of the important socio-economic factors that refers to easy access to health services in a cost-effective manner. On the other hand, ethnic diversity can affect access to energy from both social and economic viewpoints that can be treated under the energy justice and equity frameworks. Delivering continuous energy not only possesses positive social aspects but also its cost-effective access can improve the well-being of consumers. This issue directly influences the condition of fair energy distribution even for those consumers who suffer from income inequality.

In the grid modernization, socio-economic plans require empowering the system in a way to improve the accessibility index to public health. As accessibility is key for such plans in the primary step, free access to such services is also targeted as the secondary stage. By considering ethnic diversity, the mentioned plans will be in line of delivering different services under equity conditions. Indeed, inputting this factor into the development phase of socio-economic plans allows for promoting the most equitable strategies in serving energy to various consumers. The other key layer of this advancement relates to taking the well-being factor into account by benefiting a set of capable ways for the continuous energy supply. Such ways will also be most effective and complete when they simultaneously consider the income inequality in uninterrupted energy serving by adopting purposeful energy schemes and tariffs.

In grid modernization, sustainability is recognized as a critical issue, especially for those energy networks that are penetrated with fully renewable systems to benefit both economic and environmental achievements while bringing public satisfaction and well-being [26]. To this end, the development factors for modernizing future energy networks are scheduled to be aligned with the realization of sustainability in renewable-dominant energy structures. In this regard, taking appropriate development parameters for grid sustainability can directly affect the status of energy poverty and promote fair energy supply and access for various groups of consumers in different geographical regions. Herein, various development parameters that affect energy poverty are demonstrated in Figure 1.3 [23, 27–31].

According to the information presented in Figure 1.3, energy poverty has severe impacts on critical domains of development. The dependency of most daily activities on diverse carriers of energy has become its accessibility and continuously availability a critical requirement for the subjective wellbeing. Indeed, its affordable availability brings satisfactory conditions for consumers and avoids physical violence that can come from inappropriate completing commitments due to the energy shortages or cuttings. On the other hand, energy poverty can directly threaten public health in terms of physical and mental as due to the existence of high reliance between the well-being factors and accessibility to energy. This is

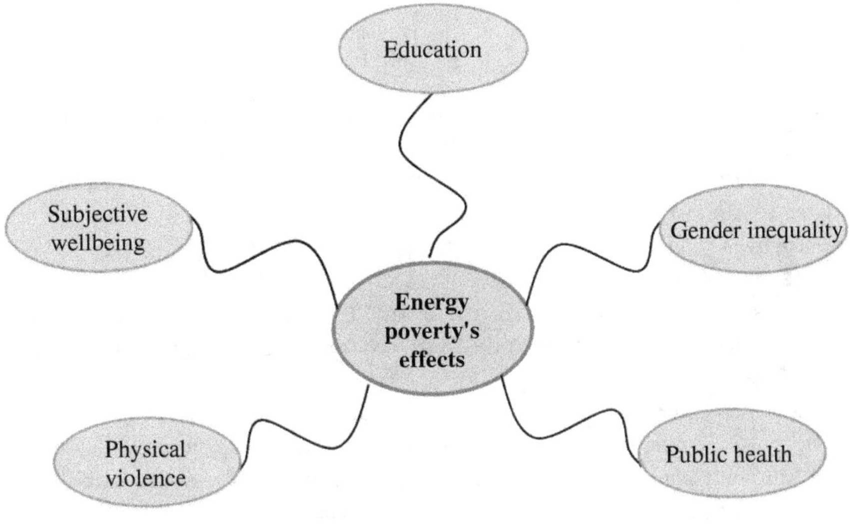

Figure 1.3 Energy poverty's effects on development parameters. Source: Adapted from [23, 27–31].

while its influence scope is not restricted to the above-mentioned areas, and energy poverty can affect education and gender inequality. In other words, its justice availability can assist in growing the education quantity and quality as well as equality supply, avoiding possible inequalities in access to energy.

1.4 Summary

Energy justice and equity have attracted substantial attention in the sustainable development of modern energy grids due to the importance of delivering fair and affordable multi-energy to various groups of end users in different geographical regions. To discuss this, the current chapter analyzes the effects of energy justice and equity in the sustainable development of modern energy networks. Economic, environmental, and social rights were assessed as three main domains of energy justice and equity in meeting sustainable development goals as embedded rights. To clarify various aspects of the relationships between energy and poverty, the chapter delved into the energy-poverty nexus in energy justice and equity by focusing on socio-economic factors in dealing with energy poverty and making energy poverty's effects on development parameters clear. The main achievement is the clarification of how energy justice and equity affect the sustainable development goals and drive them to meet the grid modernization goals in supplying fair and affordable energy to all consumers with various conditions.

References

1 Daneshvar, M., Mohammadi-Ivatloo, B., and Zare, K. (2022). A fair risk-averse stochastic transactive energy model for 100% renewable multi-microgrids in the modern power and gas incorporated network. *IEEE Transactions on Smart Grid* 14 (3): 1933–1945.

2 Burdyny, T. and Dam, B. (2023). Fossil-fuel funding could speed energy transition. *Nature* 613 (7945): 629.

3 Nations, U. (2015). *Transforming Our World: The 2030 Agenda for Sustainable Development*, vol. 1, 41. New York: United Nations, Department of Economic and Social Affairs.

4 Association internationale pour l'évaluation du rendement scolaire (2017). *Energy Access Outlook 2017: From Poverty to Prosperity*. IEA, https://scholar.google.com/scholar?hl=en&as_sdt=0%2C5&q=Energy+Access+Outlook+2017%3A+From+Poverty+to+Prosperity.+IEA&btnG=.

5 Daneshvar, M., Mohammadi-Ivatloo, B., and Zare, K. (2023). An innovative transactive energy architecture for community microgrids in modern

multi-carrier energy networks: a Chicago case study. *Scientific Reports* 13 (1): 1529.

6 Chang, Jeffrey. (2018) "The role of the Clean Development Mechanism in contributing to energy justice." PhD diss.

7 Saalfrank, Lara (2023), "Energy Justice in Offshore Grid Connection Planning Processes A Case Study on the 'Seetrassen 2030'". Master thesis, The University of Groningen.

8 Stumpf, R. (2021). Can automated smart-homes increase energy efficiency and grid flexibility?-A case study of Stavanger, Norway investigating barriers and justice implications. Master thesis. Uis.

9 Islar, M., Brogaard, S., and Lemberg-Pedersen, M. (2017). Feasibility of energy justice: exploring national and local efforts for energy development in Nepal. *Energy Policy* 105: 668–676.

10 Ciplet, D. (2021). From energy privilege to energy justice: a framework for embedded sustainable development. *Energy Research & Social Science* 75: 101996.

11 Barrie, J. and Cruickshank, H.J. (2017). Shedding light on the last mile: a study on the diffusion of Pay As You Go Solar Home Systems in Central East Africa. *Energy Policy* 107: 425–436.

12 Daneshvar, M., Ivatloo, B.M., Zare, K. et al. (2020). A stochastic transactive energy model for optimal dispatch of integrated low-carbon energy hubs in the incorporated electricity and gas networks. *2020 International Conference on Smart Grids and Energy Systems (SGES)*, 568–573. IEEE.

13 Thukral, E.G. (1996). Development, displacement and rehabilitation: locating gender. *Economic and Political Weekly* 31 (24): 1500–1503.

14 Lerner, S. (2006). *Diamond: A Struggle for Environmental Justice in Louisiana's Chemical Corridor*. MIT Press.

15 Sze, J. (2006). *Noxious New York: The Racial Politics of Urban Health and" Environmental Justice*. MIT Press.

16 Biswas, S., Echevarria, A., Irshad, N. et al. (2022). Ending the energy-poverty nexus: an ethical imperative for just transitions. *Science and Engineering Ethics* 28 (4): 36.

17 Miller, C.A., Richter, J., and O'Leary, J. (2015). Socio-energy systems design: a policy framework for energy transitions. *Energy Research & Social Science* 6: 29–40.

18 Jenkins, K., McCauley, D., Heffron, R. et al. (2016). Energy justice: a conceptual review. *Energy Research & Social Science* 11: 174–182.

19 González-Eguino, M. (2015). Energy poverty: an overview. *Renewable and Sustainable Energy Reviews* 47: 377–385.

20 Pachauri, S., Mueller, A., Kemmler, A., and Spreng, D. (2004). On measuring energy poverty in Indian households. *World Development* 32 (12): 2083–2104.

21 Lewis, J., Hernández, D., and Geronimus, A.T. (2020). Energy efficiency as energy justice: addressing racial inequities through investments in people and places. *Energy Efficiency* 13: 419–432.

22 Bouzarovski, S., Petrova, S., and Sarlamanov, R. (2012). Energy poverty policies in the EU: a critical perspective. *Energy Policy* 49: 76–82.

23 Pan, L., Biru, A., and Lettu, S. (2021). Energy poverty and public health: global evidence. *Energy Economics* 101: 105423.

24 Nguyen, C.P. and Nasir, M.A. (2021). An inquiry into the nexus between energy poverty and income inequality in the light of global evidence. *Energy Economics* 99: 105289.

25 Churchill, S.A., Smyth, R., and Farrell, L. (2020). Fuel poverty and subjective wellbeing. *Energy Economics* 86: 104650.

26 Daneshvar, M., Mohammadi-ivatloo, B., and Anvari-Moghaddam, A. (2023). Risk-oriented operational model for fully renewable cooperative prosumers in a modern water-energy nexus structure. *IEEE Transactions on Sustainable Energy* 14 (4): 2006–2019.

27 Nawaz, S. (2021). Energy poverty, climate shocks, and health deprivations. *Energy Economics* 100: 105338.

28 Hailemariam, A., Sakutukwa, T., and Yew, S.L. (2021). The impact of energy poverty on physical violence. *Energy Economics* 100: 105336.

29 Bridge, B.A., Adhikari, D., and Fontenla, M. (2016). Household-level effects of electricity on income. *Energy Economics* 58: 222–228.

30 Churchill, S.A. and Smyth, R. (2020). Ethnic diversity, energy poverty and the mediating role of trust: evidence from household panel data for Australia. *Energy Economics* 86: 104663.

31 Churchill, S.A. and Smyth, R. (2021). Energy poverty and health: panel data evidence from Australia. *Energy Economics* 97: 105219.

2

Importance of Energy Justice and Equity in Grid Modernization

Reza Gharibi[1], Behrooz Vahidi[2], and Rahman Dashti[1]

[1] *Clinical-Laboratory Center of Power System & Protection, Faculty of Intelligent Systems Engineering and Data Science, Persian Gulf University, Bushehr, Iran*
[2] *Department of Electrical Engineering, Amirkabir University of Technology (Tehran Polytechnic), Tehran, Iran*

2.1 Introduction

2.1.1 Motivation

Along with the continual development of energy systems and their critical sustainability, the immediate need for improvements to energy networks has become essential. This imperative extends beyond a mere technological response; it is a strategic alignment guiding societies toward a future of efficient energy systems [1]. Modernizing energy networks necessitates a fundamental shift beyond traditional energy models. At the center of this transformation is the basic idea of integrating various energy sources. Seamless integration of diverse energy sources lies at the heart of modernization efforts, enabling the harmonious coexistence of renewable, traditional, and emerging energies that collectively shape the forefront of progress [2]. The ongoing search for cleaner, more efficient energy solutions has far-reaching societal implications and has become a pressing global issue. As society moves toward a more efficient energy landscape, the profound impact of renewable energies and the parallel transformation of energy networks on the foundation of social infrastructure becomes increasingly clear [3]. Research and discussions on energy systems often focus on efficiency and environmental impacts. While these issues are undeniably significant, one dimension warranting serious attention is the issue of justice within the evolving energy paradigms. Examining the complexities of energy network modernization reveals that the transformative potential of these developments can either exacerbate existing inequities or serve as a powerful catalyst for positive change [4]. This chapter

Communities for Clean Energy Justice and Equity in Grid Modernization, First Edition.
Edited by Mohammadreza Daneshvar, Behnam Mohammadi-Ivatloo, and Amjad Anvari-Moghaddam.

is aligned with the broader objectives of the book, which seeks to explore the role of clean energy, justice, and equity in the process of grid modernization (GM). The chapter contributes to the book's overarching goal by providing a detailed examination of how modernized energy systems can address historical and emerging inequities in energy access and distribution. By exploring different facets of energy justice (EJ), distributive, procedural, and recognition justice, this chapter lays the foundation for understanding how GM can be leveraged to create more equitable energy systems for diverse communities, particularly those that have historically been marginalized.

2.1.2 Background

Incorporating EJ and equity into the process of upgrading energy networks is considered a crucial foundation in research and discussions about the evolution and future prospects of energy. As societal structures and regulations evolve, the role of equality- and justice-focused approaches within the complex framework of modern networks becomes increasingly critical. This recognition underscores the need to rethink energy approaches, extending beyond technological advancements to encompass a commitment to both social and environmental justices [5].

In this context, a body of research and projects highlights the critical significance of EJ in constructing fair and sustainable systems. In an in-depth discussion, the authors of [1] highlighted the integration of smart systems and the development of smart grids (SGs), as well as the global evolution of energy systems and the importance of modernizing multi-carrier energy networks for the efficient adoption of advanced technologies. They further explored the crucial importance of GM and examined various aspects of future networks. Additionally, they examined the roles of electricity, natural gas, heat, and water networks in the modernization process, offering insights into the future of energy markets through the lens of modern power systems. While these studies provide a comprehensive overview of the technological aspects of GM, they often lack critical focus on the socioeconomic implications of these transitions for disadvantaged communities. This chapter seeks to fill that gap by incorporating justice frameworks, especially in relation to access, affordability, and equity in energy transitions.

In addition, [6], the Global Energy Assessment (GEA) involved over 300 international researchers and 200 peer reviewers who conducted a thorough and policy-relevant study of global energy concerns. It investigates the challenges of sustainable development, evaluates existing technology, projects future energy systems, and maps out the policies required for transformative change toward sustainable energy futures. Unlike previous research, GEA offers a comprehensive study that applies to developing industrialized and emerging economies. In parallel with the researchers' work on the relevance of GM, the concept of

EJ was also explored. The authors of [7] emphasized fundamental ideas for EJ, specifically distributive, procedural, and recognition justice. These principles argue that the design of energy systems must ensure that individuals have equitable and continuous access to energy, considered a basic good.

In keeping with other principles of EJ, such as lowering greenhouse gas emissions and guaranteeing access to clean and inexpensive energy, [8] stressed the critical need to transition to a low-carbon energy system. Additionally, this study examined the rapid growth of the field of EJ, ongoing innovations in the structures, processes, and criteria of a just transition (JT), and emphasized the importance of an interdisciplinary approach, combining social, natural, and engineering sciences in EJ. Moving to sustainable energy sources poses considerable hurdles for places that rely on coal, oil, and natural gas for jobs and economic activities. In [9], various suggestions were presented, focusing on facilitating a fair energy transition, improving energy equity and flexibility, reducing environmental damage, promoting clean energy innovation, and supporting economic and workforce development in disadvantaged communities through policy. In addition to technological advancements, policy reports and interdisciplinary studies, such as [10], highlight the importance of social science perspectives in achieving EJ. Incorporating these perspectives ensures a more comprehensive understanding of GM's impact on vulnerable communities.

In [11] emphasized the significance of considering environmental justice implications related to renewable energy (RE) technologies while developing global policies. While the move to RE is crucial for decarbonizing the economy and combating climate change, some technologies may have an impact on vulnerable populations' health and livelihood. In this context, the environmental justice implications were explored, and the importance of integrating EJ considerations into these policies was emphasized. In this regard, [12] addresses general and thorough considerations regarding EJ in relation to the location of RE sources. Furthermore, the concept of self-awareness of justice was used to encourage discussions about the importance of justice in RE decision-making. While equity concerns were rarely addressed overtly, comments implicitly stressed distributive justice by mentioning local effects, aesthetics, community, economics, sustainable energy, and climate change.

Finally, a review of previous research highlights the critical significance of EJ in the evolution of the energy sector. This requires an integrated approach that focuses on topics, such as global energy systems, SGs, and the transformation of multi-carrier energy grids. It is also critical to analyze the concepts of EJ and procedure, with a focus on equitable and uninterrupted access to energy. The challenges of transferring to a low-carbon energy system require specific consideration. Additionally, the reviewed literature suggests that one key factor to consider is the impact of energy GM on the concepts of EJ. Recent developments in energy policy have increasingly focused on integrating justice frameworks into RE deployment and GM. Studies from 2022 and 2023, such as [13, 14], emphasize the need for equitable energy transitions and the necessity of incorporating social and economic dimensions into GM. This chapter builds on these insights to further explore how emerging technologies can drive more equitable outcomes.

2.1.3 Contributions

The primary objective of this chapter is to explore the importance of EJ and equity in the context of GM. By focusing on key elements, such as SGs, RE integration, and energy storage systems (ESSs), the chapter aims to assess how modernizing energy networks can serve as a catalyst for promoting EJ and equity. Another goal is to examine the challenges and opportunities associated with the transition to a low-carbon energy system, with a particular focus on distributive, procedural, and recognition justice. This chapter addresses EJ and equity in the context of GM. The chapter begins with an overview of GM, emphasizing key components such as SG, RE integration, ESS, and data analytics. This basic concept lays the groundwork for research on EJ and GM, where basic components, such as distributive justice, identification justice, and procedural justice, are defined. The evolution of the EJ movement is also explored, including its historical context and its intersection with environmental movements, as covered in the sections on transition justice. Addressing the transition to EJ, this chapter thoroughly examines issues related to affordability, new technology adoption, and job transitions. A key contribution in the sixth section is the introduction of modern energy networks as a catalyst for promoting EJ. Politicians, utilities, and GM stakeholders can gain actionable insights by examining democratic access to energy resources, energy poverty, environmental justice, and community empowerment. The main contributions and aims of this chapter are summarized as follows:

- Comprehensive investigation into GM and important components, such as SG, RE integration, ESS, and data analytics.
- Examining the fundamental components, such as distributive justice, recognition justice, and procedural justice, in order to address the conceptual framework of EJ.
- A historical overview of the EJ movement and how it intersects with environmental movements through the concept of JT.
- Examining the problems of shifting to EJ, such as affordability, acceptance of new technologies, and job changes.
- Discuss how to modernize energy networks as a catalyst for improving EJ.
- Investigating the effect of EJ on specific aspects of GM.
- Investigating the relationship between EJ and the adoption of innovative technologies.

2.1.4 Chapter Organization

In this chapter, attention is paid to five basic aspects. Section 2 addresses GM, including the SG, RE integration, ESS, and data analytics. Section 3 focuses on EJ and examines distributive, recognition, and procedural aspects. Section 4 examines the historical development of the EJ movement. Section 5 discusses the challenges of transitioning to EJ, including affordability and technology adoption. Finally, Section 6 describes GM as a transformative force that impacts democratic access to energy, energy poverty reduction, environmental justice, and community empowerment. Section 7 discusses the impact of EJ on GM, whereas Section 8 discusses EJ and the adoption of new technology. Section 9 concludes the chapter by emphasizing the need of incorporating EJ ideals into the current energy transition.

2.2 Grid Modernization

As a result of rapid technology developments and the growing demand for sustainable energy solutions, the modernization of energy networks has emerged as a significant requirement. Energy GM, which involves a comprehensive upgrade of infrastructure and operational frameworks, is a transformative step toward a more flexible, efficient, and sustainable energy system. As countries face the challenges of the shifting energy landscape, such as rising demand and a desire to reduce environmental impact, the need to modernize energy networks has never been more critical [1].

Various definitions of GM have been provided, with one of the most comprehensive coming from the California Public Utilities Commission (CPUC). In their decision, they described GM as follows. "A modern grid allows for the integration of Distributed Energy Resources (DERs) while maintaining and improving safety and reliability. It facilitates the efficient integration of DERs into all stages of electric power system planning and operations to fully utilize the capabilities that the resources offer, without undue cost or delay. This allows markets and customers to more fully realize the value of the resources, to the extent cost-effective to ratepayers, while ensuring equitable access to the benefits of DERs. A modern grid achieves safety and reliability through technological innovation to the extent that it is cost-effective to ratepayers relative to other legacy investments of a less modern character" [15]. According to this definition, energy network modernization involves upgrading and improving the infrastructure that supplies electricity to homes, businesses, and industries. The goal of this evolution is to enhance the efficiency, reliability, and capacity of energy systems to integrate new technologies. ESS, SGs, and advanced metering infrastructure (AMI) are all examples of significant achievements that have been accomplished. By utilizing these technologies, utilities are able to monitor the flow of energy in real time, maximize the utilization of energy, and react rapidly to any faults or disruptions that may occur. Through the utilization of data analytics and prediction algorithms, SGs are able to effectively balance supply and demand, hence enabling the incorporation of RE sources that are intrinsically variable, such as solar and wind production.

To ensure that the benefits of modernizing energy networks are shared equally across all segments of society, it is essential to emphasize principles of justice and equity. Traditionally, marginalized groups, including low-income neighborhoods and rural areas, have often suffered a disproportionate burden of negative consequences resulting from traditional energy systems, such as pollution and exorbitant energy expenses. As energy systems progress and become more sustainable, it is crucial to address these disparities and ensure that all communities can participate in this transformation. The development of microgrids and DERs contributes to modernization initiatives. Microgrids can operate independently of the main grid, providing localized power solutions during outages, whereas DERs such as rooftop solar panels and electric vehicles allow customers to create and store their own energy, resulting in a more decentralized and resilient energy system.

Achieving EJ and equity requires both equitable access to clean energy technologies and the empowerment of communities to actively participate in decision-making processes related to energy infrastructure development. This involves addressing structural challenges such as limited access to funding and technical knowledge, which may hinder marginalized populations from participating in the transition to modernized energy networks. Innovative technologies, such as machine learning (ML) and Internet of Things (IoT) sensors, enable proactive maintenance and predictive modeling, consequently enhancing grid dependability and reducing downtime. These developments ensure that modernized grids can better serve communities by lowering operational inefficiencies and increasing grid resilience. By prioritizing equity in GM efforts, we can create a more sustainable, inclusive, and environmentally sound energy future that benefits everyone [16–18]. The implementation of GM covers numerous aspects, which are listed below.

2.2.1 Smart Grid

The establishment and expansion of the SG is one of the most critical components driving network modernization. This advanced and interconnected network is implemented to enhance the efficiency, reliability, and stability of energy distribution. SGs act as platforms for digital communication and advanced sensors, enabling real-time information exchange between producers and consumers. With features like AMI, integration of DERs, and automated control systems, SGs enable active monitoring, rapid response to faults, and improved resilience. Through two-way communication, integration of RE sources, and support for demand response programs, SGs empower consumers to make informed energy consumption decisions while optimizing grid operations. The overarching goal of the SG is to create a dynamic and adaptive energy infrastructure that meets societal needs and aligns with modernization and environmental goals [19]. This transformative system comprises three main components: sensors, communication, and control systems, which work together in an integrated fashion to ensure efficient and reliable grid operations.

- **Sensors and Measurement devices:** Comprehensive data collection across the entire network is essential for modernizing energy systems. In SGs, critical metrics such as voltage, current, air temperature, and equipment must be gathered across the network. This data helps identify changes in current parameters and network behavior, enabling the implementation of corrective measures to ensure network reliability [20].
- **Communication networks**: The data collected by sensors is transmitted to centralized or decentralized control centers via communication networks, which form the backbone of SG operations. These networks facilitate seamless information exchange between sensors, smart meters, and control systems, enabling real time decision-making and automated responses to network conditions. For example, if a fault occurs, communication networks relay data to control systems, which can instantly deploy corrective measures, such as isolating affected sections and rerouting power. Communication networks also help in coordinating DERs and managing demand response programs, thereby enhancing the grid's overall flexibility [21].
- **Control systems**: Control systems are essential in processing data received from sensors via communication networks. They analyze the data to optimize power flow, balance supply and demand, and implement operational changes when necessary. In practical applications, control systems monitor and manage interactions between various energy sources, such as the main electrical grid, renewables, and storage devices. This real-time optimization ensures smooth transitions between grid-connected and islanded modes, making control systems pivotal in shaping decentralized and sustainable energy practices [22].

These components work together to create a responsive and intelligent energy infrastructure. For example, when sensors identify a grid irregularity,

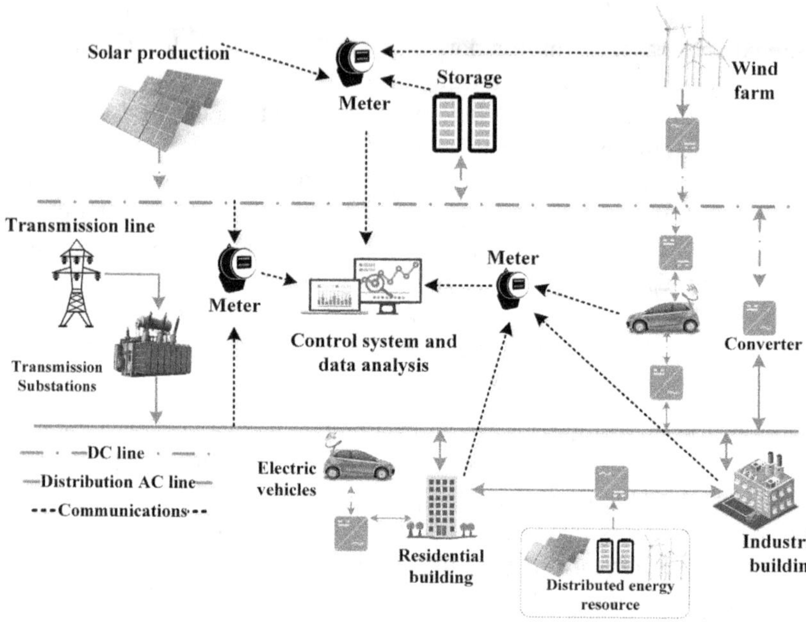

Figure 2.1 An example of a smart grid.

communication networks transmit the information to control systems, which assess the data and take appropriate action such as rerouting energy, changing supply levels, or initiating demand response programs. The combination of sensors, communication networks, and control systems allows the SG to function as a cohesive system capable of real-time adaptation and optimization, increasing the grid's resilience and efficiency (Figure 2.1).

2.2.2 Renewable Energy Integration

Integrating RE is vital for modernizing energy infrastructure and building a resilient, adaptable energy network. With society increasingly prioritizing environmental goals and the transition to cleaner energy, renewable sources like solar and wind play a crucial role in significantly reducing carbon emissions. The addition of renewables enhances the grid by diversifying the energy mix, reducing its dependence on fossil fuels, and encouraging a decentralized and distributed energy framework. Since renewable resources often have irregular output patterns, modern technologies like ESS and SG solutions are essential for effective management. For instance, solar energy generation peaks during daylight hours, whereas wind energy may fluctuate based on weather conditions.

By integrating these sources with ESS, such as batteries, surplus energy generated during peak periods can be stored and then employed when production declines, so providing a consistent energy supply. SGs are essential for coordinating the distribution of stored energy and optimizing system efficiency. A modern network that seamlessly incorporates renewable energies not only advances environmental sustainability but also enhances system reliability, flexibility, and overall efficiency. It adeptly meets the evolving requirements of a cleaner, more sustainable energy landscape [23]. The integration of RE involves three key components: renewable power plants, inverters and converters, and energy management systems (EMSs), each fulfilling a specific role.

- **Renewable Power Sources:** Renewable power plants are essential to sustainable energy networks, harnessing renewable sources to generate electricity with minimal environmental impact. These facilities utilize a range of renewable resources, including sunlight, wind, water, and biomass, through diverse technologies. Solar power plants convert sunlight into electricity through photovoltaic cells or solar thermal systems, while wind farms capture wind energy to generate power. Unlike solar and wind power plants, which rely on the weather, hydroelectric power plants have the unique capacity to provide energy consistently. Hydroelectric facilities function as natural storage systems, capable of discharging stored energy at moments of peak demand. Biomass facilities convert organic material into energy, while geothermal power plants utilize the Earth's inherent heat. One of the main characteristics of these power plants is their carbon-free status, which makes them a cleaner and more sustainable alternative to traditional energy sources. Their integration into the grid diversifies energy sources, reduces carbon emissions, and supports the development of a flexible, modern power infrastructure [24].
- **Inverters and converters**: Inverters and converters are essential for integrating RE into the power grid. Renewable sources, such as solar panels and wind turbines, primarily provide direct current (DC), which must be converted into alternating current (AC) for usage in electrical networks. Their adaptability allows for integrating a wide range of renewable sources, contributing to a flexible and adaptive energy infrastructure that aligns with the changing needs of the sustainable energy landscape. During peak solar generation, inverters facilitate the effective conversion and distribution of power to the grid or its storage in ESS for future utilization. Wind turbines utilize converters to regulate variations in wind velocity and provide stable power generation [25].
- **Energy management systems**: The distribution and consumption of energy are crucial in renewable systems, with EMSs shouldering this responsibility. In the realm of renewables, EMS efficiently balances supply and demand, manages ESS, and ensures grid stability by integrating variable sources like solar and wind. Through monitoring of energy generation and consumption

patterns, the EMS enhances the charging and discharging of ESS, ensuring the availability of stored energy during periods of decreased renewable output. The EMS collaborates with SGs to implement real-time adjustments to energy flow, hence improving efficiency and reliability. In modern grids, EMS serves as an intelligent control system, utilizing real-time data and advanced algorithms to optimize the entire energy infrastructure's performance. By monitoring and controlling components such as renewable generators, ESS, and loads, EMS enhances reliability in energy management, facilitating the transition to a more flexible and RE landscape [26].

2.2.3 Energy Storage System

Renewable energies are a crucial component of modern networks. The network faces several challenges in the integration of RE. One of these challenges is the intermittent nature of energy production from various sources. The generation from these sources can sometimes conflict with periods of high or low energy demand. ESS is a solution that plays a critical role in addressing this issue in modern systems. Various ESS exist, each with unique characteristics, benefits, and constraints, making them appropriate for different scenarios in GM [27]. Additionally, ESS can improve network flexibility. ESS in a modern grid can mitigate fluctuations, ensuring a more stable power flow. The network's reliability depends on ESS's ability to respond rapidly to sudden demand spikes or disturbances. With GM and the goal of producing 100% RE, these systems play an increasingly important role. Storage devices that are designed to work with current networks can be divided into several categories. These devices can store energy in several ways, including electrical, hybrid, mechanical, thermal, and electrochemical [28].

- **Electrical storage**: Electrical energy storage technologies, such as supercapacitors and superconducting magnetic energy storage (SMES), provide quick response capabilities and are optimal for short-term energy retention and grid stabilization. Supercapacitors are efficient in providing rapid bursts of power; yet, they are constrained by their overall energy storage capacity. SMES may retain substantial energy quantities without deterioration; yet, they can be expensive and complex to maintain. These technologies are very effective in mitigating power fluctuations and maintaining voltage stability [28].
- **Electrochemical storage**: The most widely used electrochemical storage technologies are batteries, especially flow and lithium-ion batteries. Lithium-ion batteries provide elevated energy density and are appropriate for applications necessitating frequent cycling, such as the support of solar and wind energy

systems. Nonetheless, they may deteriorate over time and necessitate meticulous monitoring of charge cycles. Flow batteries provide extended lifespans and are more appropriate for large-scale energy storage, but possessing lower energy density relative to lithium-ion batteries [29].

- **Mechanical storage**: Mechanical storage techniques, including pumped hydro storage (PHS) and compressed air energy storage (CAES), are optimal for extensive energy storage solutions. PHS is a well-established and extensively utilized technology, providing high efficiency and long-duration storage capacities; yet, it necessitates particular geographical circumstances, including elevation variations. Conversely, CAES stores energy as compressed air in underground caverns, offering substantial storage capacity, albeit with lower efficiency and significant initial investment costs [29].
- **Thermal storage**: Thermal energy storage, including molten salt and ice storage, involves the retention of energy as heat or cold. Molten salt storage is commonly employed alongside concentrated solar power (CSP) facilities to retain heat and discharge it during periods of lack of sunlight, ensuring a dependable energy source. Nonetheless, it necessitates complex infrastructure and extensive maintenance. Ice storage, primarily utilized in HVAC systems, stores energy in the form of thermal cooling during off-peak periods and discharges it during peak periods, providing economical cooling solutions, but with restricted application for widespread grid utilization [28].
- **Hybrid systems**: Hybrid systems integrate various technologies, such as batteries with supercapacitors or mechanical storage with thermal storage, to optimize the advantages of each component. These systems improve grid flexibility by utilizing the advantages of several technologies, while they may be more expensive and complex to implement [29].

Integrating a variety of ESS technologies enables contemporary grids to more effectively handle the fluctuations of RE sources, improve system reliability, and facilitate the shift toward a sustainable energy future. The appropriateness of each technology is dependent on criteria like grid requirements, scalability, and geographical limitations. As GM advances, meticulous evaluation of the benefits and drawbacks of these technologies will be crucial for enhancing energy storage options [28] (Figure 2.2).

2.2.4 Data Analytics

As the elements that make up modern networks become more intricate and manageable, intercommunication among them becomes increasingly vital. Establishing a connection among these components in modern networks necessitates platforms such as smartphones, the IoT, telecommunications networks, or other information transmission methods. The quantity of manageable components,

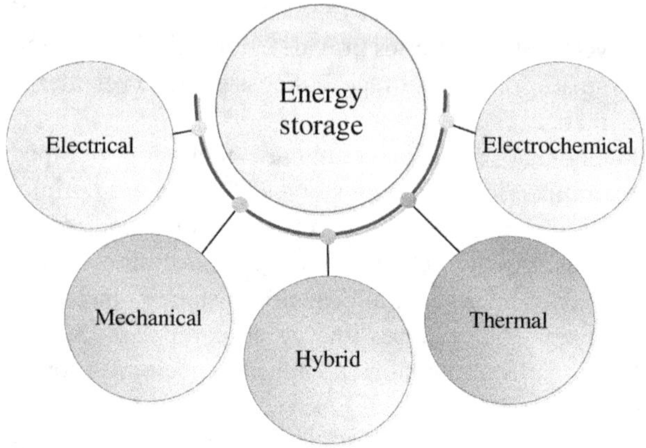

Figure 2.2 Different types of energy storage systems.

coupled with data from monitoring systems such as AMI and other network sources, produces an extensive amount of information. To achieve the objective of enhancing the efficiency of modern networks, it is essential for these networks to utilize the extensive data available for effective component interaction and optimal performance [30]. The volume of data gathered from network and monitoring components is important, yet not all of it is beneficial. The large amount of both pertinent and extraneous data generates challenges in data transit and storage, adversely affecting system efficiency. Consequently, primary data analysis, the selection of important data, and the finding of relevant patterns and relationships are essential. Modern systems employ diverse ML methods for the analysis of enormous amounts of data to tackle these difficulties [31]. For example, supervised learning methods like random forests and decision trees are used to identify abnormalities in real-time data and forecast energy consumption. Unsupervised learning methods, such as clustering, are employed to categorize analogous data patterns, enabling the identification of load profiles and the segmentation of users for demand response initiatives. Reinforcement learning is utilized in grid management to enhance energy distribution by continual learning from system feedback. These algorithms enhance decision-making, optimize grid performance, forecast faults, and increase energy distribution efficiency [32].

2.3 Energy Justice

Energy, specifically electrical energy, is extremely important in modern society, and it is related to almost every facet of societal advancement. Energy essential for the functioning of societies and various industries. For instance, industrial operations and basic needs like lighting and heating are all dependent on energy.

Despite its essential role in societal survival and growth, energy faces several challenges, such as environmental concerns and the depletion of fossil fuel resources, necessitating a shift toward clean and sustainable energy. Energy systems consist of several components, including production, energy conversion, distribution, consumption, and energy services [33]. The concept of EJ focuses on achieving equity in three basic energy sectors: energy production, distribution, and consumption. In addition, EJ delves into broader principles of environmental justice. According to EJ, energy poverty is caused not just by a lack of cost-effective energy but also by a failure to provide for society's core energy demands [7].

According to these descriptions and the importance of energy, the necessity of transitioning to a low-carbon and more equitable energy system is clear. These goals and approaches to energy in modern networks are critical components of energy strategies [34]. EJ addresses three major issues: distributive justice, which focuses on the geographic distribution of benefits and resources; recognition justice, which examines the characteristics of affected individuals; and procedural justice, which focuses on equitable decision-making processes [7]. Given these perspectives, EJ has become increasingly important within the JT framework, particularly in the context of transitioning to a low-carbon system. This "just energy transition" highlights the inclusion of the impact of energy on all societies and social groups in the transition to a low-carbon future [35]. In general, EJ requires a focus on people's interactions with energy systems and policy design to ensure a fair distribution of costs and benefits [33] (Figure 2.3).

2.3.1 Distributive Justice

Energy systems exhibit geographic disparities in the location of their sources, access to the energy produced, and the types of energy generated [36]. The uneven distribution of energy in some places leads to a lack of reliable energy access, leading to energy poverty. Such uneven distribution can hinder economic development, limit educational opportunities, and negatively affect the overall

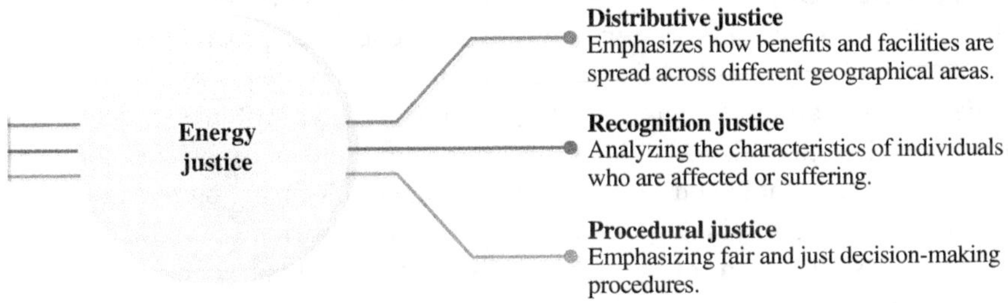

Figure 2.3 Some aspects of the energy justice.

quality of life and industrial growth in those areas. Moreover, the environmental impacts of energy production are unevenly distributed, causing some communities to face disproportionate levels of pollution and climate change effects. These instances and a lack of equitable distribution of energy resources are frequently encountered in socially excluded communities [37]. The distribution of energy production sources and related components extends beyond fossil fuels. The growth of low-carbon energy in a fair and balanced manner in various locations is directly linked to the reduction of resources related to fossil fuels, and naturally, environmental conditions improve in different regions. For instance, locations wealthy in renewable resources, like wind or solar power, may experience economic growth and cleaner energy, while regions depending on coal or oil may see economic decline if their shift to alternative energy sources is not actively promoted. This disparity highlights the necessity for targeted interventions that provide all communities, especially those economically dependent on fossil fuels, with the requisite tools and infrastructure to transition to low-carbon energy solutions [37].

Building on these concepts, distributive justice in energy underscores the need for all communities, regardless of geographic or socioeconomic factors, to have access to clean and affordable energy sources. Achieving proportional distribution depends on overcoming economic, infrastructural, or regulatory constraints and actively empowering marginalized communities. This entails including remote or rural regions, frequently overlooked in infrastructure investment, in energy planning. Moreover, in urban regions, low-income communities disproportionately face energy disparity, characterized by elevated energy expenses and diminished service reliability. Distributive justice acknowledges that achieving true equality in energy access necessitates not only technological advancements but also reforms in regulations, education, and societal attitudes. These reforms must incorporate direct community involvement in energy planning and decision-making procedures, guaranteeing that local requirements are understood and addressed. Furthermore, the distribution of subsidies, investments, and incentives must prioritize regions with the greatest need, rectifying existing disparities and guaranteeing equitable access to developments in energy infrastructure. Such measures ensure that the benefits of progress and energy access are equitably distributed across all segments of society [3, 38].

The modernization of energy networks significantly contributes to achieving distributive justice goals. Traditional infrastructure limits marginalized communities' access to reliable and plentiful energy sources. Focusing on technological advancements and developing smart, decentralized energy networks can enhance efficiency, flexibility, and equity in energy distribution systems. For example, local control over energy production and consumption is provided by decentralized RE systems, like community solar or microgrids. These systems give disadvantaged populations the ability to manage their energy resources and lessen their reliance on centralized, frequently costly. By promoting localized energy generation, these systems enable communities to enhance their resilience against external factors and rising energy expenses, so further advancing distributive justice. These modern grids respond to changing energy demands by integrating RE sources, ESS, and digital monitoring. In essence, advancements in energy networks contribute to distributive justice by removing barriers, increasing flexibility, and providing a more accessible and equitable energy landscape for both affluent and economically disadvantaged regions [6].

2.3.2 Recognition Justice

Recognizing the inequality produced by energy systems in specific segments of society and regions is basically the first step toward achieving EJ. EJ relies on the accurate identification of inequities, requiring that those affected by these injustices be recognized and addressed. This method goes beyond a simple analysis of inequities and necessitates a thorough assessment of the social, economic, and cultural variables that contribute to these injustices. Hence, recognition justice plays an important role in the larger EJ framework by highlighting the fact that attaining justice necessitates more than just allocating resources; it also entails making sure that all opinions are heard and valued during the decision-making process. Only through such comprehensive analyses can effective strategies and policies be developed to address the specific needs of affected communities, thereby paving the way for a more equitable energy framework [7]. In this context, recognition justice in EJ requires a thorough analysis of how individuals and communities are acknowledged within energy systems, along with the resulting impacts. This goes beyond simple economic concerns, attempting to identify and solve the social and cultural components of inequities caused by the distribution of energy-related benefits and costs. By connecting recognition justice with distributive and procedural justice, it is guaranteed that marginalized populations receive equitable access to energy while also being respected and empowered in decision-making processes. This idea operates from a "post-distribution" viewpoint, highlighting the importance of post-distribution planning and requiring a thorough investigation to discover and reveal inequities linked to various energy sectors [12]. After accurately identifying these injustices, policymakers and decision-makers might unintentionally overlook the true impacts of neglected sectors, resulting in recognition injustice. Nancy Fraser defines cognitive justice as addressing cultural dominance, nonrecognition, and disrespect. These concepts are applicable in contexts such as analyzing and addressing power imbalances between indigenous populations and extractive industries. It encourages decision-makers to broaden their perspectives, move beyond traditional vulnerability indicators, and consider marginalized groups such as the elderly, disabled individuals, and students. As a result, recognition justice ensures that all communities are equitably considered, which is crucial to the larger objective of maintaining energy equity. In summary, recognition justice promotes critical awareness of the social impacts of energy policies and practices and advocates for building more inclusive energy systems that respect and acknowledge diverse identities and experiences [39].

As energy systems expand to incorporate modern technology and methods, it is critical to address and include diverse populations, identities, and stakeholders. Recognition justice emphasizes equality and consideration of all individuals and groups within the energy landscape, ensuring that their perspectives, needs, and contributions are acknowledged and taken into account. This links recognition justice directly to overarching EJ objectives, as the energy transition cannot be just without the involvement and strengthening of marginalized voices. Modernizing grids enables us to overcome past inequalities and amplify marginalized groups' voices in energy grid decision-making processes. To promote a more equitable energy transfer, advanced technologies should be integrated into GM by taking into account various social, cultural, and economic factors [40].

2.3.3 Procedural Justice

Procedural justice is the third key principle in the EJ framework, relating to justice in energy-related decision-making and processes. It is a vital element of the EJ concept, aiming to ensure equitable distribution of the energy system's benefits and resources among diverse social and economic groups. Procedural justice focuses on how decisions about energy policy, resource allocation, and the level of participation of stakeholders in the energy supply process are made [7].

Different indicators and attributes of procedural justice may be articulated depending on its definition and concept. The first indicator is comprehensive decision-making, which ensures that all stakeholders involved in the energy-related decision-making process are prioritized. This includes not just government officials but also industry representatives, community members, and advocacy groups with different points of view. Comprehensive decision-making ensures that the final decision is made with all perspectives in mind and is as equitable as possible.

Procedural justice is essential in promoting other elements of EJ, such as distributive and recognition justice, by guaranteeing that the processes governing energy allocation are equitable and inclusive. Equitable processes (procedural justice) guarantee that underprivileged populations not only gain from energy access (distributive justice) but also participate in the formulation of policies that impact them. This is crucial as historically marginalized groups frequently encounter unequal energy distribution, and their contributions are essential for formulating policies that rectify these disparities. In this sense, procedural justice ensures that the energy transition does not exacerbate existing inequities by actively engaging populations most susceptible to energy poverty and exclusion. This connection underscores the notion of distributive justice, which emphasizes equitable outcomes. Furthermore, by acknowledging the unique cultural and socioeconomic requirements of various groups (recognition justice), procedural justice guarantees their appropriate representation in the decision-making process. This promotes not only equitable results but also just methods of achieving those results [41, 42].

Another indication is nondiscrimination, which, in line with procedural justice, demands a decision-making process free from discrimination. This emphasizes the importance of providing all persons and groups, regardless of social position, ethnicity, or gender, with an equal opportunity to participate in and make decisions about energy systems. This equitable strategy aligns with the ideas of recognition justice, guaranteeing that historically excluded or marginalized populations are actively involved in conversations and choices affecting their energy access and future. Transparency is another important aspect to consider in procedural justice. Transparency in energy-related decision-making procedures is another crucial aspect of procedural justice. Providing clear and accessible information about projects and effective procedures that align with this characteristic creates trust among the community and stakeholders. Equal access to decision-making processes for all stakeholders is essential to procedural justice. Examples of procedural justice include providing information in multiple languages for accessibility and holding meetings in locations that are convenient for all stakeholders [42].

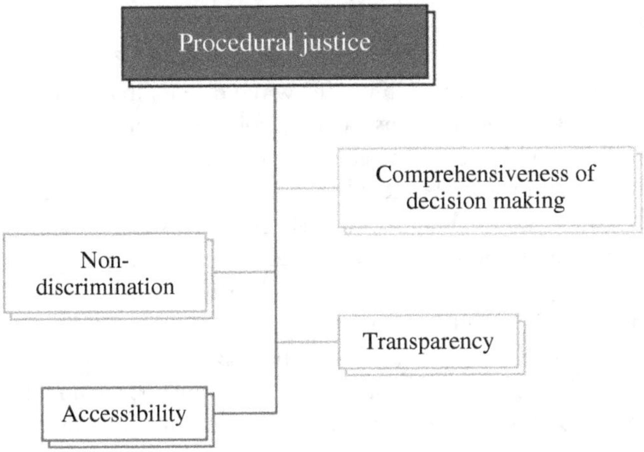

Figure 2.4 Particular aspects of procedural justice.

These strategies enhance recognition justice by ensuring that marginalized populations are included in essential discussions regarding energy systems.

Ultimately, procedural justice guarantees that energy policy development is democratic and inclusive, allowing all groups to participate in decisions that will influence the future of energy access and distribution. Alongside distributive and recognition justice, it establishes an integrated framework for achieving equality in energy systems (Figure 2.4).

2.4 Just Transition: From Environmental to Energy Justice

Moving from environmental justice to EJ requires a significant shift in the framework of sustainability issues and the resolution of associated challenges. This movement focuses on fundamental environmental concepts, primarily emphasizing ecosystem preservation and minimizing environmental damage. It represents a significant step forward in our understanding of the complexities of environmental, social, and economic challenges. The concept of EJ, as the unifying thread of these complexities, goes beyond conventional discussions of environmental protection and addresses the underlying disparities in the energy system. EJ reveals that marginalized groups suffer disproportionately from the adverse impacts of energy production while receiving fewer economic benefits or access to energy. These conditions demand comprehensive decisions that not only address environmental concerns but also confront structural and systemic inequalities, integrating environmental and energy issues into decision-making processes. EJ is about empowering and supporting those who are disproportionately affected. It also supports equal access to energy resources for all members of society and emphasizes cleanliness and affordability. This movement seeks to redress environmental injustices while developing inclusive policies that favor marginalized populations. This is where frameworks like community energy initiatives and energy subsidies for low-income populations may be implemented to guarantee equal access to sustainable energy. Moreover, regulatory frameworks such as the establishment of EJ committees can guarantee the inclusion of excluded communities in energy decision-making processes, thereby addressing both procedural and distributive justice issues. Beyond environmental considerations, the discourse on EJ encompasses significant social and economic effects. This movement further underscores that within the concept of a "JT," individuals in fossil-fuel-related industries must be adequately supported in transitioning to renewable sources, ensuring minimal economic repercussions for any specific segment of society. In this context, just energy transition funds, intended to retrain workers in the fossil fuel sectors and offer economic assistance throughout transitions, represent an essential framework for concurrently addressing environmental and EJ concerns [5].

In general, transitioning from environmental justice to EJ requires a comprehensive and integrated approach. This principle demonstrates knowledge and understanding of the intricate relationships between environmental sustainability, social equality, and economic equality. Recognizing these connections enables us to envision a future in which environmental sustainability, energy, and economic justice are inseparably interconnected, aiming for a more balanced relationship with the environment. Mechanisms such as integrated energy policies, which prioritize both sustainability and equity, serve as a foundation for achieving this vision. These policies may encompass mandates for RE production in disadvantaged regions and financial incentives for RE adoption by marginalized communities. This requires a framework and system in which decision-making processes address environmental and energy challenges simultaneously.

2.4.1 Just Transition and Environmental Movements

The effort to mitigate climate change involves a pivotal shift from fossil fuel consumption to a zero-carbon trajectory, necessitating substantial changes in key industries and energy production sources. This transition has profound implications for workers' lives and the economy, emphasizing the critical need to recognize the social dimensions, particularly the well-being of workers, in the pursuit of zero-carbon objectives. The focus on justice for the working class can be traced back to the late 1970s with the emergence of the term "JT," advocating for the support of workers and vulnerable communities in political decisions [43]. Originally rooted in economic considerations, the concept of JT has evolved significantly. Scholars now advocate not only for the seamless transition of workers and communities from declining industries but also for justice that guarantees the economic security of those involved, enabling them to retire with dignity. This expanded perspective goes beyond employment considerations, emphasizing broader societal progression. In this context, specific frameworks like JT Funds have been established to offer financial and economic assistance to areas impacted by the cessation of fossil fuels. These grants support retraining initiatives for workers, equipping them with the requisite skills for employment in the RE sector, so ensuring they are not marginalized as sectors shift toward sustainable energy sources. Government initiatives that foster workforce development for green energy employment, along with specific financial incentives, can mitigate the economic shocks faced by these areas [44].

Scholars believe that JT is intrinsically related to climate goals, with a successful shift toward environmental sustainability requiring the equitable alignment of societal needs with environmental objectives. This fundamental principle underlines the significance of creating a JT framework that systematically faces the different demands of the entire society. The environmental justice movement began in the 1980s when it was dubbed "environmental racism" to call attention to minority racial populations that were disproportionately located in environmentally damaged areas. This movement emphasizes the need of marginalized populations to fully participate in environmental decision-making processes and for the equal distribution of environmental harms and benefits [45]. Contrary to prevailing perceptions, environmental justice advocates assert that endeavors aimed at climate change mitigation may inadvertently contribute to an exacerbation of air pollution unless these endeavors are comprehensively evaluated through the lens of human health. These apprehensions, as articulated by Mendez [46], have garnered acknowledgment and consideration among the United States climate policymakers. The recognition of this nuanced interplay between climate change mitigation efforts and potential collateral environmental impacts reflects a heightened awareness within policymaking circles of the intricate connections between environmental justice, public health, and climate resilience. To guarantee that JT policies encompass both environmental and EJ issues, procedural frameworks have been established to assure the representation of marginalized populations in decision-making processes related to energy projects. Community energy boards and participatory governance approaches are employed to empower individuals historically marginalized in energy decision-making, hence advancing procedural justice.

The recognition of carbon dioxide as a pollutant shows the pivotal connection between fossil fuels and its profound environmental consequences. To underscore the intricate interplay between environmental justice and climate concerns, the conceptualization of "climate pollution" was introduced, thereby illuminating shared apprehensions regarding substances contributing to climate change. This heightened understanding prompted the categorization of greenhouse gases as a substantive component of pollution within a grant program, explicitly terming them "air pollutants." The discourse on climate justice, which gained prominence in the late 1990s, delves into the repercussions of climate change. This discourse accentuates the notion that nations with historically minimal contributions to climate change bear the brunt of its impact and consequences [47]. Recent data from 2022 underscores the disparity in greenhouse gas emissions, revealing that less developed countries collectively contribute a mere 3.5%, while countries like the United States singularly shoulder the responsibility for approximately 11.5% of all greenhouse gas emissions [48]. In response to these inequities, environmental and climate justice movements are challenging prevailing political decisions that have failed to adequately address these disproportionate and unjustifiable harms. In order to ensure that less developed countries do not bear an undue financial burden during the transition period, mechanisms like equitable carbon pricing models and global climate accords are being adopted to hold higher-polluting countries accountable. This dynamic landscape necessitates a reevaluation of global policies to rectify the disproportionate burdens borne by vulnerable nations in the face of climate change, and it underscores the urgency of a more equitable and inclusive approach to environmental governance. These developing frameworks demonstrate the increasing acknowledgment that tackling environmental justice is fundamentally connected to attaining EJ. In the future, integrated policy frameworks that combine environmental sustainability, social equality, and economic resilience will be crucial for attaining a JT for everyone.

2.4.2 History of the Energy Justice Movement

Throughout the 1980s and 1990s, justice-oriented movements emerged, focusing on environmental and economic justice problems [43]. Following the environmental justice movement, the EJ movement emerged in response to the unequal distribution of environmental and social effects throughout all energy systems, including production, distribution, and consumption. The EJ movement aims to address injustices in access to affordable and sustainable energy as well as engage all stakeholders in decision-making processes. The EJ movement addresses a wide range of concerns, including environmental racism, energy poverty, and the harmful effects of fossil fuel extraction on marginalized populations [5].

In the 2000s, the EJ movement started focusing on energy issues, such as the negative consequences of coal mining, fossil fuel power plants, and other types of energy production on marginalized groups. Activists and scholars in these fields said that low-income and minority groups frequently experience an excessive amount of environmental and health problems associated with fossil fuel extraction and consumption. This approach and study revealed that, in addition to damaging the environment, conventional energy systems increase social inequities [49]. Following the definition of justice in the field of energy and scholars' attention to this subject, another significant aspect of the EJ movement was the development of EJ principles. Scholars and activists have attempted to develop a framework that extends beyond environmental concerns to address wider social and economic issues. The primary objective of these changes was to ensure that the distribution of gains and harms is equitable with respect to production and consumption. Community energy ownership models and local decision-making councils are crucial for promoting equitable participation in the design and decision-making of energy systems. These frameworks are essential to the advancement of EJ and facilitate the consideration of both environmental and social implications in energy policies. The main topics included ensuring equal access to clean and affordable resources, increasing community engagement in decision-making, and resolving past inequalities in energy policies [50].

The development of these concepts was critical to the movement's advancement since it enabled the integration of social, economic, and environmental justice. One of the emergences of relevant and fundamental principles related to EJ can be found in the article [4], which presents and expresses insights related to sustainability and environmental movements.

After the principles and visions of EJ became clear, there was an increase in the anti-fossil-fuel movement. Environmentalists criticized fossil fuels and coal plants, which had significant negative effects on certain segments of society. These movements aimed to address the historical and structural injustices linked to the extraction and use of fossil fuels for a variety of purposes [51]. Along with attempts to limit fossil fuels, activists and supporters of the transition to cleaner energy sources gained traction in advocating for increased use of renewable energies within the framework of EJ. The EJ movement recognized and accepted RE as a solution to environmental injustices and an expression of distributive justice. Furthermore, promoters of RE said that, in addition to environmental benefits, the decentralized nature of RE promotes the economic development of marginalized communities. Decentralized energy systems like microgrids and community-owned RE projects are frameworks that have contributed to addressing both energy poverty and environmental justice concerns. These methods guarantee localized energy generation and management, empowering communities and diminishing dependence on detrimental fossil fuels. Finally, renewable energies have been recognized as providing an important part in EJ, contributing to the development of EJ and the promotion of inclusive policies in the transition to clean energy [52].

As efforts to reduce fossil fuel consumption and promote RE increased, the EJ movement grew and gained global prominence. The worldwide community

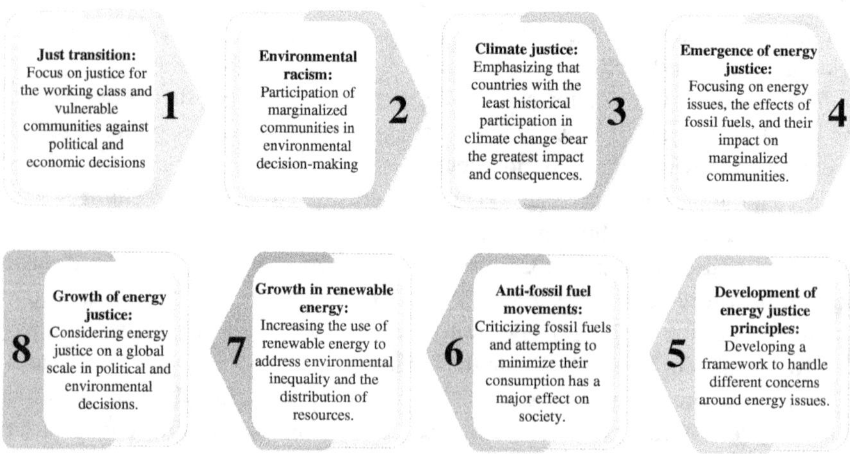

Figure 2.5 A view on the history of the energy justice movement.

demonstrated knowledge of environmental concerns, social justice, and their ties to energy systems. EJ gained momentum by supporting marginalized communities with equitable and sustainable solutions to the impacts of energy production. According to the global EJ concept, the effects of energy choices transcend political boundaries, and environmental degradation and social inequality can be interconnected on a planetary scale. In this regard, people and organizations from all around the world began to form groups to work on developing fair solutions and sharing their ideas and resources in order to progress toward a more equitable and sustainable energy future [34] (Figure 2.5).

2.5 Challenges of Transition to Energy Justice

Achieving sustainability and EJ requires societies to fundamentally alter their relationship with traditional energy systems. Fundamental changes in traditional systems present a number of challenges that must be addressed and resolved in order to achieve EJ. The complexities of EJ reveal three primary obstacles: the affordability of energy, the acceptance of new technologies and ideas, and the transformation of traditional job roles. These challenges are interconnected, forming a cycle in which resolving one issue may concurrently mitigate or intensify others. Affordable energy options can promote the integration of innovative technology while creating new job possibilities can assist a more seamless transition for workers displaced from conventional energy industries. A complete framework is essential to address these concerns holistically, ensuring that solutions to one do not exacerbate inequality in other areas. These challenges will be addressed separately in the parts that follow.

- **Affordability:** Different ways of energy supply result in significant disparities in the affordability of the obtained energy for different members of society. The affordability of energy for people in any society indicates the ability to pay their energy bills. Those facing multiple daily barriers experience this issue more acutely, as it compounds other difficulties and exacerbates inequality. This imbalance results in unequal access to energy, forcing some individuals to work harder to meet their energy needs and struggle with costs. These inequities result in issues such as excessive energy bills and a lack of reliable energy. These challenges place these populations at greater risk compared to others [53]. The geographical distribution of various energy production sources is one of the significant variables influencing energy affordability from the consumer's viewpoint. In some areas, underdevelopment or limited access to RE sources may force communities to rely on fossil fuels. This, in turn, increases the financial burden caused by energy costs. This issue has a dual effect: it reduces energy affordability in these areas and contributes to pollution and negative environmental impacts. The result is an increase in environmental injustice, with these communities suffering an increasing burden of negative environmental effects. The negative effects of this double problem extend beyond economic considerations. This challenge is closely linked to social and environmental inequities, illustrating the complex relationship between energy choices, socioeconomic development, and environmental justice. Overall, this challenge underscores the need for a comprehensive approach to energy policies that addresses economic sustainability, environmental integrity, and social equity [54]. In addition to geographical dispersion, a lack of energy efficiency in residential areas and different industries increases the challenges of energy affordability. Old and inefficient homes with poor insulation lead to increased energy use, causing residents, often low-income families, to incur higher energy costs. Addressing this part of the EJ issue requires targeted policies and investments in energy-efficient technology to ensure that all members of society have access to affordable, dependable, clean, and sustainable energy [55].

- **Acceptability and acceptance of new technologies**: As the energy landscape develops, the utilization of innovative solutions and technologies is critical. However, the success of technologies is strongly dependent on their acceptability by society in the areas where they are used. These technologies might involve the deployment of a solar-powered energy production system, the development of smart and modern networks, and novel storage systems. The success of these improvements and modernizations in the energy system closely depends on society's willingness to incorporate them into daily life. Many variables play a role in the complexities of societal acceptance. Among these considerations are the levels of public awareness of this technology, a comprehension of its perceived benefits, and concern about the consequences of not having these technologies. According to these assessments, achieving EJ requirements requires not only the continual advancement of new technologies but also an effort to interact with communities on a deep level in order for them to accept related technologies appropriately. The involvement includes an educational component designed to inform communities about the numerous benefits of these advancements. This approach actively empowers communities and fosters understanding, helping them commit to and emphasize a sustainable energy future. This strategy establishes the framework for achieving EJ by taking social concerns into account and promoting the harmonious integration of revolutionary technologies into a broader environment that aims for a sustainable and equitable energy future [56, 57].
- **Transformation of employment positions:** A major challenge in transitioning to EJ is the need to overhaul outdated jobs dependent on fossil fuels and traditional energy systems. Modernization in the energy sector, especially in low-carbon industries, increases concerns about the displacement of workers

and other stakeholders that play critical roles in these industries. As the green economy and EJ gain traction, certain job sectors may lose ground, and communities that have long relied on old energy sources will be severely impacted and transformed. This transition requires reskilling and upskilling the labor force from traditional sectors and raises socioeconomic issues, including the potential loss of stable, high-quality jobs. Along with addressing job demands, this challenge emphasizes the need for equitable opportunities in the growing green economy. Transitioning from traditional to modern energy systems involves more than just technological or commercial changes; it constitutes a social transformation that requires thoughtful and principled leadership. Addressing the issue and the concerns of the impacted people and communities requires a comprehensive strategy that considers not only the economic effects but also the social and cultural aspects. Achieving EJ requires achieving a balance between environmental sustainability and worker well-being, as well as empowering communities affected by the transformation [58, 59].

The challenges of affordability, acceptability of new technologies, and workforce transformation are inextricably linked. High-energy prices, for example, do not exist in isolation; they affect and are influenced by the other two concerns. For example, communities facing affordability challenges are more likely to resist new technology, especially if the expenses of implementing these innovations are thought to be prohibitively expensive. This resistance inhibits the general adoption of RE systems. Furthermore, when these same areas are economically insecure owing to job displacement caused by the loss of fossil fuel sectors, the situation deteriorates. A worker who has lost their job as a result of the transition to RE may find it difficult to bear the higher energy costs, making them less likely to support or adopt new technologies, even if those technologies provide long-term benefits such as reduced emissions or energy efficiency. In the same way, a lack of adoption of new technologies may impede efforts to transform job opportunities. If communities are not informed of the benefits of RE and modern energy networks, they may resist job retraining attempts or fail to recognize the worth of switching from traditional energy jobs to green energy positions. This can lead to a loop in which new technologies are not adopted, job potential in green industries is unexplored, and cost remains a significant issue.

To address these interconnected challenges, a comprehensive, equitable transition framework is required. This framework should include financial incentives and subsidies to make RE more cheap, public education campaigns to increase adoption of new technologies, and retraining and job creation programs to assist workers impacted by the energy transition. Policymakers must prioritize collaboration across governments, industry, and communities to develop long-term job prospects in the green energy sector, ensuring that no group is left behind in the

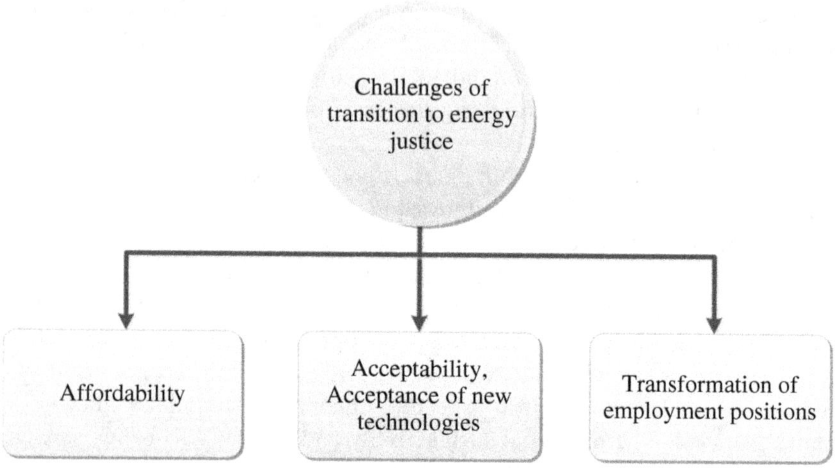

Figure 2.6 Some challenges of transition to energy justice.

transition to a low-carbon economy. By combining these solutions affordability programs, educational outreach, and labor reskilling we can overcome the barriers to achieving EJ while promoting equality and sustainability throughout society (Figure 2.6).

2.6 Modernizing Energy: A Catalyst for Energy Justice

Following the EJ vision, energy GM emerges as a transformative component of the energy system. Beyond addressing existing gaps, the primary goal is to propel EJ toward achieving its broader objectives. EJ, as a component of sustainable development, requires a thorough and comprehensive strategy to ensure that the benefits of the energy system reach every sector of society. Understanding GM's significance in its broad impact highlights its influence across sectors and its alignment with various EJ goals. The important interplay between modernizing energy networks and promoting an inclusive and equitable energy future emphasizes the interconnectedness of their activities. As energy networks continue to evolve, their efficiency improves while simultaneously creating opportunities for equitable distribution of benefits. Technological advancements in energy networks have the potential to dismantle barriers, ensuring that the benefits of a burgeoning energy landscape are accessible to all. In the following, we delve into the importance and varied effects of GM on various sectors and goals of EJ, explaining how technological progress in energy networks can facilitate and pave the way for an inclusive and fair energy future.

2.6.1 Democratizing Access to Energy Resources

At the heart of EJ is the principle that all members of society, irrespective of their economic status, should have equal access to energy resources. This concept is more than just a conceptual ideal; it serves as a rallying point for a transformative movement set to revolutionize the energy landscape. GM is a key driver of this movement, with significant potential to democratize energy access. Modern networks reduce barriers to energy access by adopting decentralized energy production and distribution. Modern networks integrated with RE sources usher in a new era, empowering communities to take control of their energy future. These communities are no longer dependent on centralized and often unfair energy systems, and they can rely on their own ability to generate and consume energy on a local scale. Moving away from traditional top-down energy systems and toward decentralized and community-based models can help empower marginalized groups. The benefits of this development extend beyond energy generation, as it causes these communities to manage and control resources, which, if managed effectively, promotes their social and economic revolutions. Individuals and communities, according to this viewpoint, become not only consumers of energy but also producers, which reduces their environmental impact. The decentralization of energy resources enables local economies to provide jobs and growth in previously ignored areas [1, 60].

2.6.2 Addressing Energy Poverty

Modern energy networks play a crucial part in reducing energy usage, which is an important issue in marginalized areas. New infrastructure developments pave the way for the development of energy-efficient technologies, resulting in significant savings in energy usage and household expenses. Alongside advancements in energy network technologies, the integration of energy sources is reducing costs, easing the burden of high energy expenses for those who have long shouldered them. GM addresses energy poverty by increasing efficiency while also promoting EJ ideals [61]. The integration of multiple energy sources in modern networks offers additional benefits, such as the ability to switch between sources when one becomes insufficient. This integration also increases management capabilities, as well as economic and environmental efficiencies, and allows for the provision of energy access to end consumers [62].

2.6.3 Improvement of Environmental Justice

EJ is not only an economic issue but also an environmental one, especially for societies that have suffered the effects of environmental pollution and destruction under the conventional energy system. Restoring energy networks is a vital step closely linked to the transition toward cleaner and RE sources, with ongoing efforts in this direction. These significant changes not only reduce the environmental impact of energy generation but also safeguard vulnerable communities from pollution. However, this process goes beyond basic protection and establishes the conditions for an equitable distribution of the advantages of a cleaner environment. Transforming energy networks into modern, sustainable systems provides a foundation for environmental justice within the broader framework of EJ. This transformation will help rectify historical environmental inequities and bring societies closer to a future where everyone benefits from a cleaner, greener planet [49].

2.6.4 Engagement and Increased Community Empowerment

Modernizing energy networks enhances community engagement and empowerment, emphasizing the social dimensions of EJ. By involving the local community in the development, implementation, and maintenance of modern

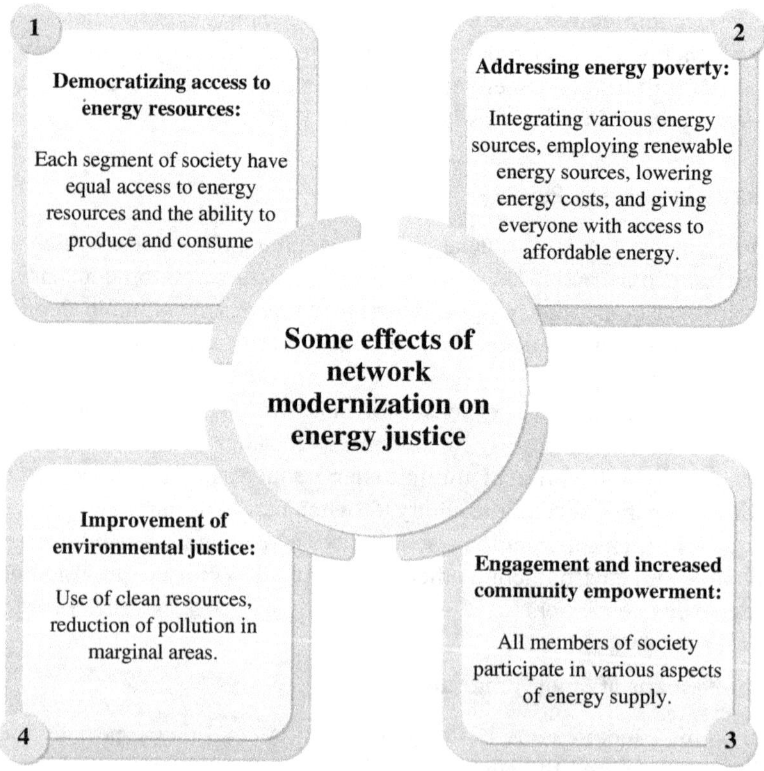

Figure 2.7 Some impacts of modernizing energy networks on energy justice.

energy infrastructure, technological advancements are more equitably distributed. Educational activities focused on new technologies provide opportunities for community members to actively participate in the energy transition movement. This community-based strategy not only empowers individuals but it also strengthens the social sector and provides the groundwork for developing a more equitable and adaptable energy economy. With the modernization of energy networks and their decentralization, community members can actively participate in the planning and execution processes, increasing their sense of ownership over local energy projects. This approach enhances the social, economic, and environmental dimensions of energy systems, improving their overall societal impact. Overall, improving energy networks allows for a wider range of community participation, which not only maintains the beneficial impacts of EJ but also gives a flexible framework for the continued evolution of sustainable energy practices [63, 64] (Figure 2.7).

2.7 Impact of Energy Justice on Grid Modernization Aspects

The influence of EJ on GM underscores the connection between the vital goal of equitable energy access and the pressing need to upgrade and optimize energy networks. EJ is a concept that focuses on ensuring a just and equal allocation of energy resources and benefits. EJ has become central to discussions on sustainable development and climate action. As communities work to improve their energy infrastructure through GM programs, it is essential to consider the potential impacts on energy equity. Examining various aspects of GM, such as hardware upgrades, security improvements, redundancy, performance optimization, and system monitoring through the lens of EJ, provides a clearer understanding of how these initiatives can either promote or hinder EJ. The following sections will explore how these features specifically impact justice and energy equality [1, 16].

2.7.1 Hardware Upgrades

Hardware upgrades are a critical component of network modernization, often incorporating energy-efficient devices. However, addressing EJ is essential in this context, particularly in ensuring the equitable distribution of costs associated with these upgrades. Marginalized groups, who are already constrained by financial limitations, may encounter disproportionate difficulties in obtaining energy-efficient solutions, thus worsening current injustices. In order to tackle these concerns, it is crucial to give priority to making energy-efficient hardware upgrades affordable and accessible. This will ensure that not only are these upgrades economically feasible but they are also available to a wide range of people. To prevent further marginalization and promote a JT toward energy-efficient infrastructure, stakeholders should implement policies and initiatives that improve affordability and remove financial barriers. This approach is in line with the principles of EJ, which seek to protect the rights of all communities to access sustainable and technologically advanced solutions without any form of discrimination or exclusion [1, 65].

2.7.2 Security Enhancements

The implementation of security measures can have a substantial impact on the dynamics of EJ. Enhancing network security measures has the potential to protect communities from cyberattacks while ensuring the reliability and integrity of energy supplies. Nevertheless, it is essential to acknowledge that the introduction of certain security protocols can unintentionally worsen preexisting inequalities, especially those affecting marginalized groups. Security protocols that use a significant amount of energy or depend on costly technology may create accessibility difficulties for socioeconomically disadvantaged communities. Achieving a balance between the need for security and the importance of EJ requires a careful and subtle approach. This entails giving priority to the implementation of cost-effective and energy-efficient security measures that not only enhance the resilience of the network but also reduce the negative impact on vulnerable populations. By giving priority to inclusivity and sustainability in efforts to strengthen security, stakeholders may guarantee that the advantages of increased security are distributed equitably among all members of society in accordance with the principles of EJ [66].

2.7.3 Redundancy and High Availability

Integrating redundancy and high availability measures is essential for improving energy resilience and reliability in GM efforts. While these approaches have the potential to strengthen energy systems against disruptions, it is important to assess their impacts from the perspective of EJ. Financial accessibility for marginalized populations may be compromised if the expenses of establishing redundant infrastructure or backup energy sources become unaffordable. Therefore, it is vital to prioritize the accessibility and affordability of energy resilience measures for all communities to uphold the principles of EJ in GM efforts. Emphasizing the affordability and accessibility of redundancy and high availability measures allows stakeholders to ensure equitable access to reliable energy infrastructure, advancing the goals of EJ in modernized networks [67, 68].

2.7.4 Performance Optimization

Optimizing network performance presents an opportunity to improve energy efficiency by minimizing wasteful resource usage. However, in terms of EJ, it is important to carefully analyze the potential trade-offs between maximizing efficiency and ensuring equitable access to network resources. Although prioritizing efficiency is important, it is equally vital to prevent any unintended negative effects that could exacerbate service inequalities or hinder access to essential services, particularly for marginalized communities. Therefore, finding an appropriate balance between improving performance and ensuring accessibility is crucial. By giving priority to solutions that not only maximize performance but also provide fair access to network resources, stakeholders may address concerns about EJ and promote a network infrastructure that is more inclusive and sustainable [69].

2.7.5 Monitoring and Management

Providing monitoring and management tools is essential for enhancing energy efficiency in GM efforts. These tools can significantly reduce wasteful energy use by identifying and addressing inefficiencies in network operations. However, in the context of EJ, it is crucial to focus on ensuring equitable access to these resources. Disparities may arise when certain communities do not have the necessary resources or infrastructure to properly use monitoring and management systems or when specific groups are unfairly burdened with the job of overseeing energy consumption. To address these challenges, it is crucial to provide support and resources for community-driven programs focused on overseeing and managing energy consumption. By providing communities with the necessary means and expertise to successfully monitor their energy consumption, stakeholders may promote inclusivity and justice in efforts to modernize the energy network, thus furthering the concepts of EJ [70, 71].

2.8 Energy Justice and the Adoption of Innovative Technologies

EJ and equity are crucial for fostering the broad acceptance of innovative technologies and RE sources. EJ advocates for equitable access to sustainable energy solutions despite socioeconomic status, fostering an environment where marginalized populations can enjoy the advantages of RE. This acceptance addresses not only environmental concerns but also economic inequities. Communities with equitable access to RE sources, such as solar or wind power, benefit from reduced energy costs, improved health outcomes, and increased resilience to energy-related emergencies. In addition, advocating for EJ promotes the establishment of localized RE efforts, empowering communities to exert more authority over their energy generation and usage. This, in turn, promotes a feeling of ownership and involvement in the shift toward cleaner energy sources [56].

EJ programs have also been extremely important in stimulating innovation in the field of RE. In order to sidestep the high initial costs that are connected with the installation of individual solar panels, community-owned solar projects have shown to be an effective means of providing marginalized communities with access to sustainable energy. Through the sharing of ownership of solar power plants, communities can collectively lower their energy bills and benefit from RE, hence leading to increased economic resilience in the local community. The implementation of projects such as microgrids in remote areas has made it possible for marginalized populations to acquire RE that is both dependable and affordable. This has resulted in an increase in energy security and the promotion of local governance of energy systems. Through the implementation of these projects, environmental justice programs have demonstrated their capacity to effectively encourage technological innovation while simultaneously addressing socioeconomic disparities [63].

Furthermore, energy equity stimulates innovation in the RE sector. Incorporating several kinds of viewpoints and opinions in the decision-making processes related to energy infrastructure results in the creation of more efficient and enduring solutions. Emphasizing equitable access to RE technology motivates policymakers and industry leaders to allocate resources for research and development that addresses the needs of marginalized communities. For instance, campaigns that focus on energy-efficient housing for families with low incomes have led to advancements in house retrofitting technology, which have made energy efficiency more accessible and inexpensive. In addition, the proliferation of energy cooperatives makes it possible for communities to combine their resources and make investments in RE infrastructure, such as wind and solar farms, and then reap the financial and environmental benefits of these investments. These cooperatives have been successful in empowering local communities, fostering a collective shift toward sustainable practices, and guaranteeing that even economically disadvantaged populations are able to participate in the transition to a more sustainable energy system. Moreover, promotion of energy equity encourages collaboration among government agencies, private sector stakeholders, and grassroots organizations, creating an environment conducive to technological advancement and implementation. EJ and equity programs assist bridge the gap between technological innovation and practical use in places that are disadvantaged. This is accomplished by ensuring that all voices are engaged in the processes of energy planning. This helps to create an environment in which cutting-edge clean energy technology may be evaluated and put into practice in an efficient manner, which will have positive effects on society over a long period of time. Integrating the concepts of EJ and equality into energy policy and planning can expedite the transition toward a sustainable and inclusive energy system [56, 72].

2.9 Conclusion

This chapter offers a detailed analysis of the relationship between modernizing energy networks and fostering a more equitable energy landscape. GM is unquestionably important since it serves as a basis for a sustainable and efficient energy infrastructure. As described in previous sections, modernizing energy networks improves dependability, resilience, and availability, consequently contributing to energy system sustainability. This technical advancement lays the groundwork for cleaner, more sustainable energy sources. EJ emerges as a critical lens through which the influence of energy policies on various socioeconomic groups is investigated. It is clear that the benefits and drawbacks of the energy transition are unevenly distributed, with vulnerable populations often bearing the greatest environmental and social impacts. The concept of "JT" highlights the importance of addressing inequality and developing a fair framework for all stakeholders. Addressing the transition to EJ requires a multifaceted approach. From policy formation to community engagement, concerted efforts are required to prevent negative consequences on vulnerable populations while also ensuring that the advantages of modernization are shared equitably. In this context, modernizing energy networks can effectively drive progress toward achieving EJ. Enhancing infrastructure not only boosts energy system efficiency but also helps address historical injustices and support community-based initiatives. The advances in technology today allow humans to create networks that empower people, overcome socioeconomic disparities, and lead the path to a more sustainable and equitable energy future. In the modern dynamic energy transmission landscape, GM and EJ are not only desirable but also critical. Moving forward, it is essential to consider diverse perspectives and develop policies that prioritize fairness and equity. The areas discussed here serve as a foundation for ongoing research, policy creation, and collaborative action to create a more equitable and sustainable energy ecosystem for current and future generations. The main topics and reviewed results addressed in this chapter are summarized in the following list:

- GM: The process of modernizing energy networks includes enhancing infrastructure and operating frameworks to efficiently incorporate dispersed energy resources, enhance safety and dependability, and optimize the benefits of these resources for customers.
- EJ: The significance of energy in modern society emphasizes the necessity to shift toward low-carbon and equitable energy systems. Research has examined the concept of EJ, which prioritizes equitable distribution, identification of affected individuals, and fair decision-making processes, particularly within the context of transitioning to low-carbon alternatives.

- JT: To transition from environmental to EJ, addressing disparities in the energy system and empowering marginalized communities is vital, along with integrating environmental and energy considerations into decision-making. This shift emphasizes fair access to clean energy, inclusive policies for fossil fuel industry workers and aims for a balanced relationship between environmental sustainability, social equity, and economic justice.
- Challenges of transition: Achieving sustainability and EJ necessitates a substantial departure from conventional energy systems, resulting in issues pertaining to affordability, adoption of new technologies, and change in the jobs. This underscores the significance of adopting a well-rounded and inclusive strategy.
- Catalyst for EJ: The modernization of energy networks, as promoted by the concept of EJ, has a significant impact on promoting equitable access to energy and achieving sustainable development objectives. It involves a thorough plan to ensure that all sections of society benefit from the energy system, in line with various EJ goals and promoting an inclusive and equitable energy future through technical advancements.

Abbreviations

EJ	energy justice
ESS	energy storage system
JT	just transition
RE	renewable energy
DERs	distributed energy resources
DC	direct current
EMS	energy management systems
GM	grid modernization
GEA	global energy assessment
SG	smart grid
CPUC	california public utilities commission
AMI	advanced metering infrastructure
AC	alternating current

References

1 Daneshvar, M., Asadi, S., and Mohammadi-Ivatloo, B. (2021). *Grid Modernization–Future Energy Network Infrastructure*. Springer.

2 Yadollahi, Z., Gharibi, R., Dashti, R., and Jahromi, A.T. (2024). Optimal energy management of energy hub: a reinforcement learning approach. *Sustainable Cities and Society* 102: 105179.

3 Lippert, I. and Sareen, S. (2023). Alleviation of energy poverty through transitions to low-carbon energy infrastructure. *Energy Research and Social Science* 100: 103087.

4 Carley, S., Engle, C., and Konisky, D.M. (2021). An analysis of energy justice programs across the United States. *Energy Policy* 152: 112219.

5 Healy, N. and Barry, J. (2017). Politicizing energy justice and energy system transitions: fossil fuel divestment and a "just transition". *Energy Policy* 108: 451–459.

6 Team, G.E.A.W. (2012). *Global Energy Assessment: Toward A Sustainable Future*. Cambridge University Press.

7 Van Uffelen, N., Taebi, B., and Pesch, U. (2024). Revisiting the energy justice framework: doing justice to normative uncertainties. *Renewable and Sustainable Energy Reviews* 189: 113974.

8 McCauley, D., Ramasar, V., Heffron, R.J. et al. (2019). Energy justice in the transition to low carbon energy systems: exploring key themes in interdisciplinary research. *Applied Energy* 233: 916–921.

9 Raimi, D., Barone, A., Carley, S. et al. (2021). Policy options to enable an equitable energy transition. Resources for the Future. https://www.rff.org/publications/reports/policyoptions–to-enable-an-equitable-energy-transition (accessed 7 January 2025).

10 Osička, J., Szulecki, K., and Jenkins, K.E.H. (2023). Energy justice and energy democracy: separated twins, rival concepts or just buzzwords? *Energy Research and Social Science* 104: 103266.

11 Levenda, A.M., Behrsin, I., and Disano, F. (2021). Renewable energy for whom? A global systematic review of the environmental justice implications of renewable energy technologies. *Energy Research and Social Science* 71: 101837.

12 Bailey, I.G. (2017). Self-recognition and the integration of energy and environmental-social justice into renewable energy siting. *Local Environment* 10 (13549839): 1418848.

13 Volodzkiene, L. and Streimikiene, D. (2024). Towards energy equity: understanding and addressing multifaceted energy inequality. *Energies (Basel)* 17 (17): 4500.

14 Adelekan, O.A., Ilugbusi, B.S., Adisa, O. et al. (2024). Energy transition policies: a global review of shifts towards renewable sources. *Engineering Science & Technology Journal* 5 (2): 272–287.

15 Cooke, A.L., Homer, J.S., and Schwartz, L.C. (2018). *Distribution System Planning-State Examples by Topic (No. PNNL-27366)*. Richland, WA (United States): Pacific Northwest National Laboratory (PNNL).

16 Salter, R., Gonzalez, C.G., and Warner, E.A.K. (2018). Energy justice: frameworks for energy law and policy. In: *Energy Justice* (ed. R. Salter, C.G. Gonzalez, and E.A. Kronk Warner), 1–11. Edward Elgar Publishing.

17 Munro, P. (2023). Energy justice, modernity and transitions: more-than-modern energy for all in the Global South. In: *Handbook on Energy Justice* (ed. S. Bouzarovski, S. Fuller, and T. Reames), 213–225. Edward Elgar Publishing.

18 Sovacool, B.K., Carley, S., Kiesling, L., and Heleno, M. (2024). Energy justice and equity: applying a critical perspective to the electrical power grid for a more just transition in the United States. *IEEE Power and Energy Magazine* 22 (4): 18–25.

19 Chong, A.T.Y., Mahmoud, M.A., Lim, F.-C., and Kasim, H. (2020). A review of smart grid technology, components, and implementation. *2020 8th International Conference on Information Technology and Multimedia (ICIMU)*, 166–169.

20 Mitra, S., Chakraborty, B., and Mitra, P. (2024). Smart meter data analytics applications for secure, reliable and robust grid system: survey and future directions. *Energy* 289: 129920.

21 Mochinski, M.A., Biczkowski, M., Chueiri, I.J. et al. (2024). Developing an intelligent decision support system for large-scale smart grid communication network planning. *Knowledge-Based Systems* 283: 111159.

22 Elmouatamid, A., Ouladsine, R., Bakhouya, M. et al. (2020). Review of control and energy management approaches in micro-grid systems. *Energies (Basel)* 14 (1): 168.

23 Khalid, M. (2024). Smart grids and renewable energy systems: perspectives and grid integration challenges. *Energy Strategy Reviews* 51: 101299.

24 Sebestyén, V. (2021). Renewable and sustainable energy reviews: environmental impact networks of renewable energy power plants. *Renewable and Sustainable Energy Reviews* 151: 111626.

25 Mahafzah, K.A., Obeidat, M.A., Mansour, A. et al. (2024). A new smart grid hybrid DC–DC converter with improved voltage gain and synchronized multiple outputs. *Applied Sciences* 14 (6): 2274.

26 Darvishi, A., Ranjbar, B., Gharibi, R. et al. (2024). Multi-objective optimization of a socio-economic energy hub with demand response program and considering customer satisfaction. *Journal of Energy Storage* 100: 113624.

27 Olabi, A.G., Onumaegbu, C., Wilberforce, T. et al. (2021). Critical review of energy storage systems. *Energy* 214: 118987.

28 Bomma, R.R., Jayakumar, J., and Bogaraj, T. (2020). Different types of energy storage systems: a literature survey. *Innovations in Electrical and Electronics Engineering: Proceedings of the 4th ICIEEE 2019*, 515–540.

29 Abdi, H., Mohammadi-ivatloo, B., Javadi, S. et al. (2017). Energy storage systems. *Distributed Generation Systems* 7: 333–368.

30 Athanasiadis, C.L., Papadopoulos, T.A., Kryonidis, G.C., and Doukas, D.I. (2024). A review of distribution network applications based on smart meter data analytics. *Renewable and Sustainable Energy Reviews* 191: 114151.

31 Oad, A., Ahmad, H.G., Talpur, M.S.H. et al. (2023). Green smart grid predictive analysis to integrate sustainable energy of emerging V2G in smart city technologies. *Optik (Stuttg)* 272: 170146.

32 Entezari, A., Aslani, A., Zahedi, R., and Noorollahi, Y. (2023). Artificial intelligence and machine learning in energy systems: a bibliographic perspective. *Energy Strategy Reviews* 45: 101017.

33 Apergi, M., Eicke, L., Goldthau, A. et al. (2024). An energy justice index for the energy transition in the global South. *Renewable and Sustainable Energy Reviews* 192: 114238.

34 Jenkins, K.E., Stephens, J.C., Reames, T.G., and Hernández, D. (2020). Towards impactful energy justice research: transforming the power of academic engagement. *Energy Research and Social Science* 67: 101510.

35 Hosan, S., Sen, K.K., Rahman, M.M. et al. (2024). Energy innovation funding and social equity: mediating role of just energy transition. *Renewable and Sustainable Energy Reviews* 197: 114405.

36 Wood, N. (2023). Problematising energy justice: towards conceptual and normative alignment. *Energy Research and Social Science* 97: 102993.

37 Opoku, E.E.O. and Acheampong, A.O. (2023). Energy justice and economic growth: does democracy matter? *Journal of Policy Modeling* 45 (1): 160–186.

38 Sovacool, B.K., Hess, D.J., and Cantoni, R. (2021). Energy transitions from the cradle to the grave: a meta-theoretical framework integrating responsible innovation, social practices, and energy justice. *Energy Research and Social Science* 75: 102027.

39 Fraser, N. (2008). Social justice in the age of identity politics: redistribution, recognition, and participation. In: *Geographic Thought* (ed. G. Henderson and M. Waterstone), 72–89. Routledge.

40 Milchram, C., Künneke, R., Doorn, N. et al. (2020). Designing for justice in electricity systems: a comparison of smart grid experiments in the Netherlands. *Energy Policy* 147: 111720.

41 Heffron, R.J. (2022). Applying energy justice into the energy transition. *Renewable and Sustainable Energy Reviews* 156: 111936.

42 Stojilovska, A. (2023). Energy poverty and the role of institutions: exploring procedural energy justice–Ombudsman in focus. *Journal of Environmental Policy & Planning* 25 (2): 169–181.

43 Pai, S., Harrison, K., and Zerriffi, H. (2020). *A Systematic Review of the Key Elements of a Just Transition for Fossil Fuel Workers*. Canada, ON: Smart Prosperity Institute Ottawa.

44 Cha, J.M. (2020). Just transition: tools for protecting workers and their communities at risk of displacement due to climate policy. *Putting California on the high road: A jobs and climate action plan for* 2030: 149–175.

45 Figueroa, R.M. (2022). Environmental justice. In: *The Routledge Companion to Environmental Ethics* (ed. B. Hale, A. Light, and L. Lawhon), 767–782. Routledge.

46 Méndez, M. (2020). *Climate Change from the Streets: How Conflict and Collaboration Strengthen the Environmental Justice Movement*. Yale University Press.

47 Birkmann, J., Liwenga, E., Pandey, R. et al. (2022). Poverty, livelihoods and sustainable development. In: *Climate Change 2022: Impacts, Adaptation and Vulnerability. Contribution of Working Group II to the Sixth Assessment Report of the Intergovernmental Panel on Climate Change* (ed. H.-O. Pörtner, D.C. Roberts, M. Tignor, et al.), 1171–1274. Cambridge, UK and New York, NY, USA: Cambridge University Press https://doi.org/10.1017/9781009325844.010.

48 Monica, C., Diego, G., Federico, P. et al. (2023). *GHG Emissions of All World Countries*. Belgium: Publications Office of the European Union https://doi.org/10.2760/235266, ISBN: 978-92-68-07550-0 (online),978-92-68-07551-7 (print).

49 Bullard, R.D. (2018). *Dumping in Dixie: Race, Class, and Environmental Quality*. Routledge.

50 Jenkins, K., McCauley, D., Heffron, R. et al. (2016). Energy justice: a conceptual review. *Energy Research and Social Science* 11: 174–182.

51 Fitzgerald, L.M. (2023). Tracing the development of Anti-Fossil Fuel Norms: insights from the Republic of Ireland. *Climate Policy* 23 (9): 1101–1114.

52 Hanke, F., Guyet, R., and Feenstra, M. (2021). Do renewable energy communities deliver energy justice? Exploring insights from 71 European cases. *Energy Research and Social Science* 80: 102244.

53 Antunes, M., Teotónio, C., Quintal, C., and Martins, R. (2023). Energy affordability across and within 26 European countries: insights into the prevalence and depth of problems using microeconomic data. *Energy Economics* 127: 107044.

54 Singh, S. and Ru, J. (2022). Accessibility, affordability, and efficiency of clean energy: a review and research agenda. *Environmental Science and Pollution Research* 29 (13): 18333–18347.

55 González-Torres, M., Pérez-Lombard, L., Coronel, J.F. et al. (2022). A review on buildings energy information: trends, end-uses, fuels and drivers. *Energy Reports* 8: 626–637.

56 Hazrati, M. (2023). Social acceptance for renewable energy technologies: the role of the energy justice framework. In: *The Power of Energy Justice & the Social Contract* (ed. R.J. Heffron and L. de Fontenelle), 83–91. Springer.

57 Ellis, G., Schneider, N., and Wüstenhagen, R. (2023). Dynamics of social acceptance of renewable energy: an introduction to the concept. *Energy Policy* 181: 113706.

58 Sievers, L. (2020). Regionale Verteilungswirkung der Energiewende. PhD Thesis. Forthcoming., Universität der Bundeswehr, München.

59 Sovacool, B.K., Martiskainen, M., Hook, A., and Baker, L. (2019). Decarbonization and its discontents: a critical energy justice perspective on four low-carbon transitions. *Climatic Change* 155: 581–619.

60 e Silva, F.A.C., da Costa, V.B.F., dos Santos, I.F.S., and Bonatto, B.D. (2024). Electric sector modernization in Brazil: milestones, challenges, and prospects. *Utilities Policy* 90: 101793.

61 Welton, S. (2016). Grid modernization and energy poverty. *NCJL & Technology* 18: 565.

62 Gharibi, R. and Vahidi, B. (2022). Coordinated planning of thermal and electrical networks. *Coordinated Operation and Planning of Modern Heat and Electricity Incorporated Networks* 449–479.

63 Hogan, J.L. (2024). Why does community ownership foster greater acceptance of renewable projects? *Investigating energy justice explanations. Local Environ* 29 (9): 1221–1243.

64 Jenkins, K.E.H. (2019). Energy justice, energy democracy, and sustainability: normative approaches to the consumer ownership of renewables. *Energy Transition: Financing Consumer Co-Ownership in Renewables* 1: 79–97.

65 Shabalov, M.Y., Zhukovskiy, Y.L., Buldysko, A.D. et al. (2021). The influence of technological changes in energy efficiency on the infrastructure deterioration in the energy sector. *Energy Reports* 7: 2664–2680.

66 Canaan, B., Colicchio, B., and Ould Abdeslam, D. (2020). Microgrid cyber-security: review and challenges toward resilience. *Applied Sciences* 10 (16): 5649.

67 Netes, V. (2020). Modern network technologies and dependability. *2020 International Scientific and Technical Conference Modern Computer Network Technologies (MoNeTeC)*, 1–10.

68 Zengin, Z.O. (2023). The energy justice imperative for clean energy storage alternatives. In: *The Power of Energy Justice & the Social Contract* (ed. R.J. Heffron and L. de Fontenelle), 217–225. Springer.

69 Vågerö, O. and Zeyringer, M. (2023). Can we optimise for justice? Reviewing the inclusion of energy justice in energy system optimisation models. *Energy Research and Social Science* 95: 102913.

70 Bonilla, D., Samaniego, M.G., Ramos, R., and Campbell, H. (2018). Practical and low-cost monitoring tool for building energy management systems using virtual instrumentation. *Sustainable Cities and Society* 39: 155–162.

71 Riyaz, W. and Sood, E.S. (2024). A smart grid monitoring system. In: *Latest Trends in Engineering and Technology* (ed. S. Singh and S. Kaur), 340–344. CRC Press.

72 Finley-Brook, M. and Holloman, E.L. (2016). Empowering energy justice. *International Journal of Environmental Research and Public Health* 13 (9): 926.

3

Decision-Making Tools for Advancing Clean Energy Justice and Equity with Supply Chain and Project Management

Monica Bhatia[1] *and Pradyumn Chaturvedi*[2]

[1] *School of Business Management, NMIMS (Deemed to be University), Indore, India*
[2] *EED, VNIT, Nagpur, India*

Acronyms

CfDs	contracts for differences
DSS	decision support system
MCDA	multi-criteria decision analysis
SWOT	strengths, weakness, opportunity, threats

3.1 Introduction

In the ongoing pursuit of sustainable development, the integration of clean energy solutions has emerged as a paramount goal for nations, industries, and communities worldwide. However, the transition to clean energy is not merely a matter of technological advancement; it is equally about ensuring justice and equity throughout the supply chain and project management processes. As we navigate the complexities of this transition, decision-making tools play a pivotal role in shaping the trajectory toward a future where clean energy is not only accessible but also justly distributed [1, 2]. The book titled *Communities for Clean Energy Justice and Equity in Grid Modernization* emphasizes the critical role of communities in achieving equitable and just access to clean energy as part of modernizing energy grids. It focuses on the social, economic, and environmental impacts of grid modernization, ensuring that all stakeholders, particularly underrepresented and marginalized groups, benefit from the clean energy transition. The transition to clean energy is critical for mitigating climate change and fostering a sustainable future. However, achieving this transition equitably

Communities for Clean Energy Justice and Equity in Grid Modernization, First Edition.
Edited by Mohammadreza Daneshvar, Behnam Mohammadi-Ivatloo, and Amjad Anvari-Moghaddam.

requires addressing the social and economic challenges that disproportionately affect marginalized and vulnerable communities. The linkage between clean energy and justice is essential for ensuring that the benefits of decarbonization, such as reduced pollution and green jobs, are shared by all, while the burdens, such as economic displacement or environmental degradation, are not unfairly concentrated.

The chapter titled "Decision-Making Tools for Advancing Clean Energy Justice and Equity with Supply Chain and Project Management" aligns with the book's broader objectives by providing a practical approach to achieving clean energy justice and equity. The chapter contributes to the book's discourse on grid modernization by exploring how supply chain and project management can be leveraged to ensure fair distribution of clean energy benefits. This integration aligns with the book's aim to create energy systems that serve all communities, making sure that the technical aspects of grid modernization do not overshadow the social imperative of equitable energy access.

3.1.1 Motivation

The urgency of addressing climate change and environmental degradation has brought clean energy to the forefront of global agendas. From solar and wind to hydro and geothermal, renewable energy sources offer promising alternatives to traditional fossil fuels. Yet, the deployment and management of clean energy infrastructure entail multifaceted challenges, particularly concerning equity and justice. This introduction sets the stage for exploring decision-making tools tailored to address these challenges, facilitating the advancement of clean energy justice and equity in both supply chain operations and project management [3].

At the heart of the clean energy transition lies the concept of justice, encompassing environmental, social, and economic dimensions. Environmental justice demands that the burdens and benefits of energy production and consumption are equitably distributed among all communities, regardless of socioeconomic status or geographical location. Social justice requires inclusive decision-making processes that empower marginalized groups and ensure their voices are heard in shaping energy policies and projects. Economic justice entails fostering opportunities for equitable participation in the clean energy economy, thereby reducing disparities and promoting shared prosperity. In the context of clean energy supply chains, decision-making tools play a critical role in promoting justice and equity. Supply chains for renewable energy technologies, such as solar panels and wind turbines, span across multiple stages, from raw material extraction to manufacturing, transportation, and installation. At each stage, decisions regarding sourcing, production methods, logistics, and labor practices

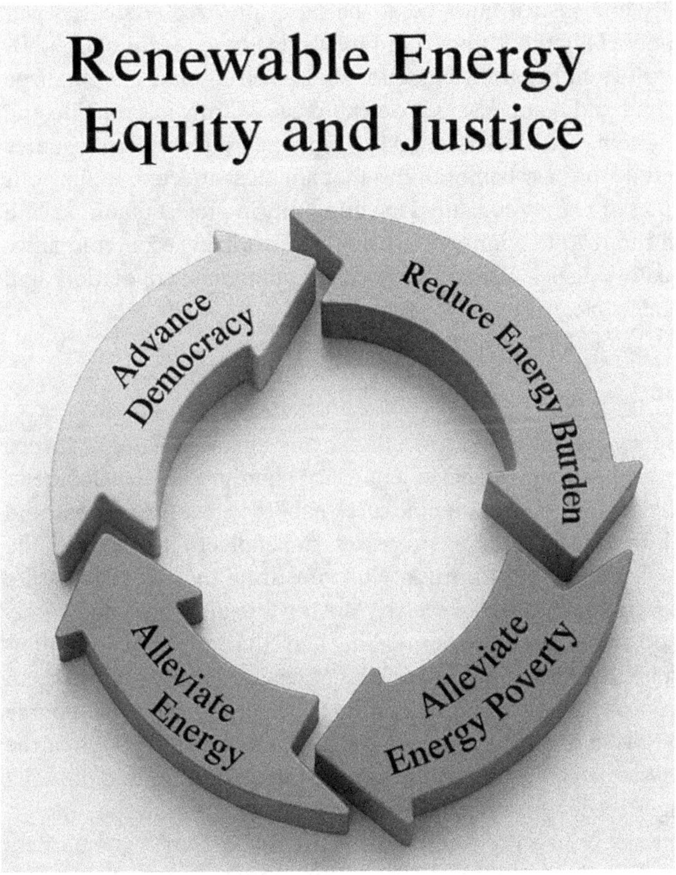

Figure 3.1 Framework of renewable energy equity and justice.

have far-reaching implications for environmental sustainability, social equity, and economic development [4]. Figure 3.1 shows the framework of renewable energy equity and justice. The framework of renewable energy equity and justice encompasses principles and strategies aimed at ensuring fair and inclusive access to renewable energy resources and benefits while addressing historical inequalities and social disparities. This framework is grounded in the recognition that the transition to renewable energy must prioritize the needs and rights of marginalized communities, including low-income households, communities of color, indigenous peoples, and other vulnerable populations.

The clean energy transition offers enormous potential to combat climate change, but it cannot be achieved at the expense of the world's most vulnerable communities. Equity and justice must be at the heart of these strategies, particularly in the supply chains that underpin renewable energy technologies. By addressing the inequities in resource extraction and ensuring that marginalized groups are empowered and compensated, we can create a truly sustainable and just global energy system. The decision-making processes surrounding resource extraction often exclude the very communities that are most affected. Indigenous and local communities are rarely consulted meaningfully in project planning, and when they are, their input is often ignored in favor of maximizing economic gains. This lack of participatory justice further marginalizes vulnerable populations and fuels resentment against clean energy projects.

3.1.2 Background

Decision-making tools are indispensable for advancing clean energy justice and equity in both supply chain management (SCM) and project development. By integrating principles of environmental sustainability, social inclusion, and economic fairness into decision-making processes, stakeholders can navigate the complexities of the clean energy transition while ensuring that its benefits are shared equitably among all members of society. Moving forward, concerted efforts are needed to develop and deploy decision-making tools that empower communities, promote transparency, and uphold principles of justice throughout the clean energy value chain. Lots of researchers work in the field of renewable energy equity and justice. Nelson et al. [5] analyzed social equity considerations in the renewable energy system. This study contributes to our understanding of social fairness concerning electricity network fees, investments in renewable energy, and household earnings. While previous studies have focused on social fairness in electricity access, there has been limited exploration into how various network tariff structures, particularly those associated with "Contracts-for-Difference" (CfDs) for renewable energy investments, affect individuals with low incomes. To address this gap, we delve into the implications of CfDs, specifically examining whether they disproportionately burden low-income groups when the costs are recuperated through network fees. Through an analysis of energy consumption and network fee structures in Australia, we discover that CfDs can exacerbate inequality, serving as regressive taxation mechanisms used by state governments to finance renewable energy initiatives. We elucidate the differential impacts of CfDs on various energy consumers and propose policy reforms aimed at enhancing social equity in renewable energy strategies. Govindaranjan et al. [6] assessed the concept of energy governance and environmental justice. Promoting renewable energy and transitioning to a low-carbon economy worldwide requires addressing the intricate dynamics of energy governance and environmental justice. To facilitate this integration, we propose a morphological analysis framework comprising six dimensions and twenty-five variations. This framework can effectively assist policymakers across different nations in navigating their transition to a low-carbon economy through renewable energy initiatives. Currently, environmental justice discourse primarily resides within the realm of justice literature. However, achieving a comprehensive understanding of environmental justice necessitates a reinvigorated research focus on emerging energy governance issues. This focus is essential to ensuring global justice while transitioning to low-carbon energy sources through the deployment of renewable energy systems. Montanes et al. [7] analyzed enabling and centering equity and justice in clean energy transition research. Throughout history, energy researchers have concentrated on exploring ways in which power systems can deliver affordable, environmentally friendly, and dependable energy. Although these aspects are pivotal for transitioning to clean energy, the energy justice movement advocates for an equally essential emphasis on fairness and justice. In recent times, scholarly inquiry has aimed to comprehend how decision-making processes and infrastructure planning can facilitate equitable socio-technical transformations within energy systems, spanning from generation to consumption to decommissioning. The chapter provides specific tools and methodologies for decision-making that can guide project managers, policymakers, and community leaders in aligning their efforts with the goals of clean energy justice and equity. This supports the book's larger vision by offering actionable frameworks that can be applied to community-based energy projects, ensuring decisions are made with fairness, inclusivity, and transparency. While the book focuses on communities' involvement in grid modernization, the chapter enhances this by addressing how clean energy initiatives can be managed from a supply chain perspective to support long-term sustainability. The decision-making tools discussed in the chapter help ensure that the grid modernization efforts are not only technically sound but also socially responsible, fostering trust and engagement within communities.

3.1.3 Research Gap

One key challenge in clean energy supply chains is the sourcing of raw materials, such as rare earth metals for solar panels and lithium for batteries. Extractive industries often operate in regions inhabited by indigenous communities and vulnerable populations, raising concerns about land rights, environmental degradation, and social conflicts. Decision-making tools that incorporate principles of responsible sourcing, stakeholder engagement, and impact assessment can help mitigate these risks and promote sustainable supply chain practices [8].

Furthermore, ensuring fair labor practices and worker rights throughout the supply chain is essential for advancing clean energy justice. From the mines and factories where raw materials are extracted and processed to the construction sites where renewable energy infrastructure is installed, workers must be treated ethically and provided with safe working conditions, fair wages, and opportunities for collective bargaining. Decision-making tools, such as ethical sourcing guidelines, labor standards audits, and supplier codes of conduct, can help companies uphold these principles and hold their supply chain partners accountable.

In addition to supply chain considerations, decision-making tools are instrumental in promoting equity and justice in the management of clean energy projects. Project development and implementation involve a multitude of stakeholders, including government agencies, private developers, local communities, and environmental organizations. Balancing competing interests and ensuring equitable outcomes require transparent decision-making processes that prioritize community engagement, participatory planning, and inclusive governance.

Community-based renewable energy projects offer a promising model for advancing clean energy justice at the local level. By involving residents in the planning, ownership, and benefits sharing of renewable energy installations, these projects empower communities to take control of their energy future while addressing local needs and priorities. Decision-making tools, such as participatory mapping, consensus-building techniques, and community benefit agreements, can facilitate meaningful engagement and collaboration among stakeholders, leading to more equitable and sustainable outcomes.

Furthermore, integrating principles of equity and justice into project siting and design is essential for avoiding environmental injustice and social inequities. Historically, marginalized communities, including low-income neighborhoods and communities of color, have disproportionately borne the burdens of pollution and industrial development. To avoid replicating these injustices, decision-makers must proactively assess the potential environmental and social impacts of clean energy projects and prioritize siting criteria that minimize harm and maximize benefits for affected communities.

3.1.4 Contributions

The author's contribution to the research on "Decision-Making Tools for Advancing Clean Energy Justice and Equity with Supply Chain and Project Management" lies in developing innovative approaches to address the intersectionality of clean energy, justice, and equity. By synthesizing principles from SCM, project management, and social justice frameworks, the author aims to offer actionable insights and tools to promote fair and inclusive clean energy transitions. Through empirical analysis and practical case studies, the author contributes to bridging the gap between theory and practice, offering tangible solutions for stakeholders involved in clean energy initiatives.

3.1.5 Objectives

- Identify key challenges and opportunities at the intersection of clean energy, justice, and equity within the supply chain and project management contexts.
- Develop decision-making tools that integrate principles of justice and equity into supply chain and project management practices for clean energy projects.
- Evaluate the effectiveness of these tools through empirical analysis and case studies, assessing their impact on promoting fairness, inclusivity, and community engagement.
- Provide actionable recommendations for policymakers, industry stakeholders, and community organizers to advance clean energy justice and equity through improved decision-making processes through the SWOT analysis.
- Foster interdisciplinary collaboration and knowledge-sharing among researchers, practitioners, and policymakers to drive sustainable and equitable clean energy transitions.
- The chapter's focus on decision-making tools for supply chain and project management provides the necessary operational and strategic framework to achieve the broader goals of clean energy justice and equity in the modernization of energy grids, directly supporting the central theme of the book.

3.2 Renewable Energy Equity and Justice

Renewable energy equity and justice entail ensuring that the benefits and burdens of transitioning to renewable energy sources are distributed fairly across society. This involves addressing disparities in access to clean energy technologies, as well as considering the social, economic, and environmental impacts of renewable energy deployment.

Equity in renewable energy involves ensuring that all individuals and communities, regardless of socioeconomic status or geographic location, have access to affordable and reliable clean energy options. This includes initiatives to make renewable energy technologies more accessible to marginalized communities, such as through incentives, subsidies, or community-owned renewable energy projects. Justice in renewable energy extends beyond access to include considerations of fairness and accountability in decision-making processes, as well as addressing historical and systemic inequalities. This encompasses ensuring that the benefits of renewable energy development are shared equitably while also mitigating any negative impacts on vulnerable communities, such as displacement or environmental degradation [9].

Efforts to promote renewable energy equity and justice may involve policy interventions, community engagement, and collaborative planning processes that prioritize the needs and perspectives of all stakeholders. By prioritizing equity and justice in renewable energy transitions, societies can work toward more inclusive and sustainable energy systems that benefit everyone. Creating a mathematical formulation for renewable energy equity and justice involves quantifying factors related to access, distribution, fairness, and accountability in renewable energy deployment [10].

Let us define

- E_i is the level of access to renewable energy for the community i.
- P_i is the level of participation or engagement of community in renewable energy projects or decision-making processes i.
- B_i is the distribution of benefits from renewable energy projects within the community i.
- C_i is the distribution of costs or burdens associated with renewable energy projects within the community i.
- S_i is the level of sustainability or environmental impact of renewable energy projects within the community i.

Then, a basic formulation for renewable energy equity and justice could be:

1) Equity Index

$$E_{\text{index}} = \frac{\sum_i E_i}{N} \tag{3.1}$$

2) Participation Index

$$P_{\text{index}} = \frac{\sum_i P_i}{N} \tag{3.2}$$

3) Benefit Distribution Index

$$B_{\text{index}} = \frac{\sum_i B_i}{N} \tag{3.3}$$

4) Cost Distribution Index

$$C_{\text{index}} = \frac{\sum_i C_i}{N} \tag{3.4}$$

5) Sustainability Index

$$S_{\text{index}} = \frac{\sum_i S_i}{N}, \tag{3.5}$$

where N is the total number of communities or regions under consideration. These indices can then be used to assess and compare the level of equity and justice in renewable energy deployment across different communities or regions. Adjustments and weighting factors can be incorporated based on specific considerations and priorities within each context. Additionally, qualitative factors and stakeholder perspectives should also be taken into account to ensure a comprehensive evaluation of equity and justice in renewable energy initiatives [11].

3.3 Types of Decision Tools

Decision-making tools are instruments, methodologies, or frameworks designed to facilitate the decision-making process by providing structure, analysis, and insights. These tools help individuals or groups make informed decisions by organizing information, evaluating alternatives, and assessing potential outcomes. The primary purpose of decision-making tools is to improve the quality of decisions, enhance understanding, and mitigate risks associated with uncertainty. Decision-making tools come in various forms, each suited to different contexts and objectives. Following are some common types of decision tools:

- **Decision Trees:** These are graphical representations of decisions and their possible consequences, including probabilities. They are useful for analyzing decisions with multiple possible outcomes. In the context of renewable energy equity and justice, decision trees can assist in identifying the most appropriate course of action for maximizing benefits while ensuring equitable distribution and fair outcomes.
- **Matrix Diagrams:** These tools help to analyze complex decisions by comparing different criteria and options in a matrix format. The matrix diagram, also known as the matrix chart or matrix grid, is a management tool used for organizing and displaying data in a matrix format. It is commonly employed in quality management, project management, and various other fields where complex relationships need to be visualized. In the context of renewable energy equity and justice, a matrix diagram can help stakeholders understand how various dimensions, such as accessibility, affordability, socioeconomic inclusion, and environmental sustainability, intersect and influence each other. This understanding can inform decision-making processes and guide efforts to prioritize actions that promote equity and justice.
- **Pareto Analysis:** This technique helps identify the most important factors among a set of options. It is based on the Pareto Principle, which states that roughly 80% of effects come from 20% of causes. In renewable energy equity and justice, Pareto Analysis can assist in identifying the key barriers or opportunities that disproportionately affect marginalized communities. By focusing resources and interventions on addressing these critical factors, stakeholders can work toward more equitable and just renewable energy systems [12].
- **SWOT Analysis:** SWOT stands for Strengths, Weaknesses, Opportunities, and Threats. This tool helps evaluate the internal and external factors affecting a decision or a situation. SWOT analysis helps identify the strengths, weaknesses, opportunities, and threats associated with renewable energy equity and justice initiatives. By understanding these factors, stakeholders can develop strategies

that leverage strengths, mitigate weaknesses, capitalize on opportunities, and address threats to promote more inclusive and just renewable energy transitions.

- **Decision Matrices:** Similar to matrix diagrams, decision matrices allow for comparing different options against multiple criteria to facilitate decision-making. A Decision Matrix is a tool used to evaluate and prioritize a list of options based on a set of criteria. Each option is scored against each criterion, and weights may be assigned to reflect the importance of each criterion. The scores are then aggregated to identify the best option. It is widely used in business and engineering for rational decision-making
- **Cost–Benefit Analysis:** This tool compares the costs of a decision with its expected benefits to determine if it is economically viable. In the context of equity and justice, cost–benefit analysis can help assess the distributional impacts of renewable energy initiatives on different communities and identify strategies to maximize benefits for underserved populations while minimizing costs and negative externalities [13].
- **Scenario Analysis:** It involves creating multiple scenarios based on different assumptions to assess the potential outcomes of a decision under various conditions. In renewable energy equity and justice, scenario analysis can help stakeholders anticipate and prepare for different social, economic, and environmental changes that may affect the distribution of benefits and burdens associated with renewable energy transitions. By exploring alternative scenarios, stakeholders can develop robust strategies that are resilient to uncertainty and promote more equitable outcomes.
- **Fishbone Diagrams (Ishikawa or Cause-and-Effect Diagrams):** These diagrams help identify potential causes of a problem or an outcome by organizing factors into categories. Fishbone diagrams help identify the root causes of problems or barriers to equity and justice in renewable energy initiatives. By categorizing factors such as policy, technology, economics, and social aspects, stakeholders can visualize the interconnectedness of various factors influencing equity and justice.
- **Force Field Analysis:** This technique examines the driving and restraining forces behind a decision to understand what factors support or hinder its implementation. Force field analysis helps stakeholders identify driving forces that promote equity and justice in renewable energy and restraining forces that hinder progress. By analyzing the balance between these forces, stakeholders can develop strategies to strengthen driving forces and mitigate or overcome restraining forces.
- **Multi-Criteria Decision Analysis (MCDA):** MCDA techniques allow for the comparison of alternatives against multiple criteria, often involving weighting each criterion based on its importance. MCDA refers to a set of techniques

or methods used to evaluate and compare alternatives when multiple, often conflicting, criteria are involved. It provides a structured framework to support complex decision-making by combining qualitative and quantitative data. Examples include the Analytic Hierarchy Process (AHP) and Technique for Order Preference by Similarity to Ideal Solution (TOPSIS).

- **Decision Support Systems (DSS):** These are computer-based tools that provide analytical support to decision-makers by processing data and presenting insights or recommendations. A Decision Support System is a computer-based system that supports business or organizational decision-making activities. DSS helps users make informed decisions by integrating data, sophisticated analytical models, and user-friendly interfaces. It is commonly used in areas like healthcare, finance, and supply chain management.
- **Monte Carlo Simulation:** This technique involves running multiple simulations of a decision using random variables to assess the range of possible outcomes and their probabilities. Monte Carlo simulation enables stakeholders to assess the uncertainty associated with renewable energy equity and justice initiatives. By modeling the impact of uncertain factors such as policy changes, economic fluctuations, and social dynamics, stakeholders can better understand the range of potential outcomes and develop robust strategies that are resilient to uncertainty.
- **Game Theory:** Particularly useful in competitive decision-making contexts, game theory models interactions between decision-makers to analyze optimal strategies. Game theory provides a framework for analyzing strategic interactions among different stakeholders with competing interests. In the context of renewable energy equity and justice, game theory can help identify potential conflicts, negotiate agreements, and design cooperative strategies that balance the needs and preferences of various stakeholders while promoting fairness and inclusivity in decision-making processes.
- **Heuristic Methods:** These are simple, practical decision-making strategies or rules of thumb used to solve problems efficiently, though they may not always guarantee optimal solutions [14]. A Heuristic Method is a problem-solving technique that uses practical, experience-based strategies to find good-enough solutions quickly when exact methods are impractical. Though not guaranteed to be optimal, heuristics are useful for solving complex or large-scale problems efficiently. Examples include greedy algorithms and rule-of-thumb approaches.
- **Analytics and Business Intelligence Tools:** These encompass a range of software applications and techniques for collecting, analyzing, and visualizing data to support decision-making processes. Table 3.1 shows the comparative analysis of different decision-making tools for advancing clean energy justice and equity.

Table 3.1 Comparative analysis of different decision-making tools for advancing clean energy justice and equity.

Decision-making tool	Description	Advantages	Challenges
Decision Trees	Decision Trees are graphical representations of decision-making processes, showing different decision options and their potential outcomes based on various criteria and probabilities.	Provides a systematic framework for evaluating complex decisions and identifying the most favorable outcomes. Helps visualize and communicate decision pathways and uncertainties.	Requires accurate data and assumptions to construct reliable decision trees, which may be challenging in the context of clean energy justice and equity. Can oversimplify complex decision-making processes and overlook qualitative factors.
Matrix Diagrams	Matrix Diagrams organize information into rows and columns to analyze relationships between different factors, such as criteria and alternatives, in decision-making.	Facilitates systematic comparison and prioritization of decision criteria and alternatives. Helps identify patterns and correlations among different factors to inform decision-making.	May become unwieldy and difficult to interpret when analyzing a large number of criteria and alternatives. Limited in capturing dynamic and interdependent relationships among decision factors.
Pareto Analysis	Pareto Analysis prioritizes problems or issues based on their frequency or impact, focusing efforts on addressing the most significant contributors to an outcome.	• Helps identify key drivers of inequality and injustice in clean energy access and distribution. • Enables targeted interventions to address the most pressing challenges and maximize impact.	Assumes that addressing a few high-priority issues will lead to significant improvements in clean energy justice and equity, overlooking systemic and structural barriers. May neglect less frequent but equally important issues that contribute to overall inequities.

SWOT Analysis	SWOT Analysis assesses the strengths, weaknesses, opportunities, and threats associated with a particular decision, strategy, or project, providing a comprehensive overview of internal and external factors.	• Facilitates strategic planning and risk management by identifying internal strengths and weaknesses and external opportunities and threats. • Helps stakeholders anticipate potential challenges and leverage opportunities to advance clean energy justice and equity.	• Can be subjective and prone to biases, depending on the perspectives and priorities of stakeholders involved in the analysis. • May overlook nuanced interactions between different factors and fail to capture emergent trends or uncertainties.
Cost–Benefit Analysis	Cost–Benefit Analysis quantifies the costs and benefits of different alternatives to determine the most economically efficient option, comparing monetary values of outcomes to inform decision-making.	• Provides a rigorous framework for evaluating trade-offs between economic efficiency and equity considerations in clean energy projects. • Helps decision-makers allocate resources efficiently and maximize societal welfare.	• Challenges in assigning monetary values to intangible benefits and social costs, such as environmental impacts and social equity. • Limited in capturing nonmarket values and distributional impacts that are central to clean energy justice and equity.
Scenario Analysis	Scenario Analysis explores different future scenarios and their potential implications on clean energy justice and equity, considering uncertainties and alternative pathways.	• Helps decision-makers anticipate and prepare for future changes and challenges, including technological, economic, and social shifts. • Enables stakeholders to test the robustness of strategies and policies under different plausible futures.	• Requires robust data and modeling capabilities to develop credible scenarios and assess their impacts accurately. • May overlook wild card events and low-probability, high-impact risks that could disrupt clean energy justice and equity efforts.

Source: [15]/Emerald Group Publishing.

3.4 Data Collection and Assessment of Different Countries

Data collection allows researchers and policymakers to identify disparities in renewable energy access and distribution across different demographic groups, geographic regions, or socioeconomic levels. By analyzing these disparities, decision-makers can better understand where inequities exist and formulate targeted interventions to address them. Data-driven insights inform the development of policies and initiatives aimed at promoting renewable energy justice and equity. Policymakers rely on robust data analysis to design targeted interventions, incentives, and regulations that ensure equitable access to renewable energy resources and benefits for all communities, including marginalized groups. Ongoing data collection and analysis are essential for monitoring the implementation and outcomes of renewable energy initiatives over time [16]. By tracking key indicators and performance metrics, stakeholders can assess progress toward equity goals, identify emerging challenges, and adapt strategies accordingly to ensure that benefits are distributed equitably. Data-driven evidence serves as a powerful tool for advocacy efforts aimed at advancing renewable energy justice and equity. By highlighting disparities, documenting impacts, and advocating for policy changes based on empirical findings, advocates can hold decision-makers and stakeholders accountable for addressing inequities in renewable energy development and distribution. Table 3.2 presents data on various countries across several indicators related to renewable energy access, energy poverty, socioeconomic inequality (as measured by the Gini Coefficient), and community engagement. This table provides a snapshot of how different countries compare across these key indicators, offering insights into their renewable energy adoption, energy poverty levels, socioeconomic inequality, and community involvement in energy-related initiatives [17].

The renewable energy access percentage refers to the proportion of the population or the geographical area that has access to renewable energy sources for meeting their energy needs. This percentage is calculated by dividing the number of people or the area with access to renewable energy by the total population or area, respectively, and then multiplying the result by 100 to express it as a percentage. Energy poverty rate refers to the proportion of the population that lacks access to modern energy services or struggles to afford adequate energy for basic needs. Energy poverty is often characterized by households or communities relying on traditional, inefficient, and polluting energy sources such as biomass, kerosene, or coal for cooking, heating, and lighting, which can have adverse effects on health, safety, and well-being. The Gini Coefficient is a widely used socioeconomic indicator that measures income inequality within a population. It provides

Table 3.2 Data set of renewable energy justice and equity.

Country	Renewable energy access (percentage)	Energy poverty rate (percentage)	Socioeconomic indicator (Gini Coefficient)	Community engagement (scale: 1–10)
United States	25	10	0.45	7
Brazil	40	15	0.55	6
India	30	20	0.35	8
South Africa	20	25	0.60	5
Germany	50	5	0.30	9
China	35	12	0.50	7
Nigeria	15	30	0.65	4
Australia	60	8	0.25	8
Mexico	28	18	0.48	6
Japan	45	7	0.32	8
Canada	55	6	0.28	7
United Kingdom	48	9	0.42	8
France	42	11	0.38	6
Spain	38	13	0.47	7
Italy	32	16	0.36	6
Argentina	22	22	0.58	5
Indonesia	18	28	0.70	4
Saudi Arabia	12	35	0.75	3
Sweden	58	4	0.22	9
Russia	20	14	0.52	5

Source: Adapted from [11].

a numerical representation of the distribution of income or wealth among individuals or households within a country or region. The Gini Coefficient provides insights into how income or wealth is distributed among different segments of the population, highlighting disparities and inequalities [18]. Table 3.3 provides statistical summary information for three variables: renewable energy access rate, energy poverty rate, and Gini Coefficient. The mean (average) value for each variable across all observations. For example, the mean renewable energy access rate is 34.65%, the mean energy poverty rate is 15.4%, and the mean Gini Coefficient is 0.45. Figure 3.2 shows a normal probability plot of residuals of the Gini Coefficient. This plot is a graphical tool used to assess whether the residuals (the differences between observed and predicted values) of a statistical model involving the Gini

Table 3.3 Descriptive analysis of different parameters of clean energy equity and justice.

	Renewable energy access rate (%)	Energy poverty rate (%)	Gini Coefficient
Count	20	20	20
Mean	34.65	15.4	0.45
Standard deviation	14.77	8.82	0.15
Median	33.5	13.5	0.46
Minimum	12	4	0.22
Maximum	60	35	0.75
Range	48	31	0.53
Variance	218.34	77.93	0.022

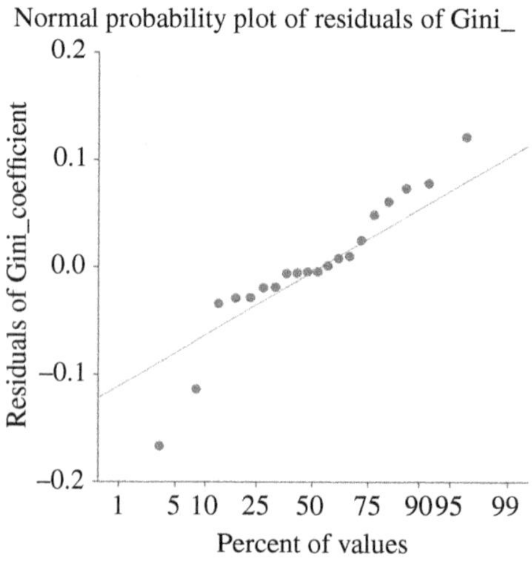

Figure 3.2 Normal probability plot of residuals of Gini Coefficient.

Coefficient follow a normal distribution. The Gini Coefficient is a measure of income or wealth inequality within a population, with values ranging from 0 (perfect equality) to 1 (perfect inequality). Figure 3.3 shows a histogram of residuals of the Gini Coefficient. In this case, the histogram of residuals of the Gini Coefficient shows the frequency distribution of the differences between observed and predicted Gini Coefficient values. It provides insights into the shape and spread of the residuals, helping to assess the model's performance and identify any patterns or outliers. Figure 3.4 shows a plot between residuals of the Gini Coefficient and energy poverty. Figure 3.5 shows a plot between residuals of the Gini Coefficient and renewable energy access. Renewable energy access measures the availability

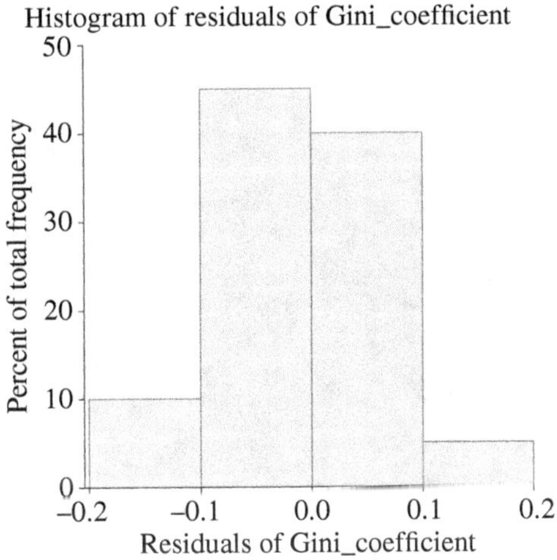

Figure 3.3 Histogram of residuals of Gini Coefficient.

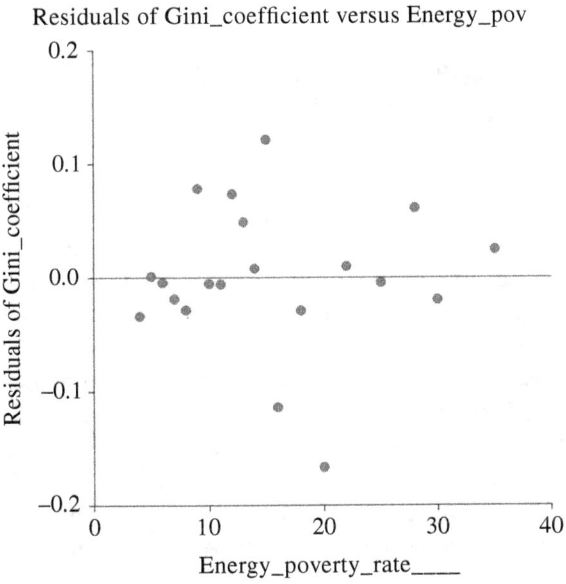

Figure 3.4 Plot between residuals of Gini Coefficient and energy poverty.

and affordability of renewable energy sources, such as solar, wind, or hydro power. By plotting the residuals against renewable energy access, researchers can explore whether there is any association between the model's errors and the degree of renewable energy access across different areas or demographics.

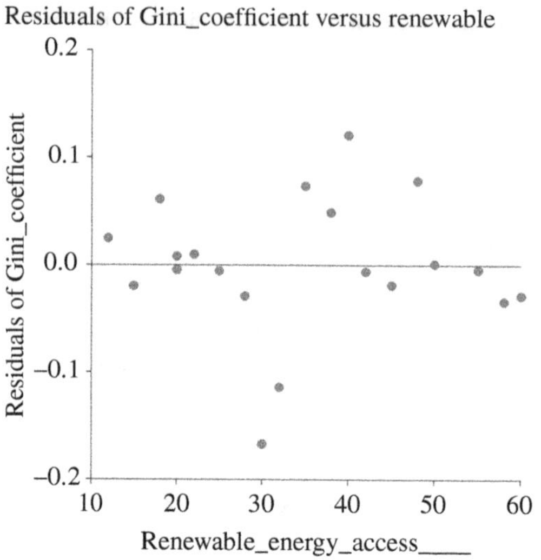

Figure 3.5 Plot between residuals of Gini Coefficient and renewable energy access.

3.5 Energy Equity and Justice of Each Renewable Energy Source in India

Energy equity and justice in the context of renewable energy sources in India refer to ensuring fair and equitable distribution of clean energy resources across different socioeconomic groups and geographical regions, with a focus on addressing disparities and promoting inclusivity in access to sustainable energy solutions. Achieving energy equity and justice in India's renewable energy sector requires a holistic approach that considers social, economic, environmental, and governance dimensions, with a focus on promoting inclusivity, fairness, and sustainability in the distribution and utilization of clean energy resources [19]. Following are the energy equity and justice of different renewable energy sources in India:

Solar Energy: Solar energy has the potential to provide decentralized and off-grid energy solutions, improving access to electricity for remote and underserved communities. However, equitable distribution of solar energy systems is essential to ensure that all regions and socioeconomic groups benefit from this renewable resource. Solar energy projects can create local employment opportunities and stimulate economic development, but there is a need to ensure that benefits reach marginalized communities and that they have the capacity to participate in solar value chains. Income inequality and affordability issues may hinder equitable access to solar energy solutions. Addressing socioeconomic

Table 3.4 Different parameters of the energy justice and equity of solar energy system.

State	Population (millions)	Access to solar energy (percentage)	Energy poverty rate (percentage)	Gini Coefficient	Community engagement (scale: 1–10)
Uttar Pradesh	220	30	25	0.35	6
Maharashtra	120	45	15	0.40	7
Bihar	110	20	35	0.38	5
West Bengal	100	25	30	0.42	6
Madhya Pradesh	90	35	20	0.37	6
Rajasthan	80	50	10	0.45	8
Tamil Nadu	75	40	18	0.39	7
Karnataka	70	45	12	0.41	8
Gujarat	65	55	8	0.36	7
Andhra Pradesh	60	35	22	0.34	6

disparities and implementing inclusive financing mechanisms are crucial for promoting energy justice. Solar energy has minimal environmental impacts compared to conventional energy sources. However, land use issues, particularly in densely populated areas, and concerns about the disposal of solar panels at the end of their lifespan need to be addressed sustainably. Table 3.4 provides data on various states in India across several indicators related to population, access to solar energy, energy poverty rate, Gini Coefficient, and community engagement. It offers insights into how different states in India compare across these key indicators, providing valuable information for policymakers, researchers, and stakeholders involved in energy planning and development initiatives [20]. Table 3.5 shows the descriptive statistics of solar energy parameters. Figure 3.6 shows the histogram of residuals of the Gini Coefficient. Figure 3.7 shows the normal probability plot of residuals of the Gini Index.

Wind Energy: Wind energy projects are often located in rural areas, providing economic opportunities to local communities. However, ensuring equitable access to wind energy benefits, such as job creation and revenue generation from land leases, is essential. Equity considerations include fair compensation for land use, transparent project development processes, and opportunities for local community ownership or participation in wind energy projects. Income inequality and land tenure issues may affect the distribution of wind energy benefits. Addressing these challenges through community engagement, capacity building, and revenue-sharing mechanisms is crucial for promoting energy

Table 3.5 Descriptive statistics of solar energy parameters.

	Gini Coefficient	Access to solar energy (percentage)	Energy poverty rate (percentage)
Count	10	10	10
Mean	0.387	38	19.5
Standard deviation	0.0340098	11.10555	8.746428
Median	0.385	37.5	19
Minimum	0.34	20	8
Maximum	0.45	55	35
Range	0.11	35	27
Variance	0.001156667	123.3333	76.5

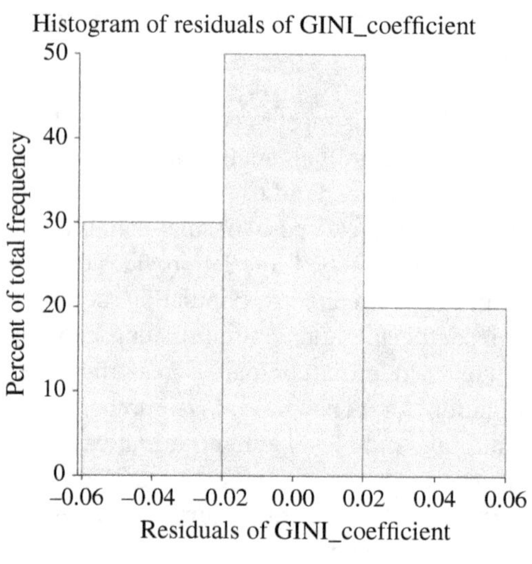

Figure 3.6 Histogram of residuals of Gini coefficient.

justice [21]. Wind energy is considered relatively environmentally friendly, but concerns about wildlife impacts, land use conflicts, and visual pollution need to be addressed through effective planning and stakeholder engagement [22]. Table 3.6 shows the energy justice and equity of the wind energy system. Table 3.7 shows the descriptive analysis of wind energy parameters. Figure 3.8 shows the histogram of residuals of Gini Coefficient of wind energy system. Figure 3.9 shows the normal probability plot of residuals of the Gini Index of the wind energy system.

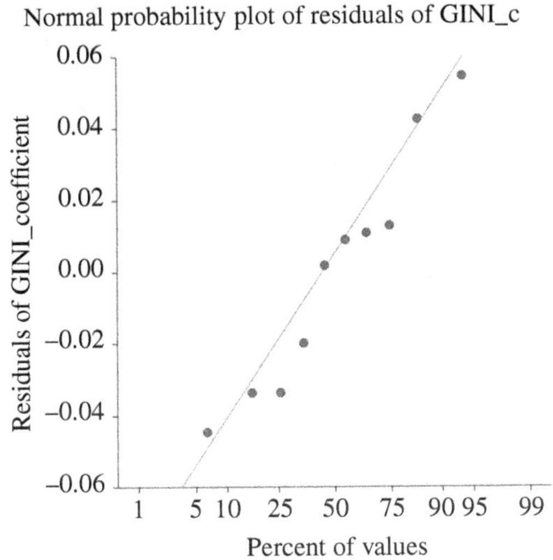

Figure 3.7 Normal probability plot of residuals of the Gini Index.

Table 3.6 Energy justice and equity of wind energy system.

State	Population (millions)	Access to wind energy (percentage)	Energy poverty rate (percentage)	Gini Coefficient	Community engagement (scale: 1–10)
Gujarat	65	40	8	0.36	7
Tamil Nadu	75	50	15	0.39	8
Maharashtra	120	30	20	0.40	6
Rajasthan	80	45	12	0.45	8
Andhra Pradesh	60	25	22	0.34	5
Karnataka	70	35	18	0.41	7
Madhya Pradesh	90	20	25	0.37	6
Uttar Pradesh	220	15	30	0.35	5
West Bengal	100	10	35	0.42	4
Bihar	110	5	40	0.38	3

Biomass Energy: Biomass energy is widely used in rural areas for cooking and heating. However, the reliance on traditional biomass sources such as firewood and crop residues can contribute to indoor air pollution and health disparities, particularly among women and children. Promoting cleaner and more efficient biomass energy technologies can improve indoor air quality and reduce health risks, benefiting marginalized communities that depend on biomass for cooking and heating. Energy poverty and lack of access to modern energy services are key

Table 3.7 Descriptive analysis of wind energy parameters.

	Gini Coefficient	Access to wind energy (percentage)	Energy poverty rate (percentage)
Count	10	10	10
Mean	0.387	27.5	22.5
Standard deviation	0.0340098	15.13825	10.1571
Median	0.385	27.5	21
Minimum	0.34	5	8
Maximum	0.45	50	40
Range	0.11	45	32
Variance	0.001156667	229.1667	103.1667

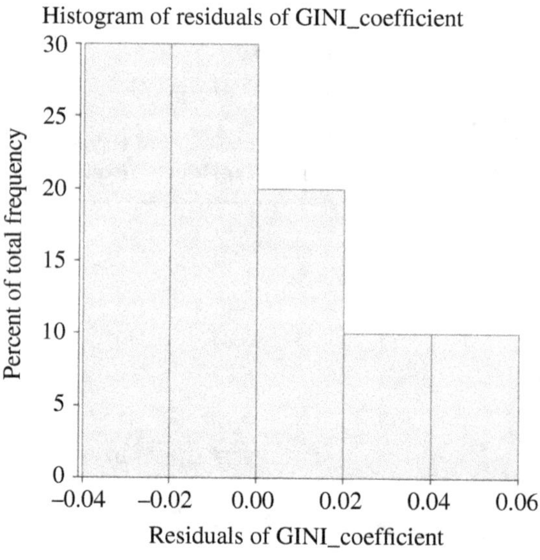

Figure 3.8 Histogram of residuals of Gini Coefficient of the wind energy system.

equity concerns in biomass-dependent communities. Providing access to clean cooking solutions and promoting sustainable biomass management practices are essential for energy justice [23]. Unsustainable biomass harvesting practices can lead to deforestation, soil degradation, and loss of biodiversity. Promoting sustainable biomass production and consumption practices is crucial for minimizing environmental impacts and ensuring intergenerational equity [24]. Table 3.8 shows the energy justice and equity of the biomass energy system. Table 3.9 shows the descriptive statistics of different parameters of the biomass energy system.

Figure 3.9 Normal probability plot of residuals of Gini Coefficient of the wind energy system.

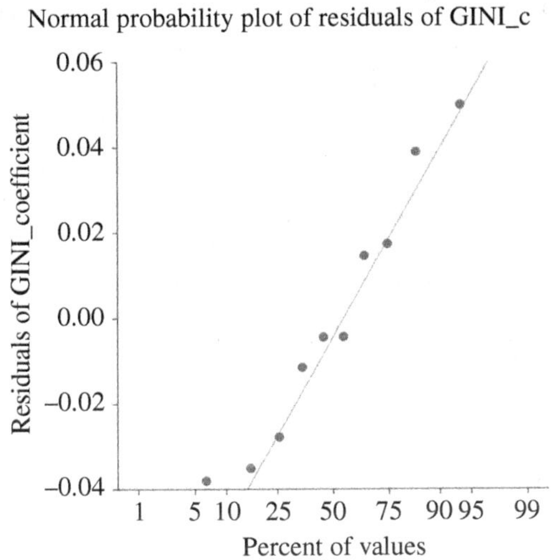

Table 3.8 Energy justice and equity of biomass energy system.

State	Population (millions)	Access to biomass energy (percentage)	Energy poverty rate (percentage)	Gini Coefficient	Community engagement (scale: 1–10)
Uttar Pradesh	220	60	30	0.35	6
Bihar	110	70	40	0.38	5
Madhya Pradesh	90	50	25	0.37	6
West Bengal	100	75	35	0.42	7
Odisha	50	40	20	0.36	6
Assam	40	65	28	0.39	5
Jharkhand	35	55	33	0.43	4
Chhattisgarh	30	45	22	0.41	6
Rajasthan	80	30	15	0.45	7

Figure 3.10 shows the histogram of residuals of Gini Coefficient of biomass energy system. Figure 3.11 shows the normal probability plot of residuals of the Gini Index of the biomass energy system.

Hydro Energy: Large hydroelectric projects can provide grid-connected electricity to urban and rural areas, but concerns about displacement and loss of livelihoods among affected communities need to be addressed. Equitable compensation, resettlement, and livelihood restoration measures are essential for ensuring that communities affected by hydroelectric projects benefit

Table 3.9 Descriptive statistics of different parameters of the biomass energy system.

	Access to biomass energy (percentage)	Energy poverty rate (percentage)	Gini Coefficient
Count	9	9	9
Mean	54.44444	27.55556	0.3955556
Standard deviation	14.67235	7.891627	0.03395258
Median	55	28	0.39
Minimum	30	15	0.35
Maximum	75	40	0.45
Range	45	25	0.1
Variance	215.2778	62.27778	0.001152778

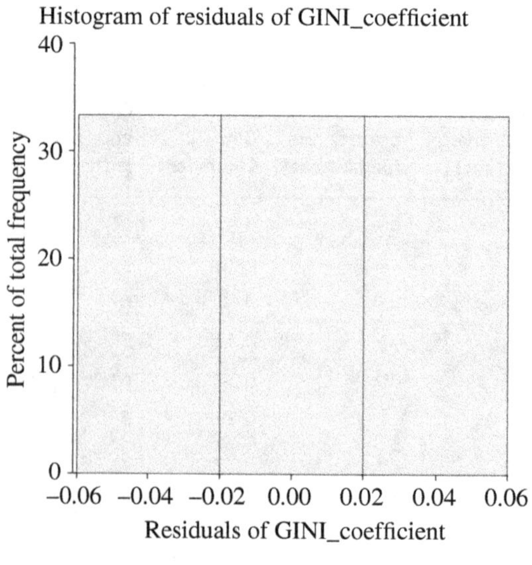

Figure 3.10 Residuals of Gini Coefficient of the biomass energy system.

from electricity generation and associated development activities. Hydropower projects can exacerbate socioeconomic disparities if not implemented inclusively. Inclusive decision-making processes, benefit-sharing mechanisms, and community-driven development approaches are essential for promoting energy justice. Large hydroelectric projects can have significant environmental impacts, including habitat destruction, altered river ecosystems, and greenhouse gas emissions from reservoirs. Environmental impact assessments and mitigation measures are necessary to safeguard ecosystems and promote environmental

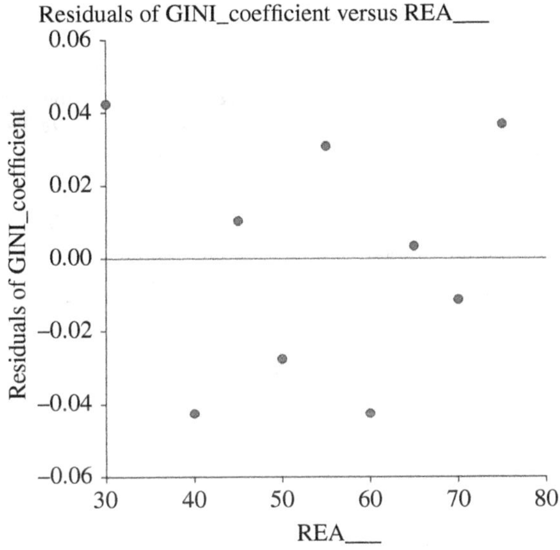

Figure 3.11 Residuals of Gini Coefficient vs. access to biomass energy.

Table 3.10 Energy justice and equity of hydro energy system.

State	Population (millions)	Access to hydro energy (percentage)	Energy poverty rate (percentage)	Gini Coefficient	Community engagement (scale: 1–10)
Himachal Pradesh	10	70	15	0.34	8
Uttarakhand	12	60	18	0.36	7
Arunachal Pradesh	1.5	90	5	0.32	9
Sikkim	0.7	80	8	0.30	8
Jammu and Kashmir	13	50	20	0.38	6
Kerala	35	30	25	0.42	5
Karnataka	70	40	18	0.41	6
Maharashtra	120	20	30	0.40	4
West Bengal	100	15	35	0.43	3
Assam	40	25	28	0.39	5

justice. By considering these factors, policymakers, developers, and civil society organizations can work together to promote energy justice and equity in renewable energy development in India, ensuring that the transition to clean energy benefits all segments of society. Table 3.10 shows the energy justice and

Table 3.11 Statistical analysis of different parameters of the hydro energy system.

	Access to hydro energy (percentage)	**Energy poverty rate (percentage)**	**Gini Coefficient**
Count	10	10	10
Mean	48	20.2	0.375
Standard deviation	26.26785	9.51957	0.04377975
Median	45	19	0.385
Minimum	15	5	0.3
Maximum	90	35	0.43
Range	75	30	0.13
Variance	690	90.62222	0.001916667

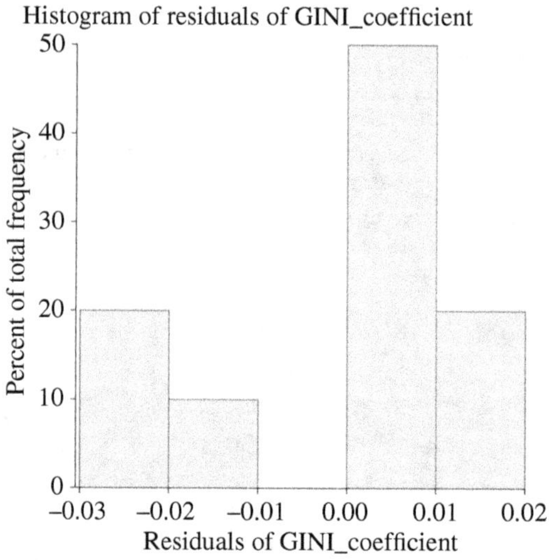

Figure 3.12 Histogram of residuals of Gini Coefficient of the hydro energy system.

equity of hydro energy systems [25]. Table 3.11 shows the statistical analysis of different parameters of the hydro energy system. Figure 3.12 shows the histogram of residuals of the Gini Coefficient of hydro energy system. Figure 3.13 shows the normal probability plot of residuals of the Gini Index of hydro energy system.

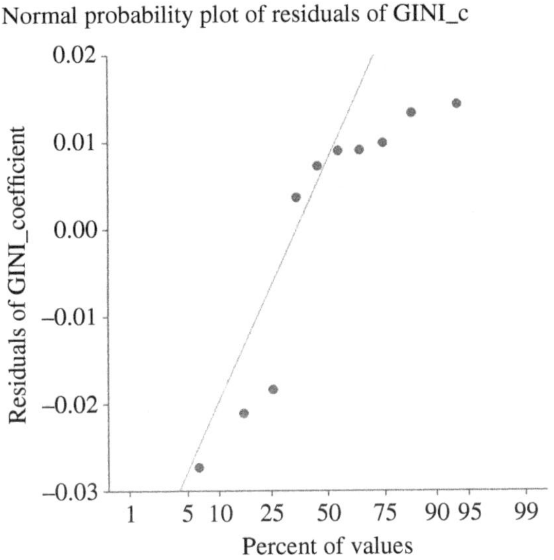

Figure 3.13 Normal probability plot of residuals of Gini Coefficient of hydro energy.

3.6 Role of Supply Chain Management in the Renewable Energy Equity and Justice

In the global pursuit of sustainable development, the integration of renewable energy sources stands as a beacon of hope for mitigating climate change and reducing environmental degradation. Yet, the realization of a truly sustainable and just energy future extends beyond the mere adoption of clean technologies; it necessitates a holistic approach that addresses issues of equity and justice throughout the entire renewable energy supply chain. At the nexus of these imperatives lies the pivotal role of SCM in shaping the equity and justice outcomes of renewable energy deployment [26].

The transition to renewable energy represents a paradigm shift in the way societies produce and consume power. From solar and wind to hydro and geothermal, renewable energy sources offer promising alternatives to fossil fuels, with the potential to reduce greenhouse gas emissions, enhance energy security, and create new economic opportunities. However, the benefits of this transition must be distributed equitably across communities, ensuring that no group is disproportionately burdened or excluded from its advantages [27]. Figure 3.14 shows the role of supply chain in renewable energy equity and justice.

SCM, as a discipline focused on the efficient and sustainable flow of goods and services from raw material extraction to end-user consumption, plays a central

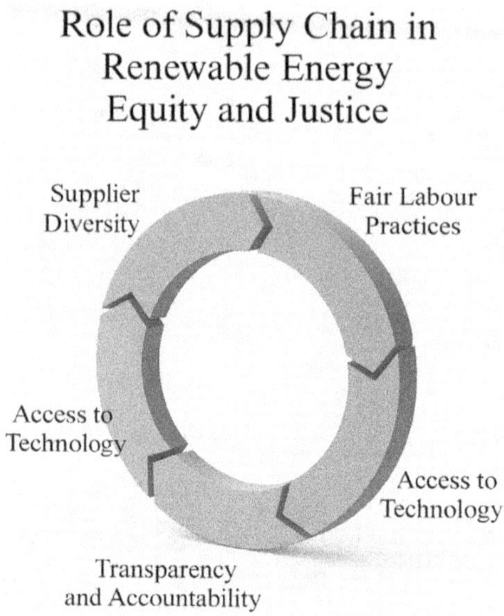

Figure 3.14 Role of supply chain in renewable energy equity and justice.

role in determining the equity and justice outcomes of renewable energy deployment. By encompassing sourcing, manufacturing, distribution, and disposal, the renewable energy supply chain embodies a complex web of interdependencies and stakeholders, each with its own set of interests, incentives, and power dynamics [28, 29].

One of the key challenges in achieving equity and justice in the renewable energy supply chain lies in the sourcing of raw materials and components. Renewable energy technologies, such as solar panels, wind turbines, and batteries, rely on a diverse array of materials, including rare earth metals, silicon, and lithium. The extraction and processing of these materials often occur in regions characterized by environmental vulnerability, social inequality, and geopolitical tensions, raising concerns about resource exploitation, land rights, and community displacement [30].

SCM practices that prioritize responsible sourcing, stakeholder engagement, and transparency can help mitigate these risks and promote equitable outcomes. By establishing robust supplier codes of conduct, conducting due diligence assessments, and fostering partnerships with local communities and Indigenous peoples, renewable energy companies can ensure that their supply chains adhere to ethical and environmental standards, respect human rights, and contribute to sustainable development [31].

Furthermore, SCM plays a crucial role in addressing labor rights and social equity issues within the renewable energy industry. From the construction of solar farms and wind parks to the operation and maintenance of renewable energy infrastructure, a diverse workforce is needed to drive the clean energy transition forward [32]. However, labor conditions in the renewable energy sector vary widely, with reports of exploitative practices, unsafe working conditions, and inadequate wages in some regions.

To promote social equity and justice, SCM strategies must prioritize fair labor practices, worker rights, and diversity and inclusion initiatives. By implementing labor standards audits, promoting fair wages and working conditions, and investing in workforce development programs, renewable energy companies can create employment opportunities that benefit local communities, empower marginalized groups, and contribute to inclusive economic growth [22].

Moreover, SCM can serve as a vehicle for fostering community engagement and empowerment throughout the renewable energy project lifecycle. From project siting and permitting to construction and operation, renewable energy projects often elicit opposition from local communities due to concerns about environmental impact, land use conflicts, and loss of cultural heritage. Effective stakeholder engagement and participatory decision-making processes are essential for building trust, addressing community concerns, and fostering social license to operate [33].

SCM practices that prioritize community consultation, consent, and benefit-sharing can help ensure that renewable energy projects are developed in a manner that respects the rights and interests of affected communities. By involving stakeholders in project planning, design, and implementation, renewable energy companies can cocreate solutions that maximize local benefits, minimize adverse impacts, and enhance the overall sustainability and acceptance of renewable energy projects.

The role of SCM in advancing equity and justice in the renewable energy sector cannot be overstated. As the world transitions toward a cleaner and more sustainable energy future, SCM practices must evolve to address the social, environmental, and ethical dimensions of renewable energy deployment. By prioritizing responsible sourcing, fair labor practices, and community engagement, SCM can help ensure that the benefits of renewable energy are equitably distributed, empowering communities, promoting social inclusion, and advancing environmental justice. Moving forward, concerted efforts are needed to integrate equity and justice considerations into the core of renewable energy SCM practices, thereby fostering a more sustainable and just energy transition for all [34].

3.7 Role of Project Management in the Renewable Energy Equity and Justice

In the global quest for sustainable energy solutions, the integration of renewable energy sources has emerged as a critical pathway toward mitigating climate change, reducing pollution, and promoting energy security. However, the success of renewable energy initiatives extends beyond the mere deployment of technology; it hinges on the equitable distribution of benefits and the just treatment of all stakeholders involved. At the forefront of ensuring equity and justice in the renewable energy sector lies the indispensable role of project management. This introduction sets the stage for an exploration of how project management practices can influence and shape equity and justice outcomes in the realm of renewable energy [35].

Renewable energy technologies, ranging from solar and wind to hydro and biomass, offer promising alternatives to fossil fuels, providing clean and sustainable sources of power. Yet, the transition to renewable energy entails complex challenges, including socioeconomic disparities, environmental justice concerns, and community resistance. Addressing these challenges requires a holistic approach that goes beyond technical expertise, encompassing social, ethical, and governance considerations. Project management, as a discipline focused on planning, executing, and controlling projects to achieve specific goals, plays a pivotal role in orchestrating the diverse array of activities involved in renewable energy initiatives [36].

One of the key challenges in achieving equity and justice in renewable energy projects lies in the distribution of benefits and burdens across communities. Historically, marginalized communities, including low-income neighborhoods and communities of color, have borne a disproportionate share of the environmental and health impacts associated with energy production and consumption. In many cases, renewable energy projects have inadvertently perpetuated these injustices by siting facilities in environmentally vulnerable areas or by excluding local communities from decision-making processes [37].

Project management practices that prioritize equity and justice can help mitigate these disparities and ensure that renewable energy projects deliver tangible benefits to all stakeholders. By adopting inclusive decision-making processes, conducting comprehensive impact assessments, and engaging with affected communities from the outset, project managers can identify and address potential social, environmental, and economic concerns early in the project lifecycle. Moreover, by establishing mechanisms for transparent communication, accountability, and community participation, project managers can build trust, foster collaboration, and empower communities to shape the outcomes of renewable energy projects.

Furthermore, project management plays a critical role in addressing environmental justice concerns in renewable energy development. While renewable energy technologies offer significant environmental benefits compared to fossil fuels, they are not without their own environmental impacts. From land use changes and habitat disruption to wildlife mortality and water consumption, renewable energy projects can pose risks to ecosystems and biodiversity, particularly when poorly planned or implemented [38].

Effective project management practices can help minimize these environmental impacts and ensure that renewable energy projects are developed and operated in a manner that protects and preserves natural resources. By conducting thorough environmental assessments, adhering to best practices in project siting and design, and implementing robust monitoring and mitigation measures, project managers can reduce the ecological footprint of renewable energy projects and safeguard sensitive habitats and ecosystems. Moreover, by incorporating principles of sustainability and ecosystem stewardship into project planning and decision-making processes, project managers can promote the long-term resilience and viability of renewable energy systems [39].

In addition to addressing environmental and social concerns, project management plays a crucial role in promoting economic equity and justice in the renewable energy sector. Renewable energy projects have the potential to create jobs, stimulate economic growth, and enhance local prosperity, particularly in rural and underserved communities. However, realizing these benefits requires intentional efforts to prioritize local hiring, workforce development, and equitable contracting practices throughout the project lifecycle [40].

Project management practices that prioritize economic equity and justice can help maximize the positive socioeconomic impacts of renewable energy projects and ensure that they contribute to inclusive economic development. By partnering with local stakeholders, supporting small and minority-owned businesses, and investing in skills training and capacity-building initiatives, project managers can create opportunities for meaningful employment, entrepreneurship, and wealth generation in communities hosting renewable energy projects. Moreover, by fostering partnerships with local governments, educational institutions, and industry associations, project managers can leverage the collective expertise and resources needed to address systemic barriers to economic opportunity and promote equitable access to the benefits of renewable energy.

The role of project management in advancing equity and justice in the renewable energy sector is multifaceted and far-reaching. By integrating principles of equity, justice, and sustainability into project planning, execution, and evaluation processes, project managers can help ensure that renewable energy projects deliver positive social, environmental, and economic outcomes for all stakeholders involved. Moving forward, concerted efforts are needed to institutionalize equity and justice considerations within the core of project management practices, thereby fostering a more inclusive, resilient, and sustainable renewable energy transition for present and future generations.

Effective project management involves comprehensive planning, stakeholder engagement, risk assessment, and performance monitoring. In the context of renewable energy initiatives, it can:

- **Ensure Inclusive Decision-Making:** Project managers can lead efforts to include underrepresented voices, particularly those from marginalized communities, in decision-making processes. This helps ensure that the needs and concerns of these communities are considered from the outset.
- **Mitigate Environmental and Social Disparities:** Through proactive risk assessments and careful planning, project managers can foresee potential **environmental and social impacts** and design mitigation strategies, minimizing harm to vulnerable populations.
- **Distribute Benefits More Equitably:** Inclusive project management ensures that economic and social benefits—such as jobs, community investments, and access to clean energy—are distributed fairly rather than being concentrated in wealthier areas or among corporate stakeholders.

3.8 Impact of Management Program on the Clean Energy Equity and Justice

The impact of a management program on clean energy equity and justice can be profound, influencing various aspects of policy implementation, resource allocation, community engagement, and decision-making processes. Following are some potential impacts:

Equitable Resource Allocation: A well-designed management program can ensure that resources for clean energy projects are distributed equitably, prioritizing communities that have historically been marginalized or underserved. By incorporating principles of equity and justice into project planning and funding allocation, the program can help address disparities in energy access and affordability.

Community Empowerment: Effective management programs engage local communities in the development and decision-making processes of clean energy projects. By providing opportunities for meaningful participation, such as community meetings, consultations, and partnerships, these programs empower residents to voice their needs, concerns, and preferences, ultimately leading to more inclusive and responsive project outcomes.

Social and Economic Benefits: Management programs that prioritize equity and justice can maximize the social and economic benefits of clean energy projects for all stakeholders, particularly disadvantaged communities. This may include job creation, workforce development, local procurement, and revenue-sharing arrangements that ensure communities directly benefit from renewable energy investments.

Quantifying social impacts focuses on assessing how management programs affect communities, particularly marginalized groups. Following are some key metrics and methods for evaluation:

- **Community Engagement and Participation:**
  - **Metrics:** Number of community meetings held, percentage of community members engaged, diversity of participants (age, gender, socioeconomic background).
  - **Method:** Surveys and interviews with community members to gauge awareness and involvement in clean energy projects.
- **Social Equity and Inclusion:**
  - **Metrics:** Changes in representation of marginalized groups in decision-making processes, the percentage of local community members employed by the project, and satisfaction levels among community members.
  - **Method:** Analyze employment statistics and conduct satisfaction surveys to assess whether management practices have improved equity.
- **Quality of Life Indicators:**
  - **Metrics:** Access to clean energy, improvements in health outcomes (reduced respiratory issues, etc.), and changes in educational opportunities resulting from clean energy projects.
  - **Method:** Collect data on energy access before and after project implementation and analyze health and educational statistics in the affected communities.

Environmental Justice: A management program focused on clean energy equity and justice considers the environmental impacts of energy projects on vulnerable communities, particularly those already burdened by pollution or environmental degradation. By conducting thorough environmental assessments, implementing mitigation measures, and promoting renewable energy technologies with minimal environmental footprints, the program can help prevent further environmental injustices and promote environmental health and resilience.

Quantifying environmental outcomes is crucial for understanding how management programs contribute to sustainability and ecological health.

- **Reduction in Greenhouse Gas Emissions**
  - **Metrics:** Measured reductions in CO_2 and other greenhouse gas emissions from project implementation compared to baseline levels.

- **Method:** Use emission modeling tools and data from local environmental agencies to assess the carbon footprint reduction resulting from clean energy projects.
- **Biodiversity and Ecosystem Health**
  - **Metrics:** Changes in local biodiversity indices, restoration of natural habitats, and assessments of land use changes.
  - **Method:** Conduct ecological surveys and habitat assessments before and after project implementation to quantify impacts on local ecosystems.
- **Water and Air Quality Improvements**
  - **Metrics:** Changes in air quality index (AQI) and water quality parameters (e.g. chemical contaminant levels, turbidity) over time.
  - **Method:** Monitor environmental quality indicators through partnerships with local environmental agencies and collect pre- and post-project data.

Policy and Regulatory Reform: Management programs play a critical role in advocating for policy and regulatory reforms that advance clean energy equity and justice at the local, regional, and national levels. This may include advocating for equitable energy policies, incentive programs for disadvantaged communities, regulatory measures to address energy poverty, and mechanisms to promote community ownership and control of energy resources.

Capacity Building and Education: A management program committed to clean energy equity and justice invests in capacity building and education initiatives to empower communities with the knowledge, skills, and resources needed to participate effectively in the clean energy transition. This may involve providing training programs, technical assistance, and educational resources tailored to the needs of diverse stakeholders, including community members, policymakers, and industry professionals.

3.9 SWOT Analysis

A SWOT analysis (Strengths, Weaknesses, Opportunities, Threats) for renewable energy justice and equity can provide insights into its current state and potential avenues for improvement.

- **Strengths**

  Environmental Sustainability: Renewable energy sources like solar, wind, and hydro are inherently sustainable and have lower environmental impacts compared to fossil fuels.

  Social Inclusion: Renewable energy projects can promote social inclusion by providing opportunities for community involvement, ownership, and empowerment.

Economic Opportunities: The renewable energy sector can create jobs and stimulate local economies, especially in rural areas where resources are abundant.

Technological Innovation: Advances in renewable energy technologies continue to drive down costs and increase efficiency, making them more accessible to a wider range of communities.

- **Weaknesses**

  Access Barriers: Initial costs and infrastructure requirements may create barriers to access for marginalized communities, exacerbating existing energy inequalities.

  Grid Integration Challenges: Integrating renewable energy into existing grids can be complex and costly, leading to delays and inefficiencies in deployment.

  Land Use Conflicts: Renewable energy projects may face opposition due to concerns about land use, biodiversity impacts, and conflicts with other land uses.

  Funding Limitations: Limited funding and financial resources may constrain the implementation of renewable energy projects in underserved areas.

- **Opportunities**

  Policy Support: Favorable policies and regulations can incentivize investment in renewable energy projects that prioritize justice and equity considerations.

  Community Partnerships: Collaborating with local communities and stakeholders can help identify needs, preferences, and solutions that promote justice and equity.

  Education and Awareness: Increasing public awareness and understanding of renewable energy benefits can build support for inclusive energy transition initiatives.

  Technological Advancements: Continued innovation in renewable energy technologies can drive down costs and improve accessibility, particularly for off-grid and remote communities.

- **Threats**

  Political Resistance: Opposition from vested interests and political agendas may hinder efforts to prioritize justice and equity in renewable energy planning and implementation.

  Socioeconomic Inequalities: Socioeconomic disparities may limit the ability of marginalized communities to benefit from renewable energy projects without targeted support and interventions.

  Policy Instability: Changes in government policies and regulations can create uncertainty and undermine long-term investment in renewable energy justice and equity initiatives.

Environmental Risks: Poorly planned renewable energy projects can pose environmental risks, such as habitat destruction, pollution, and water scarcity, which may disproportionately affect vulnerable communities.

By conducting a SWOT analysis, stakeholders can identify strategic priorities and develop targeted interventions to address challenges and capitalize on opportunities for advancing renewable energy justice and equity.

3.10 Renewable Energy Equity and Justice Matrix

Creating a matrix diagram for renewable energy equity and justice involves plotting the various aspects related to these concepts along two axes. Figure 3.15 shows the renewable energy equity and justice matrix.

In this matrix, justice is plotted along the horizontal axis, with factors such as accessibility, affordability, participation, and fair distribution represented. Equity is plotted along the vertical axis, with factors such as socioeconomic inclusion, community engagement, empowerment, and environmental sustainability represented. Each quadrant of the matrix represents different intersections between equity and justice considerations in renewable energy initiatives. Stakeholders can use this matrix to identify specific areas where efforts are needed to improve

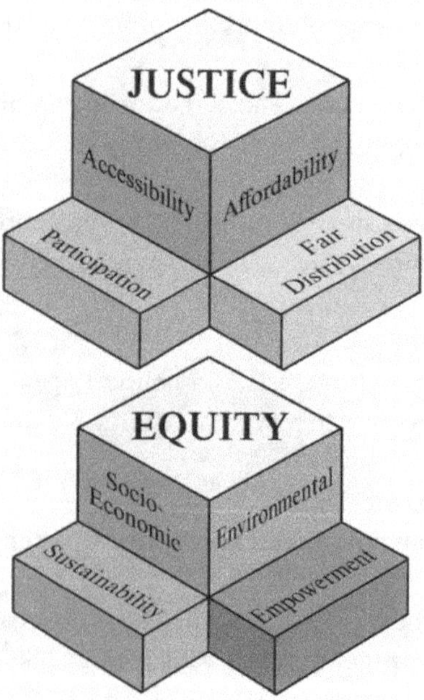

Figure 3.15 Renewable energy equity and justice matrix.

both equity and justice in renewable energy planning and implementation. This matrix provides a visual representation that can help stakeholders assess the current state of renewable energy initiatives in terms of equity and justice and develop targeted strategies for improvement.

3.11 Conclusion

The integration of decision-making tools within the supply chain and project management processes is paramount for advancing clean energy justice and equity. By leveraging these tools effectively, stakeholders can navigate complex challenges, address systemic barriers, and promote inclusive and equitable outcomes throughout the clean energy transition. SCM tools, such as supply chain mapping and life cycle assessment, enable stakeholders to identify and mitigate environmental and social risks associated with renewable energy projects. This ensures that clean energy initiatives are implemented in a sustainable manner, minimizing adverse impacts on communities and ecosystems.

Project management tools, including Gantt charts, Critical Path Method (CPM), and risk management frameworks, facilitate efficient planning, execution, and monitoring of clean energy projects. By incorporating equity and justice considerations into project management processes, stakeholders can prioritize actions that benefit marginalized communities, promote community engagement, and ensure equitable distribution of benefits. Furthermore, decision-making tools such as SWOT analysis, Pareto Analysis, scenario analysis, and game theory provide valuable insights into the strengths, weaknesses, opportunities, and threats associated with clean energy justice and equity initiatives. These tools empower stakeholders to develop strategic interventions, anticipate potential challenges, and negotiate equitable solutions that address the diverse needs and priorities of all stakeholders involved.

Incorporating Fishbone diagrams, force field analysis, and Monte Carlo simulation enhances decision-making by facilitating the identification of root causes, understanding of driving and restraining forces, and assessment of uncertainty associated with clean energy justice and equity efforts. These tools enable stakeholders to develop robust strategies that promote fairness, inclusivity, and resilience in the transition to clean energy systems. In conclusion, the effective use of decision-making tools within supply chain and project management processes is essential for advancing clean energy justice and equity. By integrating equity and justice considerations into decision-making frameworks, stakeholders can foster sustainable, equitable, and inclusive clean energy transitions that benefit present and future generations.

Acknowledgment

This work is financially supported by the Department of Science and Technology (DST), Government of India, under DST-SERB Project File No. IPA/2021/000048.

References

1 Khare, V., Nema, S., and Baredar, P. (2017). Optimization of the hybrid renewable energy system by HOMER, PSO and CPSO for the study area. *International Journal of Sustainable Energy*, Taylor and Francis 36: 326–343.

2 Caiado, R.G.G., Scavarda, L.F., Azevedo, B.D. et al. (2022). Challenges and benefits of sustainable industry 4.0 for operations and supply chain management—a framework headed toward the 2030 agenda. *Sustainability* 14 (2): 830.

3 Cao, Y. (2022). Design, dynamic simulation, and optimal size selection of a hybrid solar/wind and battery-based system for off-grid energy supply. *Renewable Energy* 187: 1082–1099.

4 Wang, F. (2022). Optimal design of solar-assisted steam and power system under uncertainty. *Journal of Cleaner Production* 336: 130294.

5 Nelson, T. (2023). Contracts-for-difference: an assessment of social equity considerations in the renewable energy transition. *Energy Policy* 183: 1–10.

6 Govindarajan, H.K. (2022). Integrating energy governance and environmental justice: role of renewable energy. *Renewable Energy Focus* 43: 24–36.

7 Montanes, C.C. (2023). Enabling and centering equity and justice in clean energy transition research. *Joule* 7: 437–441.

8 Benti, N.E. (2022). Techno-economic analysis of solar energy system for electrification of rural school in Southern Ethiopia. *Cogent Engineering* 9: 2021838.

9 Khare, V. (2019). Prediction, investigation, and assessment of novel tidal–solar hybrid renewable energy system in India by different techniques. *International Journal of Sustainable Energy*, Taylor and Francis 38: 447–468.

10 Bhave, A.G. (1999). Hybrid solar–wind domestic power generating system—case study. *Renewable Energy* 17: 355–358.

11 Borowy, B.S. and Salameh, Z.M. (1996). Methodology for optimally sizing the combination of a battery bank and PV array in a wind/PV hybrid system. *IEEE Transactions on Energy Conversion* 11: 367–375.

12 Borowy, B.S. and Salameh, Z.M. (1997). Dynamic response of stand-alone wind energy conversation system with battery energy storage to a wind gust. *IEEE Transactions on Energy Conversation* 12: 73–78.

13 Batista, L., Bourlakis, M., Smart, P., and Maull, R. (2018). In search of a circular supply chain archetype–a content-analysis-based literature review. *Production Planning and Control* 29 (6): 438–451.

14 Bazmi, A.A. and Zahedi, G. (2011). Sustainable energy systems: role of optimization modelling techniques in power generation and supply a review. *Renewable and Sustainable Energy Reviews* 15 (8): 3480–3500.

15 Belhadi, A., Kamble, S., Gunasekaran, A., and Mani, V. (2022). Analyzing the mediating role of organizational ambidexterity and digital business transformation on industry 4.0 capabilities and sustainable supply chain performance. *Supply Chain Management An International Journal* 27 (6): 696–711.

16 Bounbumroong, U., Pratinthong, N., Thepa, S. et al. (2011). Particle swarm optimization for AC-coupling stand alone hybrid power systems. *Solar Energy* 85: 560–569.

17 Khare, V., Nema, S., and Baredar, P. (2016). Solar–wind hybrid renewable energy system: a review. *Renewable and Sustainable Energy Reviews*, Elsevier 58: 23–33.

18 Khare, V., Nema, S., and Baredar, P. (2013). Status of solar wind renewable energy in India. *Renewable and Sustainable Energy Reviews*, Elsevier 27: 1–10.

19 Khare, V., Nema, S., and Baredar, P. (2016). Optimization of hydrogen based hybrid renewable energy system using HOMER, BB-BC, and GAMBIT. *International Journal of Hydrogen Energy*, Elsevier 41: 16743–16751.

20 Chong, W.T., Nagavi, M.S., Poh, S.C., and Mahlia, T.M.I. (2011). Techno economic analysis of a wind-solar hybrid renewable energy system with rain water collection feature for urban high rise application. *Applied Energy* 88: 4067–4077.

21 Carter, C.R. and Easton, P.L. (2011). Sustainable supply chain management: evolution and future directions. *International Journal of Physical Distribution and Logistics Management* 41 (1): 46–62.

22 Cristobal, J.M.S. (2011). A multi criteria data envelopment analysis model to evaluate the efficiency of the renewable energy technologies. *Renewable Energy* 36: 2742–2746.

23 Diaf, S., Diaf, D., Belhamel, M. et al. (2007). A methodology for optimal sizing of autonomous hybrid PV/wind system. *Energy Policy* 35: 5708–5718.

24 Davies, J. and Joglekar, N. (2013). Supply chain integration, product modularity, and market valuation: evidence from the solar energy industry. *Production and Operations Management* 22 (6): 1494–1508.

25 Dawson, P. and Daniel, L. (2010). Understanding social innovation: a provisional framework. *International Journal of Technology Management* 51 (1): 9–21.

26 Dufo-Lopez, R., Bernal-Agustin, J.L., and Yusta-Loyo, J.M. (2011). Multi-objective optimization minimizing cost and life cycle emissions of stand-alone PV–wind–diesel systems with batteries storage. *Applied Energy* 88: 4033–4041.

27 Elhadidy, M.A. and Shaahid, S.M. (2005). Decentralized/stand-alone hybrid wind-diesel power systems to meet residential loads of hot coastal regions. *Energy Conversion and Management* 46 (2005): 2501–2513.

28 Delabre, I., Alexander, A., and Rodrigues, C. (2020). Strategies for tropical forest protection and sustainable supply chains: challenges and opportunities for alignment with the UN sustainable development goals. *Sustainability Science* 15 (6): 1637–1651.

29 Dutta, G., Kumar, R., Sindhwani, R., and Singh, R.K. (2020). Digital transformation priorities of India's discrete manufacturing SMEs–a conceptual study in perspective of industry 4.0. *Competitiveness Review: An International Business Journal* 30 (3): 289–314.

30 Khare, V., Nema, S., and Baredar, P. (2019). Reliability analysis of hybrid renewable energy system by fault tree analysis. *Energy & Environment*, SAGE Publication 30: 542–555.

31 Jain, A., Balasubramanian, R., and Tripathy, S.C. (2011). Reliability analysis of wind embedded power generation system for Indian scenario. *International Journal of Engineering, Science and Technology* 3: 93–99.

32 Brown, H.E., Suryanarayan, S., and Natarajan, S.A. (2012). Improving reliability of islanded distribution system with distributed renewable energy resources. *IEEE Transactions on Smart Grid* 3: 2028–2038.

33 Madlener, R., Antunes, C.H., and Dias, L.C. (2006). Assessing the performance of biogas plants with multicriteria and data envelopment analysis. *European Journal of Operational Research* 197: 1084–1094.

34 Jose, L., Agustin, B., and DufoLopez, R. (2009). Simulation and optimization of standalone hybrid renewable energy system. *Renewable and Sustainable Energy Reviews* 13: 2011–2018.

35 Kharrich, M. (2023). An effective design of hybrid renewable energy system using an improved archimedes optimization algorithm: a case study of Farafra, Egypt. *Energy Conversion Management* 283: 116907.

36 Hasan, S. (2023). "Techno-economic feasibility analysis of an electric vehicle charging station for an International Airport in Chattogram", Bangladesh. *Energy Conversion and Management* 293: 117501.

37 Mansouri, A. (2023). Nonlinear observer with reduced sensors for WECS involving Vienna rectifiers theoretical design and experimental evaluation. *Electric Power Systems Research* 225: 109847.

38 Hassan, Q. (2023). Techno-economic assessment of green hydrogen production by an off-grid photovoltaic energy system. *Energies* 16: 744–750.

39 Akundi, A., Euresti, D., Luna, S. et al. (2022). State of industry 5.0—analysis and identification of current research trends. *Applied System Innovation* 5 (1): 27.

40 Alves, J., Lima, T.M., and Gaspar, P.D. (2023). Is industry 5.0 a human-centered approach? A systematic review. *Processes* 11 (1): 193.

4

Multi-energy Community Structure for Energy Justice and Equity

Peyman Afzali[1], *Sahar Mobasheri*[2], *Banafsheh Alipour*[2], *Sogand Hosseinalipour*[2], *Masoud Rashidinejad*[2], *and Sobhan Dorahaki*[3]

[1] *Department of Energy (AAU Energy), Aalborg University, Aalborg, Denmark*
[2] *Department of Electrical Engineering, Shahid Bahonar University of Kerman, Kerman, Iran*
[3] *Department of Electrical Engineering, Qatar University, Doha, Qatar*

Nomenclature

Acronyms

CAES	compressed-air energy storage
CHR	Cumulative Heat Release
CEP	Clean Energy Package
CHP	combined heat and power
CO_2	carbon dioxide
CVaR	conditional value at risk
DRP	demand response programs
DSO	distribution system operator
EDR	electrical demand response
EH	energy hub
EHP	electrical heat pump
ENS	energy not supplied
EV	electric vehicle
HV	heat value
IGDT	information gap decision theory
MEC	multi-energy community
P2G	power-to-gas
P2P	peer-to-peer
PEV	plug-in electric vehicle
PV	photovoltaic

Communities for Clean Energy Justice and Equity in Grid Modernization, First Edition.
Edited by Mohammadreza Daneshvar, Behnam Mohammadi-Ivatloo, and Amjad Anvari-Moghaddam.

SEH	smart energy hubs
TDR	thermal demand response
TES	thermal energy storage

Indices

t	index of time periods, from 1 to T
s	index of scenarios, from 1 to S

Parameters

π_s	probability of the generated scenarios
TS^{max}	maximum charging and discharging ranges of the TES
$TC^{max/min}$	maximum/minimum stored energy of TES
$\rho_t^{Max,e/th}$	maximum percentage of increased demand
α	confidence level
β	risk coefficient
HV	heat value of the natural gas
$V^{c/exp}$	variable maintenance and operation costs of the compressor/expander
MC^{CHP}	CHP unit maintenance cost coefficient
$P^{CHP}_{max/min}$	upper/lower limits of the CHP unit outputs
η_{HE}	efficiency of the heat exchanger
HPR	heat to power ratio of the CHP unit
λ_t^{Gas}	the price of the natural gas
η^{CHP}	electrical efficiency of the CHP unit
$P^{Ramp-up/Ramp-down}$	ramp-up/ramp-down rates of the CHP
$e^{i/p}$	the efficiency of injected/produced power of the CAES system
$V^{pumped}_{max/min}$	maximum/minimum level of pumped air from storage to combustion chamber
$V^{injected}_{max/min}$	maximum/minimum level of injected air into CAES
$PS_{max/min}$	maximum/minimum level of CAES

Variables

PS	the amount of stored energy in the CAES
CB_s	cost of buying electricity from local grid
RS_s	income of selling electricity to local grid
$OC_s^{CHP/Boiler/CAES}$	operation cost of CHP/boiler/CAES
$ENS^E_{t,s}/ENS^T_{t,s}$	amount of electrical and thermal ENS

$P^{CHP}_{t,s}/T^{CHP}_{t,s}$	electricity/thermal generation of the CHP
PC_s	penalty cost
SUC/SDC	the startup and shutdown cost of CHP
$V^{injected/pumped}_{t,s}$	the energy equivalent of injected/pumped air to CAES/ combustion chamber
$P^{cons.gen.}_{t,s}$	the power consumed/output by/of CAES
$TS^{ch/dch}_{t,s}$	thermal input and output of the TSS
$T^{BoilerEHP}_{t,s}$	thermal output of the boiler/EHP
$P^{EHP}_{t,s}$	electrical input of the EHP
$P^{buy/sell}_{t,s}$	amount of imported/exported electricity
$D^{DR,e}_{t,s}$	the amount of demand after applying demand response program
$DR^{DR,th}_{t,s}$	the amount of demand after applying demand response program
$\phi^{e/th}_{t,s}$	the shifted demand from other hours
DRP	the participation factor of demand response program
$\rho^{e/th}_{t,s}$	the percentage of increased demand.
κ_s	auxiliary variable to assess the CVaR
ζ	value at risk
P^{wind}	power generation of wind turbine

4.1 Introduction

Multi-energy community (MEC) framework emerges as a pivotal model for promoting energy justice and equity within the realm of sustainable electricity development. Given the significant environmental challenges and the large-scale transition toward clean energy, it is imperative to ensure that the benefits of this transition are equitably shared among communities. As we confront climate change and the need for cleaner, more sustainable energy sources, it is crucial that these advantages are efficiently distributed. Before exploring the MEC model, it is important to understand the concepts of energy justice and equity. Energy justice refers to the fair distribution of both the benefits and burdens associated with energy production and consumption, considering environmental impacts, economic opportunities, and access to affordable, clean energy. Equity, in contrast, focuses on fairness and impartiality, stressing the need for just and unbiased treatment during the energy transition. The MEC framework adopts a holistic approach to energy planning and development that moves beyond focusing on a single energy source. It integrates various energy vectors—such as electricity, heat, and mobility—acknowledging the interconnectedness of different energy aspects within a community. This model adjusts to the unique characteristics of each community, considering geographical, social, and economic factors to customize energy solutions to their specific needs. The concept of an energy community goes beyond technical, financial, and environmental aspects to include behavioral and social considerations, taking into account the preferences and impacts on energy end users. Within these communities, smart buildings play a crucial role, with each building's energy scheduling managed by an energy management system. Prosumers in these communities engage in energy trading through peer-to-peer (P2P) platforms with neighbors and with the upstream grid through interactions with the energy network manager. Smart building appliances are categorized as interruptible (e.g. washing machines, vacuum cleaners) and uninterruptible (e.g. lights), enabling prosumers to participate in demand response programs (DRPs) to reduce interruptible demand during peak hours, earning incentives from DRP aggregators or the distribution system operator (DSO). Additionally, prosumers may own photovoltaic (PV) systems covering their energy needs through self-consumption or buying from the upstream grid. Excess energy generated can also be sold back to the grid, facilitated by interactions with the energy community manager. As these MECs develop, their ability to incorporate initiatives for energy justice and equity becomes crucial. The discussion not only delves into the technical complexities of grid modernization but also highlights the coordination and implementation of energy justice and equity initiatives within sustainable and clean energy communities, especially those with a significant share of renewable energy resources. Through this exploration, the chapter aims to offer insights into achieving a harmonious balance between modernizing energy grids and ensuring fair and equal energy access for all.

4.1.1 Motivation and Background

The contemporary energy landscape is experiencing a profound transformation driven by the imperatives of sustainability, resilience, and social inclusivity. This shift is spurred by the urgent need to mitigate the impacts of climate change and transition toward cleaner, more sustainable energy sources. Traditional energy systems often perpetuate disparities, disproportionately affecting marginalized communities and exacerbating existing social and economic inequalities. Moreover, traditional energy systems have long been criticized for their centralized nature, reliance on fossil fuels, and lack of inclusivity, often exacerbating social, economic, and environmental disparities. These systems tend to disproportionately impact marginalized communities, subjecting them to higher energy costs, unreliable access, and exposure to environmental hazards such as pollution from fossil-fuel-based energy generation. Furthermore, the centralized grid structure is vulnerable to disruptions and lacks the flexibility needed to integrate renewable energy sources efficiently. MECs address these issues by decentralizing energy generation and empowering local communities to participate in energy production and decision-making. By integrating diverse renewable energy sources such as solar, wind, and biomass, MECs enhance system resilience and sustainability while simultaneously reducing environmental impact. This decentralized approach fosters greater energy equity by allowing communities to control their energy resources and directly benefit from cleaner, more affordable energy, ultimately addressing the shortcomings of traditional systems. Through these advancements, MECs play a crucial role in promoting both energy justice and a sustainable transition to renewable energy.

In response to these challenges, the motivation behind this chapter is rooted in the belief that energy communities, empowered with diverse and decentralized energy sources, can serve as catalysts for positive change. The foundation of this study is built on the evolving discourse on energy justice and equity, recognizing that the benefits and burdens associated with energy production and consumption should be fairly distributed across diverse communities. As societies navigate the complexities of transitioning to cleaner energy, the MEC structure emerges as a holistic framework that moves beyond the traditional focus on a single energy source. This approach acknowledges the interconnected nature of energy use within communities and lays the groundwork for a transformative approach to energy planning and development. Historically, energy systems have been criticized for their environmental impact, economic disparities, and lack of universal access. The growing awareness of these issues has led to a paradigm shift, emphasizing the need for a more inclusive, resilient, and sustainable energy model. The chapter's motivation lies in addressing the historical shortcomings of energy systems and harnessing the potential of MECs to not only modernize energy grids but also champion energy justice and equity. Furthermore, the MEC structure is motivated by a desire to transcend the limitations of traditional energy frameworks, which often prioritize a single energy source over the diverse needs of communities. This method recognizes that a combination of renewable energy sources, such as solar, wind, biomass, and hydropower, can enhance resilience, reduce environmental impact, and contribute to a more sustainable energy portfolio. The goal is to explore a comprehensive energy planning model that integrates electricity, heat, and mobility, tailored to the specific characteristics of each community, addressing geographical, social, and economic factors.

In the context of this chapter, it is essential to distinguish between the concepts of energy justice and energy equity, as both are integral to the development of fair and sustainable energy systems. Energy justice emphasizes the fair distribution of both the benefits and burdens associated with energy production and consumption. It encompasses three main pillars: recognition justice, procedural justice, and distributive justice, each aiming to address systemic inequalities and ensure that marginalized communities are actively included in energy decision-making processes. Energy equity, on the other hand, focuses more specifically on ensuring that all individuals and communities have equal access to affordable, clean, and reliable energy resources. While energy justice is a broader framework addressing structural and procedural fairness, equity targets the outcomes, ensuring that no group disproportionately bears the negative effects of energy systems or is left behind in the transition to renewable energy. Together, these concepts lay the foundation for an inclusive, just, and equitable transformation of the energy landscape.

The motivation for this chapter is deeply embedded in the conviction that MECs represent a pivotal force in shaping a fair, just, and sustainable energy future. It seeks to provide insights into not only the technical intricacies of grid modernization but also the practical implementation of energy justice and equity principles within the dynamic context of communities embracing a significant share of renewable energy sources. Through this exploration, the chapter aims to contribute meaningful perspectives and strategies to achieve a harmonious balance between modernizing energy grids and ensuring fairness and equality in energy access for all stakeholders. Numerous investigations have concentrated on the MEC. In [1], a thorough economic assessment is presented for a flexible MEC, emphasizing the intricate coordination of electrical and heating networks. The hub is employed as a centralized coordinator for renewables and storage devices. Notably, the unscented transformation method is adeptly utilized to model uncertainty, effectively reducing computational overhead. In [2], an innovative algorithm for scheduling multiple MECs is introduced, with a focus on efficiency and privacy preservation. The proposed approach leverages federated learning and matching learning, utilizing twin-delayed deep deterministic policy gradients for adaptive control of energy conversion in response to varying prices and diverse demands of electricity, heat, and natural gas. Shifting the focus to optimal economic programming of an MEC in the power system while addressing uncertainties related to renewable resources, [3] presents a new model efficiently estimating consumption-related uncertainty of electric vehicles (EVs) and uncertainties related to renewable energy sources using information gap decision-making theory. In [4], a virtual MEC that combines smart energy grids, hubs, and virtual power plants is introduced. The virtual MEC optimizes operations in day-ahead and intraday markets, showcasing improved profitability and heightened risk tolerance for a coalition of smart energy hubs (SEHs). Introducing a decentralized energy market design in [5], featuring SEHs, utility grids, consumers, and an aggregator, addresses renewable energy uncertainty. The model reveals the influence of pricing methods on MECs energy contribution during peak hours. Results indicate the potential for collaborative efforts among SEHs, showcasing mutual benefits and reduced energy bills through cooperative interactions. Contrary to traditional development programs specifying the size, location, and optimal time for adding a new resource, some researchers have focused on the optimal operation of MECs instead of traditional development programming. MECs, considering cooperation between different types of energy, provide a new degree of freedom in energy supply, presenting an opportunity to enhance system performance [6]. The MEC concept represents a generalization or expansion of network nodes in an electrical system [7]. Combining different types of energy can enhance energy efficiency. Using a converter and energy storage, an MEC receives, transforms, stores, and delivers energy, making it a crucial concept in SEHs. An MEC can encompass thermal power plants, diesel generators, electric and thermal exchangers, and boilers. These converters enable MECs to use different energy carriers, resulting in greater reliability compared to traditional systems [8]. Many research has been conducted on the MEC concept. The fundamental idea of the MEC is discussed in [6], emphasizing their crucial role in future power systems. Authors in [9] developed a management model for an energy hub (EH), suggesting various ways electricity can be supplied, including diesel generators, electricity and heat cogeneration units, wind turbines, or through purchasing from electricity markets or bilateral contracts. Modeling the natural gas network in [10] considered hourly and daily constraints on pipelines, areas, power plants, and generating units. An integrated model in [11] evaluated the interdependence of electricity and gas networks on power system reliability by considering gas network constraints. Several studies have explored the interaction between different energy carriers in optimal MECs [12, 13]. [12] determined the installation capacity of various equipment for an MEC, addressing issues such as uncertainty in price forecasts and long-term energy purchase decisions [13] introduced an approach for evaluating financial investment in MECs capable of converting, storing, and managing demand side, aiming to review economic power flow and optimal power distribution problems related to energy delivery [14] established a long-term multi-district development program for natural gas and electricity, incorporating the amount of natural gas passing through pipelines from supply to customer use [15] proposed a probabilistic model for the economic power flow problem in systems with multiple energy carriers, utilizing optimization techniques to find the absolute optimal point [16] implemented a residential MEC based on decision technologies from the smart grid to minimize costs and peak loads, considering user preferences. [17] introduced a dynamic linear programming model for the efficient scheduling of combined heat and power (CHP) units, demonstrating adaptability across various operational scenarios [18] explored a system integrating diverse energy resources to minimize the annual energy cost for a commercial building. Performance and strategic planning of EHs have been investigated in numerous studies. In [8], an exhaustive linear model aims to optimize MEC utilization while addressing reliability constraints [19] develops a two-step hybrid approach for the MEC, including renewable energies, providing an effective way to combine renewable energy systems. In [20], the authors propose a power-flow model for MECs and develop an optimal optimization method for solving the nonlinear MEC problem, incorporating power-flow for natural gas and electricity.

4.1.2 Challenges and Opportunities

4.1.2.1 Challenges

Integrating energy justice and fairness initiatives into the MEC framework presents several challenges. A primary hurdle is balancing the advancement of carbon-free energy production with ensuring equitable energy access. The shift toward cleaner energy often involves substantial upfront costs, which can disadvantage economically marginalized communities. Addressing this financial barrier and ensuring that the benefits of clean energy reach all socioeconomic levels are significant challenges that demand innovative solutions.

Additionally, achieving inclusive energy access within the MEC framework is a complex task. Overcoming energy poverty and ensuring that marginalized communities have equitable opportunities to benefit from the energy transition requires targeted interventions. Tailoring energy solutions to the varied needs of communities, while accounting for differences in geographical, social, and economic contexts, adds another layer of complexity. Striking a balance between the goals of environmental sustainability and social inclusivity requires a nuanced understanding of the specific challenges faced by different types of communities.

Moreover, the chapter explores the complexities of community engagement and empowerment. While active participation in decision-making processes is crucial for the success of an MEC structure, ensuring meaningful involvement and representation from all community members poses a challenge. Overcoming cultural, educational, and institutional barriers to foster a sense of ownership and inclusivity demands careful navigation.

4.1.2.2 Opportunities

Despite the challenges, integrating energy justice and fairness within the MEC framework offers numerous opportunities. Diversification of energy sources, a cornerstone of this model, not only enhances resilience by reducing reliance on a single energy type but also promotes a more sustainable energy portfolio. The advocacy for a variety of renewable sources such as solar, wind, biomass, and hydropower opens doors for local economic development, job creation, and the stimulation of related industries. Additionally, the incorporation of advanced energy storage systems and smart grid technologies presents the potential for efficient management of energy resources. This not only improves the reliability of the energy supply but also facilitates the seamless integration of intermittent renewable resources into the energy mix. These advancements contribute significantly to the stabilization of the grid and ensure that energy generation can be more predictable and controlled, aligning with the goals of energy justice by providing consistent and reliable energy access to all community members. The potential to manage energy demand and deliver in a smart and adaptive way opens avenues for optimizing energy intake patterns, decreasing wastage, and improving universal electricity efficiency. The community engagement and empowerment element of the MEC shape gifts an opportunity to transform energy customers into lively members and members. Empowering nearby residents via training, schooling, and involvement in energy tasks fosters an experience of ownership, ensuring that the blessings of smooth electricity are shared equitably. This shift closer to an extra participatory version can lead to the creation of vibrant, resilient, and self-enough communities. Inclusive energy gets admission to, every other key guideline of the MEC shape, offers opportunities to bridge socio-financial gaps. By presenting established and inexpensive get right of entry to smooth energy, the framework seeks to deal with historical disparities in energy provision. This entails now not simply making sure that marginalized communities have get entry to easy energy however additionally developing opportunities for them to actively take part in and enjoy the broader energy transition. While demanding situations exist, the MEC shape opens up a wealth of possibilities to create an extra resilient, inclusive, and sustainable energy future. Through strategic and coordinated efforts, those opportunities can be harnessed to drive superb trade and make contributions to the broader dreams of energy justice and fairness.

4.1.3 Contributions

This chapter emphasizes the importance of a holistic approach to energy planning, highlighting how MECs integrate electricity, heat, and mobility aspects. This comprehensive approach moves beyond single-source energy solutions and recognizes the interconnected nature of energy consumption within communities. In addition to technical and economic considerations, the chapter underscores the significance of behavioral and social aspects in energy community development. It highlights the importance of understanding energy end users' behaviors, preferences, and their impact on energy supply and consumption patterns. Moreover, this chapter explores how MECs offer a unique opportunity to integrate energy justice and equity programs into modern energy grids. It recognizes the importance of ensuring equitable access to clean energy services and emphasizes the role of MECs in delivering user-friendly services to diverse community members. Also, a fundamental question regarding the simultaneous advancement of carbon-free energy production and distribution while embracing energy justice and equity initiatives. It acknowledges the complexities involved in balancing environmental goals with social equity objectives within the context of energy community development. By proposing innovative approaches, the chapter aims to contribute practical insights into the coordination and implementation of energy justice and equity principles within sustainable and clean energy communities. It seeks to bridge the gap between theoretical understanding and practical application, offering guidance for policymakers, practitioners, and community stakeholders.

This chapter delves into the transformative potential of MECs for promoting energy justice and equity within the broader framework of sustainable development. The chapter aims to explore how MECs can integrate various energy sources, i.e. electricity, heat, and mobility, to ensure that the benefits of a modernized energy grid are distributed equitably among all community members. It underscores the necessity of considering social and behavioral aspects in energy planning, ensuring that solutions are not only technically feasible but also socially inclusive. The chapter argues for a holistic approach, where energy justice and equity are integral to the design and operation of energy systems, potentially aligning with global sustainable development goals and addressing the specific needs of diverse communities.

4.2 Energy Communities

Energy communities represent a paradigm shift in power governance, fostering decentralized and participatory approaches to energy production and consumption. Within the MEC framework, they play a pivotal role in shaping a sustainable and equitable energy future. This section explores various facets of energy communities, including their definition, implementation strategies, control mechanisms, and the complex interplay of socio-techno-economic-environmental factors. It also delves into understanding the behavior and preferences of end users within these communities, which is crucial for tailoring effective energy solutions.

Energy communities signify a departure from traditional centralized energy models, embracing a decentralized and participatory ethos. At the core of their perspective is the recognition that each community possesses unique characteristics shaped by geographical, social, and economic factors. Acknowledging these diversities is key to customizing energy solutions that align with the specific needs and preferences of each community.

In the context of the MEC framework, community engagement is not just a procedural step but a cornerstone of success. Active involvement of community members in decision-making processes, project planning, and implementation is essential to the model's effectiveness. Empowering local residents through education and training initiatives ensures a deeper understanding of the energy transition and fosters a sense of ownership, which, in turn, encourages commitment to the sustainable development of the community's energy landscape.

Smart buildings play a pivotal role in this perspective, acting as dynamic nodes within the energy network. The energy scheduling of each building, managed through an energy management system, exemplifies the integration of technology to optimize energy consumption. Prosumers—individuals who both produce and consume energy—engage in energy trade through P2P platforms within the community and communicate with the energy network manager to interact with the broader energy grid. This two-way communication facilitates the seamless integration of renewable energy sources, DRPs, and energy storage technologies.

The concept of prosumers highlights the behavioral and social aspects inherent in energy communities. Their engagement in energy trade via P2P platforms and participation in DRPs exemplifies the community's collective agency in shaping its energy future. The categorization of smart building appliances as interruptible and uninterruptible allows prosumers to actively participate in DRPs during peak load hours, contributing to grid reliability and earning incentives.

Moreover, the ownership of PV systems by prosumers epitomizes the decentralization of energy production. They can meet their energy needs through self-consumption or purchase energy from the upstream grid, and any excess energy generated can be sold back to the grid through interactions with the energy network manager. This dynamic interaction reflects the evolving role of individuals within the energy landscape, transforming them from passive consumers to active participants in the overall energy environment.

As these MECs evolve, their capacity to integrate energy justice and equity initiatives becomes increasingly pivotal. The chapter recognizes that the perspective of energy communities extends beyond technical, financial, and environmental considerations. It encapsulates a vision where communities actively contribute to and benefit from the sustainable development of their energy landscape. This exploration not only aims to delineate the technical intricacies of grid modernization but also to emphasize the coordination and implementation of energy justice and equity schemes within these dynamic, sustainable, and clean energy communities. Through this comprehensive exploration, the chapter seeks to provide valuable insight into achieving a harmonious balance between modernizing energy grids and ensuring equity and equality in energy access for all.

4.2.1 Definition

The idea of energy communities is foundational to the transformative paradigm of the MEC structure. In its essence, an energy community represents a departure from conventional electricity models with the aid of fostering a decentralized and participatory technique to power manufacturing and consumption. Within the purview of this chapter, an energy community is a dynamic entity, normally comprising people, households, or agencies within a specific geographic place, united via a shared goal of actively collaborating within the era, consumption, and change of energy sources. At its center, the definition of an electricity community incorporates a shift from passive electricity purchasers to lively contributors who play a twin function as both power purchasers and producers. Unlike traditional centralized energy structures, wherein energy flows unidirectionally from large-scale producers to cease-customers, energy communities introduce a dynamic where local residents grow to be prosumers engaging no longer best in the consumption of energy however also in its era through localized, frequently renewable, assets. This fundamental redefinition of roles emphasizes the democratization of electricity selection making, selling a feel of possession and empowerment among community individuals.

Furthermore, the geographic scope of an energy network is crucial in defining its boundaries and the scope of its impact. Whether a community, a rural village, or a city district, the particular traits of the network form the dynamics of energy production and intake. This localized approach permits an extra tailored and context-specific electricity solution, accounting for the specific sociocultural, financial, and environmental factors inherent to each community. The overarching goal of defining a power community within the MEC shape is to emphasize its position as a collaborative and self-sustaining entity. These communities frequently leverage digital systems and verbal exchange technology to facilitate collective selection making and foster an experience of concord among members. This collective company is essential to the fulfillment of energy communities, contributing to the creation of a resilient, sustainable, and inclusive power destiny. In essence, the definition of an energy network encapsulates now not most effective a shift within the technical elements of power systems however, more significantly, a cultural and societal transformation toward an extra equitable and participatory power panorama.

4.2.2 Implementation

The implementation of energy communities involves a multifaceted procedure that integrates technological innovations, regulatory frameworks, and community engagement strategies. At its core, a success implementation pivots on the deployment of renewable power sources, advanced technologies, and the status quo of collaborative decision-making systems in the community. Technological implementation starts with the mixing of various renewable power assets, together with sun panels, wind generators, biomass, and hydropower, tailored to the unique energy desires and geographic characteristics of the community. These assets collectively shape a resilient and sustainable electricity portfolio, decreasing reliance on a unique power type and contributing to a decrease environmental impact. The adoption of smart grid technologies performs a pivotal position in optimizing energy distribution, enabling actual-time tracking, and facilitating manner conversation among electricity manufacturers and consumers. Regulatory frameworks are instrumental in developing allowing surroundings for energy communities. Policies that help decentralized energy manufacturing, incentivize renewable power adoption, and ensure honest reimbursement for energy contributions from network contributors are essential. Governments and regulatory bodies play a pivotal role in imparting a legislative foundation that fosters the boom of power communities, permitting them to thrive and contribute to broader electricity transition dreams. Community engagement is a linchpin for a hit implementation. The established order of network-pushed governance systems, facilitated through virtual systems and conversation gear, empowers citizens to actively take part in decision-making approaches related to electricity initiatives. Education and recognition campaigns inside the network are essential to make certain that residents recognize the blessings of the energy transition, their role as prosumers, and the capability financial and environmental advantages of nearby electricity projects. Moreover, the scalability and replicability of energy network fashions across numerous geographic and cultural contexts are crucial elements of implementation. Case research and first-rate practices from distinct areas provide treasured insights into adapting these fashions to varying socioeconomic situations, ensuring that the blessings of electricity communities are handy to a wide spectrum of communities, regardless of their size or region.

4.2.3 Socio-Techno-Economic-Environmental Aspects

The integration of power communities into the multi-energy structure features a complicated interplay of socio-techno-economic-environmental components, reflecting a holistic approach that extends beyond technical considerations.

Sociocultural Dynamics: Energy communities are deeply embedded in the sociocultural cloth of the communities they serve. Cultural values, social norms, and network choices play an important function in shaping the popularity and adoption of clean energy practices. Understanding and respecting these dynamics is essential for tailoring electricity answers that resonate with the network's identity. Community engagement initiatives, cultural recognition applications, and localized schooling campaigns are instrumental in fostering a feel of collective responsibility and ownership.

Technological Implications: The technological dimension of electricity communities encompasses the deployment of renewable electricity sources, clever grid technology, and energy control structures. The preference and integration of these technologies have far-reaching implications for the network's energy resilience and sustainability. Moreover, the technological infrastructure impacts the diploma of community empowerment, shaping how citizens interact with and advantage of their nearby electricity surroundings. Striking stability among cutting-edge technologies and user-friendly interfaces is vital to ensuring the seamless integration of technological answers inside diverse communities.

Economic Considerations: The economic elements of power communities expand past the immediately value-gain analysis of energy initiatives. Localized energy tasks have the capacity to stimulate monetary development by using developing task possibilities, fostering the increase of related industries, and contributing to the overall nicely being of the community. The financial empowerment of network contributors as prosumers, participating in energy trade packages and contributing to demand reaction efforts, provides a layer of monetary resilience and inclusivity to the power community model. Furthermore, information on the financial disparities inside a community is critical for designing financial fashions that make certain equitable get admission to the benefits of the energy transition.

Environmental Impact: The environmental measurement is critical to the ethos of energy communities, emphasizing the shift toward sustainable and occasional-impact energy practices. Diversifying energy sources, incorporating renewables, and decreasing dependence on fossil fuels contribute to mitigating ecological footprints. The environmental effect of localized energy projects, along with issues of land use and conservation, is a crucial factor that demands cautious assessment. A comprehensive environmental technique also involves tracking and minimizing any potential poor effects, ensuring that the power community version aligns with broader ecological sustainability dreams.

4.2.4 Behavior and Preference of End Users

Understanding the conduct and choices of give up users inside energy communities is paramount to the achievement and sustainability of the MEC structure. This subsection delves into the intricacies of how people and communities engage with, devour, and make a contribution to energy structures, emphasizing the importance of tailoring answers that align with various options and behaviors.

Behavioral Dynamics: The behavior of give up users is shaped via a myriad of things, such as cultural influences, lifestyle choices, and ancient energy intake patterns. Recognizing that energy consumption is a deeply ingrained aspect of daily life, behavioral technological know-how turns into integral to designing powerful electricity solutions. Behavioral economics concepts, inclusive of nudging and incentive structures, may be leveraged to inspire sustainable energy practices. Community-particular studies and nonstop tracking are critical to adapting strategies as behaviors evolve through the years.

Preferences and User-Centric Design: End customers inside energy communities are not homogeneous; choices and wishes range extensively. Some people may also prioritize fee savings while others may emphasize environmental effect or energy independence. User-centric design principles, which involve actively concerning give up customers within the layout method, are crucial for developing energy solutions that resonate with their choices. This method guarantees that technologies are user-friendly, reachable, and align with the values and priorities of the network participants.

Active Participation through Prosumer Models: The emergence of prosumer models, where people transition from passive purchasers to energetic individuals in electricity manufacturing and trade, marks a paradigm shift in user behavior. Prosumers engage in energy change in the community through P2P structures, participate in call for reaction packages, or even make a contribution surplus electricity back to the grid. The energetic participation of end users is not most effective a technical innovation however also a behavioral transformation, fostering a sense of empowerment and ownership.

DRPs and Incentives: Understanding the responsiveness of give up customers to demand reaction packages is critical for grid reliability and height load management. Interruptible and uninterruptible appliances in smart homes allow prosumers to actively interact in demand response projects in the course of top load hours. Incentive structures, whether or not economic rewards or nonmonetary blessings, play a vital role in motivating cease-customers to regulate their energy intake conduct. Recognizing the various desires and motivations of stop-users is pivotal for designing incentive applications that attract an extensive spectrum of community participants.

Community Empowerment: Beyond character behaviors, network-level dynamics notably effect the success of power communities. Empowering neighborhood citizens via training, training, and involvement in electricity initiatives fosters an experience of collective ownership. This communal empowerment not best ensures the success of localized energy projects but additionally creates supportive surroundings for the broader dreams of energy justice and equity.

4.3 Multi-energy Communities

MECs represent an innovative and holistic strategy in response to the intricate demands of contemporary energy infrastructures. By leveraging a mix of numerous energy assets, MECs strive to streamline and decorate the whole process of energy era, utilization, and dispersal. These communities harness a diverse array of energy carriers, encompassing energy, heating, cooling, herbal gasoline, solar, wind, water, hydrogen, and carbon dioxide (CO_2), ensuring a complete and sustainable method of electricity control. Central to the functioning of MECs are EHs, which function pivotal conduits for orchestrating and high-quality-tuning the difficult web of electricity flows within the community [21]. Through the strategic use of EHs, MECs endeavor to optimize resource allocation and enhance the overall efficiency of the communal energy grid, thus paving the way for a more resilient and environmentally conscious energy landscape [22].

In today's dynamic energy landscape, MECs are emerging as the preferred approach to deliver energy in diverse forms. While their operation may be more complex than conventional single-carrier systems, MECs offer significant advantages in terms of stability, resilience, and robustness, owing to their access to a wider range of energy sources [23]. These systems utilize electricity, heating, cooling, natural gas, solar, wind, water, hydrogen, and CO_2 energies simultaneously. Moreover, they are equipped with a variety of energy converters that can transform these energies into one another, adapting to changing demands and market conditions [24, 25].

Moreover, MECs are designed with a focus on equity and inclusivity, ensuring that the distribution of energy resources and benefits is fair across all members, including marginalized groups. One of the key ways MECs achieve this is by decentralizing energy generation and distribution, which allows for local control over resources, ensuring that vulnerable communities have direct access to clean and affordable energy. MECs often incorporate mechanisms such as tiered pricing or subsidies to make energy costs more equitable, prioritizing low-income households and those historically underserved by traditional energy systems. Furthermore, stakeholder participation and transparency are integral to the MEC model. Through participatory governance structures, MECs actively involve community members in decision-making processes, ensuring that the voices of all stakeholders, particularly those from underrepresented groups, are heard and accounted for. This includes organizing community meetings, offering accessible platforms for engagement, and ensuring clear communication about energy decisions and policies. To address the underrepresentation or misrepresentation of vulnerable communities, MECs implement targeted outreach and engagement strategies to ensure that these groups are not only included but also empowered to influence energy policies. By embedding equity and participation into their core structure, MECs create a more just and inclusive energy system that better reflects the needs and priorities of diverse community members.

The MEC serves as a multi-enter/output framework encompassing a couple of power converters, power transmission lines, energy garage units, and numerous load necessities. This integrated gadget performs a crucial role in enhancing the reliability, flexibility, and efficiency of power distribution. In essence, the EH efficaciously manages and optimizes energy generation and intake by means of seamlessly integrating and coordinating a couple of energy companies [26, 27]. Electrical and natural gas networks are the primary sources of energy for MEC systems. These integrated electricity–gas systems, also known as integrated power–heat systems, offer enhanced resilience and reliability by providing energy backup from two parallel sources. They can cater to a wide range of energy demands, including electrical, heating, and cooling loads [28]. Renewable energy sources, such as solar PV, solar water heating, wind turbines, digester units for biogas production, oscillating seawater column converters, and anaerobic digestion reactors for hydrogen production from sewage flow, are increasingly employed in MEC systems to diversify energy sources and reduce reliance on fossil fuels [29, 30]. Energy storage plays a critical role in optimizing MEC system performance. Storage technologies effectively shift energy from off-peak to peak periods, alleviating grid imbalances and enhancing the utilization of renewable sources. Common energy storage options in MEC systems include batteries, hybrid battery–capacitor systems, pumped hydro, thermal storage, and compressed air storage. CHP systems are the primary energy converters in MEC systems, converting natural gas into electricity and heat simultaneously, leading to increased efficiency and reduced energy consumption. Other common energy converters include gas boilers, fuel cells, electrical and absorption chillers, and power-to-gas (P2G) systems [31]. Despite their promising benefits, MEC structures face challenges related to parametric uncertainty, mainly fluctuations in renewable energy technology, power fees, and load needs. To cope with these uncertainties, various optimization strategies, together with stochastic programming, factor estimate technique, strong optimization, fuzzy programming, information gap decision theory (IGDT), and danger-constrained optimization approach, had been developed. DRPs emerge as effective strategies to decorate community flexibility, flatten load curves, stability masses, and minimize operational prices. DRPs incentivize customers to shift their adjustable hundreds from top periods to off-peak hours, thereby lowering strain on the grid and optimizing power consumption [32]. Environmental pollutants, specifically greenhouse fuel emissions, stay a good-sized challenge related to MEC structures. To mitigate these emissions, carbon taxes, carbon capture, and garage technologies were proposed to reduce the environmental footprint of these structures. In precis, MEC structures offer a promising approach to cope with the complexities and challenges of present-day energy structures. Their ability to integrate multiple energy sources, utilize energy converters and incorporate energy storage and DRPs paves the way for a more resilient, efficient, and sustainable energy future [33].

4.3.1 Components of a Multi-career System

- **CHP Unit:** The CHP unit, also known as the gas turbine, plays a crucial role in the MEC system as it serves as the link between the electrical and natural gas networks.
- **Boiler Unit:** Another common heat source in MEC systems is the boiler. This unit utilizes fuel to generate heat. The operational expenses of a boiler include fuel, maintenance, and pollution costs.
- **Energy Storage Unit:** All energy storage systems store energy when energy prices are low or when there is an excess of energy and release it to the system when energy prices are high or when there is a shortage of energy. The hybrid battery capacitor is an especially efficient method for managing uncertainties related to renewable energy.
- **Chiller Unit:** In MEC systems, cooling energy is generated by electric and absorption chillers, which transform electric and heating power into cooling energy.
- **P2G Unit:** The primary goal of MEC systems is to achieve low-carbon economic dispatch. This means that in order to maintain a certain level of system reliability, excess renewable energy should not be wasted on electrical loads. P2G technology is an innovative approach with an efficiency ranging from 60% to 80%. It is adept at utilizing surplus renewable energy to generate profit, minimize costs, and reduce pollution. P2G works by converting electrical power into hydrogen and methane gases. Figure 4.1 shows the structure of the P2G system, which comprises two main components: the water electrolyzer and the methanation process. The water electrolyzer breaks water into oxygen and hydrogen using three different technologies: alkaline water electrolysis, acidic proton exchange membrane electrolysis, and solid oxide electrolysis. The resulting hydrogen can be stored or used directly in the gas turbine, fuel cell, or methanation process. In the methanation process, the hydrogen is combined with CO_2 under the Sabatier reaction, which is highly exothermal. The heat produced by this reaction can be utilized in the boiler or carbon capture unit.

4.3.2 Structure of the Multi-energy Community

The structure of the MEC involves two stages main level and lateral level. The main level includes the selection of energy carrier types in the system, such as (electrical, thermal, gas, and water and utilizes), energy conversion, generation, and determining the network components includes storage equipment such as electrical transformers, boilers, CHP units, wind turbines, desalination units, electrical and thermal storages, water reservoirs, and P2G storage systems, planning model such as linear or nonlinear and selection of objective function. The lateral level includes some critical energy democracy programs such as electrical and thermal

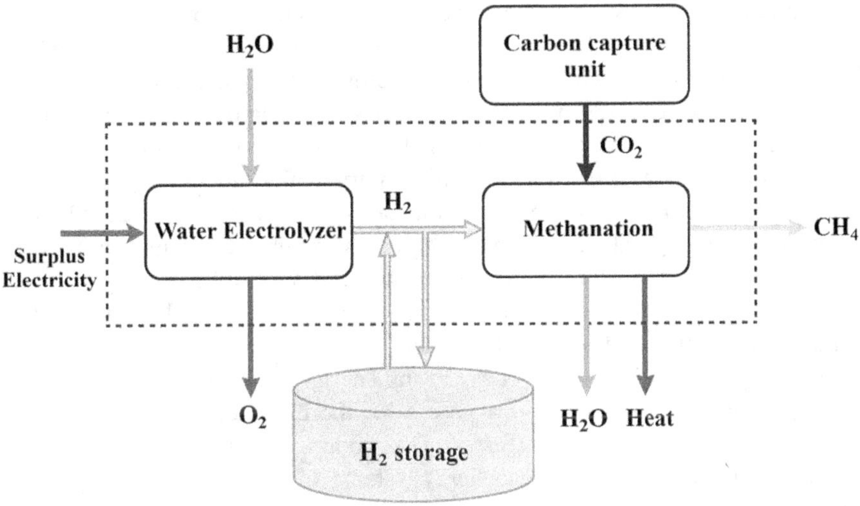

Figure 4.1 Different energy carrier types in the system such as electrical, thermal, gas, and water and utilizes. Source: [34]/with permission of Elsevier.

DR programs that is a management tool that deals with load side rather than the generation side, uncertainty management, and energy resilience.

4.4 Energy Justice and Equity in Multi-energy Communities

The integration of electricity justice and fairness standards within MECs represents a pivotal and visionary method for the transformation of power systems. This section delves into the nuanced panorama of ensuring equity, inclusivity, and social justice in the context of the MEC shape. At the center of energy justice lies the dedication to the honest distribution of benefits and burdens related to power manufacturing and intake. In the context of MECs, this interprets into ensuring that the benefits of localized power projects are available to all community participants, no matter socio-monetary popularity. Simultaneously, efforts must be made to avoid disproportionately burdening susceptible or marginalized businesses within the network. Equitable access to smooth electricity, job possibilities, and the socio-financial blessings of electricity projects become essential to attaining energy justice inside these dynamic communities.

Community-driven decision-making in MECs is a powerful tool for addressing the unique and varied needs of diverse communities. Unlike top-down approaches in traditional energy systems, where decisions are often made by external entities with limited knowledge of local circumstances, community-driven models place decision-making power in the hands of the community members themselves. This localized approach ensures that energy strategies are tailored to the specific social, economic, and geographic characteristics of each community. For instance, some communities may prioritize energy affordability, while others might focus on reducing their environmental impact or increasing energy security and resilience. By engaging directly with community members through participatory processes, such as public consultations, workshops, and local advisory boards, MECs can identify these priorities and develop energy solutions that are aligned with the community's goals and values. Additionally, this model allows marginalized or vulnerable groups to have a direct say in the energy planning process, ensuring their specific challenges—such as access to affordable energy or protection from energy insecurity—are addressed. This bottom-up approach creates more responsive, inclusive, and effective energy systems that reflect the diversity of the communities they serve.

Energy justice advocates for prevalent and less costly get admission to smooth energy, addressing the historic disparities in energy provision. In the context of MECs, this entails a dedication to eradicating energy poverty and ensuring that marginalized communities have the possibility to actively participate in and enjoy the energy transition. The implementation of inclusive electricity entry involves targeted interventions to triumph over limitations inclusive of monetary constraints, academic gaps, and technological literacy, fostering an absolutely inclusive electricity panorama. Active community participation is not handiest a means to a cease however an essential factor of energy justice in MECs. Empowering local citizens through education, education, and significant involvement in selection-making methods fosters a feel of possession and ensures that the advantages and burdens of energy tasks are shared equitably. By actively engaging communities in the planning, implementation, and control of energy structures, the MEC structure will become a vehicle for democratizing energy governance, aligning with the standards of electricity justice. The transition to an MEC structure gives unique possibilities for activity creation and financial development, contributing to the wider goals of social and financial fairness. Localized energy initiatives stimulate a boom in related industries and offerings, creating employment opportunities inside the community. By fostering financial improvement, energy justice goals to ensure that the blessings of the energy transition are not only reachable but contribute to the overall nicely being of the community, especially in regions historically disadvantaged by way of conventional electricity fashions. Achieving power justice within MECs requires a delicate stability among embracing technological innovation and upholding social equity. While present-day technologies beautify energy efficiency and grid management, a comprehensive approach must ensure that those improvements do not inadvertently exclude or downside positive network participants. User-pleasant interfaces, inclusive design ideas, and ongoing community engagement efforts are essential to bridge the gap among technological innovation and social fairness.

This section addresses the complicated and multifaceted landscape of energy justice and fairness within MECs. By weaving those concepts into the material of the MEC shape, the segment pursuits to clarify the strategies, demanding situations, and opportunities related to fostering fairness, inclusivity, and social justice in the ongoing evolution of energy structures. Through a comprehensive exploration of those subject matters, the segment contributes to the broader discourse on reshaping energy systems to not most effective meet technical and environmental goals but additionally to enhance societal nicely being and fairness for all stakeholders.

4.4.1 Definition of Energy Justice and Equity

The European Union has introduced the clean energy package (CEP) to promote system-wide changes in energy policies and climate to transition to a low-carbon energy system [35]. This transition extends beyond decarbonization and includes energy production, distribution, and consumption changes. MECs are an important component of the CEP. These communities focus on utilizing local renewable energy sources to meet the energy needs of end users [36]. They provide a sustainable alternative for large power plants dependent on fossil fuels. The development of MECs in Europe is driven by factors such as energy technologies, carriers, regional conditions, and emerging regulatory developments. The goal is to integrate distributed energy resources and multiple energy carriers at various scales, facilitating the optimal design and operation of these communities [37]. The "Clean Energy for All" policy aims to address issues of energy justice, ensuring that all community members have access to affordable, reliable, and sustainable energy services for their basic needs. It also emphasizes the importance of community participation in governance and the equitable sharing of benefits and burdens [38]. The European Union expects energy community initiatives to be fair in the European energy transition. Member states are required to establish a legally binding "enabling framework" for renewable energy communities [39]. This policy framework is designed to support and encourage the development of community energy initiatives. The anticipation is that these initiatives will bring benefits to energy consumers and contribute to the promotion of energy justice. Furthermore, energy justice requires affordability because high costs can worsen inequality. Therefore, it is important to consider strategies that advance cost-effective energy and alleviate the economic challenges faced by vulnerable communities. Energy justice embodies principles related to equity and fairness principles within the energy sector, focusing on three key aspects: distributive, procedural, and recognition [40].

Distributive justice focuses on the consequences of decision-making, including the distribution of benefits and costs, risks, and responsibilities. These outcomes can manifest as material entities or negative impacts or societal benefits. It means everyone should have reliable, affordable, and sustainable energy and that communities should not disproportionately suffer from the negative impacts of energy production, like climate change or pollution [41].

Procedural justice refers to the fairness of energy decision-making processes, which ensures meaningful stakeholder participation and transparent, accountable implementation of energy policies. It involves top-down or bottom-up decision-making, viewed from both ethical and political perspectives. Equitable procedures engage all stakeholders in a nondiscriminatory manner, providing access to decision-making. Fair decision-making is crucial for the legitimacy of rules and stakeholder cooperation, and it is achieved through unbiased decision-makers, meaningful participation, and early engagement. Information sharing in consultation processes is vital, as it impacts social acceptance, public trust, and the adoption of ethical and sustainable energy choices [42].

Recognition justice differs from procedural justice by focusing on the processes involving disrespect, insult, and degradation, which diminish specific individuals and place identities. This concept underscores the importance of fair representation and recognizes the structural underrepresentation of vulnerable communities, along with the unequal distribution of transition costs and benefits within energy communities. Furthermore, recognition justice delves into the repercussions of economic and social inequality on participation opportunities, prompting dialogue on the prerequisites for fostering inclusive participation [39]. The research associated with recognition justice investigates which segments of society are misunderstood, ignored, or misrepresented to construct inclusive energy decision-making processes that honor diverse perspectives. It places value on all individuals, comprehends disparate impacts, and endeavors to rectify injustices to promote fairness and equity in the energy system. Notably, recognition justice synergizes with distributional and procedural justice, forming an integral part of a comprehensive approach [43].

In the context of social justice and environmental sustainability, communities have paid much attention to energy equity in recent years. Although equity and justice have some similarities, they have different meanings in relation to fairness, behavioral economics, and social welfare [44]. Essentially, justice is a broad framework of fairness and moral rightness, whereas equity is a specific mode of operation within that framework and emphasizes the fair distribution of resources, rights, and opportunities [44].

In MECs, energy equity involves ensuring that all community members have access to reliable, affordable, and clean energy services, regardless of their geographical location or income. Fair treatment for all energy consumers, irrespective of their socioeconomic status or location, is of considerable importance. By promoting equal access to energy resources, it can contribute to poverty energy, improved outcomes, and social welfare. By simultaneously developing equity and justice, a just and sustainable energy system can be created that benefits every member of the community [44].

The critical features of the energy structure for energy justice and equity include the integration of electricity justice and fairness standards within MECs, which represents a pivotal and visionary method for the transformation of power systems [45].

These critical features of the energy structure for energy justice and equity can be summarized as follows:

Equitable Distribution: Ensuring a fair sharing of energy resources and benefits among diverse groups and regions.

Accessibility: Providing fair access to reliable and affordable energy sources for all, irrespective of income or location.

Affordability: Making energy services affordable for all, especially for low-income and marginalized groups, to prevent energy poverty.

Sustainability: Ensuring that energy systems are sustainable over the long term, taking into account social, environmental, and economic factors to meet the needs of both present and future generations.

Clean Energy: Transitioning toward clean and renewable energy sources is essential for mitigating environmental harm and promoting public health [46].

Community Engagement: Involving communities, especially marginalized ones, in decision-making processes regarding energy projects in order to address their specific needs and concerns [47].

4.4.2 Energy Poverty

The use of dependable and low in cost energy sources is essential for sustained development and a higher excellent of life. Clean, reliable, and sustainable energy is crucial for accomplishing numerous sustainable improvement dreams. Lack of access to lower-priced and dependable power no longer only hinders financial development but additionally has serious effects on schooling, fitness, and social welfare. Energy justice offers a framework that aligns with distribution, availability, and affordability. Energy poverty typically serves because the analytical boundary for discussions on power justice. In addition to thinking about earnings and affordability, policymakers and researchers need to also cope with the social and environmental dimensions of power poverty. This consists of reading the effect of electricity selections on prone populations, together with low-profits households, marginalized communities, and areas disproportionately stricken by electricity-associated environmental degradation. Enhancing the sector of electricity poverty studies allows for an extra complete knowledge of the obstacles and opportunities for accomplishing power justice. To successfully fight electricity poverty, it is important to undertake an interdisciplinary method that includes insights from economics, sociology, environmental research, and coverage analysis. This interdisciplinary approach enables a holistic comprehension of the interrelated factors influencing electricity poverty and gives feasible solutions to mitigate its effects. It is important to ensure that proposed interventions are contextually appropriate and equitable. Also, via integrating insights from behavioral economics, policymakers can work toward mitigating these dimensions of injustice and enhancing electricity justice consequences for individuals and communities affected by power poverty. Energy poverty is strongly related to one-of-a-kind styles of injustice, which include distributive injustice, procedural injustice, and recognition injustice. Distributive injustice entails the choppy allocation of sources or burdens inside a network. In the context of electricity poverty, this consequences in disparities within the availability of sufficient power offerings. Marginalized or inclined companies, as a result, face a disproportionate loss of entry to critical energy assets. The fairness and transparency of the selection-making procedures are the primary issues about procedural inequity. Procedural inequity within the context of energy poverty may also result from a lack of inclusive and transparent energy device control processes. This consists of the advent and implementation of guidelines related to energy access and affordability. As a result, marginalized communities will not be covered in selection-making approaches that significantly affect their usage and access to power assets [41]. Recognition injustice happens while individuals and communities aren't stated or respected for their identities, experiences, and desires. In the context of energy poverty, reputation injustice is seen, while the unique challenges faced by way of prone communities in accessing and the usage of energy offerings are not stated. This can also contain ignoring the cultural, social, and financial factors that make a contribution to electricity poverty, thereby perpetuating inequalities in power accessibility and utilization. The effect of power poverty on choice-making and the perpetuation of energy injustices may be better understood through the lens of behavioral economics. By analyzing how people make selections and behave in approaches that cross in opposition to rational choice theory, policymakers can develop extra powerful strategies to deal with energy poverty. Nudging and boosting strategies, derived from behavioral economics, can be used to influence individuals' electricity-related behaviors without proscribing their picks or converting monetary incentives. This technique can help people make extra best energy use selections and decrease energy poverty. Consumers may be endorsed to expand energy-saving behaviors and make extra optimum decisions about how much electricity they use through nudging strategies that originate from behavioral economics. By providing statistics, feedback, and social norms, stimulation can nudge individuals closer to power-green alternatives [48].

Boosting techniques can be employed to enhance individuals' competencies and decision-making abilities related to energy use. By providing skills, knowledge, and decision tools reinforcement can empower individuals to take action to improve their energy access and usage.

Behavioral economics provides a lens through which the decision-making processes influencing energy consumption patterns in economically vulnerable communities can be examined. Socioeconomic factors further compound the relationship between energy poverty and behavioral economics. Limited financial resources, a common characteristic of energy poverty households, often result in suboptimal energy usage decisions. For instance, individuals may resort to energy-intensive practices due to a perceived lack of alternatives or awareness of more energy-efficient options.

By incorporating behavioral economics methods into energy poverty alleviation measures, policymakers can empower individuals to make more informed energy choices and take actions that contribute to overall energy justice.

4.4.3 Energy Democracy

Energy democracy is deciding how energy is produced, distributed, and used in communities. It aims to create a more inclusive and fair system of resource distribution and decision-making. The goal is to enable communities and individuals to participate in the generation, distribution, and consumption of energy by democratizing energy systems. Energy democracy aims to challenge traditional centralized energy systems dominated by large corporations and government agencies by promoting decentralized, community-led initiatives that prioritize social and environmental justice. More specifically, energy production must not pollute the environment or harm people. Based on the original idea, energy democracy leads to greater public participation in decision-making [43]. Various forms and practices are used to engage in public participation. This is because the current hierarchical and expert-centered approach in the energy political sector limits its power and influence. This approach favors those in control and is not effective for policy-making and problem-solving. The participatory governance approach argues that decision-making should include non-state stakeholders like the business sector and civil society to improve governing capabilities, achieve social objectives, and solve problems. Energy democracy should result in the generation of surplus value and employment, involving advantages for the creation of employment and capital. In 2018, Szulecki offered a definition of energy democracy in which citizens, both as producers and consumers, play a key role in shaping energy sector policies. The process of making decisions involves including and being transparent, and encouraging the participation of knowledgeable and responsible people, all to achieve general well-being [49]. The concept of energy democracy has a lot in common with energy justice, as they both focus on the distributional effects, procedural procedures, environmental aspects, and social impact of energy production and consumption. Energy democracy aims to address disparities in energy distribution by decentralizing production and empowering communities for equitable access. It advocates for inclusive decision-making structures, ensuring that diverse stakeholders, including marginalized communities, shape energy policies. Prioritizing sustainability, it promotes renewables to mitigate environmental burdens on vulnerable communities. Social justice is upheld by providing tools for sustainable energy access and fostering cohesion, resilience, and community autonomy. These principles align with distributive, procedural, environmental, and social justice in energy governance [50]. Energy democracy aims to rectify injustices. However, the concept has limits, as it does not adequately address critical issues such as energy extraction operations and indigenous community rights.

4.5 Clean Energy Initiatives in Multi-energy Communities

The transition to a sustainable energy future is intricately linked to the adoption of clean energy initiatives within the MEC structure. This section explores the diverse resources of clean energy available to these communities, delves into the integration of renewable energies, and envisions the development of carbon-free MECs.

4.5.1 Resources of Clean Energy

Clean energy projects within MECs draw from numerous sets of sources that now not best meet the power wishes of the network but additionally align with sustainability goals. One of the number one and extensively adopted sources is solar energy, harvested through PV structures. Solar panels convert daylight into energy, imparting a dependable and considerable supply of easy energy. The decentralized nature of the sun energy era allows communities to harness energy domestically, lowering transmission losses and promoting energy independence.

Wind energy stands as another prominent resource, tapped via wind generators strategically located in wind-wealthy areas. The kinetic energy of the wind is transformed into energy, supplying a scalable and environmentally friendly alternative to traditional energy resources. Biomass, derived from organic materials along with agricultural residues or dedicated energy vegetation, offers a flexible and renewable energy supply. Biomass energy technology no longer only mitigates waste disposal demanding situations but additionally contributes to the discount of greenhouse gas emissions [51].

Hydropower, generated by harnessing the energy of flowing water, stays a steadfast resource within MECs. Whether through small-scale run-of-river structures or large hydropower plants, communities can leverage the consistent and controllable nature of hydropower for a steady deliver of smooth energy. The various blend of these sources guarantees resilience inside the face of variable energy production, enhancing the reliability of the general energy system.

Furthermore, emerging technologies along with tidal and wave energy make contributions to the increasing repertoire of easy electricity assets. These technologies harness the electricity generated by the motion of tides and waves, offering a continuous and predictable source of power, especially in coastal regions. By tapping into a number of clean energy sources, MECs can diversify their energy portfolio, minimizing dependency on a single supply and growing basic gadget resilience. The assets of easy energy within MECs encompass a rich tapestry of renewable resources. The synergistic use of solar, wind, biomass, hydropower, and rising technology bureaucracy the inspiration of a sustainable and environmentally accountable power atmosphere. As communities retain to explore and innovate with these sources, they pave the way for a resilient, decentralized, and smooth energy future.

4.5.2 The Integration of Renewable Energies in Energy Communities

The integration of renewable energies within energy communities is a transformative technique that is going beyond the adoption of character clean electricity resources. It includes a complete method to harnessing the overall ability of solar, wind, biomass, and hydropower, with a focal point on seamless integration, technological innovation, and network engagement.

Smart Grids and Energy Management Systems: At the coronary heart of powerful integration lies the deployment of clever grids and energy management systems. Smart grids facilitate the bidirectional go with flow of energy, taking into account the integration of renewable electricity sources and green distribution within the network. Energy control systems play an essential function in real-time tracking, call for forecasting, and optimizing the utilization of renewable electricity. By offering a dynamic framework for energy trade, smart grids empower communities to balance supply and demand, improving ordinary grid reliability.

Localized Energy Storage: The intermittent nature of renewable energy resources necessitates the incorporation of power storage systems inside energy communities. Batteries and different energy storage technology save extra energy generated for the duration of height production durations, making it to be had at some point of durations of low or no era. This localized storage capacity no longer best complements grid stability however additionally allows the efficient usage of renewable energy, minimizing wastage and maximizing the community's self-sufficiency.

Community-Driven Decision-Making: The hit integration of renewable· energies in energy communities is intricately tied to community-pushed selection-making methods. Engaging citizens in decisions associated with the sort, scale, and area of renewable energy tasks fosters a sense of ownership and ensures that answers align with the precise traits and wishes of the network. This collaborative technique not most effective complements the social reputation of renewable power projects however also leverages the collective awareness of community members for effective undertaking implementation.

Case Studies and Best Practices: Examining case studies and nice practices from various power communities offers treasured insights into hit integration strategies. By studying the stories of others, communities can adapt validated fashions, technologies, and community engagement processes to their particular context. This understanding-sharing fosters a collaborative ethos in the broader smooth electricity surroundings, accelerating the transition in the direction of sustainable and integrated power structures.

Addressing Variability and Grid Resilience: The variable nature of renewable energy resources provides a venture in retaining grid stability. Strategies to cope with this variability consist of a diverse mix of renewable resources, complemented by means of superior forecasting techniques and responsive call for-facet management. Integrating various renewable energies guarantees a more regular energy supply, decreasing reliance on an unmarried supply and improving general grid resilience.

In the end, the mixing of renewable energies within power communities is a dynamic and multifaceted process. By combining smart grid technologies, localized energy garage, community-driven selection making, and classes from successful case studies, communities can create strong and adaptive systems. This integration not only contributes to environmental sustainability however additionally positions electricity communities at the leading edge of the smooth energy transition, showcasing the ability for decentralized, network-centric strategies to redefine the destiny of electricity.

4.5.3 Carbon-Free Multi-energy Communities

The vision of carbon-unfastened MECs represents a formidable and imperative purpose in the pursuit of a sustainable and environmentally responsible future.

This subsection explores the strategies, challenges, and blessings associated with envisioning and knowing communities that not only embrace smooth energy however actively paintings closer to getting rid of carbon emissions across diverse aspects of their energy systems.

Comprehensive Energy Efficiency Measures: A foundational detail in the adventure toward carbon neutrality involves imposing complete power efficiency measures inside MECs. This encompasses adopting electricity-efficient technology in buildings, infrastructure, and transportation, thereby lowering typical electricity demand. Energy-efficient practices, starting from clever constructing design to optimized transportation systems, play a pivotal position in minimizing the carbon footprint of the network.

Electrification of Transportation: A vast contributor to carbon emissions is the reliance on fossil gas-powered transportation. Carbon-unfastened MECs prioritize the electrification of transportation by means of transitioning to EVs and investing inside the vital charging infrastructure. By promoting using easy energy sources for energy transportation, these communities make a contribution to both decreased neighborhood air pollutants and reduced carbon emissions, aligning with broader sustainability desires.

Carbon Offset Initiatives: To offset residual emissions that may be tough to cast off entirely, carbon-unfastened MECs interact in carbon offset initiatives. These projects involve investing in projects that seize or prevent the release of an equal amount of carbon dioxide someplace else. Afforestation initiatives, renewable electricity initiatives in areas with excessive emissions, and investments in carbon seize and storage technology are examples of tasks that make contributions to the network's usual carbon neutrality.

Community Awareness and Behavioral Shifts: Achieving carbon neutrality within MECs calls for a vast shift in network conduct and attention. Educational campaigns, workshops, and network engagement packages play a crucial role in raising focus about the environmental impact of electricity intake and inspiring sustainable practices. By fostering a collective commitment to lowering carbon emissions, communities can create a shared experience of responsibility, using behavioral shifts toward more environmentally conscious picks.

Technological Innovation and Research: Continuous technological innovation and studies are crucial additives to the transition to carbon-free MECs. This involves exploring modern technology, improving current clean energy answers, and investing in studies and improvement initiatives that push the limits of carbon reduction opportunities. By staying at the forefront of technological improvements, communities can make sure the nonstop development of their carbon-loose projects and hold resilience within the face of evolving electricity landscapes.

Challenges and Opportunities: While the imaginative and prescient of carbon-loose MECs is laudable, they come with a set of demanding situations and opportunities. Overcoming the reliance on traditional energy assets, addressing economic considerations, and navigating the complexities of technological transitions are demanding situations that require strategic planning and collaboration. On the other hand, the possibilities lie in job creation, monetary improvement, and positioning the network as a leader in sustainability, doubtlessly attracting investments and citizens who prioritize environmental obligation.

4.6 Problem Formulation

The proposed MEC consists of a CHP unit, a boiler, a compressed-air energy storage (CAES) system, a thermal energy storage (TES) facility, and an electrical heat pump (EHP) system, along with connections to the local electricity and natural gas grids.

A CHP unit generates electricity and captures the heat that would otherwise be wasted to provide useful thermal energy. This dual output makes CHP units highly efficient, often exceeding 80% efficiency. The CHP unit in the MEC is pivotal because it provides both electrical and thermal energies directly to the microgrid, reducing reliance on external energy sources and enhancing overall energy efficiency. The boiler in the MEC is primarily used to produce steam or hot water for heating purposes. It can serve as a backup or supplementary source of thermal energy, particularly useful during high-demand periods or when the CHP unit is offline. Boilers can be powered by various fuels, including natural gas, which might be supplied by the local natural gas grid. CAES systems store excess electrical energy by compressing air in an underground cavern or pressure vessel. When energy demand peaks, the compressed air is released and used to drive a turbine to generate electricity. In the MEC, the CAES acts as an energy buffer, smoothing out fluctuations in energy supply and demand and enhancing grid stability. TES systems store thermal energy by heating or cooling a storage medium (like water, molten salt, or phase-changing materials) during periods of low energy demand. This stored energy can then be used for heating or cooling during periods of high demand. Integrating TES into the MEC allows for better management of the thermal energy produced by the CHP unit and boiler, reducing waste and increasing the system's responsiveness to changes in thermal demand. The EHP is an efficient way to transfer heat between the outside air, ground, or water sources to the building. It can be used for both heating and cooling. In the context of the MEC, an EHP can utilize excess electrical energy to provide additional TES in the TES or direct climate control, further optimizing energy usage. The effectiveness of the MEC hinges on a sophisticated control system capable of dynamically managing the operations of all components based on current and predicted demand for electricity and heat, pricing signals from energy markets, and the operational status of each component. This system must optimize for cost efficiency, energy conservation, and system reliability, often using predictive analytics and real-time data processing. By combining these various technologies and systems, the MEC aims to provide a resilient, efficient, and cost-effective solution for energy management at a local scale. This integration not only supports sustainability by maximizing the use of renewable resources and waste heat but also enhances the grid's adaptability to varying energy loads and generation capacities.

The primary aim of the envisioned MEC model is to minimize the overall cost specified in Eq. (4.1).

$$\min \cdot \sum_{s=1}^{S} \pi_s \times \left(OC_s^{CHP} + OC_s^{Boiler} + OC_s^{CAES} + OC_s^{PEV} + CB_s - RS_s + PC_s\right) + SUC + SDC \tag{4.1}$$

where OC_s^{CHP}, OC_s^{Boiler}, OC_s^{CAES}, and OC_s^{PEV} are the operation cost of the CHP unit, boiler, CAES system, and plug-in electric vehicle (PEV) parking lot, respectively. CB_s and RS_s are the cost and revenue of buying/selling power from/to main grid, respectively. SUC and SDC are the startup and shutdown costs of CHP unit, respectively.

The cost of energy not supplied (ENS) as a penalty cost can be calculated by Eq. (4.2).

$$PC_s = \sum_{t=1}^{T} VOLL^{EL} \times ENS_{t,s}^{E} + VOLL^{TH} \times ENS_{t,s}^{T}, \tag{4.2}$$

where $ENS_{t,s}^{E}$ and $ENS_{t,s}^{T}$ are amount of electrical and thermal ENS, respectively. $VOLL^{EL}$ and $VOLL^{TH}$ are the electrical and thermal loss of load, respectively.

CHP units, often hailed for their energy efficiency, seamlessly integrate the generation of electricity and useful thermal energy. The fuel consumption of CHP units is intricately tied to their operation and maintenance costs, as outlined by a dynamic interplay of factors denoted by Eqs. (4.3)–(4.5). This symbiotic relationship underscores the significance of optimizing fuel efficiency to curtail overall operational expenditures, making CHP systems a compelling solution for sustainable and cost-effective energy production.

$$OC_s^{CHP} = \sum_{t=1}^{T} \left(F_{t,s}^{CHP} + M_{t,s}^{CHP}\right) \tag{4.3}$$

$$F_{t,s}^{CHP} = P_{t,s}^{CHP} \times \left(\frac{\lambda_t^{Gas}}{HV \times \eta^{CHP}}\right) \tag{4.4}$$

$$M_{t,s}^{CHP} = P_{t,s}^{CHP} \times MC^{CHP}, \tag{4.5}$$

where $F_{t,s}^{CHP}$ and $M_{t,s}^{CHP}$ are fuel and maintenance costs of the CHP plant, respectively. The parameters HV and η^{CHP} correspond to the heat value of natural gas and the electrical efficiency of the CHP plant, respectively. MC^{CHP} represents the coefficient of CHP plant maintenance cost.

The limitations of electrical/thermal productions of the CHP plant can be specified as the constraints Eqs. (4.6) and (4.7), respectively.

$$P_{\min}^{CHP} \le P_{t,s}^{CHP} \le P_{\max}^{CHP} \tag{4.6}$$

$$T_{t,s}^{CHP} = P_{t,s}^{CHP} \times HPR \times \eta_{HE}. \tag{4.7}$$

HPR denotes the heat-to-power ratio of the CHP plant, while η_{HE} represents the efficiency of the heat exchanger.

The ramping limitations of the CHP plant have been shown by Eqs. (4.8) and (4.9).

$$P_{t,s}^{CHP} - P_{t-1,s}^{CHP} \le P^{Ramp-up} \times \gamma_{t-1} + i_t \times P_{\min}^{CHP} \tag{4.8}$$

$$P_{t-1,s}^{CHP} - P_{t,s}^{CHP} \le P^{Ramp-down} \times \gamma_t + j_t \times P_{\min}^{CHP}, \tag{4.9}$$

where $P^{Ramp-up}$ and $P^{Ramp-down}$ are the ramp-up and ramp-down rates of the CHP plant, respectively. i_t and j_t are binary variables and show the startup and shutdown status of the CHP plant, respectively. Similarly, γ_t is a binary variable and shows whether the CHP plant is on or not.

The startup and shutdown costs of the CHP plant are expressed by Eqs. (4.10) and (4.11), respectively.

$$SUC = \sum_{t=1}^{T} (SU \times i_t) \tag{4.10}$$

$$SDC = \sum_{t=1}^{T} (SD \times j_t) \tag{4.11}$$

SUC and SDC denote the startup and shutdown costs associated with the CHP plants. Furthermore, the constraints Eqs. (4.12)–(4.16) ensure the binary variables i_t, j_t, and γ_t have a valid value.

$$0 \le i_t \le \gamma_t \tag{4.12}$$

$$\gamma_t - \gamma_{t-1} \le i_t \le 1 - \gamma_{t-1} \tag{4.13}$$

$$0 \le j_t \le \gamma_{t-1} \tag{4.14}$$

$$\gamma_{t-1} - \gamma_t \le j_t \le 1 - \gamma_t \tag{4.15}$$

$$\gamma_{t-1} - \gamma_t + i_t - j_t = 0 \tag{4.16}$$

CAES technology centers around extracting stored energy from highly pressurized compressed air, converting it into an alternative energy form through the various components of the CAES system. The limitation of the air injection into CAES is presented in Eq. (4.17) [52]. Constraint Eq. (4.18) outlines the power generated by the CAES system. The inflow of air into the chamber and the injection of air into storage are contingent upon pressure constraints and valve dimensions, mathematically expressed as Eqs. (4.19) and (4.20), respectively.

To prevent the simultaneous injection and pumping states within the CAES system, binary variables $h_{t,s}$ and $f_{t,s}$ are incorporated, as outlined in Eq. (4.21). This strategy is crucial for the optimal functioning of the CAES system.

$$V_{t,s}^{injected} = e^{i} \times P_{t,s}^{cons.} \tag{4.17}$$

$$P_{t,s}^{\text{gen.}} = e^{p} \times V_{t,s}^{pumped} \tag{4.18}$$

$$V_{\min}^{pumped} \times h_{t,s} \leq V_{t,s}^{pumped} \leq V_{\max}^{pumped} \times h_{t,s} \tag{4.19}$$

$$V_{\min}^{injected} \times f_{t,s} \leq V_{t,s}^{injected} \leq V_{\max}^{injected} \times f_{t,s} \tag{4.20}$$

$$f_{t,s} + h_{t,s} \leq 1 \tag{4.21}$$

Here, $V_{\min}^{pumped}/V_{\max}^{pumped}$ represent the minimum and maximum levels of pumped air from storage to the combustion chamber, while $V_{\min}^{injected}/V_{\max}^{injected}$ denote the minimum and maximum range of injected air into the CAES system.

The CAES's operational capability is constrained by factor Eq. (4.22). The equilibrium of stored air within each temporal block is intricately connected to the inflow and outflow of injected air, a relationship that can be articulated through expression Eq. (4.23).

$$PS_{\min} \leq PS_{t,s} \leq PS_{\max} \tag{4.22}$$

$$PS_{t+1,s} = PS_{t,s} + V_{t,s}^{injected} - V_{t,s}^{pumped} \tag{4.23}$$

The heat rate denotes the quantity of natural gas consumed per unit of maximum electrical energy produced by the expander [52]. Eq. (4.24) furnishes the operational cost of the CAES system.

$$OC_{S}^{CAES} = \sum_{t=1}^{T} \left(V^{c} \times P_{t,s}^{cons.}\right) + \left(\left(V^{\exp.} + CHR \times \lambda_{t}^{Gas}\right) \times P_{t,s}^{gen.}\right) \tag{4.24}$$

The operational expenses incurred by the expander during discharging and the compressor during charging contribute to the overall operation cost of the CAES system. Variables $V^{\exp.}$ and V^{c} denote the variable maintenance and operation costs associated with the expander and compressor, respectively. Additionally, cumulative heat release (CHR) signifies the requisite heat rate for the proposed CAES system during the discharging state.

Charge and discharge constraints of PEVs' parking lot can be defined by Eqs. (4.25 and 4.26):

$$0 \leq CH_{t,s}^{PEV} \leq CH^{\max} \tag{4.25}$$

$$0 \leq DCH_{t,s}^{PEV} \leq DCH^{\max} \tag{4.26}$$

State of charge (SOC) constraint of each PEV is as follows:

$$SOC_{t,s} = SOC_{t-1,s} + \left(\frac{CH_{t,s}^{PEV} \cdot \eta_{CH}^{PEV}}{E_{PEV}^{\max}} - \frac{DCH_{t,s}^{PEV}}{E_{PEV}^{\max} \cdot \eta_{DCH}^{PEV}} \right) \tag{4.27}$$

The upper limits for charging/discharging capabilities of the TES are specified in Eqs. (4.28) and (4.29), respectively. To preclude the TES from being in a simultaneous charging and discharging state, binary variables $m_{t,s}$ and $n_{t,s}$ are incorporated into the subsequent constraints. Eq. (4.30) illustrates the constraint on the thermal energy stored in the TES.

$$0 \leq TS_{t,s}^{ch} \leq TS^{\max} \times m_{t,s} \tag{4.28}$$

$$0 \leq TS_{t,s}^{dch} \leq TS^{\max} \times n_{t,s} \tag{4.29}$$

$$TC^{\min} \leq T_{t,s}^{TES} \leq TC^{\max} \tag{4.30}$$

$$m_{t,s} + n_{t,s} \leq 1 \tag{4.31}$$

Equation (4.32) outlines the thermal energy balance within the TES.

$$T_{t,s}^{TES} = T_{t-1,s}^{TES} \times \eta_{TES} + \left(TS_{t,s}^{ch} \times \eta_{TES}^{ch}\right) - \left(TS_{t,s}^{dch} / \eta_{TES}^{dch}\right) \tag{4.32}$$

The TES exhibits efficiencies during standby and charging/discharging operations, represented by η_{TES}^{ch} and η_{TES}^{dch}, respectively.

The boiler's thermal output is restricted by Eq. (4.33):

$$T_{\min}^{Boiler} \leq T_{t,s}^{Boiler} \leq T_{\max}^{Boiler} \tag{4.33}$$

Here, $T_{\min}^{Boiler}$ and $T_{\max}^{Boiler}$ represent the lower and upper limits of the boiler's output, respectively. The operational cost of the boiler can be expressed as follows:

$$OC_s^{Boiler} = \sum_{t=1}^{T} T_{t,s}^{Boiler} \times \left(\frac{\lambda_t^{Gas}}{HV \times \eta^{Boiler}} \right) \tag{4.34}$$

In this context, HV signifies the heat value of natural gas, while η^{Boiler} represents the boiler's efficiency.

The constraints governing the EHP system are articulated in Eqs. (4.35) and (4.36). In these constraints, $T_{\min}^{EHP} / T_{\max}^{EHP}$ denote the lower/upper limits of the EHP power outputs.

$$T_{\min}^{EHP} \leq T_{t,s}^{EHP} \leq T_{\max}^{EHP} \tag{4.35}$$

$$T_{t,s}^{EHP} = P_{t,s}^{EHP} \times CP \tag{4.36}$$

Here, CP represents the coefficient of performance for the EHP system.

The restrictions on the exchange of electricity between the MEC and the local grid are expressed in Eqs. (4.37) and (4.38). In the subsequent equations, $x_{t,s}$

and $y_{t,s}$ are binary variables signifying the status of electricity export and import, respectively.

$$0 \leq P_{t,s}^{buy} \leq ECL \times x_{t,s} \tag{4.37}$$

$$0 \leq P_{t,s}^{sell} \leq ECL \times y_{t,s} \tag{4.38}$$

$$x_{t,s} + y_{t,s} \leq 1 \tag{4.39}$$

Equations (4.40) and (4.41) delineate the expenses and earnings linked to the electricity import and export with the local grid, respectively.

$$CB_s = \sum_{t=1}^{T} P_{t,s}^{buy} \times \lambda_{t,s}^{buy} \tag{4.40}$$

$$RS_s = \sum_{t=1}^{T} P_{t,s}^{sell} \times \lambda_{t,s}^{sell} \tag{4.41}$$

Meeting thermal and electrical demands necessitates the supplied energy from the local grid and MEC, formulated as Eqs. (4.42) and (4.43).

$$D_{t,s}^{DR,e} - ECL_{t,s} = \begin{pmatrix} P_{t,s}^{CHP} + P_{t,s}^{buy} + P_{t,s}^{gen.} - P_{t,s}^{sell} - P_{t,s}^{cons.} - P_{t,s}^{EHP} \\ + P_{t,s}^{wind} + P_{t,s}^{solar} + DCH_{t,s}^{PEV} - CH_{t,s}^{PEV} \end{pmatrix} \tag{4.42}$$

$$DR_{t,s}^{DR,th} - TCL_{t,s} = T_{t,s}^{CHP} + T_{t,s}^{boiler} + TS_{t,s}^{dch} + T_{t,s}^{EHP} - TS_{t,s}^{ch} \tag{4.43}$$

Implementing a DRP can serve various objectives. This paper employs the load shifting method, wherein specific loads within the MEC are moved from peak to off-peak intervals, resulting in reduced operational costs for the system. In the proposed electrical demand response (EDR) and thermal demand response (TDR) models, end users have the ability to shift a defined portion of their demand during each operational horizon.

The limit on the transferable amount of electrical and thermal demand to off-peak times is articulated in Eq. (4.44):

$$DRP_{t,s}^{e/th} \leq DRP_{t}^{\mathrm{Max},e/th} \tag{4.44}$$

Here, $DRP_{t}^{\mathrm{Max},e/th}$ represents the maximum participation factor of the DRP at time t within the scheduling horizon. Upon implementation of the DRP, the ultimate demands of the end users should align with Eq. (4.45):

$$D_{t,s}^{DR,e/th} = \phi_{t,s}^{e/th} + D_{t,s}^{e/th} \cdot \left(1 - DRP_{t,s}^{e/th}\right) \tag{4.45}$$

In the given equation, the transferred demands from various hours to the designated time in scenario s are denoted by $\phi_{t,s}^{e/th}$. It is important to acknowledge that the transferable demand quantities during scheduling intervals may vary. Additionally, the total demand of the MEC after the DRP equals the total demand prior to its application over a 24-hour horizon. Consequently, the proposed MEC is capable of fulfilling both total electrical and thermal demands. The computation and restriction of shifted demands in each operational horizon are expressed as follows:

$$DR_{t,s}^{I,e/th} = \phi_{t,s}^{e/th} - \left(D_{t,s}^{e/th} \times DRP_{t,s}^{e/th}\right) \tag{4.46}$$

$$\left|DR_{t,s}^{I,e/th}\right| \leq D_{t,s}^{e/th} \times \rho_{t,s}^{e/th} \tag{4.47}$$

$$\rho_{t,s}^{e/th} \leq \rho_{t}^{Max,e/th} \tag{4.48}$$

$$\sum_{t=1}^{T} DR_{t,s}^{I,e/th} = \sum_{t=1}^{T} D_{t,s}^{e/th} \times DRP_{t,s}^{e/th} \tag{4.49}$$

Here, $\rho_t^{\mathrm{Max},e/th}$ denotes the maximum percentage increase in demand.

Incorporating risks associated with uncertainties is imperative in addressing the optimal scheduling problem to shield the MEC operator from undesirable situations. In this study, the conditional value at risk (CVaR) model [53], derived from the VaR methodology, is employed to assess the risk within the stochastic programming. The formulation of CVaR can be expressed as follows:

$$\text{Min} \left\{ \zeta + \frac{1}{1-\alpha} \cdot \sum_{s=1}^{S} \pi_s \times \kappa_s \right\} \tag{4.50}$$

The scheduling problem for the sustainable EH, considering risk constraints, can be reformulated as follows:

$$\text{Min} \left\{ (1-\beta) \cdot Z - \beta \cdot \left(\frac{1}{1-\alpha} \cdot \sum_{s=1}^{S} \pi_s \times \kappa_s \right) \right\} \tag{4.51}$$

Subject to:

$$\zeta - Z \leq \kappa_s \tag{4.52}$$

$$\kappa_s \geq 0 \tag{4.53}$$

Here κ_s is the scenario s' cost and ζ is value-at-risk. α denotes the confidence level and β represents the coefficient of risk.

This chapter evaluates the effects of various values of the risk aversion parameter on the optimal solution of the proposed scheduling for the MEC integrated with wind energy. The CVaR confidence level is established at 90% and 60% for a thorough investigation.

4.7 Numerical Results

The effectiveness of the presented model is demonstrated through the implementation of the proposed method in several case studies. The structure of the proposed MEC is shown in Figure 4.2.

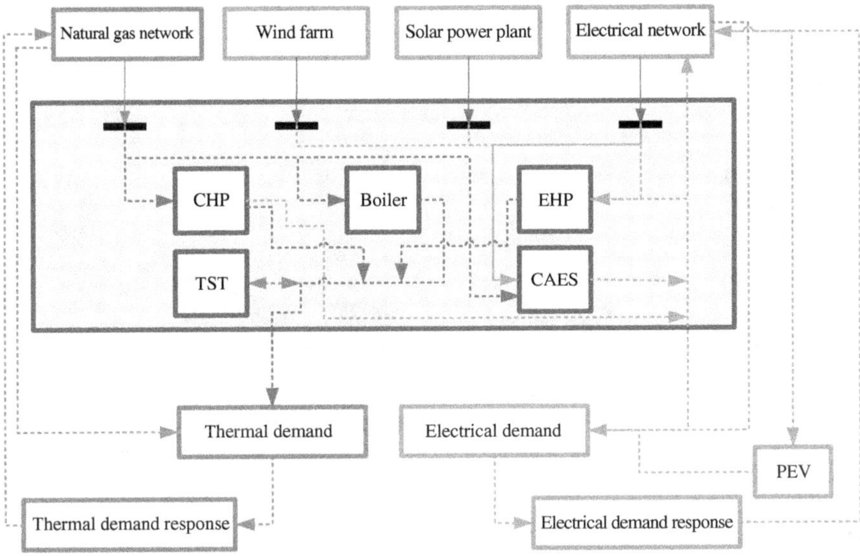

Figure 4.2 Schematic of an MEC.

The proposed MEC concept includes a CHP unit, a boiler, a CAES system, PEVs, a TES, and an EHP system, alongside connections to local electricity and natural gas grids. The inputs of the system include local grid electricity, natural gas, and wind energy. On the other hand, the outputs comprise thermal and electrical loads.

Parameters for the CHP and CAES system are detailed in Tables 4.1 and 4.2, respectively, with CAES system characteristics sourced from Ref. [52]. Additional specifications for the boiler and EHP can be found in Tables 4.3 and 4.4, respectively. Table 4.5 shows specifications for the TES. The risk aversion parameter is set at 0.15. Average VOLL is 3 \$/kWh for electrical demands and 2 \$/kWh for thermal demands. M&O costs for the compressor and expander are referenced from [54]. The price of electricity and natural gas over 24 hours are illustrated in Figure 4.3,

Table 4.1 Parameters of CHP plant.

Parameter	Value
Capacity [kW]	3,000
Electrical ramp-up/ramp-down [kW/h]	675
Thermal ramp-up/ramp-down [kW/h]	720
Startup/shutdown cost [$]	65
Maintenance cost [$/kWh]	0.260
Electrical/thermal conversion efficiency [%]	45/48

Table 4.2 Parameters of CAES system.

Parameter	Value
Maximum/minimum energy level [kWh]	500/50
$V_{\max}^{injected}$ [kW/h]	150
$V_{\min}^{injected}$ [kW/h]	5
$V_{\max}^{pumped}$ [kW/h]	150
$V_{\min}^{pumped}$ [kW/h]	5
Efficiency of injected/produced power [%]	90/90

Table 4.3 Parameters of boiler.

Parameter	Value
Capacity [kW]	1,500
Efficiency [%]	80
Startup/shutdown cost [$]	40

Table 4.4 Parameters of EHP system.

Parameter	Value
Coefficient of performance	1,500
Maximum energy level [kWh]	350
Minimum energy level [kWh]	0

Table 4.5 Parameters of the TES.

Parameter	Value
Maximum charging/discharging range [kWh]	300
Capacity [kW]	1,000
Charging/discharging efficiency [%]	95
Standby efficiency [%]	97

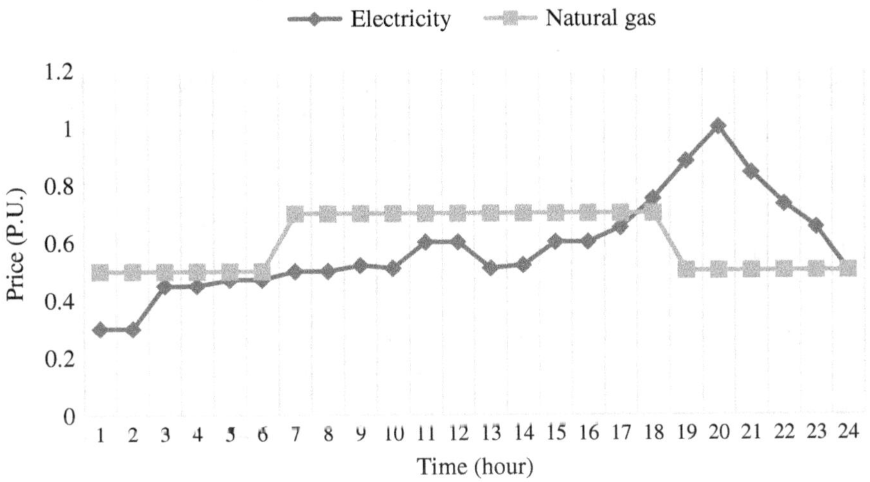

Figure 4.3 The price of electricity and natural gas.

with base values considered as 0.2 \$/kWh for electricity and 0.1 \$/m^3 for natural gas. The price of power sold to local grid $\lambda_{t,s}^{sell}$, is assumed to be $5 \times \lambda_{t,s}^{buy}$ at each scenario and time.

This paper employs a normal distribution to represent the uncertainty in electrical and thermal loads. Variations in electrical and thermal demands are illustrated in Figure 4.4, with base values of 1600 kW for electrical demands and 3000 kW for thermal demands, respectively.

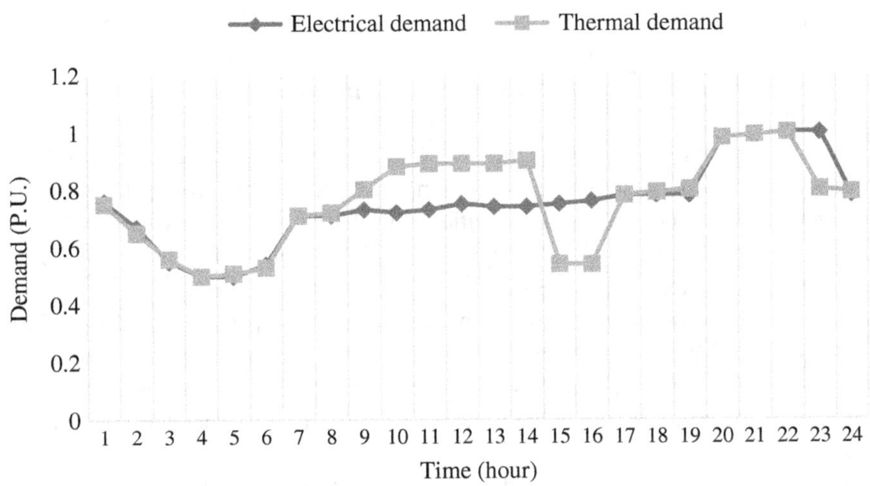

Figure 4.4 The electrical and thermal demands.

Table 4.6 Parameters of the PEV.

Parameter	Value
Maximum charge and discharge of PEVs parking lot [kWh]	400
Capacity [kW]	1,000
Minimum/maximum SoC of PEV [p.u.]	0/0.9
Charging/discharging efficiency [%]	94

Table 4.7 Probability of each scenario.

Scenario	1	2	3	4	5	6	7	8	9	10
Probability	0.041	0.055	0.244	0.151	0.090	0.063	0.088	0.080	0.096	0.092

The wind turbine has a capacity of 550 kW, and its characteristics are derived from [55]. The Monte Carlo method is employed to generate 1000 scenarios, each with equal probabilities. Parameters for the PEV parking lot are outlined in Table 4.6, and the probabilities associated with each reduced scenario are presented in Table 4.7 Under these specified conditions, the proposed stochastic MILP model is solved using the CPLEX solver within GAMS software [56].

Table 4.1 outlines the essential parameters of the CHP plant. With a capacity of 3000 kW, this table provides details on the electrical and thermal ramp-up/down rates, startup/shutdown costs ($65), and maintenance costs ($0.260/kWh). Additionally, the table specifies the electrical and thermal conversion efficiencies, standing at 45% and 48%, respectively. These parameters serve as a foundational reference for understanding the operational characteristics of the CHP plant within the MEC.

Table 4.2 presents key parameters of the CAES system. It includes the maximum and minimum energy levels (500/50 kWh), maximum and minimum levels of injected and pumped air, and the efficiency of injected and produced power, which is set at 90%. This table provides crucial insights into the storage and retrieval capabilities of the CAES system within the MEC framework.

Table 4.3 provides details on the boiler component within the MEC. With a capacity of 1500 kW and an efficiency of 80%, the table also includes the startup/shutdown cost ($40). This information is fundamental for understanding the role of the boiler in the overall energy system and its associated operational costs.

The EHP system's parameters are detailed in Table 4.4 Notably, it includes the coefficient of performance (1500) and the system's ability to store energy, with a maximum energy level of 350 kWh. While the minimum energy level is not specified, these parameters offer a foundation for evaluating the EHP system's efficiency and storage capabilities.

Table 4.5 outlines the TES system parameters. It provides information on the maximum charging/discharging range (300 kWh), system capacity (1000 kW), and efficiencies during charging/discharging (95%) and standby (97%). These details are crucial for understanding the role of TES in managing thermal energy within the MEC.

The parameters of the PEVs parking lot are detailed in Table 4.6. This includes the maximum charge/discharge capacity (400 kWh), overall capacity (1000 kW), and charging/discharging efficiency (94%). Additionally, the SoC ranges from 0 to 0.9, providing insights into the PEV system's operational characteristics.

Table 4.7 shows the probabilities of different scenarios. The provided probabilities represent the likelihood of each scenario occurring within the smart EH system, offering a basis for evaluating the potential outcomes and associated risks.

Table 4.8 presents various case studies incorporating different technologies. It provides a comprehensive overview of expected operation costs for each case study, aiding in the comparison and selection of the most cost-effective configurations. The optimal expected operation costs is obtained in Case #5 having the lowest cost at $11,620.15.

Table 4.9 displays the operation costs of the MEC under different risk conditions (β values). The expected operation cost for the boiler, CHP plant, CAES system, PEV parking lot, cost of buying electricity, revenue from selling electricity, penalty cost, and total operation cost are presented for different levels of the risk.

The analysis of the operational costs within the SEH system reveals significant trends and outcomes as the robustness parameter (β) varies. The operation cost of the boiler shows a declining trend as β increases, with the cost dropping from $2578.57 at $\beta = 0$ to $2235.09 at $\beta = 0.5$, before a slight increase to $2294.50 at $\beta = 1$. This indicates an initial cost efficiency gained with moderate robustness, which slightly diminishes at higher levels of β.

Table 4.8 Features and operation costs of different case studies for $\beta = 1$.

Case study	WT	EHP	CAES	EDR	TDR	PV	PEV	Expected operation cost [$]
#1	√	×	×	×	×	×	×	14, 604.12
#2	√	√	×	×	×	×	×	13, 965.73
#3	√	√	√	×	×	×	×	13, 634.46
#4	√	√	√	√	√	×	×	12, 512.01
#5	√	√	√	√	√	√	√	11, 620.15

Table 4.9 Operation costs of the MEC in different risk conditions.

	Value in different risk parameter (β) levels		
Expected cost	$\beta = 0$	$\beta = 0.5$	$\beta = 1$
Expected operation cost of boiler [$]	2,578.57	2,235.09	2,294.50
Expected operation cost of CHP plant [$]	5,800.76	6,191.55	6,132.98
Expected operation cost of CAES system [$]	2,605.40	2,605.40	2,605.40
Expected operation cost of PEVs parking lot [$]	5.38	43.18	24.98
Expected cost of buying electricity [$]	584	59.98	77.82
Expected revenue of selling electricity [$]	2,268.89	1,190.26	481.23
Expected penalty cost [$]	0	41.42	965.70
Expected total operation cost [$]	9,305.22	9,986.36	11,620.15

In contrast, the operation cost for the CHP plant initially increases as β is adjusted from 0 to 0.5, rising from $5800.76 to $6191.55, then slightly decreasing to $6132.98 at $\beta = 1$. This pattern suggests that increased robustness in operating the CHP plant does not linearly correlate with cost efficiency, highlighting the complexity of its operational dynamics under uncertainty.

The operation cost of the CAES system remains consistently at $2605.40 across all levels of β, indicating stability in its operational costs regardless of the robustness parameter adjustments. This stability is crucial for predictable budgeting and planning in energy systems that incorporate CAES technology.

The operation costs associated with the PEVs parking lot exhibit notable fluctuations, with the lowest cost at $5.38 at $\beta = 0$ and peaking at $43.18 at $\beta = 0.5$. This volatility underscores the sensitivity of PEVs' operational costs to changes in robustness parameters and could influence strategies on integrating PEVs within smart grids.

Electricity trading demonstrates dynamic financial outcomes. The cost of purchasing electricity peaks at $584 at $\beta = 0$ and is lowest at $59.98 at $\beta = 0.5$. Conversely, revenue from selling electricity is highest at $2268.89 at $\beta = 0$, dramatically reducing to $481.23 at $\beta = 1$. This variation highlights the critical impact of robustness on cost and revenue flows from energy trading, suggesting areas for strategic adjustments to enhance financial outcomes.

Finally, the expected penalty cost is observed to be highest at $965.70 at $\beta = 1$, reflecting the increasing financial risk associated with higher robustness settings. The total expected operation cost reflects a general decrease from $11,620.15 at $\beta = 1$ to $9986.36 at $\beta = 0.5$ and further down to $9305.22 at $\beta = 0$. This overall reduction with decreasing β underscores the potential for cost savings under lower robustness conditions, though it must be balanced against the increased risk of operational disruptions.

4.8 Conclusion

MECs offer a promising pathway toward energy justice and fairness amid the challenges of the energy transition. Their holistic approach weaves sustainability and resilience into the fabric of our energy future. By integrating diverse renewable sources like solar, wind, biomass, and hydropower, MECs establish the bedrock of clean energy initiatives. Clever grids and localized storage ensure reliable power supply, while socioeconomic-environmental considerations underscore a commitment to inclusivity, empowerment, and sustainability. The shift from passive consumers to active prosumers within MEC communities fosters a participatory energy landscape, respecting diverse preferences and behaviors. Integration of energy justice principles ensures fair distribution of benefits, inclusive access, and community engagement, positioning MECs as models of social and environmental responsibility. Looking forward, the vision of carbon-free MECs signifies a bold yet necessary goal, achieved through comprehensive efficiency measures, electrification, carbon offsetting, community awareness, and innovation. MECs embody a paradigm shift in energy conception and consumption, championing sustainability and inclusivity. As communities navigate modernization while ensuring equitable energy access, the insights presented guide the journey toward a more sustainable and just energy future. This chapter explores the role of MEC in energy justice in the energy society. In summary, the main findings of the paper are as follows:

- The decreasing cost of MECs, facilitated by energy storage systems like CAES, not only minimizes operation costs but also enhances energy justice by making clean energy more affordable and accessible to a broader spectrum of communities.
- The decreasing cost of MECs, bolstered by initiatives such as EDR and thermal DRPs, trims operational expenses while simultaneously advancing energy justice by widening the reach of affordable and sustainable energy solutions to a more diverse array of communities.
- CVaR provides insights into the potential impact of different operational decisions on energy system reliability and resilience. By managing risk effectively, MECs can enhance their ability to provide uninterrupted energy services, particularly to vulnerable communities, thus contributing to energy justice.

To fully realize the potential of MECs, policymakers must support their growth with comprehensive and inclusive policies. This includes financial incentives for renewable energy integration, regulatory frameworks that encourage local control and decision-making, and measures that ensure equitable access to affordable energy for marginalized communities. Policymakers should also promote transparency and stakeholder participation in energy planning processes, ensuring that all community voices are heard, particularly those historically underrepresented in energy policy. Additionally, risk management strategies like CVaR should be embedded in policy to bolster the resilience and reliability of energy systems, prioritizing vulnerable populations. Ultimately, the insights provided in this study offer valuable guidance for designing policies that enable MECs to champion sustainability, resilience, and energy justice.

References

1 Zhang, X., Yu, X., Ye, X., and Pirouzi, S. (2023). Economic energy management of networked flexi-renewable energy hubs according to uncertainty modeling by the unscented transformation method. *Energy* 278: 128054. https://doi.org/10.1016/j.energy.2023.128054.

2 Chen, M., Sun, Y., Xie, Z. et al. (2023). An efficient and privacy-preserving algorithm for multiple energy hubs scheduling with federated and matching deep reinforcement learning. *Energy* 284: 128641. https://doi.org/10.1016/j.energy.2023.128641.

3 Zhang, T. and Sobhani, B. (2023). Optimal economic programming of an energy hub in the power system while taking into account the uncertainty of renewable resources, risk-taking and electric vehicles using a developed routing method. *Energy* 271: 126938. https://doi.org/10.1016/j.energy.2023.126938.

4 Khouzestani, L.B., Sheikh-el-eslami, M.K., Salemi, A.H., and Moghadam, I.G. (2022). Virtual smart energy hub: a powerful tool for integrated multi energy systems operation. *SSRN Electronic Journal* 265: 126361. http://dx.doi.org/10.2139/ssrn.4080727.

5 Ebrahimi, M. and Sheikhi, A. (2023). A local integrated electricity-heat market design among multi smart energy hubs with renewable energy generation uncertainty. *Electric Power Systems Research* 218: 109217. https://doi.org/10.1016/j.epsr.2023.109217.

6 Hemmes, K., Zachariahwolf, J., Geidl, M., and Andersson, G. (2007). Towards multi-source multi-product energy systems. *International Journal of Hydrogen Energy* 32 (10–11): 1332–1338. https://doi.org/10.1016/j.ijhydene.2006.10.013.

7 Geidl, M., Koeppel, G., Favre-Perrod, P. et al. (2007). Energy hubs for the future. *IEEE Power and Energy Magazine* 5 (1): 24–30. https://doi.org/10.1109/MPAE.2007.264850.

8 Shahmohammadi, A., Moradi-Dalvand, M., Ghasemi, H., and Ghazizadeh, M.S. (2015). Optimal design of multicarrier energy systems considering reliability constraints. *IEEE Transactions on Power Delivery* 30 (2): 878–886. https://doi.org/10.1109/TPWRD.2014.2365491.

9 Najafi, A., Falaghi, H., Contreras, J., and Ramezani, M. (2016). Medium-term energy hub management subject to electricity price and wind uncertainty. *Applied Energy* 168: 418–433. https://doi.org/10.1016/j.apenergy.2016.01.074.

10 Li, T., Eremia, M., and Shahidehpour, M. (2008). Interdependency of natural gas network and power system security. *IEEE Transactions on Power Systems* 23 (4): 1817–1824. https://doi.org/10.1109/TPWRS.2008.2004739.

11 Liu, C., Shahidehpour, M., Fu, Y., and Li, Z. (2009). Security-constrained unit commitment with natural gas transmission constraints. *IEEE Transactions on Power Systems* 24 (3): 1523–1536. https://doi.org/10.1109/TPWRS.2009.2023262.

12 Geidl, M. and Andersson, G. (2006). Operational and structural optimization of multi-carrier energy systems. *European Transactions on Electrical Power* 16 (5): 463–477. https://doi.org/10.1002/etep.112.

13 Kienzle, F., Ahcin, P., and Andersson, G. (2011). Valuing investments in multi-energy conversion, storage, and demand-side management systems under uncertainty. *IEEE Trans Sustain Energy* 2 (2): 194–202. https://doi.org/10.1109/TSTE.2011.2106228.

14 Unsihuay-Vila, C., Marangon-Lima, J.W., de Souza, A.C.Z. et al. (2010). A model to long-term, multiarea, multistage, and integrated expansion planning of electricity and natural gas systems. *IEEE Transactions on Power Systems* 25 (2): 1154–1168. https://doi.org/10.1109/TPWRS.2009.2036797.

15 Moeini-Aghtaie, M., Dehghanian, P., Fotuhi-Firuzabad, M., and Abbaspour, A. (2014). Multiagent genetic algorithm: an online probabilistic view on economic dispatch of energy hubs constrained by wind availability. *IEEE Trans Sustain Energy* 5 (2): 699–708. https://doi.org/10.1109/TSTE.2013.2271517.

16 Bozchalui, M.C., Hashmi, S.A., Hassen, H. et al. (2012). Optimal operation of residential energy hubs in smart grids. *IEEE Transactions on Smart Grid* 3 (4): 1755–1766. https://doi.org/10.1109/TSG.2012.2212032.

17 Alipour, M., Zare, K., and Mohammadi-Ivatloo, B. (2014). Short-term scheduling of combined heat and power generation units in the presence of demand response programs. *Energy* 71: 289–301. https://doi.org/10.1016/j.energy.2014.04.059.

18 Marnay C., Venkataramanan G., Stadler M. et al. (2007). Optimal technology selection and operation of microgrids in commercial buildings. *In 2007 IEEE Power Engineering Society General Meeting, IEEE* (1–7 June 2007). http://dx.doi.org/10.1109/PES.2007.385847.

19 Petruschke, P., Gasparovic, G., Voll, P. et al. (2014). A hybrid approach for the efficient synthesis of renewable energy systems. *Applied Energy* 135: 625–633. https://doi.org/10.1016/j.apenergy.2014.03.051.

20 Moeini-Aghtaie, M., Abbaspour, A., Fotuhi-Firuzabad, M., and Hajipour, E. (2014). A decomposed solution to multiple-energy carriers optimal power flow.

IEEE Transactions on Power Systems 29 (2): 707–716. https://doi.org/10.1109/TPWRS.2013.2283259.

21 Cao, Y., Wei, W., Wang, J. et al. (2020). Capacity planning of energy hub in multi-carrier energy networks: a data-driven robust stochastic programming approach. *IEEE Trans Sustain Energy* 11 (1): 3–14. https://doi.org/10.1109/TSTE.2018.2878230.

22 Chen, Y., Wei, W., Liu, F. et al. (2018). Analyzing and validating the economic efficiency of managing a cluster of energy hubs in multi-carrier energy systems. *Applied Energy* 230: 403–416. https://doi.org/10.1016/j.apenergy.2018.08.112.

23 Dababneh, F. and Li, L. (2019). Integrated electricity and natural gas demand response for manufacturers in the smart grid. *IEEE Transactions on Smart Grid* 10 (4): 4164–4174. https://doi.org/10.1109/TSG.2018.2850841.

24 Wang, X., Bie, Z., Liu, F. et al. (2020). Bi-level planning for integrated electricity and natural gas systems with wind power and natural gas storage. *International Journal of Electrical Power & Energy Systems* 118: 105738. https://doi.org/10.1016/j.ijepes.2019.105738.

25 Mohamed, A., Salehi, V., Ma, T., and Mohammed, O. (2014). Real-time energy management algorithm for plug-in hybrid electric vehicle charging parks involving sustainable energy. *IEEE Trans Sustain Energy* 5 (2): 577–586. https://doi.org/10.1109/TSTE.2013.2278544.

26 Alipour, B., Abdollahi, A., Rashidinejad, M. et al. (2023). Possibilistic–probabilistic risk-based smart energy hub scheduling considering cyber security in advanced metering infrastructures. *Sustainable Energy, Grids and Networks* 36: 101159. https://doi.org/10.1016/j.segan.2023.101159.

27 Dorahaki, S., MollahassaniPour, M., Mobasheri, S., and Rashidinejad, M. (2023). Smart grids and building energy flexibility. *Building Energy Flexibility and Demand Management* 191–215. https://doi.org/10.1016/B978-0-323-99588-7.00010-9.

28 Feng, L., Dai, X., Mo, J., and Shi, L. (2019). Performance assessment of CCHP systems with different cooling supply modes and operation strategies. *Energy Conversion and Management* 192: 188–201. https://doi.org/10.1016/j.enconman.2019.04.048.

29 Mobasheri, S., Dorahaki, S., Rashidinejad, M., and MollahassaniPour, M. (2023). Overview of multi-energy interconnected systems in different energy grids. *IoT Enabled Multi-Energy Systems: From Isolated Energy Grids to Modern Interconnected Networks* 9–30. https://doi.org/10.1016/B978-0-323-95421-1.00002-1.

30 Dorahaki, S., Mobasheri, S., Zadsar, S.S. et al. (2022). Overview of the optimal operation of heat and electricity incorporated networks. In: *Coordinated*

Operation and Planning of Modern Heat and Electricity Incorporated Networks, 79–113. Wiley: https://doi.org/10.1002/9781119862161.ch5.

31 Shao, C., Wang, X., Shahidehpour, M. et al. (2017). An MILP-based optimal power flow in multicarrier energy systems. *IEEE Trans Sustain Energy* 8 (1): 239–248. https://doi.org/10.1109/TSTE.2016.2595486.

32 Mazzeo, D. (2019). Solar and wind assisted heat pump to meet the building air conditioning and electric energy demand in the presence of an electric vehicle charging station and battery storage. *Journal of Cleaner Production* 213: 1228–1250. https://doi.org/10.1016/j.jclepro.2018.12.212.

33 Li, Y., Li, Z., Wen, F., and Shahidehpour, M. (2020). Minimax-regret robust co-optimization for enhancing the resilience of integrated power distribution and natural gas systems. *IEEE Trans Sustain Energy* 11 (1): 61–71. https://doi.org/10.1109/TSTE.2018.2883718.

34 Eladl, A.A., El-Afifi, M.I., El-Saadawi, M.M., and Sedhom, B.E. (2023). A review on energy hubs: models, methods, classification, applications, and future trends. *Alexandria Engineering Journal* 68: 315–342. https://doi.org/10.1016/j.aej.2023.01.021.

35 Spasova, D. and Braungardt, S. (2022). The EU policy framework for energy communities. In: *Energy Communities*, 25–42. Elsevier https://doi.org/10.1016/B978-0-323-91135-1.00022-5.

36 Dorahaki, S., Rashidinejad, M., Fatemi Ardestani, S.F. et al. (2023). An integrated model for citizen energy communities and renewable energy communities based on clean energy package: a two-stage risk-based approach. *Energy* 277: 127727. https://doi.org/10.1016/j.energy.2023.127727.

37 Dorahaki, S., Rashidinejad, M., MollahassaniPour, M. et al. (2023). A sharing economy model for a sustainable community energy storage considering end-user comfort. *Sustainable Cities and Society* 97: 104786. https://doi.org/10.1016/j.scs.2023.104786.

38 Sovacool, B.K. and Dworkin, M.H. (2015). Energy justice: conceptual insights and practical applications. *Applied Energy* 142: 435–444. https://doi.org/10.1016/j.apenergy.2015.01.002.

39 van Bommel, N. and Höffken, J.I. (2021). Energy justice within, between and beyond European community energy initiatives: a review. *Energy Research and Social Science* 79: 102157. https://doi.org/10.1016/j.erss.2021.102157.

40 Lee, J. and Byrne, J. (2019). Expanding the conceptual and analytical basis of energy justice: beyond the three-tenet framework. *Frontiers in Energy Research* 7: https://doi.org/10.3389/fenrg.2019.00099.

41 Hanke, F., Guyet, R., and Feenstra, M. (2021). Do renewable energy communities deliver energy justice? Exploring insights from 71 European cases. *Energy Research and Social Science* 80: 102244. https://doi.org/10.1016/j.erss.2021.102244.

42 Jenkins, K., McCauley, D., Heffron, R. et al. (2016). Energy justice: a conceptual review. *Energy Research and Social Science* 11: 174–182. https://doi.org/10.1016/j.erss.2015.10.004.

43 Carley, S. and Konisky, D.M. (2020). The justice and equity implications of the clean energy transition. *Nature Energy* 5 (8): 569–577. https://doi.org/10.1038/s41560-020-0641-6.

44 Fairchild, D. and Weinrub, A. (ed.) (2017). *Energy Democracy*. Washington, DC: Island Press/Center for Resource Economics https://doi.org/10.5822/978-1-61091-852-7.

45 Opal, A. and Nathwani, J. (2023). Global energy transition risks: evaluating the intergenerational equity of energy transition costs. *Energy Democracies for Sustainable futures* 301–310. https://doi.org/10.1016/B978-0-12-822796-1.00032-2.

46 Castro-Diaz, L. et al. (2023). Evaluating equity and justice in low-carbon energy transitions. *Environmental Research Letters* 18 (12): 123003. https://doi.org/10.1088/1748-9326/AD08F8.

47 Initiative for Energy Justice (2024). Section 1—defining energy justice: connections to environmental justice, climate justice, and the just transition. Initiative for Energy Justice. https://iejusa.org/section-1-defining-energy-justice (accessed 17 April 2024).

48 DellaValle, N. and Sareen, S. (2020). Nudging and boosting for equity? Towards a behavioral economics of energy justice. *Energy Research and Social Science* 68: 101589. https://doi.org/10.1016/j.erss.2020.101589.

49 Szulecki, K. (2018). Conceptualizing energy democracy. *Env Polit* 27 (1): 21–41. https://doi.org/10.1080/09644016.2017.1387294.

50 Droubi, S., Heffron, R.J., and McCauley, D. (2022). A critical review of energy democracy: a failure to deliver justice? *Energy Research and Social Science* 86: 102444. https://doi.org/10.1016/j.erss.2021.102444.

51 Daneshvar, M., Mohammadi-Ivatloo, B., Zare, K., and Anvari-Moghaddam, A. (ed.) (2023). From isolated energy grids to modern interconnected networks. In: *IoT Enabled Multi-Energy Systems*. Elsevier.

52 Daneshi H., Daneshi A., Tabari N. M., and Jahromi A. N. (2009) Security-constrained unit commitment in a system with wind generation and compressed air energy storage. *In 2009 6th International Conference on the European Energy Market*, IEEE (1–6 May 2009). http://dx.doi.org/10.1109/EEM.2009.5207208.

53 Dolatabadi, A. and Mohammadi-Ivatloo, B. (2017). Stochastic risk-constrained scheduling of smart energy hub in the presence of wind power and demand response. *Applied Thermal Engineering* 123: 40–49. https://doi.org/10.1016/j.applthermaleng.2017.05.069.

54 Shafiee, S., Zareipour, H., Knight, A.M. et al. (2016). Risk-constrained bidding and offering strategy for a merchant compressed air energy storage plant.

IEEE Transactions on Power Systems 1–1. https://doi.org/10.1109/TPWRS.2016.2565467.

55 Wen, S., Lan, H., Fu, Q. et al. (2015). Economic allocation for energy storage system considering wind power distribution. *IEEE Transactions on Power Systems* 30 (2): 644–652. https://doi.org/10.1109/TPWRS.2014.2337936.

56 Ren, H. and Gao, W. (2010). A MILP model for integrated plan and evaluation of distributed energy systems. *Applied Energy* 87 (3): 1001–1014. https://doi.org/10.1016/j.apenergy.2009.09.023.

5

Equity-Oriented Energy Allocation in Mixed and Underprivileged Communities

Akhtar Hussain[1] *and Petr Musilek*[2]

[1] *Department of Electrical and Computer Engineering, Laval University, Quebec City, Quebec, Canada*
[2] *Department of Electrical and Computer Engineering, University of Alberta, Edmonton, Alberta, Canada*

Nomenclature

Abbreviations

CMS community management system
H_x Home x
IEA International Energy Agency
PV photovoltaic
RER renewable energy resource

Indices

t index for time interval (1 to T).
m index for renewable energy-based homes (1 to M).
n index for grid-dependent homes (1 to N).
k index for all homes (1 to K).

Parameters and Variables

$P_{t,k}^{\mathrm{Lrat}}$ load ratio for home k at time t.
$P_{t,k}^{\mathrm{load}}$ load of home k at time t.
$P_{t,k}^{\mathrm{Epr}}$ amount of energy allocated to home k by proportional rule.
P_{t}^{Cpv} available energy at time t in the community resource.
$P_{t,k}^{\mathrm{Eee}}$ amount of energy allocated to home k by equal energy rule.
$P_{t,k}^{\mathrm{Eel}}$ amount of energy allocated to home k by equal loss rule.

Communities for Clean Energy Justice and Equity in Grid Modernization, First Edition.
Edited by Mohammadreza Daneshvar, Behnam Mohammadi-Ivatloo, and Amjad Anvari-Moghaddam.

$P^{\text{Epi}}_{t,k}$ amount of energy allocated to home k by priority rule.
J^x_t Jain's fairness index at time t for rule x.
$R^{\text{x}}_{t,k}$ ratio of allocated energy to the energy demand for rule x.
$P^{\text{x}}_{t,k}$ energy allocated to home k under rule x.
$P^{\text{exc}}_{t,m,n}$ amount of power exchanged between homes m and n at t.
$P^{\text{pv}}_{t,m}$ amount of PV power available at home m at t.
$P^{\text{buy}}_{t,m}$ amount of power bought from the grid by home m at t.
$P^{\text{sell}}_{t,m}$ amount of power sold to the grid by home m at time t.
$P^{\text{exe}}_{t,m}$ amount of power sent (exchanged) by home m at t.
$P^{\text{exe}}_{t,n}$ amount of power received (exchanged) by home n at t.
$P^{\text{Lrat}}_{t,n}$ load ratio for grid-dependent home n at time t.
$P^{\text{Erat}}_{t,m}$ excess power ratio for renewable energy-based home m at t.

5.1 Introduction

Climate change is considered one of the major threats to humanity in the upcoming years. Carbon emissions due to increased energy usage are significant factor contributing to climate change [1]. This is primarily due to the use of fossil-fuel-based energy resources to meet the energy demand, leading to carbon emissions. According to a report by the International Energy Agency (IEA) [2], the energy sector contributes more than three-quarters of the total carbon emissions. Renewables are considered a viable solution to reduce emissions from the energy sector. Therefore, several incentives have been provided by governments to facilitate the adoption of renewable energy resources (RERs). Due to these incentives and awareness of environmental benefits, the penetration of RERs is increasing globally. Recently, wind and solar photovoltaic (PV) have particularly gained momentum. It is expected that the penetration of these technologies will further increase in the near future. For example, the [3] estimates that wind and solar capacity will double by 2028 compared to 2022. Similarly, it was reported in [4] that solar energy accounted for 45% of all new electricity-generating capacity added to the US grid in the first half of 2023.

Meanwhile, several studies have reported a direct relation between socio-economic status and the disparity in energy insecurity. For example, it is reported in [5] that low-income families and people of color are most likely to live in energy-inefficient homes. Additionally, these underprivileged households experience three times more energy burden than the national average [6]. It has also been reported in [7] that these communities are exposed to more pollution than they produce and are at higher risk during grid outages. Due to the distributed nature of RERs, they can potentially mitigate these inequalities when combined with weatherization and energy efficiency measures [5]. However, improper incentivization and installation of RERs can potentially further exacerbate these disparities. For example, currently, most RERs are deployed in wealthier communities, with very limited to no installations in communities with lower socioeconomic status [8].

Several studies have recently focused on the need for energy justice and proposed different methods to enhance energy equity in different communities. The first group deals with the definition and needs of energy justice and equity. For example, it is noted in [6] that lower-income households receive far fewer tax credits for green energy compared to higher-income homes. It is emphasized in [9] that both benefits and burdens need to be equally shared among different community members, which is not the case for RES as of now. Recent disasters are analyzed in [10], and disproportionate harm to underprivileged communities is observed, emphasizing the focus on energy equity in policy discussions. The importance of considering equity impacts on underprivileged communities during low-carbon energy transitions is analyzed in [11], and an analytical framework is proposed to evaluate these impacts across key transition areas. It can be concluded that transitioning to cleaner, more equitable energy is vital for addressing climate change, energy access, economic development, equity, and environmental conservation.

In the second group of studies, different measures are suggested to enhance energy equity among different communities and community members. The energy trilemma—balancing economic, political, and environmental aims through energy justice—is proposed in [12], where an energy justice metric is introduced to guide decision-making on energy policy with a focus on equity and fairness. The role of renewable-powered microgrids in advancing energy justice is evaluated in [13] by considering economic, environmental, health, and resilience benefits. The impact of community energy storage on promoting equitable outcomes in the power system is analyzed in [14] by considering principles of energy justice. This study highlights the need to integrate these principles into project design and policy for future research and implementation. A quantitative framework is developed in [15] for policy decision-making on equitable energy interventions, utilizing a linear programming model to optimize energy investments for minimizing the energy burden of energy-insecure households.

Energy justice encompasses a broader range of principles and considerations, including equity, fairness, sustainability, and accountability within the energy system. The energy-just framework is built upon three key elements: distributional justice, procedural justice, and justice as recognition [16]. Distributional justice refers to the fair allocation or distribution of energy resources, opportunities, benefits, and burdens within a community [17]. Procedural justice refers to the fairness and transparency of the processes and procedures used to make decisions or allocate energy resources within a system [16]. Justice as recognition refers to the acknowledgment and validation of the identities, experiences, and rights of individuals and groups within a community. The principles of energy justice, such as equity, fairness, sustainability, and accountability, can be linked to different elements of the framework. For example, fairness, sustainability, and accountability can be linked to distributional justice. Similarly, fairness and equity can be linked to procedural justice. Finally, accountability and sustainability can be linked to justice as recognition.

The focus of this study is on distributional justice, specifically on the equitable allocation of energy resources among different community members. To address this issue, two types of scenarios are discussed in this chapter. First, equitable sharing of a community energy resource (community PV) among members of a community is discussed. Equitable energy allocation becomes necessary during intervals when energy from the community PV is not sufficient to fulfill the energy needs of the whole community. In this case, it is assumed that the community resource is either supported by the government or another organization for an underprivileged community. Four well-known division rules are analyzed, and the results obtained show that the proportional allocation method results in the highest fairness. In the second scenario, a mixed community of members, both with and without solar PV, is considered. The focus of the first part is to share power from members with PV to those without PV in an equitable manner. Equitable sharing becomes challenging in two cases: first, when the total energy demand of the community is greater than the available PV power. In this case, some parts of the energy demand will not be fulfilled by the PV, and available PV power needs to be shared equitably among the members. The second case is when excess PV power is more than the energy demand. In this case, equitable acceptance of PV power from members with excess PV power is needed. This is because selling to community members results in higher profits compared to selling to the grid. Therefore, proportional allocations are proposed in this chapter to allocate resources equitably during these two extreme cases. The main objectives of this chapter can be summarized as follows:

- To analyze the impact of different energy allocation rules on equity and fairness in underprivileged communities.
- To assess the effectiveness of the proportional rule in enhancing fairness of allocating energy from community resources, especially during extreme periods of excess generation and excess load.
- To investigate the equitable allocation of energy in mixed communities with some homes equipped with PV systems.
- To provide insights and recommendations for enhancing fairness and equity in the allocation of energy resources among community members.

5.2 Equitable Energy Sharing

Equitable energy sharing is crucial for ensuring fair access to energy resources, promoting economic development, and enhancing energy security. This is particularly important with RERs in compensating systematically neglected communities, such as underprivileged communities. Therefore, this section focuses on the development of mathematical models to share energy resources equitably among different community members, especially when the demand and supply are not equal. Details about the cases discussed in this study are presented in the following sections.

5.2.1 Shared Community Energy Resources

In this part, the equitability in sharing energy from a community RER is discussed. An overview of the community with a shared PV system is shown in Figure 5.2. It can be observed that all community members can fulfill their energy needs by either buying energy from the grid or using energy from the community RER. A community management system (CMS) is responsible for managing the community. If the energy from the community RER is more than the total load of the community, no additional processing is required. The energy demand of community members is fulfilled first, and the remaining energy is sold to the grid. However, when the load is higher than the energy from the community RER, equity issues arise in sharing the limited amount of energy. This is because the energy cost of the community RER is minimal compared to the price of the grid. In this section, it is assumed that the community consists of underprivileged households, and the RER is donated by the government or other organizations. Only operation and maintenance costs are managed by billing the energy of the RER, which are minimal.

Several rules have been devised in the existing literature to distribute scarce resources among different entities, known as division rules [18, 19]. This chapter analyzes the performance of four well-known rules to distribute a community RER's energy among its members. These rules include the proportional, equal energy, equal loss, and priority rules. In this section, a summary of all four rules is presented. The impact of these rules in terms of fairness is assessed in the subsequent sections.

In the case of the proportional rule, first, a load ratio ($P_{t,k}^{\text{Lrat}}$) is computed for each household (k):

$$P_{t,k}^{\text{Lrat}} = \frac{P_{t,k}^{\text{load}}}{\sum_{k=1}^{K} P_{t,k}^{\text{load}}}, \qquad \forall t \in T, \forall k \in K \tag{5.1}$$

where $P_{t,k}^{\text{load}}$ is the load of the kth home. The amount of energy allocated to home k under the proportional rule ($P_{t,k}^{\text{Epr}}$) is computed using the load ratio and the amount of energy available in the community RER at time t (P_t^{Cpv}):

$$P_{t,k}^{\mathrm{Epr}} = P_t^{\mathrm{Cpv}} \times P_{t,k}^{\mathrm{Lrat}}, \qquad \forall t \in T, \forall k \in K \tag{5.2}$$

It can be observed that the available energy of the community RER is shared proportionally among all the homes.

The basic idea of the equal energy rule is to allocate an equal amount of energy to all homes. However, some homes might have a lower demand compared to the allocated energy. It can be realized by dividing the available energy by the total number of homes, i.e. P_t^{Cpv}/K. However, the energy demand of some homes could be lower than the allocated energy. Therefore, the allocation based on the equal energy rule ($P_{t,k}^{\mathrm{Eee}}$) can be modified as:

$$P_{t,k}^{\mathrm{Eee}} = \min\left\{\frac{P_t^{\mathrm{Cpv}}}{K}, P_{t,k}^{\mathrm{load}}\right\}, \qquad \forall t \in T, \forall k \in K \tag{5.3}$$

It should be noted that, under this rule, all homes are sorted in ascending order of energy demand, and allocations start from the lowest-demand home. The remaining energy is updated after each allocation and divided by the number of remaining homes.

The idea of the equal loss rule is to allocate equal loss to all homes, where loss refers to the amount of demand unfulfilled by the community RER's energy. It can be realized by subtracting the total demand from the available energy and dividing the difference by the total number of homes, i.e. $\left(\sum_{k=1}^{K} P_{t,k}^{\mathrm{load}} - P_t^{\mathrm{Cpv}}\right)/K$. However, the energy demand of some homes could be lower than the allocated loss. To address this issue, the allocations based on the equal loss rule can be modified as:

$$P_{t,k}^{\mathrm{Eel}} = P_{t,k}^{\mathrm{load}} - \max\left\{P_{t,k}^{\mathrm{load}} - P_t^{\mathrm{LR}}, 0\right\}, \qquad \forall t \in T, \forall k \in K \tag{5.4}$$

In this case, the homes are first sorted in ascending order of load demand. Then, energy is allocated starting from the lowest load home. The loss (difference between total load and available energy) is updated after each allocation and divided by the remaining number of homes.

In the case of the priority rule, there is a set priority, and homes are sorted based on that priority. For example, homes can be prioritized based on household income or health and safety needs. Then, energy allocation starts from the highest priority homes, and the remaining energy is updated after each allocation. This process is repeated until all available energy is allocated. It can be modeled as:

$$P_{t,k}^{\mathrm{Epi}} = \min\left\{P_{t,k}^{\mathrm{load}}, \max\left\{P_t^{\mathrm{Cpv}} - \sum_{k'=1,k<k'}^{K} P_{t,k'}^{\mathrm{load}}, 0\right\}\right\}, \qquad \forall t \in T, \forall k \in K \tag{5.5}$$

where $P_{t,k}^{\mathrm{Epi}}$ is the energy allocation based on priority rule and $k < k'$ implies that home k' has higher priority over home k.

5.3 Community Resource Allocation

In this section, the performance of the four division rules discussed in the previous section is analyzed in terms of fairness/equity in allocating energy from the community RERs. A community of three underprivileged homes with PV is considered for this section, as shown in Figure 5.1. The load profiles of the community are real load profiles of households in Vancouver [20], and PV profiles are also generated for the same locality. An overview of the total load of the community and the PV power is shown in Figure 5.2. A scheduling horizon of 24 hours with a time interval of 1 hour is considered. The developed linear programming model is simulated in CPLEX [21] by integrating it with Python.

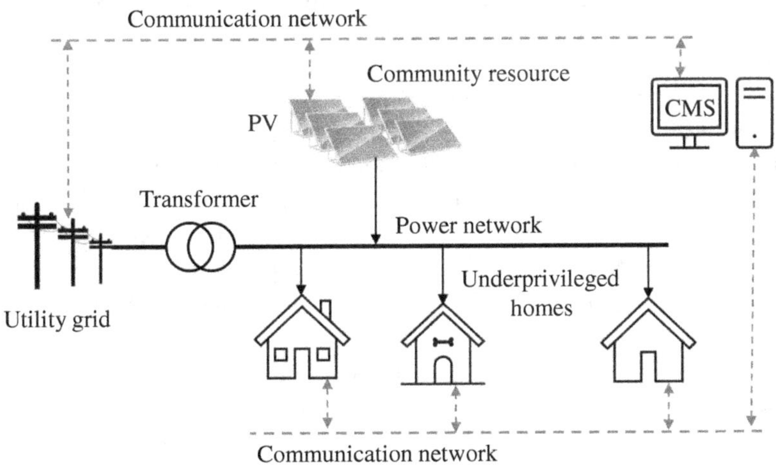

Figure 5.1 Overview of a community with shared energy resource (CMS: community management system).

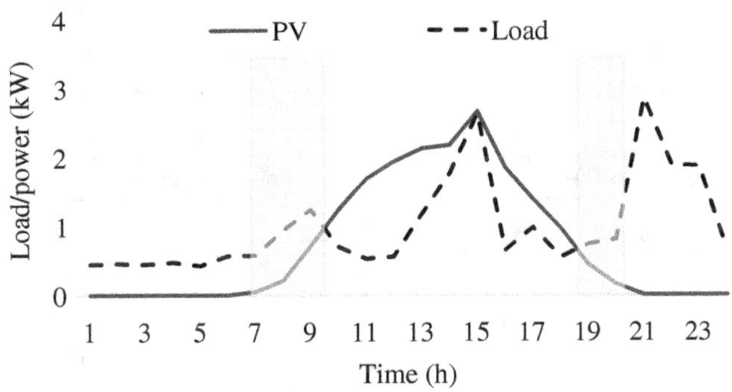

Figure 5.2 Load and PV profiles of the community under consideration.

It can be observed that during the nonzero intervals, the PV power is lower than the load during the early morning (7–9) and evening (19 and 20) hours. Equity issues arise during these intervals as a limited amount of power needs to be shared among various homes. Therefore, these intervals are discussed in the following sub-sections. During other intervals (10–18), the energy demand of all homes can be fulfilled by the RER, and thus no fairness issue arises.

5.3.1 Resource Allocation Methods

In this section, the energy allocated to different homes during shortage intervals (7–9, 19, and 20) under different rules is analyzed. The results of all four rules are summarized in Table 5.1. The priorities of homes are assumed as H2 > H1 > H3, which implies that H2 has the highest priority and H3 has the lowest.

The allocations of the proportional rule are based on the load demand, i.e. proportional to load demand. Therefore, some energy is allocated to all homes during all intervals. Conversely, the allocations of the priority rule are based on the priority. Therefore, first, the full demand of H2 is fulfilled, then that of H1, and the remaining energy is allocated to H3. This sequence is due to the defined priorities, as discussed in the previous paragraph. In the case of the equal energy rule, equal energy is allocated to all homes during most of the intervals, except interval 9. During interval 9, the load of H1 is lower, and the excess power

Table 5.1 Energy allocation to homes under different division rules.

	Proportional (kW)			Priority (kW)		
Interval (h)	**Home 1**	**Home 2**	**Home 3**	**Home 1**	**Home 2**	**Home 3**
7	0.025	0.008	0.017	0	0.05	0
8	0.083	0.047	0.080	0	0.21	0
9	0.086	0.206	0.418	0.15	0.36	0.2
19	0.101	0.214	0.125	0.08	0.36	0
20	0.043	0.036	0.081	0	0.16	0
	Equal loss (kW)			**Equal energy (kW)**		
Interval (h)	**Home 1**	**Home 2**	**Home 3**	**Home 1**	**Home 2**	**Home 3**
7	0.24	0.09	0.2	0.02	0.02	0.02
8	0.26	0.21	0.26	0.07	0.07	0.07
9	0.15	0.19	0.19	0.15	0.28	0.28
19	0.1	0.1	0.1	0.15	0.15	0.15
20	0.22	0.18	0.25	0.05	0.05	0.05

is shared among the other two homes. Conversely, the equal loss rule tries to allocate equal loss to all homes.

5.3.2 Equity in Resource Allocation

In this section, Jain's fairness index is used to evaluate the fairness/equity of all four rules in allocating community resources' energy among underprivileged homes. The fairness index at time t for rule x (J_t^x) can be computed as:

$$J_t^x = \frac{\left(\sum_{k=1}^{K} R_{t,k}^{\mathrm{x}}\right)^2}{K\mathrm{x}\sum_{k=1'}^{K}\left(R_{t,k}^{\mathrm{x}}\right)^2} \tag{5.6}$$

where K is the total number of homes in the community and $R_{t,k}^{\mathrm{x}}$ is the ratio of allocated energy to the energy demand of home k under rule x at time t. It can be computed as:

$$R_{t,k}^{\mathrm{x}} = \frac{P_{t,k}^{\mathrm{x}}}{P_{t,k}^{\mathrm{load}}} \tag{5.7}$$

where $P_{t,k}^{\mathrm{x}}$ is the energy allocated to home k under rule x at time t, and $P_{t,k}^{\mathrm{load}}$ is the load of home k at time t. It should be noted that it is a continuous index ranging between 0 and 1, where higher values refer to higher fairness and vice versa [18, 22].

The fairness results of all four rules are shown in Figure 5.3. It can be observed that the proportional rule results in the highest fairness, a value of 1 (the highest value) during most of the intervals. Conversely, the priority rules result in the lowest fairness during all intervals. This is because some of the homes do not receive any energy under this rule, as shown in Table 5.1. Both the equal loss

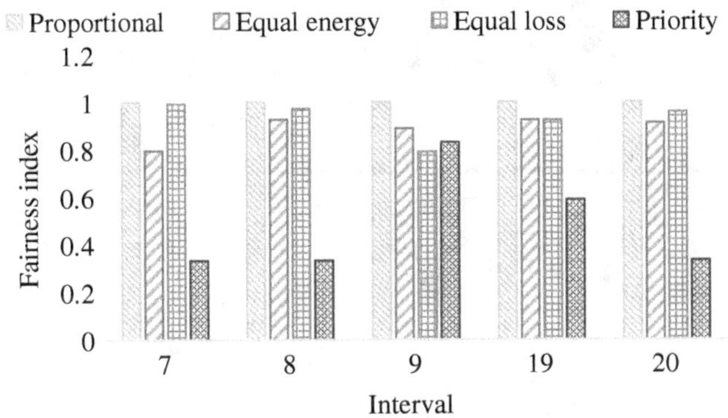

Figure 5.3 Comparative fairness performance of the four energy allocation rules over selected time intervals.

and equal energy rules have intermediate fairness levels, with the equal loss rule resulting in higher fairness during the most intervals.

This analysis shows that the proportional rule can ensure equitable energy sharing among different homes when sharing the energy of community resources. The priority rule, although resulting in the lowest fairness, can be used under specific conditions where it is necessary to prioritize one consumer over another.

5.3.3 Mixed Communities

Mixed communities refer to communities that contain both members with RERs, such as rooftop PV and members without RERs [23]. An overview of a mixed community is presented in Figure 5.4. In this figure, renewable energy-based homes refer to homes with PV, and grid-dependent homes refer to homes without PV. The CMS is responsible for managing the local energy exchange. In this study, it is assumed that the absence of PV in the homes is due to socioeconomic status, and thus the grid-dependent homes are considered as underprivileged homes. It is worth noting that local energy sharing is beneficial for both homes with and without PV. This is because the local exchange price is higher than the grid selling price and lower than the grid buying price. Therefore, both participants benefit by avoiding direct trading with the grid. This also necessitates equitable energy sharing to benefit all community members fairly.

An optimization model is developed to achieve equitable energy sharing among community members. The objective function is designed to maximize the

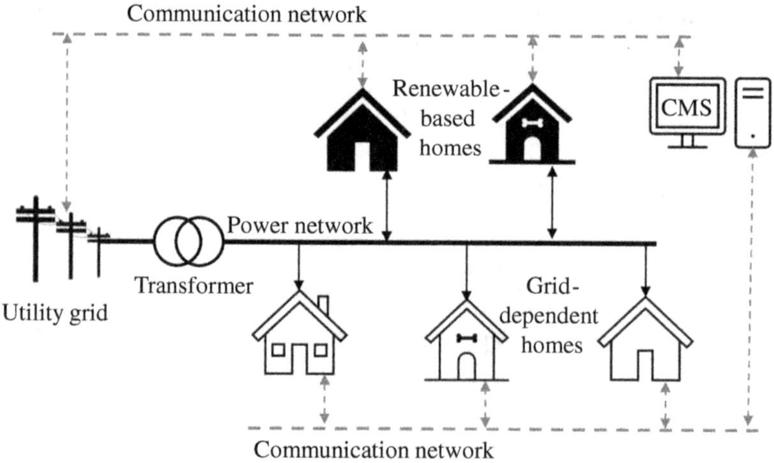

Figure 5.4 Overview of a mixed energy community.

local energy exchange ($P^{\mathrm{exc}}_{t,m,n}$) within the community. It can be mathematically modeled as:

$$\max \sum_{t=1}^{T}\sum_{m=1}^{M}\sum_{n=1}^{N} P^{\mathrm{exc}}_{t,m,n} \tag{5.8}$$

where m is the identifier for renewable energy-based homes, and n is the identifier for grid-dependent homes. Renewable energy-based homes can fulfill their energy demand ($P^{\mathrm{load}}_{t,m}$) by either buying from the grid ($P^{\mathrm{buy}}_{t,m}$) or by using energy from the local RERs ($P^{\mathrm{pv}}_{t,m}$). In addition, they can share the excess energy with other community members ($P^{\mathrm{exc}}_{t,m}$) and/or sell to the grid ($P^{\mathrm{sell}}_{t,m}$), as given by:

$$P^{\mathrm{pv}}_{t,m} + P^{\mathrm{buy}}_{t,m} - P^{\mathrm{sell}}_{t,m} - P^{\mathrm{exc}}_{t,m} = P^{\mathrm{load}}_{t,m}, \qquad \forall t \in T, \forall m \in M \tag{5.9}$$

where $P^{\mathrm{exc}}_{t,m}$ represents the total energy sent by home m at time t, i.e. $P^{\mathrm{exc}}_{t,m} = \sum_{n=1}^{N} P^{\mathrm{sen}}_{t,m,n}$. Renewable energy-based homes are designed to prioritize meeting their own energy needs using the energy generated from their resources (solar PV). Any surplus energy beyond their immediate requirements is considered excess power which can be shared with other homes (grid-dependent homes). This process enhances green energy equity among community members and maximizes profit by sharing energy locally. It can be modeled as:

$$P^{\mathrm{exc}}_{t,m} \le P^{\mathrm{pv}}_{t,m} - P^{\mathrm{load}}_{t,m}, \qquad \forall t \in T, \forall m \in M \tag{5.10}$$

Similarly, grid-dependent homes can fulfill their energy demand ($P^{\mathrm{load}}_{t,n}$) by either buying from the grid ($P^{\mathrm{buy}}_{t,n}$) and/or by using the energy received from renewable energy-based homes ($P^{\mathrm{exc}}_{t,n}$). It can be modeled as:

$$P^{\mathrm{buy}}_{t,n} + P^{\mathrm{exc}}_{t,n} = P^{\mathrm{load}}_{t,n}, \qquad \forall t \in T, \forall n \in N \tag{5.11}$$

where $P^{\mathrm{exc}}_{t,n}$ represents the total energy received by home n at time t, i.e. $P^{\mathrm{exc}}_{t,n} = \sum_{m=1}^{M} P^{\mathrm{rec}}_{t,n,m}$. Grid-dependent homes cannot receive power beyond their required energy level (load), i.e. $P^{\mathrm{exc}}_{t,n} \le P^{\mathrm{load}}_{t,n}$. Finally, the net energy exchange in the community should be zero, as given by:

$$\sum_{m=1}^{M} P^{\mathrm{exc}}_{t,m} - \sum_{n=1}^{N} P^{\mathrm{exc}}_{t,n} = 0 \qquad \forall t \in T \tag{5.12}$$

These equations are sufficient to realize energy exchange among community members. However, equity issues arise when the total surplus of renewable energy-based homes is not equal to the energy demand of grid-dependent homes [23]. There are two possible cases which raise equity issues. The first case is when the total surplus is less than the total demand. In this case, limited surplus power needs to be shared among various homes, and the remaining demand needs to be fulfilled by buying from the grid, which is more expensive. The second case is when there is more surplus than the energy demand. In this case, a limited amount of surplus power will be sold internally, and the remaining will be sold to the grid. The grid selling price is lower than the internal selling price, therefore, it raises fairness/equity issues.

Therefore, additional constraints are introduced to allocate energy equitably among different community members. To allocate energy among grid-dependent homes during intervals where excess PV is lower than the total load, a load ratio is computed as:

$$P_{t,n}^{\mathrm{Lrat}} = \frac{P_{t,n}^{\mathrm{load}}}{\sum_{n=1}^{N} P_{t,n}^{\mathrm{load}}}, \qquad \forall t \in T, \forall n \in N \tag{5.13}$$

The ratio is then used to limit the allocation of energy to each grid-dependent home. It can be modeled as:

$$P_{t,n}^{\mathrm{exc}} \leq \sum_{m=1}^{M} \max\{P_{t,m}^{\mathrm{pv}} - P_{t,m}^{\mathrm{load}}, 0\} \times P_{t,n}^{\mathrm{Lrat}}, \qquad \forall t \in T, \forall n \in N \tag{5.14}$$

where the first part on the right side is the total excess of energy in renewable energy-based homes during interval t. The second part is the load ratio determined in the previous equation. This constraint ensures that during lower excess intervals, energy is allocated proportionally and fairly to all grid-dependent homes. Proportional allocations result in the highest fairness, as discussed in the previous section.

Similarly, in the second case, where the total excess is more than the total demand, an excess power ratio is determined first:

$$P_{t,m}^{\mathrm{Erat}} = \frac{\max\{P_{t,m}^{\mathrm{pv}} - P_{t,m}^{\mathrm{load}}, 0\}}{\sum_{m=1}^{M} \max\{P_{t,m}^{\mathrm{pv}} - P_{t,m}^{\mathrm{load}}, 0\}}, \qquad \forall t \in T, \forall m \in M \tag{5.15}$$

It should be noted that in this scenario, the total surplus available in the community from renewable energy-based homes exceeds the total load. Therefore, only a fraction of the excess power can be used to fulfill the energy demand of grid-dependent homes. The excess power ratio is crucial in this context as it determines how to fairly accept or allocate a certain amount of power from all renewable energy-based homes. This ratio ensures that the allocation of excess energy is balanced among homes, optimizing the utilization of surplus energy and maintaining fairness among community members. It can be modeled as:

$$P_{t,m}^{\mathrm{exc}} \geq \sum_{n=1}^{N} P_{t,n}^{\mathrm{load}} \times P_{t,m}^{\mathrm{Erat}}, \qquad \forall t \in T, \forall m \in M \tag{5.16}$$

where the first part on the right side is the total load of the grid-dependent homes, and the second part is the excess energy ratio. This constraint ensures that energy is proportionally (fairly) accepted from all renewable homes during energy excess intervals.

5.4 Power Sharing: Mixed Communities

To analyze the performance of the developed models, a community of five homes is considered in this study, as shown in Figure 5.4. Three homes (H1–H3) are assumed to be grid-dependent homes, and the remaining two (H4, H5) are renewable energy-based homes. The load profiles of the community are real load profiles of households in Vancouver [20], and PV profiles are also generated for the same locality. An overview of the daily total load and PV power is shown in Table 5.2. It can be observed that H4 has the highest load, and H1 has the lowest load. The highest amount of PV power is in H5.

5.4.1 Benefits of Power Sharing

To analyze the benefits of sharing RER power locally, this section considers two cases. In the first case (original case), no sharing is allowed, and all community members directly trade with the grid. In the second case (with sharing), the proposed sharing mechanism is included.

An overview of the results is shown in Table 5.3. It can be observed that there is no local trading in the first case (original). Meanwhile, renewable

Table 5.2 Daily load and PV power amount in each home.

Parameter	Home 1	Home 2	Home 3	Home 4	Home 5
Load (kWh)	5.8	8.92	11.69	26.74	14.01
PV (kWh)	0	0	0	17.05	28.42

Table 5.3 Results for benefits of sharing in the original and with sharing cases.

Case	Parameter	Home 1	Home 2	Home 3	Home 4	Home 5
Original	Buy (kWh)	5.8	8.92	11.69	15.87	6.72
	Sell (kWh)	0	0	0	6.16	21.11
	Local (kWh)	0	0	0	0	0
	Cost ($)	5.699	9.146	11.113	13.173	−3.691
With sharing	Buy (kWh)	3.44	3.46	9.44	15.87	6.72
	Sell (kWh)	0	0	0	4.38	12.84
	Local (kWh)	2.36	5.46	2.25	−1.79	−8.28
	Cost ($)	4.139	5.957	9.699	13.036	−3.999

energy-based homes (Homes 3 and 4) are selling their excess power to the grid, and grid-dependent homes (Homes 1–3) are buying power from the grid. In the second case (with sharing), local exchange can be observed for all the homes. The negative values signify the sending of power, and positive values refer to receiving power. It is worth noting that renewable energy-based homes are still selling some power to the grid after fulfilling their local load demand and sending power to grid-dependent homes. This is because the available PV power during peak generation intervals is more than the total load of the community.

In terms of cost, cost reductions can be observed for all community members, where the negative cost value represents the profit. The cost reduction due to local trading is shown in Figure 5.5, where significant cost reductions can be observed for the grid-dependent homes. This is because the local trading price is much lower than the grid buying price, and the local selling price is set to be slightly higher than the price of selling to the grid. Due to this marginal difference, the cost reduction of renewable energy-based homes is lower. However, it is still profitable to share power with the community members as compared to selling to the grid.

5.4.2 Equity Analysis

In this section, two extreme cases are considered to analyze the performance of the proposed scheme in terms of equity. The first case is when the total load of grid-dependent homes is higher than the excess power of renewable energy-based homes. This generally happens at the beginning of solar generation (around sunrise) and the end of the solar generation (around sunset). Therefore, one interval is selected from each side (sunrise and sunset), and these intervals are 9 and 19 for this community.

The results of the selected intervals are shown in Table 5.4. It can be observed that the total load is higher than the surplus power in both of these intervals.

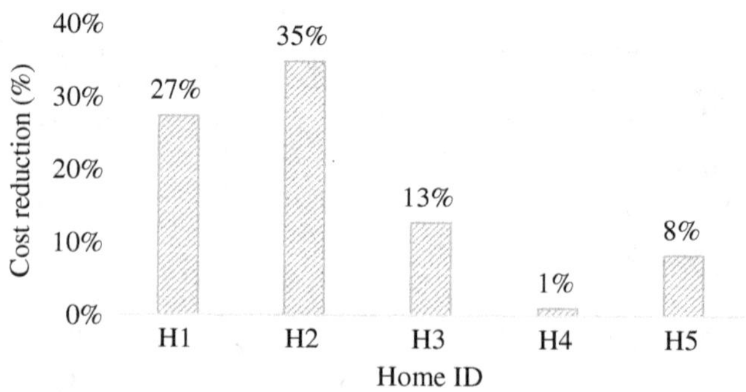

Figure 5.5 Cost reduction for each community home through local energy trading.

Table 5.4 Equity analysis during load excess intervals.

Interval (h)	Home	Buy (kWh)	Local (kWh)	Load (kWh)	Surplus (kWh)
9	Home 1	0.11	0.0378	0.15	0.312
	Home 2	0.2694	0.0906	0.36	
	Home 3	0.5462	0.1838	0.73	
	Total	0.928	0.312	1.24	
19	Home 1	0.1245	0.0455	0.17	0.198
	Home 2	0.2637	0.0963	0.36	
	Home 3	0.1538	0.0563	0.21	
	Total	0.542	0.198	0.74	

The surplus PV power is equitably (proportionally) shared among all three grid-dependent homes. For example, the allocated power (local) to the load ratio of all homes for interval 9 is 0.252, and it is 0.268 for interval 9. In addition, it can also be observed that all excess power is allocated during these intervals, i.e. the total allocated power (local) is the same as the surplus power.

Similarly, to analyze the performance of the proposed method under excess power intervals (second case), two intervals are analyzed. This generally happens during peak PV generation intervals (around noon), and intervals 12 and 13 are selected in this section.

The results of both intervals are shown in Table 5.5. It can be observed that the total PV excess power (after serving the local load) is more than the total load. The excess power is fairly (proportionally) accepted from both renewable energy-based homes. For example, the ratio of accepted power to excess power for both homes is 0.4104 for interval 12 and 0.4790 for interval 13. In addition, it can also be observed

Table 5.5 Equity analysis during PV power excess intervals.

Interval (h)	Home	Sell (kWh)	Local (kWh)	Load (kWh)	PV (kWh)	Load (kWh)
12	Home 4	0.7804	0.5431	0.62	1.9435	1.70
	Home 5	1.6622	1.1569	0.42	3.2391	
	Total	2.4426	1.70	1.04	5.1826	
13	Home 4	0.7146	0.6571	0.76	2.1317	2.22
	Home 5	1.6999	1.5629	0.29	3.5529	
	Total	2.4146	2.22	1.05	5.6846	

that the total received power from both homes (local) is equal to the load demand of the grid-dependent homes (total load) during both intervals.

This analysis shows that the proposed method allocates energy to homes fairly during lower power intervals. In addition, it also fairly accepts power from homes during excess power intervals.

5.5 Advantages and Limitations

5.5.1 Major Factors for Fairness Analysis

In this study, fairness in energy allocation is examined through the following key factors.

- The first factor is equitable green energy sharing. It ensures that all community members, including underprivileged homes, have access to green energy resources, crucial in mixed communities. Surplus energy from renewables is shared proportionally among all members, especially during excess generation and load intervals.
- The second factor is cost reduction. Analyzing different allocation rules aims to identify costs reducing methods, especially for grid-dependent homes. Local trading and sharing of renewable energy can significantly reduce costs, highlighting the need for fair and efficient allocation.
- The last factor is community benefit. Fair energy allocation reduces reliance on external sources, promoting sustainability. Considering broader impacts on community well-being and sustainability underscores the importance of fairness in energy distribution.

5.5.2 Features of Different Cases

In this chapter, two cases are analyzed. The first case examines equity and fairness in community resource allocation. The second case analyzes equity and fairness in energy sharing in mixed communities. The features and limitations of both cases are discussed in this section.

The analysis of the first case (community resource allocation) shows that the proportional rule enhances fairness in community resource allocation compared to other division rules. This ensures that all community members receive a fair share of the RER, promoting energy justice. In addition, by prioritizing proportional allocation, equity among community members is promoted, particularly for underprivileged homes. While the proportional rule improves fairness, it may not address all aspects of energy justice, such as addressing historical inequalities.

The analysis of the second case (energy sharing in mixed communities) indicates that internal energy transfer allows for a fair distribution of surplus energy among homes, particularly during intervals of excess generation and load. In addition, it can lead to cost reductions for community members, especially those reliant on the grid. However, the effectiveness of internal energy transfer depends on the availability of renewable energy generation and right mix of renewable energy-based homes and grid-dependent homes.

5.6 Conclusions

This chapter introduces different methods for allocating energy among different homes in underprivileged and mixed communities, aiming to achieve equitable distribution. In the first part, the impact of various rules on energy equity and fairness in allocating energy from a community PV resource is analyzed. Four division rules are assessed, with the proportional rule emerging as the most effective in enhancing fairness in community resource allocation. This aligns with the objective of promoting fairness and equity in energy distribution.

In the second part, a community of mixed homes (with and without PV) is considered, and the proportional rule is applied to allocate energy during two extreme cases: excess generation and excess load intervals. The results indicate that the proportional rule can equitably allocate resources during these scenarios, further supporting the objective of ensuring fair energy distribution among community members. Overall, the study's objectives of analyzing and promoting fairness and equity in energy allocation among community members are achieved through the evaluation and application of these rules.

References

1 Owusu, P.A. and Asumadu-Sarkodie, S. (2016). A review of renewable energy sources, sustainability issues and climate change mitigation. *Cogent Engineering* 3 (1): 1167990.

2 IEA (2023). Greenhouse Gas Emissions from Energy Data Explorer. *Tech. Rep.* International Energy Agency (IEA). https://www.iea.org/data-and-statistics/data-tools/greenhouse-gas-emissions-from-energy-data-explorer (accessed 3 January 2025).

3 IEA (2024). Renewables 2023. *Tech. Rep.* International Energy Agency (IEA). https://www.iea.org/reports/renewables-2023 (accessed 3 January 2025).

4 Michelle, D., Martinez, S.L., Gaston, Z. et al. (2023). Solar Market Insight Report 2023 Q3. *Tech. Rep.* SEIA/Wood Mackenzie Power & Renewables.

5 Heleno, M., Sigrin, B., Popovich, N. et al. (2022). Optimizing equity in energy policy interventions: a quantitative decision-support framework for energy justice. *Applied Energy* 325: 119771.

6 Ross, E., Day, M., Ivanova, C. et al. (2022). Intersections of disadvantaged communities and renewable energy potential: data set and analysis to inform equitable investment prioritization in the united states. *Renewable Energy Focus* 41: 1–14.

7 Kime, S., Jacome, V., Pellow, D., and Deshmukh, R. (2023). Evaluating equity and justice in low-carbon energy transitions. *Environmental Research Letters* 18 (12): 123003.

8 Sunter, D.A., Castellanos, S., and Kammen, D.M. (2019). Disparities in rooftop photovoltaics deployment in the united states by race and ethnicity. *Nature Sustainability* 2 (1): 71–76.

9 Jenkins, K., McCauley, D., Heffron, R. et al. (2016). Energy justice: a conceptual review. *Energy Research & Social Science* 11: 174–182.

10 Ballesteros-Arjona, V., Oliveras, L., Munoz, J.B. et al. (2022). What are the effects of energy poverty and interventions to ameliorate it on people's health and well-being?: A scoping review with an equity lens. *Energy Research & Social Science* 87: 102456.

11 McCauley, D., Ramasar, V., Heffron, R.J. et al. (2019). Energy justice in the transition to low carbon energy systems: exploring key themes in interdisciplinary research. *Applied Energy* 233: 916–921.

12 Heffron, R.J., McCauley, D., and Sovacool, B.K. (2015). Resolving society's energy trilemma through the energy justice metric. *Energy Policy* 87: 168–176.

13 Anderson, K., Farthing, A., Elgqvist, E., and Warren, A. (2022). Looking beyond bill savings to equity in renewable energy microgrid deployment. *Renewable Energy Focus* 41: 15–32.

14 McNamara, W., Passell, H., Montes, M. et al. (2022). Seeking energy equity through energy storage. *The Electricity Journal* 35 (1): 107063.

15 Hoffman, J., Davies, M., Bauwens, T. et al. (2021). Working to align energy transitions and social equity: an integrative framework linking institutional work, imaginaries and energy justice. *Energy Research & Social Science* 82: 102317.

16 Pellegrini-Masini, G., Pirni, A., and Maran, S. (2020). Energy justice revisited: a critical review on the philosophical and political origins of equality. *Energy Research & Social Science* 59: 101310.

17 Heffron, R.J. and McCauley, D. (2017). The concept of energy justice across the disciplines. *Energy Policy* 105: 658–667.

18 Hussain, A. and Musilek, P. (2022). Fairness and utilitarianism in allocating energy to EVs during power contingencies using modified division rules. *IEEE Transactions on Sustainable Energy* 13 (3): 1444–1456.

19 Moulin, H. (2004). *Fair Division and Collective Welfare*. MIT Press.

20 Makonin, S. (2018). HUE: the hourly usage of energy dataset for buildings in British Columbia. https://doi.org/10.7910/DVN/N3HGRN.

21 IBM (2017). User's Manual (IBM ILOG CPLEX Division, Incline Village, NV).

22 Hoßfeld, T., Skorin-Kapov, L., Heegaard, P.E., and Varela, M. (2018). A new QoE fairness index for QoE management. *Quality and User Experience* 3: 1–23.

23 Hussain, A. and Musilek, P. (2024). Equity-oriented power sharing and demand response for enhancing green energy access in mixed communities. *Energy and Buildings* 308: 114029.

6

Energy Justice and Equity in Developing a Fair Multi-energy Community and Environment

Aamer Raza[1], Sheraz Azam Mirza[2], Saqib Ali[3,4], and Muzaffar Ali[1]

[1] *Senior Engineer, Islamabad Electric Supply Company, Islamabad, Pakistan*
[2] *Senior Engineer, National Grid, Birmingham, UK*
[3] *Assistant Professor, National Fertilizer Company, Multan, Pakistan*
[4] *Postdoc Fellow, Zagreb University, Croatia*

Acronyms

AI	artificial intelligence
CTBCM	Competitive Trading and Bilateral Contract Market
DG	distributed generation
GENCOs	Generation Companies
GDP	gross domestic product
GEPCO	Gujranwala Electric Power Company
GHG	greenhouse gas emission
IESCO	Islamabad Electric Supply Company
IPPs	independent power producers
KE	Karachi Electric
LESCO	Lahore Electric Supply Company
NEPRA	National Electric Power Regulatory Authority
NG	natural gas
NTDC	National Transmission and Dispatch Company
PEPCO	Pakistan Electric Power Company
WAPDA	Water & Power Development Authority

6.1 Introduction

Energy systems play a central role in the socioeconomic development of human society. Residential comfort, commercial activities, industrial expansion, and

Communities for Clean Energy Justice and Equity in Grid Modernization, First Edition.
Edited by Mohammadreza Daneshvar, Behnam Mohammadi-Ivatloo, and Amjad Anvari-Moghaddam.

agricultural growth depend on reliable and sustainable energy systems. As a point of fact, customers do not entirely depend on a single form of energy; instead, they use multiple forms such as electricity from the national electric grid, natural gas (NG) from NG distribution system, heat from district heating system, and customer-owned renewable energy-based distributed generators (DGs) (especially solar and wind). This multiplicity of energy streams enhances reliability (probability of availability of the energy supply) and sustainability (degree of fiscal affordability and environmental cleanliness). In an energy society, there are different stakeholders, such as customers, utilities, and even the environment itself, together shaping an energy market. Actually, energy market is a commodity market that trades fuel, electricity, and heat as commodities [1]. Fuel may include oil, NG, coal, solar irradiance, wind, and nuclear energy. Now, in a marketplace, every participant has its own stakes, for instance, utilities aim to produce and distribute inexpensive energy, spend as low a capital and an operational cost on the energy infrastructure as possible, and curtail peak load to defer the network's capacity enhancement expenditures, reduce losses, improve network reliability and maximize profit earning, etc. Likewise, customers desire a low-priced, quality, safe and reliable energy supply. The third stakeholder has been the environment that expects minimum greenhouse gas (GHG) emission, to avoid extreme weather swings threatening millions of lives across the globe and health deterioration of living beings. Environment has been a stakeholder in a sense that it may positively or adversely affect the human society. In other words, the actual meaning of the environment being a stakeholder is the community and society itself as a beneficiary. Now generally, these stakes are in a state of unceasing conflict with each other, e.g. utilities intend to raise energy prices to increase their profit on one side, however, on the other hand, it would be injustice to the customers purchasing expensive energy, consequently, forcing them to limit their energy consumption quantum. In such cases, however, the loss would be bilateral, as energy consumption reduction would decrease the per capita income and human comfort for the public on one side with an unforeseen drop in the business volume of a utility on the other side. Therefore, energy justice and equity may be practiced within and between the energy communities to promote a healthy business atmosphere.

"Justice is a long-term equity; it looks to create equity within different systems as well as among the individuals in a system. Actually, justice can take equity one step further by fixing the systems in a way that leads to a long-term, sustainable and equitable access to the facilities and the amenities for the generations to come" [2]. In other words, justice and equity have been mandatory aspects to retain order in a system or in a society. The term "energy justice" may be defined as a global challenge to provide energy to the customers in an affordable, sustainable, and safe manner [3]. Since the early appearance of the term [4, 5] "energy justice," it was mainly used on distributional injustice, such as "energy poverty" under which developed parts of the globe have highly improved energy delivery services compared to the less developed or underprivileged territories.

However, now, after various research efforts [6–14], the definition of energy justice has broadened. It includes the injustice arising from: (1) the energy processes, such as intentionally or unintentionally opting for expensive or pollution-prone options of energy, (2) eliminable energy production inefficiencies (e.g. energy losses, load shedding and safety and quality issues), (3) unfair distribution (concentrating major quantum of energy in one part of the globe or country and denying easy access to the rest of the population), (4) unjust treatment of various stakeholders and groups (e.g. the power plant owners earning deceitfully/deviously/fraudulently the high amount of capacity charges from the customer), (5) energy security (uninterruptable energy supply at afford able prices), (6) energy poverty (759 million persons around the globe lack access to dependable electric supply and about 2.6 billion users use perilous and wasteful cuisine systems), and (7) undesired impact originating from inappropriate siting of the energy infrastructure. This literature shows that the concept of the energy justice is still evolving and requires more scholarly endeavors.

Another definition [15] elaborates the concept of energy justice as the "international/national energy system that justly distributes both the costs and benefits for supplied amenities to the customers" Sareen and Haarstad [16] considered energy justice as a way the customers get affected by the cost, energy-related deprivation and risks within a community and among different communities. Furthermore, references [17–20] discussed that the introduction of renewable energy-based microgrids may raise geopolitical strains over a decline in fossil fuel trade and land grabbing; therefore, these issues also require a just solution.

References [21, 22] discussed the research work conducted on energy equity and justice metrics and emphasize the energy system planners to conduct a quality analysis and find out an optimal trade-off among various stakes like energy consumption cost, emission, availability, and profitability. These variables represent the interests of stakeholders to promote just energy business. Hernández et al. [23] and Prabhakar et al. [24] tried to elucidate the history, dissimilarities, and similarities between "fuel poverty (poor segments of society having least likeliness to the required energy access), energy burden (percentage of annual earning spent on to tolerate the energy cost), energy poverty, energy vulnerability (the extent to which an energy system or entity is affected by adverse events such as economic downturn, environmental pollution, and poor governance) and energy insecurity (uninterrupted supply of energy at affordable price)" and related these terms with the principles of energy justice. Moreover, the authors emphasized the need to broaden the justice vision and apply it for wellbeing of the shareholders.

References [25–27] discussed about the application of artificial intelligence (AI) in "distributive, procedural, and recognition" justices. Authors determined that the AI techniques such as genetic algorithm, flower pollination algorithm, water cycle algorithm, and machine learning may be used to optimally trade-off the conflicting interests of various participants to bring justice among them.

Next sections of this chapter present the types and principles of energy justice, practical instances of energy injustice in different countries and their solutions, factors affecting the fairness and equity of a multi-energy community, use of AI to solve energy justice problems and conclusion and future research avenues in this area.

6.2 Axis of Energy Justice and Equity in a Multi-energy Environment

This section discusses the types of energy justice and equity and their application in the global energy market.

6.2.1 Distributional Justice

Sovacool, et al. [28] related "availability" of a commodity with the distributional justice. Authors defined availability as "the capacity of an economic system, commodity market or system/network to assure adequately available assets when desired." Distributional justice is defined as "justice in the unbiased supply of a commodity among the participants of a civilization" [27, 29]. Yenneti et al. [30] defined distributional justice as the "fairness in the distribution of goods and subsequent advantage." In fact, distributional justice identifies the inadequate distribution of benefits and their evils, and the irregular dispersal of their related duties.

McCauley et al. [3] described an instance of distributional injustice in which underprivileged public hosts objectionable energy plants compared to the well-off communities. For example, areas where affluent public resides do not have coal and biogas-run plants, whereas, localities hosting masses usually have sources that produce serious amount of air, water and soil pollution. In literature, such irregularities are also termed as "spatial inequalities."

Under the distributional justice regime, authors in [31, 32] interpreted another type of justice termed as the "access to energy services." The severity of this issue may be understood by the fact that about 746 million people around the globe do not have access to electric services. Such an energy dearth in a community is also called energy poverty. In Pakistan, where energy security remains a chronic issue, about 60 million people (25% of the total population) have no access to the national electric grid [33]. Likewise, in India 19,5000,000 households (13% of total population), and in a highly developed country like the United States, 60,000 people still do not have access to the national grid [34].

Currently, Pakistan's electric energy system has a PKR. 2.635 trillion Circular Debt [35] due to unfair capacity charges being paid to the independent power producers (IPPs), technical/system losses, energy theft, and poor revenue collection from customers. One of the major reasons of high system losses, energy theft, and inadequate revenue collection has been that the electric network has been inappropriately extended to the areas which are far-off from the national grid. Usually, longer distances and insufficient transmission and distribution capacities create voltage suppression resulting in higher than permissible energy losses. In addition to this, such far-fledged regions lack education and security. Therefore, billing and revenue collection remains a serious issue for utility's staff. By digging deep into the country's social and political history, it may be concluded that such blind network expansion was not based on any scientific and economic acumen. Mere driver of this irrational expansion was that the country's rulers politically aimed to gratify public by extending national grid to these areas with no business viability but just to take votes and enhance political influence. However, if based on these facts, electricity is not extended to these areas, the public would face distributional or spatial injustice. In other words, it may be established that sometimes, utility's lawful interests may fuel distributional injustice.

Now a question arises that how can such a scenario be avoided? In this regard, there are multiple solutions to this problem: (1) improve the law and order situation in such areas and then extend the national grid and (2) invite the local influential investors of these regions to invest in the development of an islanded microgrid. This microgrid possessing its own generation and distribution resources may provide electricity to the local public. Since the natives of the area would invest capital and operational cost, therefore, it will discourage energy theft, hurdles in the revenue collection and increase job opportunities for the local citizens.

We may witness this prevailing scenario from another angle that the financial damages, occurring due to losses from theft, voltage suppression in distant parts of the grid, and low revenue being collected in areas with inefficiencies of recovery staff and poor law order situation, are in turn recovered from the regularly paying customers. Under such a situation, few energy thieves and nonpaying customers would become a reason of loss to the paying one. Therefore, an innovative thinking such as the application of the concept of the microgrid is required to put an end to the blind expansion of the energy networks.

Another way to avoid distribution injustice described above would be that all the citizens should be involved in the process of framing the country's energy policy. This step would lead us to the concept of energy democracy, where every individual is allowed to involve in the policy-making process. At this point, a question arises that how is it possible to involve every citizen? The answer to this is quite simple that the citizens of a country have a right to elect the true leadership who makes policies in the light of the problems of public. These elected representatives would participate in making policies that favor the country and its public rather than a certain faction or a mafia. This argument clarifies that the boundaries of energy justice overlap with the social justice and fair democratic system in a society. However, unfortunately, countries like Pakistan never had a true democracy based on free and fair elections that could work for the welfare of the masses.

To meet the power demand, "Water & Power Development Authority (WAPDA), Pakistan" was established through a Parliament Act in 1958 [36]. At that time, the total generation capacity of the utility was 119 MW through hydro-thermal means. In 1965 and 1976, Mangla and Terbella Dams, respectively, were commissioned for irrigation and power generation purposes. Till 1990, WAPDA expanded its generation, transmission and distribution networks through its own finances and the Government did not spend anything. By this time, utility's network had expanded by 250 times since its inception. Nevertheless, unfortunately, most of the revenue collected by the utility went into the Government's treasury for other expenditures, and from 1976 to 1993, no earnest effort was carried out to tap a large amount of hydroelectric potential (60,000 MW), wind-based energy (346,000 MW), and coal resources (186 billion tons) available in the country [37]. It is quite astonishing to know that the total annual water inflow in the country's Indus River System is 151.38 million acre feet (MAF). Out of which 103.81 MAF, 7.3 MAF, 7.4 MAF, and 33.29 MAF are diverted to the national canal system for irrigation purposes, stored in Tarbela and Mangla Dams, and wasted into the Arabian Sea, respectively. The estimated worth of yearly ocean-dumped water (33.29 MAF) has been USD 29 billion and this too in a country that has a total of USD 130.4 billion foreign debt. Moreover, silting of the Tarbela Dam has reduced its capacity from 11.6 MAF to 7.3 MAF, requiring immediate construction of the Bhasha and Dasu Dams upstreams. These statistics reveal that proper utilization of abundantly available water, wind, and coal resources may easily alleviate the foreign debt burden and distributional injustice in the energy sector of the country.

The name, i.e. WAPDA depicts that it is a sovereign authority to predict load; expand generation, transmission, and distribution; and carry out maintenance services. However, as mentioned earlier, all the earned revenues generated were consumed by Federal Government and WAPDA practically was crippled to construct new generation and transmission projects on its own. This situation resulted in the escalation of load shedding and price surge. Moreover, the hydro projects planned by WAPDA were politicized by different political parties in the provinces. A classic example of this has been the Kala Bagh Dam, which was abandoned by merely stating that the construction of this dam would flood

Noshehra City in Khyber Pakhtunkhwa Province and bring drought to the Sindh Province, however, the studies have confirmed that these claims are erroneous. Khan et al. [38] proved that the minimum elevation of Noshehra City from sea level had been 935 feet, whereas the maximum elevation of the dam's reservoir would be 915 feet. The point of maximum elevation would be 16 km downstream of Noshehra City. Therefore, there is no chance of flooding in these areas of Khyber Pakhtunkhwa Province whatsoever. Similarly, Sindh Province gets 44.5 MAF of water annually. Indus River has 18% of the yearly water inflow in winter and 82% in summer. Excess water in summer would store in Kala Bagh Dam, which may provide additional 6.1 MAF water to the Sindh Province during dry winter season.

Lack of interest and political ill will described above finally resulted in load shedding in 1990s, thereby, destroying the business and manufacturing activities in the country. As a result of this, the investors hesitated to invest in industrial and agricultural sectors, making the country to heavily rely on the foreign imports. Since then, national gross domestic product (GDP), per capita income, job creation, and poverty alleviation have been on a continuous decline in the country. Instead of investing in the manufacturing sector, real estate sector remained the main focus of investors. As a result, the country started heavily relying on foreign imports, thus draining the dollars out of the national financial pool. Due to this, inflation increased with devaluation of the "Pakistani Rupee" against the US dollar and it was the actual starting point of the economic meltdown of the country. Pakistan's power sector is a classic example of distributional injustice originating from a lack of political acumen and ill will. During this period, the public had no real sympathetic political representation, and the country remained in hands of forces that had no democratic roots. Ill will, wickedness, and incompetence of the ruling elite, therefore, drove the public into a situation where they now have to buy unreliable and costly electricity.

In 1994, the Government in a bid to restructure WAPDA, liberalized the generation sector by introducing IPPs mainly oil-based, to reduce the load shedding. However, these IPPs proved to be an energy mafia; as their agreements (take or pay in nature) with the Government, were based on utter disloyalty; which again served the vested interests of the ruling elite who actually owned these power plants. According to these agreements, public or government had to pay unfair amount of capacity charges, which have been the major portion of circular debt till to date. Large amount of capacity charges are a main reason of price surge. These charges vary between PKR. 1500–4000 per kW in summer and winter seasons. Another surprising aspect has been that the maximum load witnessed in Pakistan in 2023 has been around 23,000 MW; however, the installed capacity has been 43,000 MW. The IPPs have been charging the capacity charges for excess 20,000 MW amounting to PKR 2.16 trillion, for year 2023–2024.

Poor public of the country has already paid PKR. 8.344 trillion, for the last 10 years (2013–2023), thus overburdening the customers under heavy tariffs.

Currently, in 2024–2025, per unit price has been PKR. 46.8. Out of which, PKR. 8, 24, and 3 are the per unit average fuel price, capacity charges being paid to the IPPs, and transmission and distribution losses (11%), respectively. Upon summing up these fractions, the total per unit cost comes out to be PKR. 35 and not PKR. 46.8. Now, a question arises that why is the Government taking excess PKR. 11.8 on a unit of electric energy from the poor public? Moreover, globally, the per unit capacity charges are 3¢ which equal PKR. 8, whereas, in Pakistan, customer has been paying PKR. 24. In short, we may say that a poor, pitiable residential customer has been paying PKR. 46.8 per unit instead of what should have been PKR. 19, i.e. paying PKR. 27.8 extra on an energy unit. Due to this cruelty, public has been under mounting stress, selling their household belongings to pay the electricity bills. Moreover, industry and businesses have been suffering, resulting in a drop in the national GDP, decline in economic growth, rise in poverty, deterioration of health and education and a surprising surge in crimes and joblessness.

There are 100 IPPs that are granted licenses by the National Electric Power Regulatory Authority (NEPRA) till 2057, among them, 51, 7, 18, 7, and 13 belong to various business groups, political personalities, international companies, Government of Pakistan and individual businessmen, respectively. We are giving just one example to show how "***Profitable White Collar Loot***" it has been. Initially, in 1994, 17 IPPs were introduced. The energy audit of 16 out of 17 IPPs till 2010 shows that the owners invested capital of PKR. 51.8 billion, however, they in return earned PKR. 415 billion, which is approximately 800% more than the initial capital investment. All of this money has either been collected from the customers or national treasury. Such a massive profiteering escalated the electricity prices over last 25 years, consequently, destroying country's economy. The severity of the matter has been so grave that IPPs have threatened the economic future and national security of the country. Amid all this, there has not been any institution or authority such as the judiciary that is willing to conduct heat and efficiency audits of these power producers and inquire from the regulator, i.e. NEPRA, that why did it allow the high-capacity payments and installation of excess 20,000 MW generation? Under all this, the Government has been craftily saying that instead of these mounting capacity charges, energy theft has been the main reason of utility's destruction, thus unveiling that it is unwilling to formulate any sympathetic framework to revise its IPP policy.

In 2007, in a bid to restructure the WAPDA, utility was unbundled into a Water Wing managing dams and four thermal generation companies (GENCOs) and a Power Wing managing 11 distribution companies, Karachi Electric (KE) (a privatized distribution company) and National Transmission and Dispatch Company (NTDC). After this, WAPDA was limited to only hydro dams and the rest, except KE, was called Pakistan Electric Power Company (PEPCO). In 2021, PEPCO was renamed as Power Planning and Monitoring Company (PPMC) and all the 11 distribution companies and NTDC have become independent entities with everyone working under their own Boards of Directors. This is how once a utility termed as the WAPDA interconnecting the millions of people from Karachi to Khyber, consisting of thousands of miles of transmission and distribution lines and providing safe, secure, reliable, and sustainable electricity, is disintegrated in the name of restructuring. Now "circular debt-ridden utility's" services are no longer secure, reliable, and affordable, and it is not in a position to drive the engine of a sickening and bewailing economy of the country.

Currently, NEPRA has shown commitment to implement the concept of energy wheeling. Under this regime, a private generator (wheeler) may be allowed to transfer power from the utility's network to a bulk consumer (greater than 1 MW load). In response to this, the generator will pay the use of system charges to the utility. Until now, proper use of system charges has not been determined yet, and if the wheeling is allowed without determining the precise charges, it would definitively result in a loss to either the utility itself or the wheeler, thus giving rise to a distributional injustice. Besides, even after a precise determination of the use of system charges, utilities will lose their customers with a consequent shrinkage in business volume. The lost ones would especially be large customers that cross-subsidize the smaller ones. Resultantly, the cross-subsidy would be recovered from smaller customers, therefore, putting an extra financial burden on them.

Apart from these issues, wheeling would introduce complications such as non-recovery of the cost of losses and investment on the network expansion and maintenance, non-recovery of stranded cost, energy banking, uncertainty of renewable energy, and energy fed to the hybrid bulk customers during peak hours, etc. It is important to mention that in the presence of energy wheeling, utility's network acts as a bank, in which generator may inject power during off-peak hours; nevertheless, the customer may draw the energy in peak hours. In this way, wheeling may overload the network, reduce its quality and reliability, increase losses, and may even force the utility to invest capital in capacity enhancement of the network. Same issues may arise when a hybrid bulk customer starts drawing power from utility during unavailability of the wheeling generator.

In recent years, NEPRA has decided to allow customers to install solar DGs of up to 1 MW on building rooftops (a concept of net metering that transforms passive consumers into prosumers). Under this facility, prosumer generates power, uses for its own load requirements and sells excess energy to the national. In case the self-generation is less than the load, the prosumer imports the power from the national grid. Under this bilateral energy trade, power imported by the prosumer is charged on retail tariff, whereas, the power injected into the grid is sold on the wholesale tariff. This mode of business seems quite attractive, however, it has some pros and cos. The issues in this mode of energy trade for the attention of energy regulators are briefly mentioned here: (1) the net metering will make heavy customers go out of the business pool of the distribution utilities. Large customers cross-subsidize the smaller ones. However, in the presence of net metering, the large prosumers will be lost and the Government will have to increase the energy prices to cover the lost cross-subsidy, and the price will be recovered from the underprivileged low-income section of society; (2) capacity charges being paid by the Government on the part of lost customers due to wheeling and net metering would worsen the picture of circular debt, and henceforth, the affordability of the per unit price; (3) solar DGs inject power back into the grid that is owned, constructed, and maintained by utilities themselves. Under this, DG owners should pay the "use of system charges" to the utility. However, these charges are not included in the tariff and utilities are facing a huge financial loss, thus giving rise to distributional injustice; (4) excess solar power injection would increase system harmonics, reduce power quality and reliability, consume the system's reactive power and increase transformer damage rate, thereby resulting in fiscal loss to the distribution utilities. Loss of reactive power may increase utility's energy losses; (5) prosumer remains independent during off-peak hours (day time), however, feeds from the national grid during peak hours (evening and night). This may overload the utility's network and force the it to procure in-optimally high amount of capacity from the national power pool; (6) solar energy is unpredictable, therefore, to ensure system stability during its unplanned outage, firm capacity has to be maintained, which will consequently increase the capacity charges even more. In short, the introduction of wheeling and net metering would overburden the poor, damage the utility business, and favor the privileged if not carefully regulated. We would like to suggest the utilities to: (1) conduct simulation studies to find the optimum injection level of solar energy from prosumer side that does not increase losses, harmonics, system's quality and stability and equipment damage rate; this injection limit (formulated in %) must never be violated to safeguard bilateral stakes of energy market participants; (2) determine proper tariff structure that compensates the capacity charges, cross-subsidy, use of system charges and rise in energy losses. These two facts may afterwards be submitted to the regulator for an open discourse and subsequent modification in the regulations provided it is deemed logical.

Privatization started since 1994 with the introduction of IPPs, when monopolized generation sub-sector was opened for private investment to reduce energy price and load shedding, however, unfortunately, both purposes could not be achieved. Concepts of net metering and energy wheeling are also aspects of privatization. Apart from all this, the Government under every distribution company, has introduced the Market Implementation and Regulatory Affairs Department (MIRAD) to promote the concept of Competitive Trading and Bilateral Contract Market (CTBCM). Under this department, the distribution companies will predict their future load and accordingly expand transmission and distribution networks and procure energy from different power plants. The scenario of power procurement would be that, e.g. Islamabad Electric Supply Company runs an optimal dispatch to obtain low energy procurement cost to feed its load. The dispatch, for instance, may show that to fulfill certain energy demands, 600 MW of hydro energy is required. In response to this, market implementation department of the Company opts to purchase this hydroelectricity from a hydroelectric plant e.g. Tarbela. However, at the same time, another distribution company may step ahead and offer Tarbela a higher per unit cost than what Islamabad Electric Supply Company is willing to pay. In such a situation, the plant may sell power to the other distribution company instead of Islamabad Electric Supply. This is how competitive energy trading would occur, thereby liberalizing or privatizing the bulk power market in Pakistan. Apparently, it seems quite attractive, but practically it is not; because whenever, a distribution company would purchase power from a power plant, the plant in turn would demand a bank guarantee from the company. Now banks would show no reluctance to provide heavy guarantees for the distribution companies that are earning profit, however, likelihood of a guarantee from any bank for the companies already running on receivables (e.g. Peshawar Electric Supply) would be quite low. Therefore, the chances of success of CTBCM in distribution companies of Islamabad, Gujranwala, Lahore, and Faisalabad seem realistic; however, rest of the distribution sector would not produce expected results.

Further, for privatization purposes, NEPRA desires to segregate the wire and supply licenses of the distribution companies. After this, companies would only own the wire side license under which load prediction, network expansion, and maintenance services would be allowed. Transmission and distribution networks would remain under the sole exclusivity of the distribution companies. Supply-side licenses would be granted to those suppliers by NEPRA who qualify the regulator-laid conditions. The duties of a supplier would be meter reading, bill distribution, recovery, installation of new connections, disconnection, and reconnection, etc. Internationally, privatization occurs in two modes: (1) outsourcing the commercial and/or technical activities or (2) selling a certain percentage of utility's assets. Now we will have to witness that which option the Govt. of Pakistan opts for.

As discussed above, there are 11 distribution companies in Pakistan, however, only Gujranwala Electric Power Company, Faisalabad Electric Supply Company, Islamabad Electric Supply Company, and Lahore Electric Supply Company are profit-earning ones and all other are under heavy loss. Nonetheless, NEPRA has announced a uniform tariff for all these companies instead of a regional tariff. Had the regional tariff been allowed, electricity prices in areas with better losses and recovery would have been low and conversely high in high-loss utilities. But unfortunately, the customers who are not involved in theft are paying the same per unit price as the ones who steal energy. This is an example of distributional injustice where a loyal customer has been paying the price of energy theft.

The current energy policy in Germany aims to reduce carbon emissions by gradually replacing fossil-fuel-based plants with renewable energy DGs [31]. To realize this strategy practically, Government has announced feed-in-tariff, which offers low electricity prices to the owners of wind and solar DGs. However, the monetary burden is shifted to the low-income class, thus, raising the concerns of distributional injustice. Moreover, the real renewable potential lies in Northern Germany, however, the industry lies in Southern Germany, and to transport this power, lengthy, complex, and expensive transmission network is required between both regions. This situation is a true manifestation of the spatial injustice related to the siting of the infrastructure.

Even the Pakistani power system has the same issue. The country's northern region relies on hydro power, which reduces in winter due to low water inflow in the Indus River System; consequently, to fulfill the energy needs, this region imports remaining power through extra high voltage tie lines from the Southern Region. Now, whenever there is a fault on any one of these four tie lines, the Northern Region's frequency collapses due to generation-load mismatch, plunging millions into darkness. To avoid this scenario, we suggest the installation of gas turbines and oil-based power plants (with high ramp rates) in the Northern Region to share the peak load and do not entirely depend on the tie lines and thermal power available in the country's south.

Germany intends to replace fossil and nuclear power plants with renewable energy to reduce the environmental dangers posed by both without aggravating the distributional injustice problem. In this regard, first endeavors could be made to replace nuclear plants with renewables, and then in the second phase, the fossil-fuel-based plants may be replaced to a degree that distributional injustice does not aggravate intolerably.

6.2.2 Procedural Justice

Procedural justice exhibits a demand for "equitable procedures" that involve all shareholders in a nondiscriminatory manner [39]. Under this type of justice, all stakeholders are involved and heard in a decision-making process, which alleviates the chance of any of them getting aggrieved. It is important to consider what procedures are utilized, who is included in the decision-making and executing guidelines, and how fair are those decisions [40]. Procedural justice is related to impartiality in making decisions, comprising the ones who play a role in the development and delivery of policies that ultimately affect customer's ability in a positive way to attain an energy service. Bojer [29] argues that "under the procedural justice, the consequence is evenhanded if the procedure is fair." Reference [31] defines procedural justice as a "need for impartial procedures, including all shareholders in an indiscriminate way."

It is evident from these definitions that due process including all participants is mandatory for fair decision-making. To achieve this objective, a strong democratic system needs to be in place that instead of taking care of the vested interests, frames policies favoring all stakeholders in a just way. Likewise, to restrict the authorities and institutions from making unfair decisions in favor of mafia, a strong criminal justice system should be in place. It can therefore be inferred that procedural justice can only be realized pragmatically by a strong democratic system backed up by actively functioning law enforcement mechanisms.

A blatant instance of procedural injustice in Pakistan has been that the Government did not adopt a comprehensive policy to produce electricity from abundantly available hydro, coal, and wind resources, which may bring energy prices down. The oil import bill for fiscal year 2022–2023 has been USD 23.318 billion, putting unbearable strains on an already collapsed economy of the country. Dollar outflow has been devaluating the local currency, raising inflation, causing unaffordable commodity prices, and decelerating much-needed economic growth. The inflation rate in 2022–2023 has been around 20%, causing severe hurdles for millions to fulfill their day-to-day needs. Likewise, the planning and procedure to introduce the IPPs with unfair capacity charges and willful oblivion, or rather intentional amnesia to install 20,000 MW excess private generation while overlooking the national and public interest, are classic instances of procedural injustice.

Under all the power distribution companies in Pakistan, a procedural failure has occurred in the form that the authority has been reluctant to recruit new staff against the retiring ones. Currently, utilities are working on 45% of the sanctioned strength, which has been causing degradation of technical, commercial and administrative services of a working unit termed as "Sub Division." Further, shortage of staff puts an extra burden on the working personnel, thereby, making them vulnerable to a fatal or nonfatal accidents. In the past few years, the accident count has alarmingly increased.

In recent times, there is unfortunate policy shift in completing Iran–Pakistan gas pipeline. Had there been a resolve in completing this project without taking any foreign dictation, there would have been a marked reduction in energy poverty and an improvement in energy sustainability in Pakistan. The effect of oil crisis in 1973 on underprivileged and aged households' capacity to satisfy their energy requirements in the United States and United Kingdom was quite significant [23]. Nevertheless, both countries have managed to come out of that predicament—while in the 1990s, the concept of poverty of fuel in the United Kingdom led to the 2001 "Fuel Poverty Strategy," whereas the United States still lacks a federal approach to avoid energy deprivation [23]. Currently, the idea of energy poverty has received quite a lot of attention in Europe by nongovernmental groups and administrations [23].

It is important to mention that an Advanced Metering Infrastructure (AMI) project (without energy management system module) has been started by Islamabad Electric Supply Company, under which, the conventional energy meters of approximately 1.3 million electricity customers, will be replaced the AMI meters. The capital expenditure of the project is PKR. 30 billion with a return period of 10 years. It is claimed that the effect of this expenditure would not profoundly increase tariff; moreover, application of the AMI meters would reduce the quantum of energy theft with multiple other lucrative benefits. Despite all this, a few questions remain unanswered: (1) Can a rational electric utility invest in an area with a return period of 10 years, when still there are the areas, where, return period could have been quite low such as 3 to 5 years? Areas with low return period may for instance be, 11/0.415 kV distribution network rehabilitation program, which may result in a significant energy loss reduction and quality as well reliability improvement. Likewise, another area could be that it has been noticed from the situation prevailing in the company, that out of these 1.3 million customers, 20% i.e. 0.26 million are running on slow and inefficient electro-mechanical meters. After investing on network rehabilitation from PKR. 30 billion, remaining amount could have been invested on the replacement of these 0.26 million non-static meters. Addition of the static meters could have increased company's sold energy against same received energy. To clarify this advantage further, we assume that if a static meter reads energy more efficiently by giving 10 additional units per customer per month, the amount of increase in the sold energy would be 31,200,000 kWh per annum for 0.26 million customers. Such a massive increase in the sold energy would have resulted in a huge revenue increase as well as meaningful loss reduction, in a fiscal year. Now, according to the current electricity price in Pakistan i.e. PKR. 46.8 per unit, company could have increased its annual revenue by PKR. 1,460,160,000. In other words, by investing in network rehabilitation and replacement of non-static meters, the company could have saved billions per annum. It is mandatory to describe that even after spending money on these two activities, a handsome amount of sum out of PKR. 30 billion, could have been saved for other issues.

Few millions out of this 30 billion if invested in recruiting and training the staff, could have invigorated our manpower. It would definitely have improved our services as energy distributor and secure our staff from fatal and non-fatal accidents.

6.2.3 Recognition Justice

McCauley et al. [3] defined recognition justice as that the "stakeholders must be impartially represented as if they are free from all types of threats and they must essentially be offered equivalent and comprehensive rights," to avoid or repair inequalities adversely affecting their energy requirements. This type of justice strives to certify the acknowledgment of the sidelined and relegated groups in an energy society. In order to seek procedural justice for fair participation in decision-making, recognition is mandatory, as, without unbiased representation, the realization of procedural justice would be impossible. Therefore, it may be concluded that the concept and efficacy of both forms of justice overlap with each other. Apart from this, in [41], authors state that fair recognition is a tool that prevents the cultural injustice and misrepresentation of individuals that originate, e.g. from different cultures, areas, religions, and languages.

The following four principles are applied while discussing recognition justice in any case:

a) **Recognizing the Specific Advantages, Disadvantages, and Circumstances for All Participants:** Recognition is based on the morality of the whole of humankind and acceptance of all types of cultural, political, age group-based, social, racial, gender-based, and ethnic differences. In terms of the concept of energy, it means identifying various drawbacks and required energy services for a community [23]. For example, the class of population related to a poor economic segment of society has the issues of energy poverty, whereas, ethnic minorities may face the situation of energy burden or energy security. Pakistan's energy society is an example of this kind of injustice, where the rich and well-off have been resorting to rooftop solar panels to avoid the expensive energy import from the national grid. Besides this, the people of poor provinces such as Baluchistan are still deprived of the NG that their own province produces.
b) **Recognizing Multiple Types of Energy Requirements**: Different groups, classes, and populations have different types of energy requirements, and these needs must be respected [23]. For example, the people residing in Northern Europe have more space and water heating needs compared to the population living in tropical weather of Africa.
c) **Recognizing the Complete Reasons of Energy Scarcity**: Complete knowledge of, and regard for, the conditions of families anguished by energy-related misery demands an acceptance of the fundamental reasons and a fair solution. Root causes may be, for example in a low-income family, appliances with poor energy-related efficiency, austerity measures, and high-energy prices [23].
d) **Non-defamation in Representing Families with Energy-related Destitution:** In this regard, it is quite evident that words such as poverty, etc., have denouncing/stigmatizing inferences, and challenging this notion requires all stakeholders to provide space to admit the physical, organizational, and operational hardships [23].

"Lawrence Weston" is acknowledged as a precinct of severe deprivation and besides, it has blatant economic, social, and topographical disparities in Bristol,

United Kingdom. As a consequence, the community of this area faces recognition injustice with continuing problems comprising poverty, crime, inferior quality housing, unemployment, and underprivileged transport means. However, the most fundamental cause of this deprivation has been that the people of Bristol did not actually realize the unjust treatment. However, a Director of the British Education Council acknowledged the issue of non-recognition and stated that the area has been facing energy deprivation, and we have a plan to revitalize the city [42]. Likewise, vast areas of Pakistan, such as the interior Sindh, Baluchistan, Khyber Pakhtunkhwa, and South Punjab, have been facing the recognition issue since its creation. Due to this, these areas do not have infrastructure, energy supply in the form of electricity and the NG, hospitals, universities, jobs opportunities and are suffering from high crime rates. Such a condition, if persists over a long-term period, may even threaten the national security. Therefore, the Government should take notice of this misrepresentation of groups and provinces and plan developmental activities to take the public out of the sense of deprivation.

6.3 Principles of Energy Justice

The principles of the energy justice system are as follows:

a) **Availability:** Availability means the ability of a market, economy, or a system that ensures adequate energy sources when required. In other words, it relates to the reliability, adequacy, and security of supply of a commodity such as energy. Integrated expansion of generation, transmission, and distribution with unbiased and fair decisions may ensure this [43].
b) **Affordability:** The second principle is the simple affordability of energy supply. Affordability covers stable costs as well as justifiable prices that do not require the low-income class to spend a disproportionate share of their income on commodities [44]. Highly obtainable energy and services are worthless unless the public cannot afford them.
c) **Due Process:** Due process ensures the involvement of all the stakeholders in the process of policymaking through energy democracy. It also forces operational alternatives through legal and administrative remedies. More precisely, the decision-making code recommends that groups must be permitted to fairly participate in deciding the issues that affect them through a strong national democratic representation [45].
d) **Good Governance:** Strong democratic system has a characteristic that politicians elected on the basis of free and fair elections impartially represent their voters. Such politicians work for the improvement of their country, ease of access to commodities, justice, education, and curbing corruption and health,

etc. Moreover, under such a scenario, there would be an abundance of transparency and information [46]. Politicians develop such an atmosphere in which both the country and the countrymen thrive as the former have to again go back to the later for votes in the next elections; and if they did not work, the public would not vote them in future. This is how social equality thrives in the countries moving from authoritarianism to full democracy, and politicians start working for their public. Nonetheless, wicked intrusion by any peripheral forces or internal departments with no political and public roots trying to seize the fair governance process would surely tumble the country in gloom, despair, and deprivation, the way Pakistan has been since its creation. According to the Foreign Ministry's report 3.275 million educated/skilled Pakistanis have left the country in the last five years (2019–2024) due to the disheartening political, economic, and social situation in the country.

e) **Sustainability:** Energy sustainability is the provision of adequacy, reliability, and affordability of energy in adherence to the environmental and the social requirements [21]. Again policymaking to achieve these aims needs political stability and strong governance.
f) **Responsibility:** This principle relates to the responsibility of a government to protect the environment from emissions and opt for the fuels that assist the low power production cost [45]. For this purpose, adoption of cheap and low-emission energy sources is mandatory.

6.4 Factors of Fairness and Equity in Developing a Multi-Energy Community

Equity has already been defined; however, fairness in literature means the "quality of making decisions that are free from prejudice and discrimination." Umpires, teachers, politicians, policymakers, and judges should strive to practice the element of fairness in making judgments. Fairness in terms of the old English dictionary also means "attraction" and "physical beauty" [26]. The factors affecting fairness and equity in developing a multi-energy community are as follows:

6.4.1 Climate Change

Climate change has been an important driver that adversely affects in developing fairness and equity-based energy communities. The geographic areas that are severely affected may face extreme weather conditions such as heat waves, low temperatures, smog, air pollution, heavy rains, floods, and droughts. These conditions result in entirely different energy consumption requirements than usual, thus drifting a community to a possibly unfair and unequal energy supply.

However, on the other hand, a community residing in an area with low to moderately affected climate change does not require an abnormally high amount of energy for comfort.

6.4.2 Structural Racism

We have witnessed over the years that whenever there is a shortage of energy, load shedding is being carried out. However, the areas chosen to shed load are usually those where the low-income class resides. Likewise, smaller cities, states, and provinces suffer the same fate. Meanwhile, the affluent communities remain intact from these difficulties.

6.4.3 Energy Prices

Recent global recession, unfriendly economic policies, and political instability have increased energy prices. The per unit price of electricity was PKR. 13.95 in 2021; however now in 2024, the same has risen to PKR. 46.8. This hike, due to alarmingly high-capacity charges and dependence on oil-based generation, has incapacitated the commoner's affordability to use electricity according to their needs.

6.4.4 Planning and Policy Failure

There are certain instances when proper planning and fair policies are not adopted in developing a market to generate, transmit, and distribute energy in light of rising demand. Choice of poor planning tools, use of inaccurate data for future load prediction, and professional inefficiencies to formulate policies may result in system instability, inadequate reliability, inaccurate network expansion, and high prices in the future. Such a situation may adversely affect a community to gain fair and equal access to multi-energy streams.

6.4.5 Political Acumen

Pakistan is a classic example of a lack of political acumen and elite capture. The country has large hydro, oil, gas, coal, and renewable energy resources. However, political instability never permitted a capable and sympathetic leadership to emerge. Due to this, precious energy sources could not be utilized resulting in energy poverty, unclean environment, high prices, and economic collapse.

6.5 Use of Artificial Intelligence in Energy Justice

Application of Artificial Intelligence (AI) algorithms in energy justice problems may open up a whole new area of research for energy specialists. However, to proceed further in this area requires mathematical modeling of energy justice scenarios which may later on be embedded in a power/energy engineering framework. Subsequently, an AI tool may be used to find an optimum solution. The remaining part of this section provides a brief literature review on the application of AI techniques for solving the problems of the energy system.

In reference [47], authors established a strategy for evaluation of the energy justice in a smart grid paradigm. The proposed strategy is founded on the "triumvirate of tenets approach." Authors take profit and cost, public funding, governance data, and knowledge sharing as distributional justice. Recognition justice is taken into account in the form of the selection of communities for participation in smart grid activities and the degree of technological accessibility in accordance with the literacy rate. Recognition justice is also incorporated in the form of the process of decision-making.

Researchers propose a framework for injustices in a smart grid of a locality [48], analyze the effect of demand response on domestic customers [49] and issues related to the nonuniform tariff [50, 51] and use of the machine learning techniques on the justice issues [52].

Mosavi et al. [53] applied various AI techniques for load prediction, wind speed forecasting, solar insolation estimation, various optimization tasks, peak demand management, and cost reduction. Results show high accuracy with low compilation time.

Reference [54] proposes an optimum voltage control for a low-voltage distributor having multiple solar DGs. Whenever the voltage increases, the active power injection from DGs is curtailed to bring the voltage back and vice versa for reverse action. Results show that AI techniques isolate DGs in less time at a low cost compared to conventional optimization techniques.

Vergara et al. [55] developed a centralized voltage control mechanism for the above-mentioned scenario in [54]. The central control takes the output of all DGs and minimizes their output to maintain the voltage within the permissible range. Machine learning technique is used, and results reveal that the proposed strategy prevents undue curtailment.

Raza et al. [56] proposed a macro energy hub control for multi-energy residential communities under the smart distribution grid paradigm. The proposed framework uses particle swarm optimization and flower pollination algorithm. Results show that later meta-heuristic algorithm outperforms many contemporary algorithms in terms of finding a global optimum. The authors succeeded in developing a justice mechanism between customers, utility, and climate change challenges by reducing energy consumption cost, peak shaving, and emission reduction, respectively.

Raza et al. [57] suggested a two-layered mathematical macro energy hub control for multi-energy commercial buildings under a smart distribution grid paradigm. The proposed framework uses a flower pollination algorithm. The authors succeeded in developing a justice mechanism between customers, utility, and climate change challenges by reducing energy consumption cost, peak shaving, and emission reduction, respectively.

Raza et al. [58] suggested a two-layered risk-neutral and risk-averse macro energy hub controls for multi-energy residential buildings, under smart distribution grid paradigm. Framework uses flower pollination algorithm and proposed two-core flower pollination algorithm. Results show that later algorithm outperforms many of the contemporary algorithms in terms of finding global optimum. The authors succeeded in developing a justice mechanism between customers, utility, and climate change challenges by reducing cost, peak shaving, and emission reduction, respectively.

Ahmed et al. [59] proposed "new Weibull Distribution-based time-coupled probabilistic generation model" for optimal sizing and siting of wind power plants. The proposed wind generator model is used to feed voltage-dependent, time-varying load. Particle swarm optimization is used, and results show that in the IEEE-33 & 69 bus systems, voltage and losses remain within permissible ranges against demand variations.

Although AI techniques are quite attractive, but there are a few problems, such as algorithmic bias, data quality issues, and transparency. Ferrara [60] discussed various reasons of bias leading to suboptimal solutions, are imprecise data, type of algorithm selected, and human decision biases. Besides, the authors proposed solutions to prevent bias like "robust data augmentation; application of counterfactual fairness alongside unbiased data collection methods." Data quality [61] plays an important role in accurate decision-making. Main reasons of imprecision may be integration of data from two sources with resulting errors, large data volume making it difficult to maintain, data security and privacy from external intrusion, and possible human errors. To detect these errors, quantitative and qualitative data analysis can be done through statistical and thematic analysis tools, respectively. Transparency [62] in AI has been vital due to bias in the data on which it is trained. Such counterchecks for instance avoid accidents in ill-trained or inaccurately designed automatic electric vehicles and auto-controls designed to handle certain processes or robots.

6.6 Conclusion and Future Research

This chapter discusses distributional, procedural, and recognition justices in various forms and shapes of energy systems. These justices call for just distribution of resources everywhere guaranteeing the entry for everybody into the decision-making process and fulfillment of fair stakeholder representation, respectively. Further, availability, affordability, due process, good governance, sustainability, and responsibility are elaborated as justice principles. Factors affecting the fair and equitable multi-energy community are discussed at length. Finally, a brief survey on the applications of AI techniques in energy justice issues reveals their worthiness in precise and rapid estimation of different problem variables.

Identified avenues for future research are: (1) digging out more literature to broaden the number of variables and philosophical vision of the problem of energy justice; (2) mathematical formulation of energy justice issues, embedding them in the existing power engineering problems and solving them with AI and machine learning techniques, would be a lucrative area of research.

References

1 Girish, G.P. and Vijayalakshmi, S. (2014). Energy trading and investing: trading, risk management and structuring deals in the energy markets-a book review. *Indian Journal of Finance* 8 (4): 49–51.

2 Kolm, S.C. (1997). *Justice and Equity*. MIT Press.

3 McCauley, D.A., Heffron, R.J., Stephan, H., and Jenkins, K. (2013). Advancing energy justice: the triumvirate of tenets. *International Energy Law Review* 32 (3): 107–110.

4 Guruswamy, L. (2010). Energy justice and sustainable development. *Colorado Journal of International Environmental Law and Policy* 21: 231–275.

5 Yumkella, K. (2010). Energy Justice Conference, October 23, Keynote address. *Colorado Journal of International Environmental Law and Policy* 1 (2009): 277–286.

6 Garavito, G.A.A. and Huebner, G. (2023). *Energy Equity in the United Kingdom*. Center for Open Science.

7 Reeves, L. (2022). *Electrification of Surrey: Energy and Equity Implications*. Simon Fraser University.

8 Forman, A. (2020). Energy and equity revisited: examining local approaches to energy justice through community renewables development in wales. Doctoral dissertation. Cardiff University.

9 Chan, W.W. (2015). *Rethinking Water and Energy Affordability in Australia: an Analysis of the Efficiency, Effectiveness and Equity of Current Policy*. The Australian National University.

10 Sovacool, B.K. (2010). *The Routledge Handbook of Energy Security*. Routledge.

11 Tomasi, S. (2024). Citizens at the core: impact evaluation and justice implications of energy transition policies. Doctoral dissertation. Free University of Bozen-Bolzano. The University of Sydney.

12 Arya, R.N. (2023). Lost in transition! an analysis of justice implications for energy transition in India. Doctoral dissertation. University of Edinburgh.

13 Hirono, K.T. 2023). Participation for health equity: a comparison of citizens' juries and health impact assessment.

14 Li, S. 2023). Equity valuation and climate change Doctoral dissertation. Open Access Te Herenga Waka-Victoria University of Wellington.

15 Sovacool, B.K., Burke, M., Baker, L. et al. (2017). New frontiers and conceptual frameworks for energy justice. *Energy Policy* 105: 677–691.

16 Sareen, S. and Haarstad, H. (2018). Bridging socio-technical and justice aspects of sustainable energy transitions. *Applied Energy* 228: 624–632.

17 Newell, P. and Mulvaney, D. (2013). The political economy of the 'just transition'. *The Geographical Journal* 179 (2): 132–140.

18 Blondeel, M., Bradshaw, M.J., Bridge, G., and Kuzemko, C. (2021). The geopolitics of energy system transformation: a review. *Geography Compass* 15 (7): e12580.

19 Vakulchuk, R., Overland, I., and Scholten, D. (2020). Renewable energy and geopolitics: a review. *Renewable and Sustainable Energy Reviews* 122: 109547.

20 Temper, L., Avila, S., Del Bene, D. et al. (2020). Movements shaping climate futures: a systematic mapping of protests against fossil fuel and low-carbon energy projects. *Environmental Research Letters* 15 (12): 123004.

21 Baker, E., Carley, S., Castellanos, S. et al. (2023). Metrics for decision-making in energy justice. *Annual Review of Environment and Resources* 48: 737–760.

22 Abdelbary, A.M., Manglicmot, L., Kanwhen, O., and Mohamed, A.A. (2024). Community-centric distributed energy resources for energy justice and decarbonization in dense urban regions. *Energy Reports* 11: 1742–1751.

23 Hernández, D., Yoon, L., and Simcock, N. (2022). Basing "energy justice" on clear terms: assessing key terminology in pursuit of energy justice. *Environmental Justice* 15 (3): 127–138.

24 Prabhakar, S.V.R.K., Shimizu, N., and Lee, S.Y. (2024). Equity and fairness in community based adaptation and disaster risk reduction. In: *Disaster Risk and Management Under Climate Change*, 369–390. Singapore: Springer Nature Singapore.

25 Noorman, M., Espinosa Apráez, B., and Lavrijssen, S. (2023). AI and energy justice. *Energies* 16 (5): 2110.

26 Apergi, M., Eicke, L., Goldthau, A. et al. (2024). An energy justice index for the energy transition in the global South. *Renewable and Sustainable Energy Reviews* 192: 114238.

27 Skare, M., Qian, Y., Xu, Z., and Gou, X. (2024). Energy justice and gaps in sustainable development: a convergence testing and clustering study. *Renewable and Sustainable Energy Reviews* 192: 114166.

28 Sovacool, B.K. and Dworkin, M.H. (2015). Energy justice: conceptual insights and practical applications. *Applied Energy* 142: 435–444.

29 Bojer, H. (2003). *Distributional Justice -Theory and Measurement*. London, New York: Routledge.

30 Yenneti, K. and Day, R. (2016). Distributional justice in solar energy implementation in India: the case of charanka solar park. *Journal of Rural Studies* 46: 35–46.

31 Jenkins, K., McCauley, D., Heffron, R. et al. (2016). Energy justice: a conceptual review. *Energy Research and Social Science* 11: 174–182.

32 Park, S., Yun, S.J., and Cho, K. (2024). Energy justice: lessons from offshore wind farm siting conflicts in south korea. *Energy Policy* 185: 113972.

33 Malik, S., Qasim, M., Saeed, H. et al. (2020). Energy security in pakistan: perspectives and policy implications from a quantitative analysis. *Energy Policy* 144: 111552.

34 Ritchie, H., Rosado, P., and Roser, M. (2024). Access to energy. *Our World in Data*.

35 Anwar, A., Arshed, N., Ahmad, S., and Hayat, N. (2023). Nexus of electricity demand, circular debt and economic progress: an evidence from Pakistan. iRASD. *Journal of Economics* 5 (1): 14–29.

36 Khalid, Z. and Iftikhar-ul-Husnain, M. (2016). Restructuring of WAPDA: a reality or a myth. *The Pakistan Development Review* 55: 349–359.

37 Abbas, Y. and Aslam, R.A. (2023). Potential of untapped renewable energy resources in Pakistan: current status and future prospects. *Engineering Proceedings* 56 (1): 108.

38 Khan, M.I., Jamil, S.M., Ali, L. et al. (2014). Feasibility study of Kalabagh Dam Pakistan. *Life Science Journal* 11 (9s): 458–470.

39 Walker, G. and Day, R. (2012). Fuel poverty as injustice: integrating distribution, recognition and procedure in the struggle for affordable warmth. *Energy Policy* 49: 69–75.

40 Trimmel, K.E., Kriechbaum, M., Lazou, R., and Brudermann, T. (2024). Between distributive and procedural justice claims: reframing patterns of discursive resistance against climate action. *Energy Research and Social Science* 109: 103424.

41 Fraser, N. (2020). From redistribution to recognition?: dilemmas of justice in a'postsocialist'age. In: *The New Social Theory Reader* (ed. A. Bianco and M. Olff), 188–196. Routledge.

42 Allan, G., Eromenko, I., Gilmartin, M. et al. (2015). The economics of distributed energy generation: a literature review. *Renewable and Sustainable Energy Reviews* 42: 543–556.

43 Yildiz, C., Köhler, K., Wasserfurth, P., and Özilgen, M. (2024). Assessment of entropy accumulation in human subjects when exposed to low energy availability. *Heliyon* 10 (17): e36792.

44 Barrella, R., Linares, J.I., Romero, J.C., and Arenas, E. (2023). Evaluating the impact of energy efficiency strategies on households' energy affordability: a Spanish case study. *Energy and Buildings* 295: 113289.

45 Heffron, R.J. and De Fontenelle, L. (2023). Implementing energy justice through a new social contract. *Journal of Energy and Natural Resources Law* 41 (2): 141–155.

46 Stojilovska, A. (2023). Energy poverty and the role of institutions: exploring procedural energy justice–ombudsman in focus. *Journal of Environmental Policy and Planning* 25 (2): 169–181.

47 Milchram, C., Künneke, R., Doorn, N. et al. (2020). Designing for justice in electricity systems: a comparison of smart grid experiments in the Netherlands. *Energy Policy* 147: 111720.

48 Knox, S., Hannon, M., Stewart, F., and Ford, R. (2022). The (in) justices of smart local energy systems: a systematic review, integrated framework, and future research agenda. *Energy Research and Social Science* 83: 102333.

49 van Bommel, N. and Höffken, J.I. (2021). Energy justice within, between and beyond European Community energy initiatives: A review. *Energy Research and Social Science* 79: 102157.

50 Calver, P. and Simcock, N. (2021). Demand response and energy justice: a critical overview of ethical risks and opportunities within digital, decentralised, and decarbonised futures. *Energy Policy* 151: 112198.

51 Neuteleers, S., Mulder, M., and Hindriks, F. (2017). Assessing fairness of dynamic grid tariffs. *Energy Policy* 108: 111–120.

52 Powells, G. and Fell, M.J. (2019). Flexibility capital and flexibility justice in smart energy systems. *Energy Research and Social Science* 54: 56–59.

53 Mosavi, A., Salimi, M., Faizollahzadeh Ardabili, S. et al. (2019). State of the art of machine learning models in energy systems, a systematic review. *Energies* 12 (7): 1301.

54 Ali, M.V.M., Nguyen, P.H., Kling, W.L. et al. (2015). Fair power curtailment of distributed renewable energy sources to mitigate overvoltages in low-voltage networks. In: *2015 IEEE Eindhoven PowerTech*, 1–5. IEEE.

55 Vergara, P.P., Salazar, M., Giraldo, J.S., and Palensky, P. (2022). Optimal dispatch of PV inverters in unbalanced distribution systems using reinforcement Learning. *International Journal of Electrical Power and Energy Systems* 136: 107628.

56 Raza, A. and Malik, T.N. (2018). Biogas supported bi-level macro energy hub management system for residential customers. *Journal of Renewable and Sustainable Energy* 10 (2): https://doi.org/10.1063/1.4996271.

57 Raza, A. and Malik, T.N. (2019). Energy management in commercial building microgrids. *Journal of Renewable and Sustainable Energy* 11 (1): https://doi.org/10.1063/1.5034352.

58 Raza, A., Malik, T.N., Khan, M.F.N., and Ali, S. (2020). Energy management in residential buildings using energy hub approach. In: *Building Simulation*, vol. 13, 363–386. Tsinghua University Press.

59 Ahmed, A., Nadeem, M.F., Sajjad, I.A. et al. (2020). Probabilistic generation model for optimal allocation of wind DG in distribution systems with time varying load models. *Sustainable Energy, Grids and Networks* 22: 100358.

60 Ferrara, E. (2023). Fairness and bias in artificial intelligence: a brief survey of sources, impacts, and mitigation strategies. *Science* 2024 (6): 3.

61 Pansara, R. (2023). Cultivating data quality to strategies, challenges, and impact on decision-making. *International Journal of Management Education for Sustainable Development* 6 (6): 24–33.

62 Mensah, G.B. (2023). Artificial intelligence and ethics: a comprehensive review of bias mitigation, transparency, and accountability in AI Systems. *Preprint* November, 10.

7

Energy Community Operation and Planning for Energy Justice and Equity

Homayoun Ghasemnejad[1], *Sogand Hosseinalipour*[1], *Masoud Rashidinejad*[1], *and Sobhan Dorahaki*[2]

[1] *Department of Electrical Engineering, Shahid Bahonar University of Kerman, Kerman, Iran*
[2] *Department of Electrical Engineering, Qatar University, Doha, Qatar*

Nomenclature

Abbreviations

CEC	Citizen Energy Community
CEP	clean energy package
CES	community energy storage
DER	distributed energy resource
DRPs	Demand Response Programs
ECM	Energy Community Manager
EEP	energy efficiency program
ESS	energy storage system
EU	European Union
HVAC	heating, ventilation, and air conditioning
P2P	peer-to-peer
PV	photovoltaic
RES	renewable energy sources
SOC	state of charge

Indices

N_g, g	index and set of shiftable appliances
N_i, N_j, i, j	index and set of prosumers
N_t, t	index and set of times
N_ϕ, ϕ	index and set of subcommunities or clusters
N_σ, σ	index and set of storages

Communities for Clean Energy Justice and Equity in Grid Modernization, First Edition.
Edited by Mohammadreza Daneshvar, Behnam Mohammadi-Ivatloo, and Amjad Anvari-Moghaddam.

Parameters

C_i	thermal capacitance (kWh/°C)
$P_{\max}^{Sell,G}, P_{\min}^{Sell,G}$	maximum and minimum sale power capacity limitation to the grid (kW)
$P_{\max}^{Sell,Pr}, P_{\min}^{Sell,Pr}$	maximum and minimum sale power capacity limitation of prosumer (kW)
$P_{\max}^{Buy,G}, P_{\min}^{Buy,G}$	maximum and minimum purchased power capacity from the grid (kW)
$P_{\max}^{Buy,Pr}, P_{\min}^{Buy,Pr}$	maximum and minimum purchased power capacity limitation of prosumer (kW)
$P_{i,t}^{pv}$	power of PV system for prosumer (kW)wtrgtweg
$P_{i,t}^{baseload}$	baseload of prosumer (kW)
$P_i^{\min}, P_i^{\max}$	maximum/minimum limitation of HVAC (kW)
$P_{\sigma,\phi}^{ch,\min}$	minimum charging limit of storage (kW)
$P_{\sigma,\phi}^{ch,\max}$	maximum charging limit of storage (kW)
$P_{\sigma,\phi}^{dis,\min}$	minimum discharging limit of storage (kW)
$P_{\sigma,\phi}^{dis,\max}$	maximum discharging limit of storage (kW)
$P_{i,\sigma,\phi}^{ch,\min}$	minimum charging power limitation of prosumer to CES (kW)
$P_{i,\sigma,\phi}^{ch,\max}$	maximum charging power limitation of prosumer to CES (kW)
$P_{i,\sigma,\phi}^{dis,\min}$	minimum discharging power limitation of prosumer to CES (kW)
$P_{i,\sigma,\phi}^{dis,\max}$	maximum discharging power limitation of prosumer to CES (kW)
R_i	thermal resistance (°C/kW)
$SOC_{\sigma,\phi}^{\min}$	minimum SOC (state of charge) limit of storage (kWh)
$SOC_{\sigma,\phi}^{\max}$	maximum SOC limit of storage (kWh)
t_g^s, t_g^f	earliest start time/latest finish time for shiftable appliances
$\pi_t^{Buy,G}$	purchase price of electricity from the grid (€/kWh)
$\pi_t^{Sell,G}$	selling price of electricity to the grid (€/kWh)
θ_t^{amb}	ambient temperature (°C)
$\theta_{i,t}^{set}$	setpoint temperature (°C)
η_i^{th}	HVAC efficiency factor
$\eta_\sigma^{storage}$	charging and discharging efficiency of storage
λ_i	tolerance degree of temperature (°C)

Variables

$D_{i,t,\phi}$	consumption of prosumer (kW)
$I_{i,t}^{Buy,\mathrm{Pr}}, I_{i,t}^{Sell,\mathrm{Pr}}$	binary status of power purchase/sale from/to the grid for prosumer
$I_{i,t,\phi}^{cooling}, I_{i,t,\phi}^{heat}$	binary status of HVAC
$I_{i,\sigma,t,\phi}^{ch}, I_{i,\sigma,t,\phi}^{dis}$	binary status of storage
$P_t^{Buy,G}$	purchased power from the grid (kW)
$P_t^{Sell,G}$	selling power to the grid (kW)
$P_{m,t}^{Buy,\mathrm{Pr}}$	power purchased from the grid by prosumer (kW)
$P_{m,t}^{Sell,Pr}$	power sold to the grid by prosumer (kW)
$P_{i,j,t,\phi}^{P2P}$	power trading (kW)
$P_{i,t,\sigma,\phi}^{ch}, P_{i,t,\sigma,\phi}^{dis}$	charging and discharging power of prosumer to CES (kW)
$P_{i,t,\phi}^{shift}$	total consumption power of shiftable loads (kW)
$P_{i,t,\phi}^{cooling}$	the cooling consumption power of HVAC (kW)
$P_{i,t,\phi}^{heat}$	the heating consumption power of HVAC (kW)
$P_{i,g,t,\phi}$	power consumption by the shiftable appliance (kW)
$P_{\sigma,t,\phi}^{ch}, P_{\sigma,t,\phi}^{dis}$	charging/discharging of storage (kW)
$SOC_{\sigma,t,\phi}$	state of charge (kWh)
t^{start}, t^{finish}	starting and finishing time of shiftable appliances
$X_{i,g,t,\phi}$	binary variable used to show the status of shiftable appliance
$\theta_{i,t,\phi}$	indoor temperature (°C)
$\theta_{i,t,\phi}^{T}$	temperature modification by HVAC (°C)

7.1 Introduction

In recent decades, environmental issues have been the biggest challenge for the energy system, which relies heavily on fossil fuels and greenhouse emissions. Distributed energy resources (DERs), including photovoltaic (PV) panels, wind generators, batteries, and electric-powered vehicles, are one of the most important solutions to conquer this impediment, as endorsed by the global environmental agencies. These resources can provide smooth, renewable, and decentralized energy to meet the demand and reduce the dependency on the grid. However, the mixing of DERs into the energy structures also brings a few management troubles, including power quality, reliability, and coordination. To cope with these issues, the energy community idea is proposed. The energy community is a collection of end users and producers who cooperate to optimize using DERs inside a local area. The energy network can be considered an important tool for reaching sustainable development in the energy systems by considering diverse alternatives together with demand response programs (DRPs), energy efficiency programs (EEPs), parking lots, peer-to-peer (P2P) energy trading, etc. [1]. These options can enhance the flexibility, resilience, and affordability of the energy system, as well as the social and environmental advantages for the network members [2]. Energy equity is a critical issue within energy communities. Energy equity and energy justice are fundamental outcomes of these communities, where prosumers, who are consumers that also produce energy, can engage in energy sharing while recognizing and addressing fairness concerns. The goals of energy equity and energy justice are to achieve fairness across both social and financial dimensions of the energy system, while also mitigating the environmental and health impacts of the power system on disadvantaged communities [3].

In this chapter, the concept and the potential of energy communities that empower end users, small groups, and local governments to produce, manage, and consume their own energy are explored. In other words, this chapter discusses how energy communities can manage the power transition by growing the proportion of renewable power, improving energy performance, reducing energy bills, creating nearby jobs, and improving social concord. Moreover, this chapter investigates the challenges and opportunities of the energy communities in achieving energy justice, which is the fair distribution of energy benefits and burdens and the inclusive participation of all stakeholders in energy decision-making. The key aims and novel concepts presented in this chapter can be summarized as follows:

1) Discussion on how energy communities serve as a suitable platform for sustainability in power systems but also face challenges.
2) Comprehensive explanation of energy communities along with their challenges and opportunities.
3) The role of energy justice in the energy communities has been investigated.
4) The role of P2P trading in the energy communities has been examined.

7.2 Energy Communities: Definition, Challenges, and Opportunities

According to the negative impacts of conventional energy sources, such as fossil fuels, on the environment, many countries are pursuing initiatives to reduce their environmental impact. These measures are geared toward diminishing greenhouse gas emissions and promoting the adoption of sustainable practices.

This strategic shift involves a dual approach: first, a heightened reliance on renewable energy sources (RESs) to replace conventional ones, and second, a simultaneous reduction in the utilization of more pollutant-laden alternatives. This intricate and dynamic transformation is commonly recognized as the process of energy transition [4]. During this transition, consumers no longer play a passive role; instead, they transform into prosumers, capable of making decisions regarding their consumption and production. Consequently, this shift poses significant socioeconomic challenges that must be effectively managed by policymakers to achieve a fair transition toward sustainable energy systems [5]. As such, terms like "community energy" and "energy communities" have been progressively attracting scholarly attention for a considerable duration. By integrating diverse social and socio-technical principles, ECs emerge as products of a transition from centralized energy systems to decentralized and democratized ones. Additionally, ECs are seen as a potential pathway toward achieving a more equitable and inclusive energy transition, empowering citizens to engage in decision-making processes and contribute to the local energy infrastructure [6].

The concept of "community" is a term with several distinct meanings utilized for various purposes, which contributes to the ambiguity surrounding the term "community energy." For this reason, the concept of community energy remains an ambiguous term, as no precise definition has been provided thus far [7]. Based on [7–9], which examined papers on social science, various definitions of community energy provided by the papers are as follows.

7.2.1 Localized

Most of the articles suggest that energy communities should be formed and run by people who are in the same location or nearby. In other words, geographical proximity is emphasized, encouraging local individuals to participate in and manage a project [10]. A common misunderstanding is that they associate this proximity with a network, but geographical proximity actually means that individuals are near each other in a certain place [11].

7.2.2 Scale

In energy communities, "scale" simply refers to the specific size or scope at which energy systems operate, encompassing all aspects related to that particular size or scope. This encompasses a diverse range of energy communities, spanning from individual households with rooftop solar panels to expansive utility-scale. This interpretation, representing 43% of the definitions, revolves around the scale of energy communities.

7.2.3 Purpose

In energy communities, "Purpose" signifies a forward-looking and sustainable approach to energy consumption and production. Communities are designed with the number one aim of achieving positive and measurable effects in phrases of environmental effect, social well-being, and financial sustainability. In the area of energy, groups focus on imposing renewable and clean strength sources, which include sun, wind, and hydropower, to lessen dependence on fossil fuels. One of the substantial results of forming electricity communities is the attainment of self-sufficiency in their power supply.

7.2.4 Participants

Citizens or entities like municipalities and companies can join the energy community as investors and take part in the production, distribution, and sale of energy resources. To participate in energy communities, essential conditions such as a sense of trust [12, 13], economic, and environmental motivations [11, 14], a feeling of involvement in energy community affairs, and possessing a trusting spirit are necessary [15]. These factors contribute to the establishment of a sustainable and participatory energy community. The studies indicate that 42% of the definitions of energy communities are associated with the concept of energy community stakeholders.

7.2.5 Social

In the realm of energy communities, three pivotal criteria play a significant role in addressing social issues. The principle of justice emerges as a key driver, emphasizing the need for a balanced distribution of benefits and burdens in energy-related matters [16]. It underscores the importance of ensuring equal access to renewable energy resources and upholding the collective right to utilize them, aligning with fundamental social and energy justice principles. Simultaneously, psychological factors, encompassing social consciousness and responsibility, wield considerable influence in steering societal transformations within the energy sector [17]. Fostering a shared sense of social unity and responsibility toward the environment and renewable energy resources contributes positively to individuals' behaviors, promoting effective energy use. Cultural factors, including shared values and beliefs within society, significantly shape the landscape of social energy considerations. Common convictions such as equal rights, entrepreneurship, collective values, traditions, and localized perspectives influence decision-making and behaviors related to energy [18]. The commitment to societal values and their acceptance within social structures is proposed as a strategic approach to enhance the social acceptance of energy initiatives [19].

7.2.6 Technology

The discourse surrounding technology within energy communities has been thoroughly examined across diverse articles [20]. These technological aspects encompass PV system technology, wind power plants, the efficient utilization of energy storage, intelligent communication infrastructures, and notably, the potential for P2P energy trading among prosumers. This exploration has spurred a multitude of research initiatives dedicated to delving into the dynamics of energy communities.

7.2.7 Comprehensive Definition

In light of the mentioned criteria, it is possible to provide a comprehensive definition for an energy community, such that it is constituted by groups composed of local citizens and investors. These individuals adhere to shared environmental and social motivations, possessing similar principles of equity and psychological perspectives. Geographically proximate to each other, these individuals have the capability to directly engage in the production, distribution, and consumption of energy. Furthermore, these communities also possess the capacity for energy storage and P2P energy trading. On the other hand, the European Union (EU) envisions that, with the introduction of the clean energy package (CEP), consumers can have ownership of energy production for their own consumption and can even benefit from energy storage, defining them as prosumers who consume their own production. This shift empowers consumers to play a more influential role in the energy market. However, it also puts individuals or those experiencing energy poverty in an unstable situation, as they lack the capacity to invest in renewable resources and storage solutions. Consequently, this prevents all members from enjoying these advantages. For this reason, the EU sought a solution to establish justice among prosumers, incorporating the three dimensions of distributive, procedural, and recognitional justice. Hence, a revision was introduced in the CEP, introducing the concept of an energy community as a framework in which justice is established among prosumers [21]. They characterize energy communities as socio-technical systems that evolve various energy resources, flexibility, connectivity of different renewable sources, and bidirectional energy flows. Furthermore, the CEP provides two definitions of energy communities:

1) Renewable Energy Communities (RECs)
2) Citizen Energy Communities (CECs).

Both of these definitions are positioned as specific approaches to collectively organize activities related to a particular energy domain.

1) **Renewable Energy Communities (RECs)**
   - Formed based on valid national regulations and requires voluntary participation.
   - Emphasizes self-sufficiency.
   - Controlled by shareholders or members participating in renewable energy projects.
   - The primary goal is to provide environmental, economic, or social benefits to stakeholders and members, rather than financial profits.
2) **Citizen Energy Communities (CECs)**
   - Established based totally on voluntary and unfastened participation.
   - Emphasizes neighborhood manipulation and self-sufficiency.
   - Controlled with the aid of members or stakeholders, such as individuals, neighborhood authorities such as municipalities, or small groups.
   - The primary goal is to provide environmental, monetary, or social benefits to members or stakeholders instead of economic income.
   - Engages in a broader range of activities, along with renewable strength technology, distribution, delivery, intake, aggregation, strength storage, power performance offerings, and charging services for electric-powered cars and energy-sharing offerings.

In other words, both CECs and renewable energy communities aim to deliver environmental, economic, and social advantages to their respective areas. However, they diverge in their organizational structures and the extent of their endeavors. CECs embrace a wider array of initiatives and draw on voluntary engagement from local governments and small enterprises, whereas renewable energy communities tend to focus on distinct renewable energy ventures with narrower scopes [22]. The question is whether energy communities can provide justice and sustainability for their members, especially those facing energy poverty. To examine this issue, it is necessary to consider the dimensions of justice in CECs and identify them. However, EU legislators need to provide a set of laws for energy communities to ensure that all members of these communities have social and political rights. To this end, the following sections will address energy justice in CECs and how to categorize members to achieve sustainability in these communities, along with a set of laws for P2P energy exchange and energy storage to achieve relative justice in CECs.

7.3 Promoting Sustainability Through Energy Communities

The concept of sustainable development, formalized in the 1987 UN report "Our Common Future," recognized the role of energy, but specifics were unclear

[23]. Sustainable energy development is a complex and multi-dimensional concept that varies based on context and user perspective. The role of energy in sustainable development has been continuously defined since the 1987 Brundtland report, now firmly on the political agenda due to environmental concerns and dwindling fossil fuel sources [23]. Researchers are actively exploring the blending of renewable and conventional sources for a transformative energy future [23].

The extensive use of fossil fuels by various industries in the world has led to severe problems such as environmental destruction, greenhouse gas production, resource depletion, and global warming. Hence, global organizations advocate for a shift toward environmentally friendly alternatives for sustainable energy production and emphasize the central concept of green energy planning [24]. Energy sustainability is intently linked to the usage of renewable power sources. This connection guarantees that future generations might not face unfavorable impacts because of resource depletion, as renewable resources are inexhaustible and without difficulty accessible. Sustainability necessitates equal consideration of three critical aspects, implying that sustainable energy resources must be economically viable, environmentally friendly, and socially equitable. The mentioned aspects are explained as follows:

7.3.1 Economic Aspect

The economic aspect of sustainability assesses whether energy supply is both financially profitable and economically feasible. Energy investments must be cost-effective to incentivize reinvestment in systems that foster sustainability. Furthermore, accessibility guarantees that energy is not merely physically accessible but also financially feasible for society. These dual aspects collaborate to uphold the sustainability of the supply infrastructure. Excessive energy provisioning costs can render it economically unfeasible for society, leading to negligible returns for suppliers. Achieving economic sustainability necessitates a system that furnishes a profitable and accessible energy supply, both presently and in the future. This encompasses evaluating the expenses linked to accessing energy reservoirs, operational costs for users, and the imperative for financial support for the system [25].

7.3.2 Environmental Aspect

The environmental dimension of sustainability is dedicated to preserving the health and productivity of natural ecosystems while mitigating the environmental footprint of energy production and consumption. Its objective is to mitigate the adverse effects of natural resource extraction for energy and human necessities while enhancing positive outcomes. Energy's significance in sustainable development lies in its potential to mitigate or prevent environmental challenges like climate change, deforestation, and pollution, primarily through international, national, and local agreements [26].

7.3.3 Social Aspect

The social dimension of sustainability examines the equity of energy distribution in society, taking into account the social acceptability, availability, and affordability of energy sources and services for all groups of people. This dimension leads to two primary subdivisions: scope and inclusion. Scope means how widely energy services are available in society. Some areas may be left out by integrated systems because they are too far away or too expensive to connect and maintain [27]. Inclusion, on the other hand, emphasizes the need to ensure that all segments of society have equitable access to energy resources and services [28].

7.4 Development of Energy Sustainability with the Energy Community

Energy communities are groups of prosumers who exchange and share energy with each other to meet their electricity demand. By enabling such transactions, energy communities help advance sustainable energy solutions. This model stimulates cooperation, creates a feeling of belonging among members, and supports the use of RES and energy-saving measures. Energy communities are thus essential for achieving a more sustainable energy future [29].

7.5 Challenges of Energy Sustainability Within Energy Communities

Energy communities are key players in the global transition to sustainable energy systems, aiming to promote sustainability, resilience, and inclusivity. As energy communities gain prominence in policy discussions and academic research, it is essential to understand the diverse challenges they face in achieving energy sustainability. Addressing the complex barriers that energy communities encounter in implementing sustainable energy practices requires insights into regulatory, technical, financial, and social aspects [30].

7.5.1 Regulatory Challenges

Energy communities face huge challenges in dealing with the complex regulatory environment. The EU has introduced new categories, such as RECs and CECs, creating more regulatory difficulties. Conforming to these regulations presents a formidable challenge for energy communities, as it necessitates the delicate equilibrium between legal mandates and community-driven energy initiatives.

7.5.2 Technical Challenges

Deploying energy communities involves integrating various technologies to harness renewable energy sources and enhance energy efficiency. However, incorporating these technologies into existing infrastructures presents significant technical challenges. Issues such as interoperability, data management, and cybersecurity are critical concerns, underscoring the urgent need for robust technological solutions to support sustainable energy practices within energy communities.

7.5.3 Financial Challenges

Obtaining enough funding is essential for energy community development, as capital availability affects the viability and growth of network electricity projects. Even though sustainable investments are more popular, getting funding for the energy community is still hard. Barriers such as limited access to capital, high upfront costs, and revenue flow doubts hinder the economic sustainability of energy communities. Overcoming these challenges requires innovative financing methods and supportive policies to encourage investment in community-led energy projects.

7.5.4 Social Challenges

The success of energy communities depends on strong community engagement and acceptance. Building trust, promoting inclusivity, and addressing social equity concerns are crucial for ensuring the long-term sustainability of energy communities. However, challenges such as stakeholder resistance, low public awareness, and socioeconomic disparities pose significant obstacles [31].

7.6 Energy Justice: Definition and Implementation

Energy justice is a key concept that addresses the fair allocation of both the advantages and disadvantages associated with energy production, distribution, and use. It emphasizes the need to ensure that everyone has access to affordable, reliable, and environmentally friendly energy, acknowledging its vital role in promoting human well-being and development. Despite this recognition, historically marginalized communities have borne the brunt of the negative impacts of energy production, such as pollution and environmental damage, while also facing obstacles in accessing clean and cheap energy sources. Stemming from the debates in the environmental justice and climate justice literature, energy justice is often defined as the creation of a global energy system that equitably distributes both the benefits and costs of energy services, along with inclusive and impartial decision-making processes [32]. Energy justice involves addressing the three principles of distributive, procedural, and recognition justice. These principles, frequently mentioned in the literature, stress the importance of ensuring fairness in the distribution of benefits and burdens among society, applying fair decision-making processes in energy issues, and respecting the rights and views of affected individuals and communities.

- Distributive justice focuses on ensuring equity in the allocation of benefits and disadvantages within society, including responsibilities and losses. In the energy context, it aims to provide equitable access to benefits such as reliable electricity and sustainable energy sources while avoiding the disproportionate imposition of energy-related costs and burdens, such as pollution, on specific communities.
- Procedural justice is concerned with fostering transparency and inclusivity in decision-making processes. This involves ensuring that all stakeholders have equal opportunities to participate in energy-related decision-making, with their voices clearly heard and decisions made based on principles of fairness, impartiality, and non-discrimination. By prioritizing these principles, the decision-making process not only yields more equitable outcomes but also enhances trust and legitimacy within society. This aspect includes scrutinizing legal processes and ensuring access to justice.
- Recognition justice underscores the significance of acknowledging and respecting cultural diversity, distinct experiences, and needs across different segments of society. In the energy context, it involves challenging prevailing assumptions and models that may marginalize certain groups. This entails ensuring that the perspectives and experiences of marginalized communities, like indigenous groups or low-income areas, are thoroughly considered in the design and implementation of energy policies and solutions, fostering fairness and inclusivity. Rectifying any injustices caused by the energy sector, such as through improved decommissioning practices, exemplifies this principle. While some view Recognition justice as part of Procedural justice, it operates as a distinct

aspect emphasizing the importance of inclusivity and cultural recognition in decision-making processes [33].

In contrast to the expansive and comprehensive perspective of energy justice, energy community scholars tend to apply this concept to analyze injustices within specific energy plans. A significant portion of this research investigates energy justice principles within a narrower context. For instance, Simcock's study prioritizes procedural justice, focusing on the fairness of processes and decision-making within university energy environments [33]. Some scholars [34], concentrate on all three principles of distributive, procedural, and recognition justice in community energy initiatives, aiming for a more comprehensive application of the energy justice framework.

7.7 The Role of Energy Communities in Advancing Energy Justice

The local residents and communities can use the benefits of energy communities in order to actively participate in some modern energy choices toward the development of an equitable energy system. In fact, energy communities enable prosumers to have a direct stake in energy generation, distribution, and consumption decisions. This empowers them to take control of their energy future, rather than relying solely on large corporations or distant authorities. Moreover, energy communities often operate on democratic principles, allowing members to have a say in how energy projects are developed and managed. This ensures that the needs and interests of the community are taken into account, promoting fairness and equity in decision-making processes [35]. Furthermore, energy communities can provide access to clean RES, such as solar, wind, or hydropower, to communities that may have been traditionally underserved or marginalized by the mainstream energy system. This helps to reduce energy poverty and improve environmental quality in these areas.

Energy projects developed by and for the community can generate various economic, social, and environmental benefits, such as job creation, revenue generation, energy cost savings, and local economic development. These benefits can help to address disparities and inequities within communities, promoting greater social justice [33]. Energy communities contribute to advancing awareness and understanding among their members, particularly regarding RES. This is because increased awareness and community engagement in addressing energy challenges can facilitate the achievement of energy justice. Optimal and more efficient use of energy resources benefits all members of society and can mitigate its adverse effects.

7.8 Examination of Energy Justice Within Energy Community

There is value in taking a broader perspective on energy justice when it comes to community energy initiatives. Using observations in the energy community literature, the analysis of community energy equity is conducted at three distinct levels: within energy community, between energy communities and external institutions, and beyond energy community [33].

7.8.1 Energy Justice Within Energy Community

Focusing on energy justice within the energy community, observed a significant emphasis on internal dynamics in the literature [35, 36]. This emphasis is understandable because the connection between energy justice and community energy is mainly based on expectations at the local level. These expectations include elements of procedural and distributive justice, such as transparent decision-making processes, fair distribution of financial profits, and community participation [37, 38]. Moreover, a fair distribution of risks, costs, and benefits within a community enhances the social acceptance of renewable energies. The benefits of energy justice within community energy initiatives often involve empowering community members through skill development, job creation, strengthening social and community resilience, and democratizing energy supply [39].

7.8.2 Energy Justice Between Energy Communities and External Institutions

Energy community interactions with external institutions and factors bring about challenges and complexities that can lead to injustices in this area. For instance, these injustices may occur concerning the approach of energy communities toward foreign investors; as these investors are inclined to engage in renewable energy investments, failure to make primary decisions in energy communities considering all members creates injustice [10]. Additionally, governments may implement policies that affect energy justice within energy communities [40].

7.8.3 Energy Justice Beyond Energy Community

So far, various forms of justice for energy communities have been discussed, all of which impact individuals within these communities. However, in addition to these, there is a type of justice related to the development of energy communities, involving inequality in opportunities for establishing these communities. This type of justice is associated with inequality in the distribution of renewable energy technology, where the benefits of these technologies are not fairly distributed among all individuals, thereby constraining the development of energy communities [33].

7.9 Challenges to Green and Equitable Energy Society

The transition toward an inexperienced and equitable energy community is vital in addressing urgent global demanding situations, such as climate change, environmental degradation, and social inequality. However, this transition is fraught with multifaceted demanding situations spanning technological, economic, social, and political dimensions. Here, the challenges are classified into five categories, and a framework of challenges for green energy transfer is proposed as follows:

Social Aspect: The difficulties faced in this domain have a direct effect on people. One significant social aspect is the shift toward green energy, presenting challenges like jeopardizing stakeholders in the existing system and the displacement of jobs tied to fossil fuels. Moreover, ensuring secure and reasonably priced energy is crucial during the transition to green energy. Energy poverty is a pressing global issue within the energy sector that requires immediate attention. A just transition in energy is essential for combating energy poverty and is fundamental for incorporating energy justice into the energy community [41].

Economic Aspect: The shift toward renewable energy encounters hurdles that affect investment, competitiveness, and economic advancement. These obstacles comprise investment uncertainties, transient dips in productivity, substantial investment requisites, subsidies favoring fossil fuels, and the expenses linked to climate change mitigation. To tackle these hurdles effectively, energy communities emerge as the optimal solution, fostering the essential incentives for investors. For instance, such communities can establish incentive schemes to encourage investment in clean and sustainable energy sources, while also furnishing infrastructure and avenues for the advancement of novel energy technologies [42].

Environmental Aspect: Encompasses challenges related to climate change, environmental pollution, depletion of natural resources, and land use. A significant challenge to the transition toward green energy is the consumption of resources, such as sand for constructing green and energy-efficient buildings, or minerals like selenium, cadmium, gallium, tellurium, lithium, and germanium used in energy storage and solar PV panels. Additionally, the management of waste in landfills and extensive land use for constructing solar and wind farms pose critical challenges to the green energy transition [43].

Institutional Aspect: In the Institutional column, there are challenges that affect decisions, policies, and strategies or are caused by them. For example, anti-innovation policies can hinder the transition to green energy, and short-term policies may also impede progress toward clean energy. Moreover, restructuring and transition always come with challenges and contradictions that may lead to resistance to change. Consequently, conflicts and differences in the development stages of policies may affect investment returns [44].

Technical Aspect: These challenges encompass obstacles to infrastructures, technology development, and technical standards. Technical aspects in transitioning to green energy are crucial as they impact the global energy budget on climate change, carbon emissions reduction, and stabilizing global average temperature. In other words, sustainable energy systems are concurrently moving toward standards, technology, and infrastructure. For instance, utilizing P2P energy exchange technologies can enhance energy efficiency on the demand side. However, the absence of technical standards can hinder the development and adoption of new low-carbon technologies, necessitating the fulfillment of global needs [43].

7.10 Proposed Framework for CEC Planning and Operation

In the realms of energy systems, two persistent issues are of paramount importance: planning and operations. These two subjects are fundamentally intertwined with energy communities. In the realm of planning, matters such as the installation of storage facilities and the establishment of renewable energy farms garner the most attention. Additionally, for the formation of energy communities, criteria such as the geographic proximity of individuals to one another are crucial [9]. Moreover, from a social perspective, factors such as income level, educational attainment, and behavioral compatibility among individuals must also be meticulously examined to shape a sustainable and successful energy community [45].

In addition, within the realm of managing energy communities, it is common to adopt centralized approaches [46]. In these approaches, an energy community manager (ECM), who is typically a member of the community, assumes the responsibility of overseeing operations. The primary goal of the ECM is to optimize the satisfaction levels of all participants involved or to minimize the overall costs associated with purchasing energy. This centralized management approach is aimed at ensuring fairness and equity for all members within the energy community, thereby promoting a sense of inclusivity and shared benefits among its constituents [47].

As stated in advance, electricity communities harness contemporary technologies like PV, centralized storage structures, and electric vehicles to offer a numerous array of amenities and offerings to their participants. These technologies not only enhance the performance and sustainability of energy production but also empower people inside the network to actively participate in power control. One of the pivotal capabilities of those communities is their ability to share surplus electricity generated locally amongst prosumers through a P2P strength alternate mechanism. This modern approach enables prosumers to exchange excess power immediately with one another, bypassing traditional electricity providers and distribution networks. By facilitating P2P energy trading, energy communities foster collaborative surroundings wherein contributors can optimize their energy intake styles and decrease reliance on outside strength sources. Furthermore, it promotes self-sufficiency and enhances the ability of the electricity network. P2P energy trading, similar to its blessings, also poses fundamental challenges. First, it introduces a heavy computational burden on energy management. With the increase in the number of members in the energy community, each individual may face a large number of peers for energy exchange, leading to increased complexity in decision-making. Second, fairness in energy exchange, which is a fundamental principle in energy communities, is challenged. If this principle is not adhered to during P2P energy exchange, individuals experiencing energy poverty may not equally benefit from the surplus energy production of other members of the energy community. To address these challenges, a comprehensive structure for the operation of energy communities is proposed, which also enhances the sustainability of the energy community and contributes to fairness in P2P energy exchanges. This structure comprises an ECM that initially divides the members of the energy community into smaller groups based on their similarities to P2P energy exchanges. This division leads to the formation of stable energy subcommunities, as prosumers prefer to engage in energy exchanges with others who share similar attitudes, promoting stability within the energy community. Additionally, other criteria exist for the formation of energy subcommunities, such as demographic factors and consumption levels. Forming energy subcommunities based on attitudes to P2P exchange similarities not only increases sustainability but also reduces the challenge of P2P energy exchanges, which previously imposed a heavy computational burden, as prosumers are restricted to exchanging energy only with members within their energy subcommunity.

In the operational phase, the ECM utilizes the energy community to reduce energy purchasing costs. Within this community, all participants possess solar panels, smart appliances with time-shift capabilities, community energy storage (CES), and heating, ventilation, and air conditioning (HVAC) systems, enabling them to engage in P2P power trading. The structural format and organization of the strength network are visually depicted in Figure 7.1, presenting a comprehensive review of its additives and functionalities. Furthermore, the mathematical

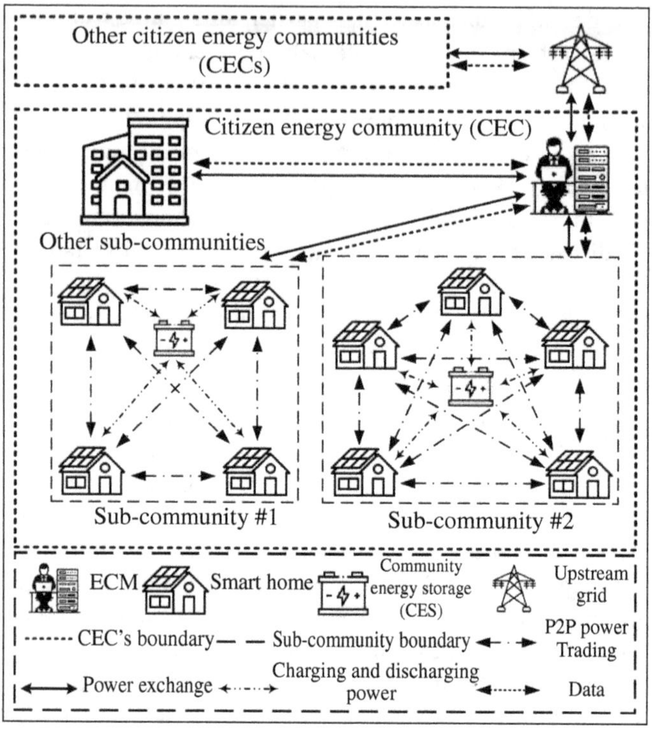

Figure 7.1 The structure of the energy community with subcommunities.

model for operating the energy community, conducted by the ECM, is as follows:

$$\min \left[\sum_{t=1}^{N_t} \left(\pi_t^{Buy,G} P_t^{Buy,G} - \pi_t^{Sell,G} P_t^{Sell,G} \right) \right] \tag{7.1}$$

The objective function in Eq. (7.1) includes two fundamental components. The first component pertains to the price of buying energy from the upstream grid, even as the second part involves the revenue generated from selling energy to the upstream grid. The constraints of the optimal operation model are as follows:

$$P_t^{Buy,G} = \sum_{i=1}^{N_i} P_{i,t}^{Buy,Pr} \tag{7.2}$$

$$P_t^{Sell,G} = \sum_{i=1}^{N_i} P_{i,t}^{Sell,Pr} \tag{7.3}$$

$$P_{\min}^{Buy,G} \leq P_t^{Buy,G} \leq P_{\max}^{Buy,G} \tag{7.4}$$

$$P_{\min}^{Sell,G} \leq P_{t}^{Sell,G} \leq P_{\max}^{Sell,G} \tag{7.5}$$

$$I_{i,t}^{Buy,\Pr} P_{\min}^{Buy,\Pr} \leq P_{i,t}^{Buy,\Pr} \leq I_{i,t}^{Buy,\Pr} P_{\max}^{Buy,\Pr} \tag{7.6}$$

$$I_{i,t}^{Sell,\Pr} P_{\min}^{Sell,\Pr} \leq P_{i,t}^{Sell,\Pr} \leq I_{i,t}^{Sell,\Pr} P_{\max}^{Sell,\Pr} \tag{7.7}$$

$$I_{i,t}^{Buy,\Pr} + I_{i,t}^{Sell,\Pr} \leq 1 \tag{7.8}$$

Constraints Eqs. (7.2) and (7.3) ensure that the amount of electricity bought or sold from/to the grid aligns precisely with the total energy transactions conducted by each prosumer, respectively. Constraints Eqs. (7.4) and (7.5) impose limits on the quantity of energy that can be purchased or offered from/to the grid. Additionally, constraints Eqs. (7.6) and (7.7) restrict the amount of power that each prosumer can buy or sell, deliberating constraint Eq. (7.8), which prevents simultaneous purchasing and selling of power.

$$P_{i,t}^{Buy,\Pr} - P_{i,t}^{Sell,\Pr} + P_{i,t}^{PV} - \sum_{j=1}^{N_j} P_{i,j,t,\phi}^{P2P} - \sum_{\sigma=1}^{N_\sigma} \left(P_{i,\sigma,t}^{ch} - P_{i,\sigma,t}^{dis} \right) = D_{i,t,\phi} \tag{7.9}$$

Constraint Eq. (7.9) illustrates the equilibrium between power usage and generation for individual prosumers. This balance is determined through various factors, such as the power consumed, electricity generated with the aid of rooftop PV, P2P energy trading, and the charging and discharging of CES. The model related to the P2P energy trading is as follows [29]:

$$P_{i,j,t,\phi}^{P2P} = -P_{j,i,t,\phi}^{P2P} \tag{7.10}$$

$$-P_{\max}^{P2P} < P_{i,j,t,\phi}^{P2P} < P_{\max}^{P2P} \tag{7.11}$$

$$\sum_{t=1}^{N_t} P_{i,j,t,\phi}^{P2P} = \sum_{t=1}^{N_t} P_{j,i,t,\phi}^{P2P} \tag{7.12}$$

Equation (7.10) shows the power trade among prosumers ith and jth, even as Eq. (7.11) represents the obstacle in the electricity exchange between these prosumers. Of greater importance is constraint Eq. (7.12), which acts as a rule ensuring fairness during energy trading. Specifically, if the prosumer *ith* transfers power to prosumer *jth*, prosumer *jth* needs to go back an equivalent amount of power to prosumer *ith* in the course of the day. This ensures that neither prosumer is disadvantaged in P2P transactions.

$$D_{i,t,\phi} = P_{i,t}^{baseload} + P_{i,t,\phi}^{shift} + P_{i,t,\phi}^{cooling} + P_{i,t,\phi}^{heat} \tag{7.13}$$

The power consumption attributed to every prosumer, as outlined in Eq. (7.13), encompasses numerous factors: the baseload of the prosumer, the electricity used by the shiftable loads appliance, and the power usage associated with HVAC systems.

$$P^{shift}_{i,t,\phi} = \sum_{g=1}^{N_g} P_{i,g,t,\phi} X_{i,g,t,\phi} \tag{7.14}$$

$$X_{i,g,t,\phi} = \begin{cases} 1 & t^{start} < t < t^{finish} \\ 0 & \text{otherwise} \end{cases} \tag{7.15}$$

$$\sum_{t^s_g < t < t^f_g - d_g} X_{i,g,t,\phi} = 1 \tag{7.16}$$

Constraints Eqs. (7.14 and 7.15) specify the designated time frames during which shiftable household appliances are allowed to operate. Moreover, once a household appliance is activated, it must sustain its operation for a designated period as outlined in Eq. (7.16). The model related to the HVAC system is as follows [48]:

$$\theta_{i,t+1,\phi} = e^{-\frac{\Delta T}{C_i R_i}} \theta_{i,t,\phi} + \left(1 - e^{-\frac{\Delta T}{C_i R_i}}\right)\left(\theta^{amb}_{i,t,\phi} - \theta^{T}_{i,t,\phi}\right) \tag{7.17}$$

$$\theta^{T}_{i,t,\phi} = R_i \eta^{th}_i \left(P^{cooling}_{i,t,\phi} + P^{heat}_{i,t,\phi}\right) \tag{7.18}$$

$$I^{cooling}_{i,t,\phi} P^{\min}_i \le P^{cooling}_{i,t,\phi} \le I^{cooling}_{i,t,\phi} P^{\max}_i \tag{7.19}$$

$$I^{heat}_{i,t,\phi} P^{\min}_i \le P^{heat}_{i,t,\phi} \le I^{heat}_{i,t,\phi} P^{\max}_i \tag{7.20}$$

$$I^{cooling}_{i,t,\phi} + I^{heat}_{i,t,\phi} \le 1 \tag{7.21}$$

$$\theta^{set}_{i,t} - \lambda_i \le \theta_{i,t,\phi} \le \theta^{set}_{i,t} + \lambda_i \tag{7.22}$$

$$\theta_{i,1,\phi} = \theta^{set}_{i,1,\phi} \tag{7.23}$$

Equation (7.17) suggests that the destiny indoor temperature is stimulated through the cutting-edge indoor temperature, ambient temperature, and temperature adjustments made via the HVAC system. The energy usage of the HVAC system is dictated by Eq. (7.18). Eqs. (7.19) and (7.20) limit the electricity consumption of the HVAC, while constraint Eq. (7.21) prevents the HVAC from simultaneously presenting heating and cooling energy. Additionally, Eq. (7.22) specifies that the set temperature ought to fall inside an acceptable variety to ensure the comfort of the prosumers, and Eq. (7.23) assumes that the real temperature is identical to the set temperature on the first time day.

$$SOC_{\sigma,t,\phi} = SOC_{\sigma,t-1,\phi} + P^{ch}_{\sigma,t,\phi} \eta^{storage}_{\sigma} - P^{dis}_{\sigma,t,\phi} / \eta^{storage}_{\sigma} \tag{7.24}$$

$$P^{ch}_{\sigma,t,\phi} = \sum_{i=1}^{N_i} P^{ch}_{i,\sigma,t,\phi} \tag{7.25}$$

$$P^{dis}_{\sigma,t,\phi} = \sum_{i=1}^{N_i} P^{dis}_{i,\sigma,t,\phi} \tag{7.26}$$

$$SOC^{\min}_{\sigma,\phi} \leq SOC_{\sigma,t,\phi} \leq SOC^{max}_{\sigma,\phi} \tag{7.27}$$

$$P^{ch,\min}_{\sigma,\phi} \leq P^{ch}_{\sigma,t,\phi} \leq P^{ch,\max}_{\sigma,\phi} \tag{7.28}$$

$$P^{dis,\min}_{\sigma,\phi} \leq P^{dis}_{\sigma,t,\phi} \leq P^{dis,\max}_{\sigma,\phi} \tag{7.29}$$

$$I^{ch}_{i,\sigma,t,\phi} P^{ch,\min}_{i,\sigma,\phi} \leq P^{ch}_{i,\sigma,t,\phi} \leq I^{ch}_{i,\sigma,t,\phi} P^{ch,\max}_{i,\sigma,\phi} \tag{7.30}$$

$$I^{dis}_{i,\sigma,t,\phi} P^{dis,\min}_{i,\sigma,\phi} \leq P^{dis}_{i,\sigma,t,\phi} \leq I^{dis}_{i,\sigma,t,\phi} P^{dis,\max}_{i,\sigma,\phi} \tag{7.31}$$

$$I^{ch}_{i,\sigma,t,\phi} + I^{dis}_{i,\sigma,t,\phi} \leq 1 \tag{7.32}$$

Equation (7.24) represents the state of charge CES, with the charging and discharging power calculated through Eqs. (7.25) and (7.26). Eqs. (7.27)–(7.32) depict limitations on the state of charge and charging/discharging power. Constraint (Eq. 7.32) prevents each prosumer from charging and discharging simultaneously.

7.11 Numerical Study

To examine and validate the structural performance proposed for an energy community in The Netherlands, this study has been conducted. In the initial stage, which involves the formation of the subcommunity, prosumers' perspectives based on data from [45] have been examined through a survey regarding their attitudes toward P2P exchange. This survey encompasses three criteria, including attitudes toward environmental issues, social perspectives, and technology. Individuals sensitive to climate change and environmental pollution offer their surplus energy to other members of the energy community to reduce energy consumption at higher levels. Those with social inclinations engage in P2P exchanges with other members of the energy community due to human-centric concerns, while individuals interested in technology are involved in utilizing technology for energy trade.

The creation of energy subcommunities, characterized by prosumers with similar perspectives, has been conducted utilizing a machine learning technique known as k-means clustering. The number of energy subcommunities is assumed to be four, which enhances stability in P2P energy trading and reduces computational burdens. Figure 7.2 illustrates the clustering results of energy subcommunities based on attitudinal similarities.

After allocating prosumers to energy subcommunities, the ECM operates this energy community. This operation includes parameters such as baseload, top

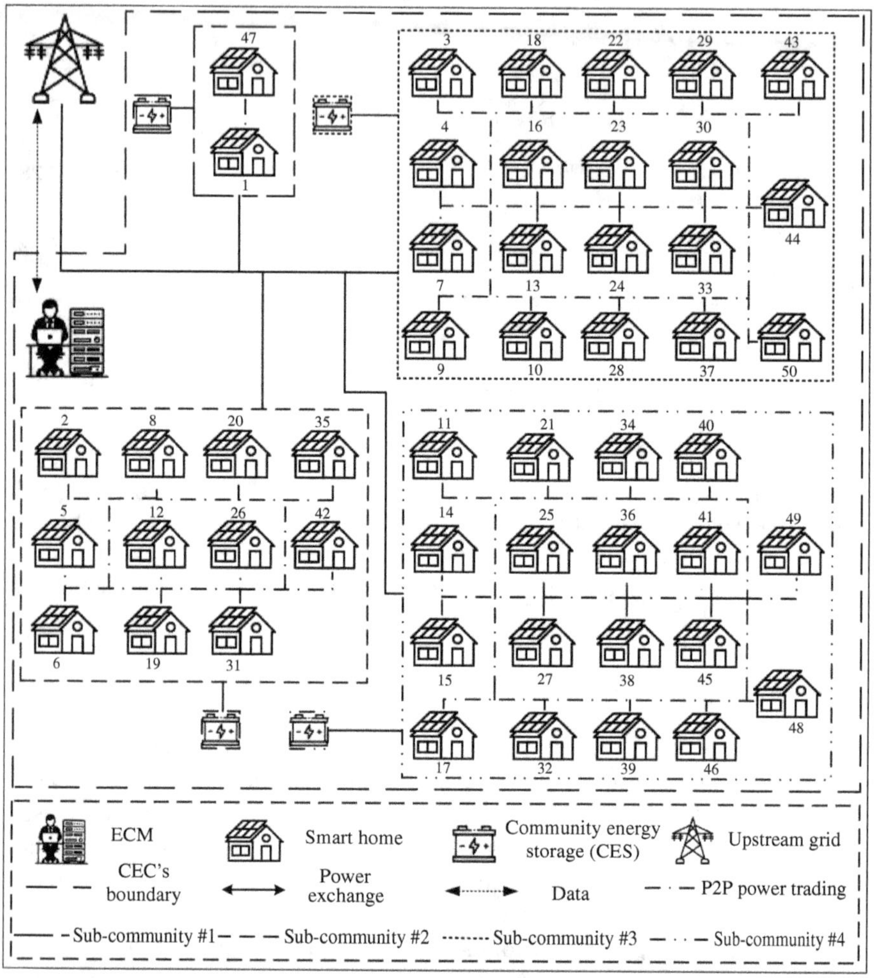

Figure 7.2 Allocation of prosumers to energy subcommunities.

roof PV production capacity, CES, and electricity prices taken from [29], HVAC parameters [48], and also household home appliance parameters taken from [49]. These parameters include: the maximum purchased power from the grid is 375 kW, and for each prosumer, this maximum limit is 5 kW. The maximum P2P trading power is 2.5 kW. The storage efficiency is 0.94, and the associated HVAC parameters are as follows: HVAC Efficiency factor is 0.02, Thermal resistance is 2°C/kW, and Thermal capacitance is 2.5 kWh/°C. The results of optimal operating costs in different scenarios are presented in Table 7.1.

Table 7.1 The operating cost of the energy community in different scenarios.

	Without P2P energy trading		With P2P energy trading	
	Total cost (€)	Savings (%)	Total cost (€)	Savings (%)
Without ESS	113.23	Based-case (NA)	107.70	4.88
With CES	69.23	38.86	66.86	40.95

As observed from Table 7.1, the operating cost is exceedingly high in scenarios where neither energy storage nor P2P exchange exists. However, when the energy community utilizes CES and P2P exchange, it leads to approximately 41% savings in operating costs, with the majority of these savings attributed to storage. In other words, the CES provides greater flexibility in operation to the CEC, resulting in reduced operating costs.

As mentioned, one of the primary objectives of energy communities is to establish justice. For this reason, in the legal modeling section, a law was proposed to enforce justice regarding P2P exchange. Table 7.2 displays the performance of this law regarding P2P exchange for two participants, labeled as 1 and 47, within the energy subcommunity 1.

Table 7.2 Power trading between prosumers 1 and 47.

Prosumer 1												
Hour (h)	1	2	3	4	5	6	7	8	9	10–	11	12
Power (kW)	1.35	−1.51	−2	−1.57	−1.71	2	0	−1.85	2	−0.52	1.3	0.9
Cumulative	1.35	−0.16	−2.16	−3.73	−5.44	−3.44	−3.44	−5.29	−3.29	−3.81	−2.51	−1.61
Hour (h)	13	14	15	16	17	18	19	20	21	22	23	24
Power (kW)	0.53	1.83	−1.3	1.57	−0.54	−0.97	−1.27	−0.48	−0.96	1.7	2	−0.5
Cumulative	−1.08	0.75	−0.55	1.02	0.48	−0.49	−1.76	−2.24	−3.2	−1.5	0.5	0
Prosumer 47												
Hour (h)	1	2	3	4	5	6	7	8	9	10	11	12
Power (kW)	−1.35	1.51	2	1.57	1.71	−2	0	1.85	−2	0.52	−1.3	−0.9
Cumulative	−1.35	0.16	2.16	3.73	5.44	3.44	3.44	5.29	3.29	3.81	2.51	1.61
Hour (h)	13	14	15	16	17	18	19	20	21	22	23	24
Power (kW)	−0.53	−1.83	1.3	−1.57	0.54	0.97	1.27	0.48	0.96	−1.7	−2	0.5
Cumulative	1.08	−0.75	0.55	−1.02	−0.48	0.49	1.76	2.24	3.2	1.5	−0.5	0

Based on Table 7.2, the "cumulative" column indicates the sum of power exchanged within the same hour as its previous hour, resulting in the total purchased and sold power between Prosumer 1 and 47 being equal to zero. This implies that if Prosumer 1 has sold power to Prosumer 47 during certain hours, Prosumer 47 has made efforts to return this power during the remaining hours. This is achieved through the implementation of the justice law in P2P power trading. Conversely, the results validate that prosumers actively contribute and hold significant importance in engaging in P2P energy trading during peak demand periods, coinciding with the peak electricity prices and surplus generation capacity of solar panels. Additionally, prosumers have a strong inclination toward P2P trading, especially when communal storage systems are in the process of charging and discharging. This result leads to efficient P2P energy trading, allowing prosumers to meet their energy needs through energy exchange. The power bought and sold to/from the grid is depicted in Figure 7.3.

The results indicate that the ECM, from 1 to 7 in the morning when electricity prices are low, attempts to purchase energy from the power grid and charge the storage. Furthermore, during the hours of 8 to 17, when solar resources are available, the purchased power from the grid significantly decreases. At times, such as during hours 10 and 11, surplus power generation occurs, leading to the sale of excess power back to the grid. Toward the end of the day, from 18 to 24, when both peak demand and electricity prices are at their highest, discharging the storage presents an opportunity to reduce the need for grid power purchases. It can be inferred that storage plays a crucial role within the energy community, alongside P2P energy trading.

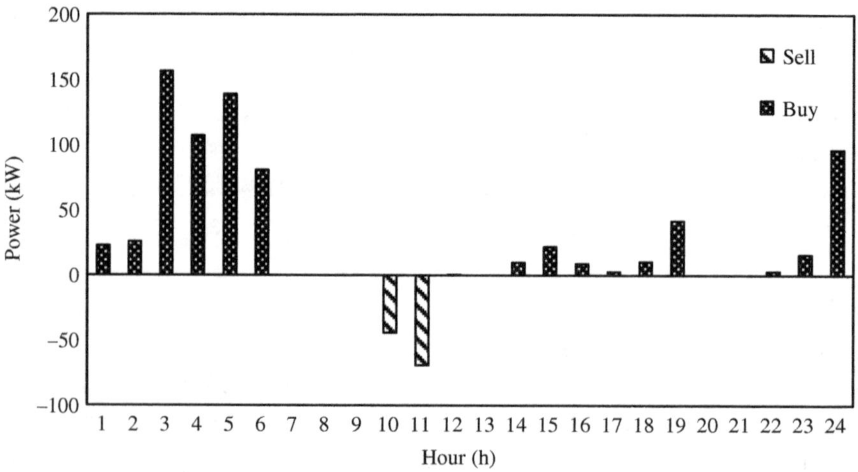

Figure 7.3 The energy exchange with the upstream network.

7.12 Conclusion

In this chapter, the concept of the energy community is introduced based on the CEP of the European Union. The study reveals that the energy community has various meanings and definitions, particularly when integrating social science aspects with technical, environmental, and economic factors. Moreover, energy communities have a significant impact on promoting sustainability and equality in society. One of the key contributions of energy communities to sustainable network development is P2P energy exchange. This study presents an optimal model for energy communities, where justice in P2P exchanges is evaluated to ensure performance accuracy. The findings demonstrate that the proposed approach can enhance community payments by enabling energy sharing among participants without compromising energy justice. Additionally, the results indicate that the CES is an effective method for achieving lower energy prices while accommodating a higher capacity of DERs. Future research will explore how prosumer behaviors influence the performance of energy communities. Moreover, understanding prosumer responses to fairness issues is a critical research topic in the energy system for future studies. The principal achievements of this study include:

- A comprehensive study was conducted on the role of the energy community in sustainability and energy justice.
- We succeeded in demonstrating energy justice in the energy community for P2P trading.
- The impact of storage on the sustainability and operation cost of the energy community was also demonstrated.

References

1 Dorahaki, S., Sarkhosh, A., Rashidinejad, M. et al. (2023). Fairness in optimal operation of transactive smart networked modern multi-carrier energy systems: a two-stage optimization approach. *Sustainable Energy Technologies and Assessments* 56: 103035. 10.1016/j.seta.2023.103035.

2 Dorahaki, S., Rashidinejad, M., MollahassaniPour, M. et al. (2023). A sharing economy model for a sustainable community energy storage considering end-user comfort. *Sustainable Cities and Society* 97: 104786. 10.1016/J.SCS.2023.104786.

3 Dorahaki, S., Muyeen, S.M., and Amjady, N. (2024). A Stackelberg game theory model for integrated community energy storage systems. *IEEE Access* 12: 152908–152920. 10.1109/ACCESS.2024.3481155.

4 Selvakkumaran, S. and Ahlgren, E.O. (2020). Impacts of social innovation on local energy transitions: diffusion of solar PV and alternative fuel vehicles in Sweden. *Global Transitions* 2: 98–115. 10.1016/J.GLT.2020.06.004.

5 Lennon, B., Dunphy, N.P., and Sanvicente, E. (2019). Community acceptability and the energy transition: a citizens' perspective. *Energy, Sustainability and Society* 9 (1): 1–18. 10.1186/S13705-019-0218-Z/FIGURES/8.

6 Hanke, F., Guyet, R., and Feenstra, M. (2021). Do renewable energy communities deliver energy justice? Exploring insights from 71 European cases. *Energy Research and Social Science* 80: 102244. 10.1016/J.ERSS.2021.102244.

7 Bauwens, T., Schraven, D., Drewing, E. et al. (2022). Conceptualizing community in energy systems: a systematic review of 183 definitions. *Renewable and Sustainable Energy Reviews* 156: 111999. 10.1016/j.rser.2021.111999.

8 Dall-Orsoletta, A., Cunha, J., Araújo, M., and Ferreira, P. (2022). A systematic review of social innovation and community energy transitions. *Energy Research and Social Science* 88: 102625. 10.1016/j.erss.2022.102625.

9 Lode, M.L., te Boveldt, G., Coosemans, T., and Ramirez Camargo, L. (2022). A transition perspective on Energy Communities: a systematic literature review and research agenda. *Renewable and Sustainable Energy Reviews* 163: 112479. 10.1016/j.rser.2022.112479.

10 Grashof, K. (2019). Are auctions likely to deter community wind projects? And would this be problematic? *Energy Policy* 125: 20–32. 10.1016/J.ENPOL.2018.10.010.

11 Seyfang, G., Park, J.J., and Smith, A. (2013). A thousand flowers blooming? An examination of community energy in the UK. *Energy Policy* 61: 977–989. 10.1016/J.ENPOL.2013.06.030.

12 Leiren, M.D., Aakre, S., Linnerud, K. et al. (2020). Community acceptance of wind energy developments: experience from wind energy scarce regions in Europe. *Sustainability* 12 (5): 10.3390/su12051754.

13 Bauwens, T. (2016). Explaining the diversity of motivations behind community renewable energy. *Energy Policy* 93: 278–290. 10.1016/j.enpol.2016.03.017.

14 Wüste, A. and Schmuck, P. (2012). Bioenergy villages and regions in Germany: an interview study with initiators of communal bioenergy projects on the success factors for restructuring the energy supply of the community. *Sustainability* 4 (2): 244–256. 10.3390/su4020244.

15 Sperling, K. (2017). How does a pioneer community energy project succeed in practice? The case of the Samsø Renewable Energy Island. *Renewable and Sustainable Energy Reviews* 71: 884–897. 10.1016/j.rser.2016.12.116.

16 Mundaca, L., Busch, H., and Schwer, S. (2018). 'Successful' low-carbon energy transitions at the community level? An energy justice perspective. *Applied Energy* 218: 292–303. 10.1016/j.apenergy.2018.02.146.

17 Tiberio, L., De Gregorio, E., Biresselioglu, M.E. et al. (2020). Psychological processes and institutional actors in the sustainable energy transition: a case-study analysis of a local community in italy. *Frontiers in Psychology* 11: 1–15. 10.3389/fpsyg.2020.00980.

18 Haf, S. and Parkhill, K. (2017). The Muillean Gaoithe and the Melin Wynt: cultural sustainability and community owned wind energy schemes in Gaelic and Welsh speaking communities in the United Kingdom. *Energy Research and Social Science* 29: 103–112. 10.1016/j.erss.2017.05.017.

19 Mjahed Hammami, S., Chtourou, S., and Al Moosa, H. (2018). A holistic approach to understanding the acceptance of a community-based renewable energy project: a pathway to sustainability for Tunisia's rural region. *Business Strategy and the Environment* 27 (8): 1535–1545. 10.1002/bse.2211.

20 Vallecha, H., Bhattacharjee, D., Osiri, J.K., and Bhola, P. (2021). Evaluation of barriers and enablers through integrative multicriteria decision mapping: developing sustainable community energy in Indian context. *Renewable and Sustainable Energy Reviews* 138: 110565. 10.1016/j.rser.2020.110565.

21 European Parliament and the Council of the European Union 2019. Directive (Eu) 2019/ 944 of the European Parliament and of the Council - of 5 June 2019 - on common rules for the internal market for electricity and amending Directive 2012/ 27/ EU.

22 Lowitzsch, J., Hoicka, C.E., and van Tulder, F.J. (2020). Renewable energy communities under the 2019 European Clean Energy Package – Governance model for the energy clusters of the future? *Renewable and Sustainable Energy Reviews* 122: 109489. 10.1016/j.rser.2019.109489.

23 Gunnarsdottir, I., Davidsdottir, B., Worrell, E., and Sigurgeirsdottir, S. (2021). Sustainable energy development: history of the concept and emerging themes. *Renewable and Sustainable Energy Reviews* 141: 110770. 10.1016/j.rser.2021.110770.

24 Bhowmik, C., Bhowmik, S., and Ray, A. (2022). Green energy sources selection for sustainable planning: a case study. *IEEE Transactions on Engineering Management* 69 (4): 1322–1334. 10.1109/TEM.2020.2983095.

25 Filer, J.E., Delorit, J.D., Hoisington, A.J., and Schuldt, S.J. (2020). Optimizing the environmental and economic sustainability of remote community infrastructure. *Sustainability* 12 (6): 2208. 10.3390/su12062208.

26 Ninno Muniz, R., Stefenon, S.F., Buratto, W.G. et al. (2020). Tools for measuring energy sustainability: a comparative review. *Energies* 13 (9): 2366. 10.3390/en13092366.

27 Vafadarnikjoo, A., Ahmadi, H.B., Hazen, B.T., and Liou, J.J.H. (2020). Understanding interdependencies among social sustainability evaluation criteria in an emerging economy. *Sustainability* 12 (5): 1934. 10.3390/su12051934.

28 Otamendi-Irizar, I., Grijalba, O., Arias, A. et al. (2022). How can local energy communities promote sustainable development in European cities? *Energy Research and Social Science* 84: 102363. 10.1016/j.erss.2021.102363.

29 Ghasemnejad, H., Rashidinejad, M., Abdollahi, A., and Dorahaki, S. (2024). Energy management in citizen energy communities: a flexibility-constrained robust optimization approach considering prosumers comfort. *Applied Energy* 356: 122456. 10.1016/J.APENERGY.2023.122456.

30 Dorahaki, S., MollahassaniPour, M., Rashidinejad, M. et al. (2025). A flexibility-oriented model for a sustainable local multi-carrier energy community: a hybrid multi-objective probabilistic-IGDT optimization approach. *Applied Energy* 377: 124678. 10.1016/j.apenergy.2024.124678.

31 Ghasemnejad, H., Dorahaki, S., Rashidinejad, M., and Muyeen, S.M. (2025). An IGDT optimization model for a prosumer-oriented citizen energy community considering hydrogen parking lots, energy sharing and thermal comfort. *International Journal of Hydrogen Energy* 101: 972–982. 10.1016/j.ijhydene.2024.12.500.

32 Heffron, R.J. (2022). Applying energy justice into the energy transition. *Renewable and Sustainable Energy Reviews* 156: 111936. 10.1016/j.rser.2021.111936.

33 van Bommel, N. and Höffken, J.I. (2021). Energy justice within, between and beyond European community energy initiatives: a review. *Energy Research and Social Science* 79: 10.1016/j.erss.2021.102157.

34 Emelianoff, C. and Wernert, C. (2019). Local energy, a political resource: dependencies and insubordination of an urban 'Stadtwerk' in France (Metz, Lorraine). *Local Environment* 24 (11): 1035–1052. 10.1080/13549839.2018.1506754.

35 van Veelen, B. and Eadson, W. (2020). Assembling community energy democracies. *Voluntary Sector Review* 11 (2): 231–249. 10.1332/204080519X15740562779512.

36 Capaccioli, A., Poderi, G., Bettega, M., and D'Andrea, V. (2017). Exploring participatory energy budgeting as a policy instrument to foster energy justice. *Energy Policy* 107: 621–630. 10.1016/j.enpol.2017.03.055.

37 Simcock, N. (2016). Procedural justice and the implementation of community wind energy projects: a case study from South Yorkshire, UK. *Land Use Policy* 59: 467–477. 10.1016/j.landusepol.2016.08.034.

38 Forman, A. (2017). Energy justice at the end of the wire: enacting community energy and equity in wales. *Energy Policy* 107: 649–657. 10.1016/j.enpol.2017.05.006.

39 Roby, H. and Dibb, S. (2019). Future pathways to mainstreaming community energy. *Energy Policy* 135: 111020. 10.1016/j.enpol.2019.111020.

40 Genus, A. and Iskandarova, M. (2020). Transforming the energy system? Technology and organisational legitimacy and the institutionalisation of community renewable energy. *Renewable and Sustainable Energy Reviews* 125: 109795. 10.1016/j.rser.2020.109795.

41 Stock, R. and Birkenholtz, T. (2021). The sun and the scythe: energy dispossessions and the agrarian question of labor in solar parks. *Journal of Peasant Studies* 48 (5): 984–1007. 10.1080/03066150.2019.1683002.

42 Gelo, T., Šimurina, N., and Šimurina, J. (2021). The economic impact of investment in renewables in croatia by 2030. *Energies* 14 (24): 10.3390/en14248215.

43 Saraji, M.K. and Streimikiene, D. (2023). Challenges to the low carbon energy transition: a systematic literature review and research agenda. *Energy Strategy Reviews* 49: 101163. 10.1016/j.esr.2023.101163.

44 Andrews-Speed, P. (2016). Applying institutional theory to the low-carbon energy transition. *Energy Research and Social Science* 13: 216–225. 10.1016/j.erss.2015.12.011.

45 Georgarakis, E., Bauwens, T., Pronk, A.-M., and AlSkaif, T. (2021). Keep it green, simple and socially fair: a choice experiment on prosumers' preferences for peer-to-peer electricity trading in the Netherlands. *Energy Policy* 159: 112615. 10.1016/j.enpol.2021.112615.

46 Dorahaki, S., Rashidinejad, M., Fatemi Ardestani, S.F. et al. (2023). An integrated model for citizen energy communities and renewable energy communities based on clean energy package: a two-stage risk-based approach. *Energy* 277: 127727. 10.1016/j.energy.2023.127727.

47 Diestelmeier, L. (2021). Citizen energy communities as a vehicle for a just energy transition in the EU – Challenges for the transposition. *Oil, Gas Energy Law Intelligence* 1: 2021.

48 Chang, H.C., Ghaddar, B., and Nathwani, J. (2022). Shared community energy storage allocation and optimization. *Applied Energy* 318: 10.1016/j.apenergy.2022.119160.

49 Bahramara, S. (2021). Robust optimization of the flexibility-constrained energy management problem for a smart home with rooftop photovoltaic and an energy storage. *Journal of Energy Storage* 36: 102358. 10.1016/j.est.2021.102358.

8

Energy Storage Systems for Energy Justice and Equity

Emre Çakır, Furkan Hanilçi, and Yiğit Çağatay Kuyu

R&D Department, Karsan Automotive, Bursa, Turkey

8.1 Introduction

Energy, serving as a fundamental pillar of contemporary civilization, critically shapes our lifestyle, industry, and technological advancements, given its indispensable role in facilitating a plethora of daily activities—from providing essential light and heat to operating vehicles and supporting intricate production processes—thereby highlighting its profound significance in not only sustaining but also advancing the modern living standards that society has come to rely upon. In today's rapidly evolving global context, the persistent escalation in energy demand has led to a marked increase in energy consumption over time, a trend propelled by factors such as exponential population growth, expansion of industrial activities, and cutting-edge technological advancements, which, in tandem, have significantly heightened the overall need for energy and profoundly impacted consumption patterns across the globe. This persistent surge in demand not only accentuates the imperative need for the energy sector to adopt more efficient and sustainable management practices but also casts a spotlight on the severe complications that arise from the finite nature of traditional energy resources coupled with persistent hurdles in achieving energy efficiency. Given the looming specter of resource depletion and the consequential environmental repercussions, the prioritization of sustainable energy production and the enhancement of energy efficiency have ascended to the forefront of strategic planning within the energy sector, establishing the effective management of energy consumption and the improvement of energy efficiency as pivotal elements crucial for bolstering societal and environmental sustainability in an increasingly resource-constrained world. [1]. Energy justice and equality are critical concepts that champion a fair and equitable approach in accessing, distributing, and

Communities for Clean Energy Justice and Equity in Grid Modernization, First Edition.
Edited by Mohammadreza Daneshvar, Behnam Mohammadi-Ivatloo, and Amjad Anvari-Moghaddam.

utilizing energy resources within the energy sector, with energy justice primarily focusing on ensuring equal and fair access to these resources for various societal segments, while energy equality extends this vision by aiming to provide access in a sustainable and inclusive manner. These pivotal concepts form the bedrock of decision-making processes and the effective utilization of resources when devising energy policies and practices, highlighting their integral role in shaping equitable energy frameworks. The introduction and implementation of energy storage systems (ESSs) play a crucial role in this context, as they have the potential to significantly enhance the efficiency and effectiveness of energy consumption by allowing for the storage of energy and its release on demand, thereby necessitating a rigorous adherence to the principles of justice and equality in their development and application. In scenarios where energy storage projects may disproportionately favor specific communities or are unevenly distributed across different regions, the risk of encroaching upon equal access to energy resources becomes pronounced, thereby potentially undermining the core principles of energy justice. This underscores the necessity of a comprehensive assessment of the societal impacts of ESSs, balancing the economic, environmental, and social outcomes that may arise from the use of these technologies. It is, therefore, essential to design and implement energy policies and projects with a keen consideration for adhering to the principles of energy justice and equality, ensuring that the benefits of the energy transition are equitably shared across a broad spectrum of society [2].

The subsequent sections are organized as follows: Section 8.2 offers a comprehensive literature review on ESSs. Section 8.3 elucidates the significance of ESSs in the context of energy justice, outlining various approaches. In Section 8.4, we delve into the utilization of data analyses and big data in relation to the aforementioned concept. Finally, Section 8.5 serves as the conclusion section, summarizing key findings and insights gleaned from the preceding discussions.

This study is meticulously designed to serve as a seminal cornerstone document within the realm of ESSs, providing a comprehensive synthesis of existing knowledge coupled with clear, actionable insights aimed at guiding future endeavors in this rapidly evolving field. Our primary objectives are multifaceted: to conduct an exhaustive literature review that captures the breadth and depth of current research, analyze the critical role that energy storage plays in achieving energy justice, and mention advanced data analysis techniques that are commonly used to significantly enhance the efficiency and accessibility of these systems, thereby facilitating a more equitable distribution of energy resources. By innovatively integrating interdisciplinary perspectives that blend the nuanced principles of energy policy, sustainability, and data science, this work endeavors to offer strategic insights into the mechanisms of equitable energy distribution and establish a robust, well-rounded framework for future research and policy formulation. As this document illuminates both the prevailing challenges and the potential future possibilities in energy storage, it seeks to serve as a guiding beacon for policymakers, industry stakeholders, and researchers alike, aiming to set a benchmark for the integration of technologically advanced energy storage solutions that are firmly aligned with the ideals of fairness and sustainability. Intended to chart the course for more equitable and sustainable energy practices, this study positions itself as a pivotal reference that ensures ESSs contribute effectively to global energy security, thereby shaping a path forward that champions both innovation and ethical responsibility.

8.2 Background

In 2014, a thorough investigation was undertaken to explore how energy storage could contribute to fostering fairness within the energy sector. The research underscored the critical importance of energy justice, a concept encapsulating principles pertaining to distribution, recognition, procedural fairness, and restoration. These principles underscore inequalities in energy access, affordability, sustainability, and accountability. Emerging developments in energy storage technologies, notably artificial photosynthesis, are being investigated as potential avenues for mitigating these disparities, with the overarching goal of cultivating a more inclusive and equitable energy framework [3].

In the same year, an essential study was conducted as part of a broader research endeavor to create a methodology and a unified tool for assessing the economic, social, and environmental impacts of different energy and fuel supply chains. This research aimed to scrutinize the adequacy of the current modeling techniques used in energy systems planning, especially in evaluating diverse energy supply options and dimensions. It covered a wide range of economic, environmental, and optimization criteria, such as minimizing costs, reducing environmental harm, and maximizing profits. The study also accounted for various constraints that shape the structure, operation, and boundaries of energy systems. Additionally, it explored how this approach could be applied to remote areas with similar energy and water system characteristics, including multiple supply sources and users, while considering various technical, environmental, social, and economic factors. However, it's worth noting that the analysis overlooked the important aspect of energy justice, which deserves attention in such assessments [4]. In 2015, a comprehensive examination and research were undertaken to explore the significance of energy justice and energy storage across four key areas. First, the focus was on identifying core values influencing the decisions made by both energy users and suppliers. Second, the study delved into crucial topics such as energy transitions and the social acceptance of low-carbon technologies, with a particular emphasis on their implications for Denmark's energy landscape. Third, it aimed to offer valuable insights for European policymakers by adopting a holistic approach involving various stakeholders, including business leaders and households, to assess the effectiveness of energy policies. Finally, the research shed light on the key challenges and opportunities facing energy policies in Denmark, underscoring the country's potential to serve as a model for others in transitioning toward low-carbon fuels. While the specific role of energy storage in promoting energy justice was not directly explored, the overall findings of the study contribute significantly to understanding the interconnectedness of energy justice and storage within the broader energy landscape [5]. Again, in 2015, this research delves into the concept of renewable energy and energy efficiency (REEE)-powering, which advocates for a holistic overhaul of Namibia's energy sector. REEE-powering entails harnessing renewable energy sources, implementing energy storage solutions, and incorporating energy-efficient technologies with the potential to foster local development. Rooted in the ideals of energy fairness, the aim is to mitigate energy poverty and guarantee widespread access to modern energy amenities. To achieve this, the study suggests establishing a framework supported by policy and regulatory initiatives to bolster energy accessibility in Namibia. By underscoring the nexus between energy justice and local socioeconomic advancement, the research underscores that energy access should not only merely fulfill basic needs but also enable productive utilization to drive societal progress [6]. The literature assesses how energy storage components function in on-grid distributed energy generation systems utilizing renewable sources. With this aim, the impact of fluctuating electricity prices on energy storage elements is explored for a hybrid renewable energy system (HRES) using an optimized fuzzy logic controller (FLC). FLC membership functions are fine-tuned based on weekly and daily forecasts to minimize the operational expenses of HRES. These predictions encompass various environmental factors like electricity price, load demand, wind speed, solar radiation, and ambient temperature. The FLC incorporates three inputs: electricity price, net power flow, and battery stack charge status. It is validated that considering electricity prices significantly influences energy storage element performance in on-grid HRES operation. Weekly and daily optimized FLCs lead to reduced operation hours for fuel cells and electrolyzers and decreased battery stack state of charge (SOC) fluctuations [7].

In 2016, this analysis highlights three instances where US energy policies clash with the principles of energy justice and disregard its tenets. First, a policy mandating only federal agencies to acquire a specific portion of renewable energy falls short of distributive energy justice principles as it fails to establish a comprehensive nationwide renewable energy strategy, thus neglecting to equitably distribute the risks of grid failures among the populace. Second, the procedural fairness principle of energy justice is undermined by the unequal representation of economic groups in policy formulation, resulting in stunted progress toward renewable energy targets and uneven distribution of associated benefits. Third, the nation's persistent reliance on fossil fuel extraction and promotion contradicts the preservative aspect of energy justice, inflicting harm on both people and the environment through physical damage. Additionally, the lack of a comprehensive federal policy on renewable energy hampers the creation of cohesive energy strategies, impeding endeavors to broaden the nation's renewable. Through the framework of energy justice, analyzing energy policies allows us to evaluate existing approaches and improve future policymaking. Championing energy justice aims to guarantee fair access, participation, and distribution of appropriate, efficient, and eco-friendly energy technologies across all societal sectors to meet their essential requirements [8]. In a 2017 study, the notion of energy democracy surfaced as a notable issue, emphasizing the exploration of alternative methods for energy provision and governance. Energy democracy entails advocating for more participatory structures in energy provision and management. The study developed a typology of energy democracy practices and policies by compiling examples from both the global North and South. This classification includes various strategies, spanning from decentralized systems that prioritize local independence to urban initiatives advocating for public or cooperative ownership of energy utilities. Additionally, it encompasses approaches promoting national energy sovereignty as alternatives to self-serving development agendas. By consolidating these varied examples, the study argues that energy democracy transcends localized and Eurocentric viewpoints, offering guidance to movements striving to cultivate fairer and more sustainable energy systems worldwide [9]. Again, in 2017, this study focused on research on energy justice and the importance of energy storage. The research highlights that energy access is a vital element for the well-being and sustainable development of modern societies. In particular, it has been stated that energy use is directly associated with economic, social, and environmental development. It has been emphasized that we are faced with challenges such as increasing global energy demand, which means that investments in infrastructure must triple. This represents significant growth and demand growth, especially for Europe's current challenging energy market. The depletion of fossil fuel reserves and the growing need for clean, accessible, and secure energy are recognized as significant priorities for all nations. In the research, quotations were made from various sources on the definition and importance of energy supply security. In particular, it was emphasized that energy supply security can be achieved by reducing dependence on energy sources and diversifying energy sources. Finally, it has been stated that investments in renewable energy sources increase energy supply security, and it is important to encourage innovation in the energy sector [10]. In a study conducted in 2017, four models were proposed to facilitate the transition from conventionally structured energy portfolios to renewable energy sources while concurrently mitigating CO_2 emissions within the framework of energy justice and sustainable development. These models underwent evaluation considering a range of technical, economic, environmental, and social parameters, encompassing aspects such as energy diversity, economic ramifications, environmental considerations, and societal impacts. Notably, the study underscored the significance of energy storage as a strategic approach to mitigate the fluctuation challenges associated with renewable energy sources. Furthermore, it emphasized the importance of social dimensions such as energy justice and the fulfillment of societal needs within this context. The focal point of this investigation was the sustainable evolution of traditional energy systems in developing nations, with specific attention given to the role of energy storage and other balancing measures in facilitating this transformation [11].

In 2018, the author underscores the intricate connection between sustainable energy transitions and social justice, stressing that these transitions encompass not only socio-technical dimensions but also carry significant implications for social equity. They highlight a prevailing tendency in the literature to treat these aspects as isolated phenomena and advocate for a holistic approach that integrates them. The author contends that such an approach, which incorporates considerations of socio-technical advancement and energy justice, must acknowledge the intertwined nature of institutional, material, and relational changes while also accounting for various spatial dimensions and normative outcomes. Furthermore, they illustrate this approach using three brief case studies focusing on the multi-scale utilization of solar energy in Portugal, demonstrating how notions of justice are deeply embedded within the strategies and policies of sustainable energy transitions. Additionally, the text explores the placement of ESSs within the broader framework of energy justice, advocating for an integrated analysis that considers both socio-technical elements and the broader societal implications of sustainable energy transitions [12]. In the same year study, there was a thorough examination of how the shift to renewable energy in energy production sectors could be realized, with a particular emphasis on the concept of energy justice. The research centered on effectively managing the transition to renewable energy generation in regions traditionally reliant on conventional energy sources, such as Queensland while ensuring alignment with principles of fairness in energy distribution. A novel conceptual framework dubbed *Just Transition Management* was introduced, amalgamating the notions of "transition management" and "just transition," to comprehensively comprehend and steer the transition process in these areas. This study delved into the political, economic, and social facets of transitioning to renewable energy, exploring strategies for achieving a transition that upholds the tenets of energy justice [13].

In a 2019 study, the significance of energy storage, a pivotal component of energy justice, was examined. According to energy justice theory, energy storage plays a crucial role in ensuring an equitable and just distribution of energy resources. Energy storage encompasses vital aspects such as cost distribution among community members, ensuring accessibility, and facilitating fair representation in decision-making processes. With energy storage capacity, energy prosumers are empowered to manage energy demands efficiently while simultaneously generating energy. This fosters a more equitable distribution of energy resources, aligning with the fundamental tenets of energy justice theory. However, achieving these objectives necessitates appropriate legal regulations and support mechanisms. This study assesses the potential of energy prosumers in advancing energy justice goals, discussing measures such as aiding low-income consumers, fostering the diversification of energy sources, and enhancing the short-term electricity market [14].

In another study conducted in 2019, there is a notable emphasis on the necessity of considering potential social impacts when assessing the environmental and economic sustainability of transitioning energy systems to low-carbon sources. The integration of a social life cycle assessment (LCA) framework emerges as crucial for comprehensively addressing equity concerns among various stakeholder groups across all stages of a low-carbon energy system's life cycle. This framework categorizes and discusses indicators pertinent to electric energy systems across four dimensions: employees, electricity consumers, local communities, and broader society. By employing this framework, disparities in fairness and equity can be pinpointed at various stages of an energy system's life cycle, not limited solely to production. The study delves into the ways in which energy justice can be fostered throughout the life cycle of electrical energy systems, thereby aiding in the design and planning processes to enhance the social sustainability of such systems [15].

A study from 2019 investigates the benefits of ESSs within microgrid (MG) operations, emphasizing advantages such as rectifying the intermittent production of renewable sources, enhancing power quality, and facilitating supplementary services like frequency regulation. Hybrid ESSs (HESSs) have emerged as a solution by amalgamating various energy storage technologies to achieve desired performance through the combination of their favorable attributes. Due to limitations such as restricted lifespan, cost, energy, and power density, as well as dynamic responsiveness, a single energy storage technology alone cannot fulfill operational requirements. Consequently, the literature proposes various configurations of HESS considering factors such as storage type, interface, control method, and provided services. The study offers a comprehensive review of the current status of HESS systems in MG applications, presenting an overview of the development within the HESS industry. Critical aspects of HESS deployment in MGs encompass methods for capacity sizing, power converter topologies for HESS interfaces, architecture, control strategies, and energy management. Additionally, an economic analysis and design methodology are provided to address HESS considerations from the perspectives of investors and distribution system engineers. Given the literature assessment and identified gaps, the future trajectory of HESS in MGs emerges as a significant area of interest [16].

In another study, a notable study delved into the pivotal concept of energy justice, specifically examining the pivotal role energy storage plays within this framework. Researchers discovered that conceptualizing energy as both a necessity and an inherent entitlement is crucial for grasping public perceptions and interpretations of energy matters. Drawing upon existing energy justice literature and prior studies on public attitudes toward energy, the researchers contend that the public universally regards energy as an essential requirement and fundamental entitlement, necessitating a distinct framework for analysis. Furthermore, they explore how energy storage technologies contribute to the realization of energy justice and propose strategies for aligning energy policies accordingly. This analysis serves as a valuable instrument for comprehending the dynamics of energy policy communication and its interface with public engagement [17].

In 2019, this research delved into the pivotal role of energy storage within the framework of energy justice. Energy storage serves to enhance energy equity by facilitating the efficient management of distributed power generation resources, particularly within systems such as peer-to-peer (P2P) energy trading. Nevertheless, the implementation of these systems may encounter significant challenges and inequalities. For instance, P2P energy trading systems necessitate access to distributed power generation resources like solar energy, which may not be readily available to economically disadvantaged communities. Thus, it is imperative for local governments to assume a pivotal role in establishing appropriate norms for the development of a fair energy trading system and in providing support for the enforcement of these norms. Within this discourse, a comprehensive assessment of research regarding the significance of energy storage in the realm of justice, and the formulation of norms within this domain, is presented [18].

In 2020, a study was conducted to examine the application of the concept of energy justice within the realm of P2P energy trading systems. The study's premise stemmed from the recognition that P2P energy trading, particularly with its predetermined technological requirements, may engender perceptions of inequity within certain communities. Consequently, the study sought to address this potential injustice by exploring avenues for integrating principles of energy justice into P2P energy trading systems and emphasizing the necessity of establishing a framework of equitable standards within this context. By analyzing the socio-technical aspects of P2P energy trading, the research endeavored to articulate the significance of energy justice and translate it into concrete norms.

Through this process, correlations were drawn between the socio-technical features of existing technologies and the principles of energy justice, resulting in the formulation of a set of norms aimed at upholding principles of fairness and equity within the P2P energy trading paradigm [19].

In the same year, this discourse delves into the intersection of "energy storage systems" with the principle of energy justice and elucidates how such systems can advance the cause of energy justice. The discussion revolves around the alignment of energy storage technologies with the tenets of energy justice, emphasizing their potential to foster a more sustainable, equitable, and efficient management of energy infrastructures. Notably, it underscores the capacity of ESSs to rectify the disparity between energy generation and consumption while facilitating a more equitable distribution of energy resources. Furthermore, the discourse explores the environmental and social ramifications of these technologies, underscoring their potential to mitigate adverse impacts. Consequently, it underscores the pivotal role that ESSs can play in promoting energy justice and fostering the development of fairer, more sustainable energy systems [20].

In 2021, this study explores the connection between ESSs and the concept of energy justice. The study explores the incorporation of ESSs within the framework of energy justice and elucidates their potential contributions toward fostering equitable energy distribution, community engagement, and enhanced energy accessibility. Furthermore, it highlighted how these systems can enable the smooth incorporation of renewable energy sources, thereby advancing the goals of energy justice. This discourse underscores the pivotal role of ESSs in promoting energy justice and underscores their potential transformative impact in this domain [21]. In 2021, this research emphasizes the crucial importance of energy justice in promoting the progress and smooth integration of energy storage technologies. Specifically, it emphasizes the necessity of evaluating and endorsing advanced solutions like pumped underground storage hydro (PUSH) within the paradigm of energy justice. PUSH presents a promising energy storage avenue, especially in repurposing idle mines for community infrastructure. However, it highlights the inadequacies of current policy, legal, and regulatory structures in acknowledging and fostering the comprehensive benefits offered by energy storage technologies. This study scrutinizes, through an energy justice lens, the existing legal and regulatory frameworks that either foster or hinder the progression of energy storage technologies. Furthermore, it proposes an alternative framework aimed at bolstering the integration of technologies such as PUSH [22]. In 2022, this study explores the importance of ESSs concerning energy justice and community advantages. It systematically evaluates the efficacy of these systems across three distinct usage paradigms: utility-level installations, community-owned initiatives, and residential-level deployments. Central to this analysis is the concept of energy access, which serves as a pivotal metric in ensuring equitable energy distribution. This metric is applied rigorously across various outage scenarios along a representative distribution line. Moreover, the study underscores the importance of exploring additional metrics such as energy burden, energy poverty, vulnerability, resilience, carbon reduction, and employment generation to comprehensively assess energy justice. It posits that ESSs possess significant potential in advancing energy justice objectives and advocates for the development of a prioritization framework to align community needs with energy infrastructure preferences [23]. In another study conducted in 2022, a notable study shed light on the concept of energy justice and the community advantages stemming from the widespread adoption of electric vehicles (EVs). EVs have transitioned into being acknowledged as valuable assets for energy storage, especially when coupled with bidirectional charging strategies. The objective of this research is to conduct a comprehensive analysis of storage adequacy for energy accessibility, thereby characterizing the societal benefits and equity associated with mobile energy storage solutions (MESSs). This analysis will encompass various usage scenarios, considering factors such as battery capacities, charging schedules, and distribution throughout the grid. The ultimate aim of this study is to provide insights that can guide decision-making processes for stakeholders involved in EV deployment, charging infrastructure development, and broader energy system enhancements. By developing a prioritization framework tailored to the specific needs of communities and energy systems, this research endeavors to offer valuable guidance for strategic planning and implementation [24].

Looking at recent studies in 2023, the discussion centers on alternative energy storage options and the concept of energy justice. Amid the ongoing energy transition, energy storage plays a crucial role as a key catalyst. Nonetheless, conventional battery technologies present inherent environmental and societal hazards, often overlooked. This inquiry delves into various energy storage options through the lens of the energy justice paradigm, assessing their alignment with principles of fairness. Drawing upon contemporary scholarship and empirical illustrations, the study scrutinizes the prospects and obstacles associated with different energy storage methodologies. By comparing traditional batteries to alternatives such as pumped hydroelectric storage (PHS), thermal energy storage (TES), and compressed air energy storage (CAES), it suggests that alternative storage solutions offer fairer results. Particularly, the application of the energy justice framework to energy storage endeavors prompts a critical examination of the widespread adoption of certain batteries known to harbor environmental risks. The main goal is to provide policy guidance that supports a fair transition within the renewable energy sector, where sustainability and fairness go beyond mere convenience [25]. Machine learning techniques are increasingly employed to address challenges in electricity systems arising from the expanding supply of energy from renewable sources. Researchers are investigating data-driven artificial intelligence (AI) technologies to predict energy usage, enhance cost-effectiveness, monitor system health, and alleviate network congestion. However, there is a growing concern about the impact of AI on energy justice, which emphasizes fair outcomes in energy-related decisions, especially during the energy transition. Despite the concept of energy justice being established for over a decade, research on energy justice in digitalized and data-driven electricity systems remains limited. Particularly, there is insufficient exploration into the ramifications of utilizing AI technology for managing electricity systems. Therefore, the main focus of this chapter is to investigate the possible impacts of integrating AI into smart electricity systems from the viewpoint of energy justice, along with its consequences for technology design and regulation [26]. In 2023, a study was conducted focusing on the development of equitable energy-sharing frameworks tailored for multiunit buildings. This investigation introduced a novel fair energy-sharing model, scrutinizing its alignment with energy justice principles. The analysis delved into the core tenets of energy justice and explored their application in ensuring equitable energy distribution during trading and distribution processes. Notably, the proposed model strategically manages energy trading endeavors through a profit-driven game theory approach, aimed at augmenting participants' financial gains. The efficacy of this model was substantiated through comparison with empirical data sourced from Austin, Texas. Furthermore, the study examined the integration of energy justice principles within the proposed energy allocation and trading mechanisms. It meticulously scrutinized the recognition, distribution, and procedural aspects of justice, observing that the devised sharing models afford accessible energy benefits to diverse groups. Within these models, equitable distribution of energy and associated benefits is achieved by factoring in ownership and priority considerations. Procedural justice is upheld by fostering an inclusive environment wherein all stakeholders partake in decision-making processes concerning energy resource utilization. Consequently, the study underscores the paramount importance of integrating principles of fairness in energy distribution into the design of energy-sharing frameworks, highlighting their versatile applicability across varying contexts [27]. In 2024, this study emphasizes that energy storage is an important tool and that there are many factors to be considered in terms of energy justice. While energy storage has an important role in carbon reduction planning, different stakeholders have diverse interests in the development and implementation of projects. The study suggests that energy storage projects should not only focus on financial returns but also consider environmental impacts and societal benefits. In particular, it draws attention to the necessity of various policy instruments for the equitable development and distribution of energy storage technologies, taking into account the environmental impacts of the implementation of energy

Table 8.1 Table of acronyms and abbreviations.

Acronyms	Abbreviation	Acronyms	Abbreviation
PV	Photovoltaic	EPC	Engineering, Procurement, and Construction
DC	Direct Current	PEMFC	Polymer Electrolyte Membrane Fuel Cell
FC	Fuel Cell	HT-PEMFC	High-Temperature PEMFC
SoC	State of Charge	DMFC	Direct Methanol Fuel Cell
EES	Electromagnetic Energy Storage	MCFC	Molten Carbonate Fuel Cell
SMES	Superconducting Magnetic Energy Storage	PAFC	Phosphoric Acid Fuel Cell
EV	Electric Vehicle	SOFC	Solid Oxide Fuel Cell
HTS	High-Temperature Superconducting	AFC	Alkaline Fuel Cell
ICCOS	Integrated Capacitor Coil Switching	MESS	Mechanical Energy Storage Systems
CES	Capacitive Energy Storage	CAES	Compressed Air Energy Storage
IES	Inductive Energy Storage	PHS	Pumped Hydroelectric Storage
EESS	Emerging Energy Storage Systems	FES	Flywheel Energy Storage
NAS	Sodium-Sulfur	CAES	Compressed Air Energy Storage
NGK	NGK INSULATORS, LTD	SESS	Spring Energy Storage Systems
TEPCO	Tokyo Electric Power Co	GES	Gravity Energy Storage
EMF	Electromotive Force	IoT	Internet of Things
LA	Lead Acid	AWS	Amazon Web Service
SLA	Sealed Lead Acid	IaaS	Infrastructure as a Service
FLA	Flooded Lead Acid	SaaS	Software as a Service
AGM	Absorbent Glass Mat	PaaS	Platform as a Service
VRLA	Valve-Regulated Lead Acid	DaaS	Data as a Service
RFB	Redox Flow Batteries	HDFS	Hadoop Distributed File System
HFB	Hybrid Flow Batteries	HPC	High-Performance Computing Clusters
VRB	Vanadium Redox Flow	GPU	Graphic Processing Unit
OCV	Open-Circuit Voltage	FPGA	Field Programmable Gate Arrays
LCA	Life Cycle Assessment	SVM	Support Vector Machine
TEA	Techno-Economic Analysis	RBF	Radial Basis Function
BESS	Battery Energy Storage Systems		

storage projects. In this context, the study emphasizes the importance of strategies such as timing, considering regional differences, and balancing the interests of different stakeholders for energy justice [28]. In the same year, this research underscores the critical role of energy storage as a pivotal tool in addressing energy justice concerns. It highlights the multifaceted considerations integral to energy storage, particularly in the context of carbon reduction initiatives. Despite its significance, the development and execution of energy storage projects involve divergent interests among stakeholders. Consequently, the study advocates for a comprehensive approach that extends beyond mere financial gains to encompass environmental ramifications and societal advantages. Moreover, it underscores the imperative need for policy frameworks aimed at fostering equitable access and distribution of energy storage technologies. Such policies should meticulously address the environmental repercussions associated with the deployment of energy storage projects. Within this framework, strategic considerations such as timing, regional disparities, and the harmonization of divergent stakeholder interests are deemed paramount for achieving energy justice [29].

In light of these studies, we see that research on energy justice and equality has been ongoing for many years. These studies are becoming richer with each passing day, thanks to advancements in new technologies and their derivatives. In today's world, energy justice has become even more crucial compared to the past, particularly with the increasing importance of clean and renewable energy concepts. ESSs, in particular, emerge as a significantly important concept in both fair energy distribution and energy efficiency fields. Undoubtedly, one of the most critical issues in today's world is information (or data). Therefore, alongside ESSs, data science studies are also of great value for energy justice and efficiency.

We can use this information for the control of energy distribution and energy storage. We need important information on how they are controlled and where these methods are used.

A compilation of studies on the algorithms used for HESSs and the methods of their utilization is shown in Table 8.1.

8.3 Energy Storage System

The significance of ESSs is rising, as they play a crucial role in addressing the expanding energy needs and optimizing the utilization of renewable energy sources. These systems store electrical energy, providing a stable power supply and enhancing the reliability of electrical grids by balancing fluctuations in energy demand. Additionally, ESSs contribute to the widespread adoption of environmentally friendly transportation, such as EVs, by being utilized in mobile applications. Within this piece, we will delve into the significance of ESSs, closely

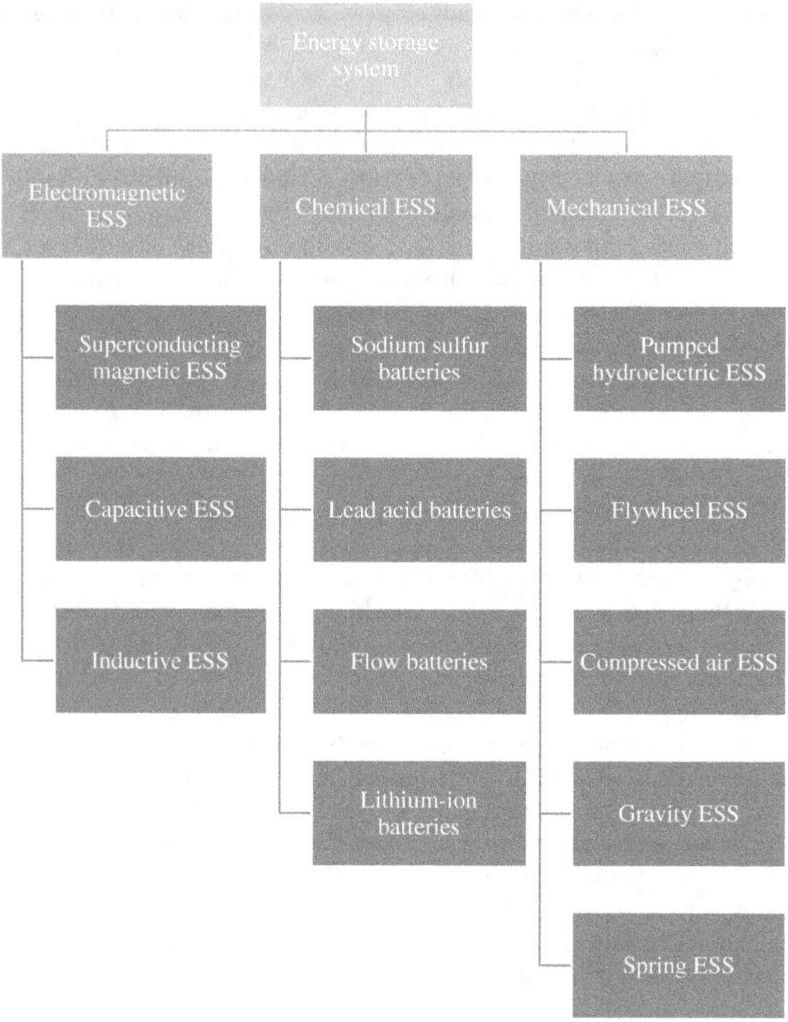

Figure 8.1 Types of energy storage systems.

examining various energy storage technologies and their applications. As depicted in Figure 8.1, we have analyzed ESSs under three main headings (Table 8.2).

8.3.1 Electromagnetic Energy Storage

The advent of electromagnetic energy storage (EES) systems represents a monumental turning point in the ongoing quest for pioneering and sustainable energy solutions. Positioned as a pivotal component, these innovative systems

Table 8.2 Summary of commonly used control algorithms.

References	Summary	Configuration		Control method	Advantages
		Energy sources	Storage technology		
[30]	In this approach, short-term charging and discharging operations are supported by supercapacitors (SCs). The strategy involves precise modeling and battery lifespan prediction in hybrid energy storage systems (HESSs). Findings indicate that prolonging the low pass filter time constant leads to a decrease in battery cycling wear.	Wind turbine	Battery/supercapacitor	Fuzzy-based control	Enhancing the longevity of batteries
[31]	A control method is introduced to distribute power among HESSs to address demand-generation imbalances. In this approach, supercapacitors (SC) provide rapid transient power when needed, while the battery handles slower dynamics.	PV & Wind turbine	Battery/supercapacitor	Fuzzy-based control	Enhances regulation of the DC grid voltage. Decreases the rates of charging and discharging.
[32]	A proposal for voltage regulation of the DC bus involves utilizing nonlinear differential flatness-based Takagi-Sugeno fuzzy logic control. This approach aims to eliminate rapid changes in battery and fuel cell (FC) currents, thereby reducing stress on their currents.	PV	Battery/supercapacitor/fuel cell	Fuzzy-based control	The strain on both the fuel cell (FC) and battery is alleviated.

(Continued)

Table 8.2 (Continued)

		Configuration			
References	**Summary**	**Energy sources**	**Storage technology**	**Control method**	**Advantages**
[33]	This source employs a blend of neural network computations and wavelet transform for controlling the system. The wavelet transform is employed to break down the net power into its fast and smooth components, while the neural network computes the battery's reference power. Additionally, supercapacitors (SCs) are employed for DC bus voltage regulation.	PV & Wind turbine	Battery/supercapacitor	Fuzzy-based control & artificial neural network	The battery's state of charge (SoC) stays within the permissible range.
[34]	The objective of the control system is to allocate sudden changes in current to the supercapacitors (SCs) and gradual current fluctuations to the battery. Model predictive control (MPC) forecasts the modulation indices for the SC and the battery's DC/DC converter, ensuring that the overall power aligns with the predetermined total power while adhering to specified constraints.	–	Battery/supercapacitor	Model predictive control	The state of charge (SoC) of the battery, the current supplied by the solar charger (SC), the current drawn from the battery, and the voltage from the solar charger are kept within acceptable limits.

[35]	This source outlines the development involves two control systems: a decentralized energy management strategy stationed at the customer's location, and a central power flow controller. The central power controller orchestrates energy distribution and power flow among the energy sources by collecting data from generators and sensors and issuing control signals to the different converters employed in managing the power system. Meanwhile, the local load strategy oversees various loads, adjusting them based on the overall energy balance of the system. It can alleviate strain on the system by temporarily disconnecting controllable loads, thereby reducing home power demand during periods of stress.	PV	Battery/fuel cell/supercapacitor	Rule-based control	Enhancing the longevity of the battery and flight controller (FC), minimizing the frequency of FC power cycles, and preventing battery deep discharge during adverse weather conditions.

play a critical role in elevating the reliability and efficiency of modern power grids and infrastructure. By harnessing the principle of storing energy within electromagnetic fields, EES systems have emerged as indispensable assets in tackling the intricate challenges associated with the integration of intermittent renewable energy sources into existing energy frameworks. Moreover, as societies worldwide continue to embrace electrification across various sectors, these advanced storage solutions are increasingly vital in meeting the escalating demands of an ever-expanding and energy-hungry global population [36]. EES technologies intricately utilize the foundational principles of electromagnetism to efficiently capture and retain energy, presenting a sophisticated mechanism that enables the rapid conversion of stored energy into electrical power whenever demanded. This transformative capability extends beyond mere technical advancement; it signifies a profound evolution in energy management, distribution, and utilization, ushering in a new era of versatility and responsiveness in our energy infrastructure. By embracing these technologies, we embark on a journey toward a more dynamic and sustainable energy landscape, where adaptability and efficiency become the cornerstones of our energy ecosystem, fostering resilience and innovation [37]. The integration of EES systems into our energy infrastructure heralds a paradigm shift toward greater sustainability and resilience. These systems offer not only enhanced reliability but also facilitate the efficient utilization of renewable energy sources, thereby reducing reliance on fossil fuels and mitigating environmental impact. As the global community strives to address the urgent challenges of climate change and energy security, the role of EES systems becomes increasingly indispensable. Their ability to store excess energy during periods of low demand and discharge it during peak hours helps to balance supply and demand, thereby stabilizing the grid and reducing the need for costly infrastructure upgrades. Furthermore, the versatility of EES technologies extends beyond grid applications, with potential implications for various sectors, including transportation, industrial processes, and residential energy management. From EVs to off-grid renewable energy systems, the integration of EES solutions holds promise for revolutionizing diverse aspects of our daily lives. In essence, the proliferation of EES systems signifies more than just a technological advancement; it represents a fundamental shift toward a more sustainable, efficient, and resilient energy future. By leveraging the transformative potential of EES technologies, we can not only address the challenges of today but also pave the way for a brighter and more sustainable tomorrow [38]. Furthermore, as the global transition toward electrification gains momentum across diverse sectors such as transportation and industrial operations, there arises an increasingly urgent demand for energy storage solutions capable of providing high power density while facilitating rapid energy discharge. EES systems, including advanced technologies like supercapacitors (SCs) and superconducting magnetic energy storage (SMES), emerge as highly suitable options to address these evolving needs. These cutting-edge solutions not only cater to requirements ranging from swift EV charging to instantaneous power backup for critical systems but also offer versatile applications across various domains where quick and efficient energy delivery is paramount [39].

The operation of EES systems is grounded in the fundamental principles of electromagnetism. Energy can be stored and retrieved through two main mechanisms. In electric field storage, within the realm of energy storage, capacitors and SCs serve as pivotal components, leveraging a method predicated upon the accumulation of energy within the electric field delineated between oppositely charged plates. This mechanism, foundational to the storage of electrical energy, is subject to meticulous scrutiny concerning its efficacy, which is contingent upon multifarious parameters. Principal among these are the surface area of the plates, the interplate separation distance, and the intrinsic characteristics of the dielectric medium employed. In tandem with the progression of materials science, notable advancements have been witnessed in the refinement of SC technology. These advancements have engendered SCs typified by heightened energy densities, thereby evincing a notable departure from conventional energy storage modalities. Importantly, these modern iterations of SCs retain the inherent advantage of rapid charge and discharge kinetics, ensuring expeditious energy transfer dynamics. The evolution of SCs constitutes a seminal milestone in the ongoing quest for proficient and sustainable energy storage solutions. The iterative refinement of materials and fabrication methodologies underscores the ceaseless endeavor to enhance energy storage capacities. In this vein, the scholarly pursuit of novel materials and the concomitant optimization of fabrication processes herald a promising trajectory toward the realization of high-performance and environmentally conscientious energy storage technologies [40]. Magnetic field storage, the utilization of ESSs, exemplified by technologies such as SMES and inductive energy storage (IES), constitutes a pivotal facet in contemporary energy management paradigms. These systems operate on the fundamental principle of capitalizing on the magnetic fields engendered by the flow of electrical currents, thereby effectuating the storage of electrical energy. Central to the efficiency and efficacy of such storage methodologies is the employment of superconducting coils, which serve to curtail energy losses inherent in conventional storage systems. The remarkable attribute of superconductivity facilitates a near-complete elimination of electrical resistance within the coil, thereby enabling the preservation of stored energy over prolonged durations. This characteristic also engenders the capability for rapid energy release, ensuring the expeditious deployment of stored energy when necessitated. Critical to note is the nuanced interplay between the choice of materials comprising the superconducting coils and the meticulous design considerations governing the construction of the storage unit. Material selection encompasses a multifaceted evaluation of parameters such as critical temperature, critical magnetic field, and critical current density, among others, to ascertain optimal superconducting performance. Additionally, the design intricacies of the storage unit encompass considerations of thermal management, mechanical stability, and electromagnetic shielding, among others, to uphold operational robustness and longevity. Consequently, the efficacy and viability of SMES and IES systems are contingent upon a judicious amalgamation of material science, electromagnetic engineering, and system design principles. The culmination of these interdisciplinary efforts culminates in energy storage solutions characterized by unparalleled efficiency, reliability, and versatility, thereby facilitating their integration into diverse applications within the contemporary energy landscape [41].

The burgeoning field of EES holds immense promise for revolutionizing contemporary energy systems, poised at the forefront of ongoing research endeavors aimed at transcending prevailing limitations and maximizing efficacy across multiple dimensions, including energy density, efficiency, and cost-effectiveness. Through concerted efforts and scientific breakthroughs in superconducting materials and dielectric technologies, the horizon of EES is steadily expanding, offering the prospect of mitigating existing constraints and broadening the applicability of such systems across diverse industrial sectors and societal domains. In the quest for enhanced energy storage solutions, the pursuit of advancements in superconducting materials has emerged as a pivotal focal point, with researchers endeavoring to unlock the full potential of these materials to facilitate unprecedented levels of energy retention and transmission efficiency. This quest involves the exploration of novel compositions, structural configurations, and fabrication techniques, all aimed at pushing the boundaries of superconductivity and harnessing its transformative capabilities within the realm of EES. Concurrently, the refinement of dielectric technologies represents another cornerstone of progress in EES research as scientists strive to optimize the performance characteristics of dielectric materials for energy storage applications. This entails the meticulous engineering of dielectric properties, such as permittivity and breakdown strength, to achieve optimal energy storage capacity while minimizing losses and ensuring operational reliability. Moreover, the integration of advanced manufacturing processes and innovative design approaches promises to further elevate the efficiency and cost-effectiveness of dielectric-based EES systems, thereby bolstering their viability for widespread deployment and adoption. As these technological advancements continue to unfold, the landscape of EES is poised for transformative change, with profound implications for the future sustainability, resilience, and efficiency of global energy systems. Through interdisciplinary collaboration and sustained research efforts, the realization of next-generation EES solutions holds the potential to catalyze a paradigm shift in how we generate, store, and utilize energy, ushering in an era of unprecedented innovation and progress in the realm of sustainable energy technologies [42].

In conclusion, EES stands as a fundamental technology pivotal in facilitating the transition toward a more sustainable, efficient, and adaptable energy paradigm. Its unparalleled capacity to rapidly and effectively store and discharge energy renders it a vital enabler in tackling the obstacles associated with the integration of renewable energy sources and the electrification of various sectors. As research endeavors push the boundaries of EES, harnessing its potential holds the key to unlocking a future characterized by enhanced energy resilience, reduced environmental impact, and heightened energy system flexibility, thus paving the way for a more sustainable and prosperous world [43].

8.3.2 Types of Electromagnetic Energy Storage

8.3.2.1 Superconducting Magnetic Energy Storage (SMES)

The fundamental operational principle underpinning SMES systems resides in the efficient storage of energy within the magnetic field induced by the circulation of direct current through a superconducting coil. Upon cooling below its critical temperature, a superconducting material enters a state of zero electrical resistance, thereby enabling the uninterrupted flow of electricity. This remarkable phenomenon facilitates the sustained generation of high currents within SMES systems, resulting in the storage of energy within the magnetic field engendered by this persistent current flow. Consequently, the stored energy can be swiftly retrieved and reintroduced into the power grid through the reversal of the current direction, endowing SMES systems with unparalleled responsiveness to the dynamic demands and fluctuations inherent in modern energy grids. A notable advancement in the realm of SMES technology encompasses the development of a pioneering high-temperature superconducting (HTS) pulse transformer-based inductive pulsed power supply circuit. This innovative endeavor integrates an enhanced integrated capacitor coil switching (ICCOS) module, thereby heralding a new era of improved efficiency and compactness in SMES design paradigms. The integration of HTS technologies not only augments the performance metrics of SMES systems but also extends their applicability to a diverse array of energy storage solutions. By overcoming traditional constraints associated with energy losses and volumetric requirements, this breakthrough holds immense promise in fortifying grid stability and optimizing energy management on a multifaceted scale. In elucidating the profound implications of these advancements, it becomes evident that the evolution of SMES technologies embodies a critical juncture in the trajectory of contemporary energy storage methodologies. Through meticulous research and iterative refinement, researchers and engineers continue to push the boundaries of innovation, thereby unlocking unprecedented capabilities and efficiencies within the realm of energy storage and grid management. Consequently, the integration of SMES systems into the fabric of modern energy infrastructure stands poised to catalyze transformative changes, ushering in a new era characterized by resilience, sustainability, and adaptability in the face of evolving energy dynamics and imperatives [44].

8.3.2.2 Capacitive Energy Storage (CES)

CES, commonly referred to as SCs or ultra-capacitors, represents a distinct paradigm in energy storage mechanisms grounded in the electrostatic accumulation of energy. Diverging from the chemical processes inherent in traditional battery technologies, CES systems leverage the physical partitioning of charge within an electric field across electrode pairs. These electrodes, typically fashioned from sophisticated materials such as porous carbon or metal oxides, are immersed in an electrolytic medium and segregated by a dielectric, which facilitates the establishment of an electric field while impeding charge transfer. The architectural attributes of electrode materials, notably surface area and porosity, serve as critical determinants in dictating the overall capacitance of SCs, exerting direct influence over the quantity of charge that can be stored. SCs exhibit exceptional prowess in swiftly charging and discharging energy, thereby furnishing elevated power outputs within compressed timeframes. This distinctive feature renders them eminently suitable for applications necessitating rapid energy surges, including but not limited to regenerative braking systems in EVs and the optimization of power quality in grid infrastructures. Recent scholarly endeavors have delved into the exploration of innovative electrode materials, particularly porous hollow high-entropy metal oxides like $(NiCoCuFeMg)_3O_4$ nanofiber anodes, with the objective of augmenting the energy density and efficiency of CES systems. These pioneering investigations have yielded promising outcomes, indicative of the latent potential for future advancements in this field. Such advancements hold the promise of not only enhancing the performance metrics of SCs but also expanding their utility across a spectrum of interdisciplinary applications, thereby ushering in a new era of energy storage technologies with far-reaching implications for sustainability and technological innovation [45].

8.3.2.3 Inductive Energy Storage (IES)

IES systems represent a pivotal domain within the broader landscape of energy storage technologies, leveraging the fundamental principles of electromagnetism to efficiently store and discharge electrical energy. At the heart of IES lies the utilization of inductors to create and manipulate magnetic fields, thus capitalizing on the phenomenon of electromagnetic induction as elucidated by Faraday's law. This principle asserts that any change in magnetic flux within a closed circuit generates an electromotive force (EMF), or voltage, within that circuit. In the context of IES, this translates to the storage of energy within the magnetic field surrounding the inductor when an electric current traverses it [46]. The operational dynamics of an IES system are intricately intertwined with the characteristics of the inductor, notably its inductance and the magnitude of the current flowing through it. Specifically, the amount of energy stored within the magnetic field is directly proportional to the square of the current and the inductance of the coil. This mathematical relationship underscores the significance of optimizing both the design parameters of the inductor and the control mechanisms governing the flow of current to enhance the overall energy storage efficiency of IES systems. One of the distinguishing features of IES systems is their remarkable durability and capability to deliver high power outputs expeditiously. This attribute renders them particularly well-suited for applications necessitating rapid energy storage and release, such as in pulsed power scenarios common in military, aerospace, and industrial domains. Furthermore, the inherent scalability of IES technology enables its deployment across a diverse array of applications, ranging from grid stabilization to renewable energy integration and beyond. The evolution of IES technology is inexorably linked to advancements in materials science and electrical engineering. In particular, the ongoing exploration of novel materials with enhanced magnetic properties, coupled with innovative design paradigms such as HTS pulse transformer-based circuits featuring improved integrated current control and power supply (ICCOS) modules, is driving significant progress in the field. These developments hold the promise of not only enhancing the efficiency and storage capacity of IES systems but also expanding the scope of their applicability across various sectors of the economy. While IES represents a formidable contender in the realm of energy storage solutions, it is by no means an isolated entity. Rather, it is part of a broader ecosystem of energy storage systems (ESSs) encompassing diverse technologies such as SMES and CES, each characterized by its unique approach to energy storage and management. The convergence of research efforts across these disparate domains, underpinned by a multidisciplinary approach encompassing physics, materials science, and engineering, holds the key to unlocking new frontiers in energy storage innovation.

In conclusion, the ongoing research and development endeavors in the field of IES are poised to yield transformative outcomes, propelling the realization of more efficient, reliable, and scalable energy storage solutions. By harnessing the power of electromagnetism and pushing the boundaries of scientific understanding, the quest for optimal energy storage technologies continues unabated, foreshadowing a future where the smooth assimilation of renewable energy sources and the electrification of diverse industries become not just feasible but inevitable.

8.3.3 A Comprehensive Comparative Analysis of Electromagnetic Energy Storage Systems

Undertaking a thorough comparative examination of EES systems requires a deep dive into the subtle differentiations among a range of metrics. These metrics include energy density, power density, efficiency, lifecycle longevity, and economic feasibility. This comprehensive exploration draws upon the latest scholarly insights to elucidate the distinguishing features and comparative benefits of the primary EES technologies. These technologies encompass SMES, CES, and IES, each playing a pivotal role within its respective domain due to its distinctive capabilities and applications. Delving into the intricacies of these systems offers valuable insights into their respective strengths and potential avenues for further development and optimization. Upon thorough analysis of the energy and power density attributes inherent in these systems, it becomes conspicuously clear that SMES systems exhibit unparalleled efficacy in delivering exceedingly high power density, a characteristic of paramount importance. This distinctive trait is particularly pivotal for applications that mandate instantaneous energy discharge and recharge cycles. SMES systems emerge as preeminent contenders in contexts necessitating rapid response capabilities, such as the management of power quality and the provision of emergency power supply systems. Their extraordinary ability to expeditiously release and absorb energy within fractions of a second assumes a pivotal role in preempting power outages and ensuring the requisite stabilization of grid operations. Within the spectrum of energy storage technologies, SMES systems stand as exemplars of high-performance solutions tailored to address the exigencies of modern power systems, thus underscoring their significance in advancing the efficacy and resilience of energy infrastructure [47]. Conversely, within the realm of ESSs, CES devices, particularly SCs, exhibit a distinctiveness stemming from their relatively elevated energy density. This defining feature empowers them to efficiently accumulate a noteworthy quantity of energy and subsequently release it with swiftness, albeit not with the instantaneous discharge capability synonymous with SMES systems. This nuanced attribute positions them favorably for applications necessitating a delicate equilibrium between sustained energy provision and rapid deployment. Such applications include but are not limited to the implementation of regenerative braking systems in EVs or the mitigation of intermittent power fluctuations inherent in renewable energy frameworks. This balance between longevity and expedited energy release underscores the pragmatic utility and versatility of CES technologies, thereby warranting further exploration and refinement within both academic discourse and practical deployment scenarios [48]. The assessment of EES encompasses multifaceted considerations, with efficiency and lifecycle longevity emerging as pivotal metrics for evaluating their overall utility and sustainability. Within this landscape, CES systems, notably SCs, emerge as exemplars of efficiency, boasting the remarkable capability to undergo millions of charge–discharge cycles without succumbing to significant degradation. This inherent resilience positions them as highly desirable options for applications where sustained reliability and minimal maintenance are paramount concerns. SCs, in particular, offer an enticing proposition owing to their ability to maintain efficiency over extensive operational periods.

Their minimal degradation profile underscores not only their robustness but also their potential for prolonged utilization, rendering them suitable for deployment across diverse sectors ranging from transportation to grid-scale energy storage. Their resilience to wear and tear aligns inherently with sustainability objectives, advocating for their integration into systems where durability and performance longevity are critical prerequisites. Moreover, the efficacy of CES systems, including SCs, extends beyond their operational efficiency to encompass broader environmental and economic benefits. Their ability to withstand numerous charge–discharge cycles translates into reduced replacement frequency and associated material consumption, thereby mitigating resource depletion and waste generation. This attribute lends credence to their role in fostering sustainable practices and promoting circular economy principles within the energy storage domain. Conversely, SMES systems also merit attention for their commendable efficiency in storing and releasing energy. However, the complexities inherent in maintaining superconductivity at cryogenic temperatures introduce substantial challenges that necessitate nuanced deliberation. The logistical intricacies and cost implications associated with cryogenic cooling infrastructure and ongoing maintenance efforts underscore the need for judicious evaluation, particularly in large-scale deployment scenarios. The interplay between technological prowess and practical feasibility underscores the imperative for comprehensive assessments when contemplating the adoption of SMES systems, especially within contexts characterized by stringent cost–benefit analyses and operational constraints. While their efficiency credentials are noteworthy, the pragmatic realities surrounding cryogenic management underscore the importance of considering a holistic array of factors, including economic viability, scalability, and operational resilience. In conclusion, the juxtaposition of CES and SMES systems highlights the intricate balance between efficiency, sustainability, and practical feasibility within the realm of energy storage technologies. While CES systems, exemplified by SCs, demonstrate remarkable efficiency and longevity, the complexities surrounding SMES systems necessitate careful scrutiny, particularly concerning their deployment at scale. Moving forward, a nuanced understanding of these dynamics will be indispensable in shaping informed decisions regarding the selection and implementation of energy storage solutions that align with overarching sustainability imperatives [49]. The economic dimensions pertinent to the deployment of EES exhibit a nuanced and multifaceted landscape, incorporating a spectrum of factors ranging from initial capital expenditure to ongoing operational and maintenance costs, culminating in the comprehensive assessment of the total cost of ownership over the system's life cycle. Within this framework, the discourse surrounding SMES systems merits particular attention. While acknowledged for their unparalleled performance in specific applications, SMES systems are inherently associated with considerable upfront investment requirements owing to the intricate materials and cooling infrastructures indispensable for achieving and maintaining superconductivity. SMES systems, characterized by their ability to store large amounts of energy with minimal losses, are distinguished by their reliance on superconducting materials, which demand sophisticated fabrication processes and advanced cooling mechanisms to sustain the requisite cryogenic temperatures. Consequently, the procurement and installation costs of SMES systems can be prohibitive, particularly in comparison to alternative EES technologies. Notwithstanding their cost implications, SMES systems offer distinctive advantages, including rapid response times, high power density, and superior efficiency, rendering them indispensable in scenarios necessitating instantaneous and reliable energy supply, such as grid stabilization and power quality enhancement applications. In contrast, capacitive energy storage (CES) technologies, exemplified by SCs, present a compelling economic alternative, demonstrating favorable cost-performance ratios across a diverse array of industrial and consumer applications. SCs, characterized by their high power density, rapid charge–discharge cycles, and extended operational lifespans, have surfaced as an economically viable option for storing energy requirements characterized by frequent cycling and short-duration energy bursts. Unlike SMES systems, SCs are distinguished by their simplified manufacturing processes, reduced maintenance needs, and absence of cryogenic cooling infrastructure, contributing to their affordability and widespread adoption in various sectors, including automotive, renewable energy integration, and portable electronics. In summary, while the economic considerations surrounding EES deployment are indeed complex, the juxtaposition of SMES and CES technologies underscores the significance of striking a delicate balance between performance attributes and cost-effectiveness. While SMES systems offer unparalleled performance capabilities, their high upfront costs necessitate careful evaluation against the economic viability and operational efficiency offered by alternative EES solutions such as SCs. Thus, informed decision-making in EES deployment mandates a comprehensive analysis encompassing not only performance metrics but also economic considerations to optimize the allocation of resources and maximize the benefits derived from energy storage investments [50]. As we cast our gaze toward the horizon of tomorrow, the terrain of EES technologies emerges as a fertile ground teeming with potential, driven by an unyielding wave of innovation in materials science, engineering progressions, and an escalating imperative for sustainable and dependable energy storage solutions. The relentless pursuit of research and development endeavors stands poised to usher in a new epoch characterized by amplified performance metrics within EES systems, concurrent reductions in associated costs, and the revelation of hitherto unexplored applications. Consequently, these advancements are poised to solidify the indispensable role of EES technologies as the cornerstone of the forthcoming generation of energy storage solutions, underscoring their pivotal significance in shaping the future landscape of sustainable energy systems.

8.3.3.1 Applications and Case Studies of Electromagnetic Energy Storage Systems

Utility-Scale Energy Storage and Grid Stabilization At the forefront of utility-scale infrastructure, EESs stand as indispensable pillars, wielding significant influence in fortifying grid stability, facilitating the seamless, smooth assimilation of renewable energy sources, and upholding stringent standards of power quality assurance. Among the myriad manifestations of EESs, SMES emerges as a prominent exemplar, harnessing the inherent properties of superconducting coils to effectuate the storage of energy within magnetic fields, thus endowing the system with the unparalleled capability for near-instantaneous energy discharge [51]. The salient attribute of SMES, characterized by its ability to swiftly release or absorb energy, assumes paramount importance within the context of contemporary power grid management. In light of the escalating prevalence of renewable energy sources, whose intermittent nature engenders considerable volatility in power generation and distribution, the imperative for responsive energy storage solutions becomes unequivocal. Consequently, SMES emerges as a vanguard technology, furnishing grid operators with a potent tool to contend with the abrupt fluctuations in power demand and supply precipitated by the stochastic nature of renewable energy inputs, notably those stemming from wind and solar sources. The efficacy of SMES and its counterparts in the realm of EESS extends beyond mere mitigation of intermittency-induced perturbations; rather, these systems serve as linchpins in the preservation of grid frequency and voltage stability. By virtue of their inherent responsiveness, SMES and analogous technologies function as dynamic buffers, swiftly modulating power flow to counteract deviations from desired frequency and voltage parameters. In doing so, these systems mitigate the deleterious effects of transient imbalances in supply and demand, thereby safeguarding the integrity of the grid and ensuring the reliable delivery of electricity to end consumers and industrial entities alike. Within the scholarly discourse surrounding power system engineering and renewable energy integration, the significance of SMES and EESS as enablers of grid resilience and sustainability commands considerable attention. Their capacity to not only store energy but also to swiftly discharge it as necessitated by fluctuating grid conditions underscores their pivotal role in effectuating a seamless transition toward a renewable energy-dominant paradigm. Through judicious deployment and integration within existing grid infrastructures, SMES and analogous EESS solutions hold the potential to engender a paradigm shift in the operational dynamics of power grids, imbuing them with unprecedented levels of flexibility, reliability, and efficiency. In conclusion, the ascendancy of

SMES and EESS within the realm of utility-scale infrastructure epitomizes a paradigmatic shift toward dynamic, adaptive energy storage solutions capable of reconciling the imperatives of renewable energy integration with the exigencies of grid stability and reliability. As the energy panorama worldwide continues to develop, driven by imperatives of sustainability and decarbonization, the centrality of SMES and EESS in shaping the contours of future power systems becomes increasingly pronounced, underscoring the urgency of sustained scholarly inquiry and technological innovation in this domain.

Renewable Energy Integration The assimilation and integration of renewable energy sources present formidable challenges attributed to their inherent intermittency and variability. Addressing these challenges necessitates the implementation of sophisticated energy storage solutions within the broader energy infrastructure. Emerging Energy Storage Systems (EESS) technologies, notably capacitive storage mechanisms such as SCs, emerge as pivotal components in this endeavor, offering robust mechanisms for mitigating the inherent intermittency of renewable energy sources.

EESS, through the utilization of SCs, functions as a crucial intermediary interface between intermittent renewable energy sources and the grid. These storage systems excel in swiftly capturing surplus energy during periods of heightened generation and subsequently discharging stored energy during instances of diminished renewable output. This capacitive storage functionality effectively serves to regulate and stabilize the energy supply, facilitating a smoother and more consistent integration of renewable energy into the grid. The deployment of EESS technologies not only optimizes the operational efficiency of renewable energy installations but also engenders a substantial reduction in reliance on conventional fossil-fuel-based peaking power plants. By mitigating the need for such carbon-intensive backup infrastructure, the transition toward a more sustainable energy landscape is markedly accelerated. Consequently, the widespread adoption of EESS technologies holds profound implications for advancing the global transition toward cleaner, more resilient energy systems, aligning with overarching sustainability objectives and climate mitigation imperatives [52].

Transportation and Electric Mobility In the ever-evolving landscape of electric mobility, energy storage and conversion systems (ESCS) represent a groundbreaking advancement in the management of energy within EVs and broader public transportation infrastructures. Among these innovations, the integration of SCs stands out as a transformative solution. SCs boast remarkable attributes such as high power density and extraordinary durability, capable of enduring millions of charge and discharge cycles with minimal degradation. These qualities render them exceptionally suited for regenerative braking systems, a critical component in enhancing the efficiency and performance of EVs [53]. Regenerative braking systems, facilitated by SCs, play a pivotal role in harnessing kinetic energy typically dissipated during braking maneuvers. By swiftly capturing and subsequently releasing this energy during acceleration, SCs effectively mitigate wastage, thereby amplifying the energy efficiency and extending the range of EVs. This not only contributes to mitigating environmental impacts by reducing reliance on conventional energy sources but also fosters a more sustainable transportation paradigm. Moreover, the application of EESS technologies extends beyond individual vehicles to encompass broader public transportation networks, including buses and trains. Embracing EESS in these contexts not only bolsters efforts to combat urban pollution but also serves as a compelling exemplar of the potential for sustainable urban mobility solutions. By integrating advanced energy storage and conversion systems into public transit, cities can significantly reduce their carbon footprint while concurrently enhancing the accessibility, reliability, and environmental sustainability of their transportation infrastructure. In essence, the advent of EESS technologies, particularly SCs, marks a significant stride toward realizing the vision of a cleaner, more efficient, and sustainable future for urban mobility. Through their application in both individual and collective transportation modes, these innovations promise to revolutionize the way energy is managed and utilized, paving the way for greener and more resilient cities.

High-Quality Power Supply for Sensitive Industries Industries reliant on sensitive electronic apparatus, such as those engaged in semiconductor fabrication, healthcare delivery, and data management within centers, exhibit an imperative need for a consistent and superior-grade power provision. Perturbations in voltage or frequency levels present formidable challenges, potentially precipitating prolonged periods of operational standstill and substantial financial repercussions stemming from equipment degradation. The advent of EESS heralds a transformative paradigm characterized by its unparalleled capacity to swiftly discern and rectify power quality anomalies, thus assuming a pivotal role in safeguarding the dependability of power dissemination across these critical sectors. The expeditious response dynamics inherent to EESS not only serve to shield delicate equipment from harm but also serve as an indispensable linchpin in preserving uninterrupted operational continuity. Consequently, the integration of EESS technologies emerges as a cornerstone strategy in fortifying the resilience of essential infrastructures against disruptions, thereby underpinning the sustainability and efficiency of industrial operations within these domains [54].

Microgrids and Decentralized Energy Systems Energy storage and management systems (EESS) constitute a fundamental technological framework within the domain of MGs and decentralized energy networks, exemplifying their indispensability in orchestrating optimal control and distribution of energy resources. In the context of remote or off-grid communities, characterized by their inherent imperative for energy self-reliance, EESS assumes a paramount role by facilitating the efficient storage and subsequent deployment of locally harnessed renewable energy reserves. This strategic utilization of EESS not only serves to bolster the reliability and sustainability of energy provision within these geographically isolated locales but also affords a notable reduction in dependency upon conventional, often pollutant-emitting diesel generators. The integration of EESS within such energy ecosystems epitomizes a paradigmatic shift toward greener and more resilient energy infrastructures. By effectively capturing and storing renewable energy surpluses, EESS enables the optimization of energy utilization patterns, thereby mitigating the sporadic nature linked with renewable energy sources like solar and wind. This, in turn, engenders a heightened degree of energy security and reliability, mitigating the susceptibility of off-grid communities to disruptions and fluctuations in energy supply. Moreover, the adoption of EESS fosters economic advantages by curtailing the reliance on diesel generators, notorious for their high operational costs and adverse environmental impacts. By circumventing the need for constant refueling and maintenance, EESS presents a compelling alternative that not only minimizes operational expenditures but also contributes to mitigating the ecological footprint associated with conventional energy generation practices. Consequently, the widespread deployment of EESS within remote or off-grid contexts marks the beginning of a significant change toward sustainable and economically feasible energy solutions, embodying the convergence of environmental stewardship and technological innovation.

In conclusion, the integration of energy storage and management systems within MGs and decentralized energy systems represents a pivotal advancement toward achieving energy autonomy and sustainability, particularly within remote or off-grid communities. Through the judicious utilization of EESS, these communities stand to benefit from enhanced energy reliability, reduced environmental impact, and improved economic viability, thereby heralding a promising trajectory toward a more resilient and sustainable energy future [55].

Consumer Electronics and Portable Devices The proliferation of consumer electronics and the concomitant demand for devices characterized by prolonged battery life and swift recharging capabilities has precipitated a pronounced integration of EESS technologies, prominently including SCs. Renowned for their expeditious charge–discharge cycling and enduring operational longevity, SCs have emerged as quintessential components within a broad spectrum of portable electronic apparatus, ranging from ubiquitous smartphones and laptops to cutting-edge wearable devices. This symbiotic relationship between technological advancements and consumer expectations has engendered significant strides in user-centric experiences, exemplified by devices that not only exhibit accelerated recharging times but also boast enhanced endurance capacities, thereby transcending the conventional boundaries of portable technology innovation. Such synergistic amalgamations epitomize a burgeoning paradigm shift within the realm of consumer electronics, underscoring the inexorable march toward increasingly sophisticated and user-friendly technological landscapes [56].

Smart Grids and Advanced Energy Management Systems The advancement of smart grid infrastructure, characterized by its refined capacity for controlling and optimizing energy flows, is substantially underpinned by the integration of energy storage and ESS technologies. These sophisticated systems play a pivotal role in orchestrating the dynamic equilibrium between energy supply and demand, facilitating the seamless assimilation of decentralized energy resources, and empowering the deployment of advanced demand response methodologies. Through the sophisticated analysis of real-time data streams and the implementation of automated control protocols, EESS technologies ensure the efficient storage, conversion, and dissemination of energy resources. This transformative capacity not only augments the operational efficiency of energy grids but also engenders greater resilience and responsiveness within energy infrastructures. Consequently, the integration of EESS technologies represents a critical milestone in the evolution of smart grids, heralding a new era of enhanced sustainability and adaptability in the domain of energy management and distribution [57].

Innovations in Emergency Power Solutions In critical infrastructures like hospitals, data centers, and emergency response services, where constant power supply is crucial, the use of advanced energy storage technologies is key to maintaining uninterrupted operations and resilience. These technologies are vital for managing modern infrastructure, providing a reliable method for storing and delivering energy when needed. They are especially crucial in environments with demanding power needs and the requirement for continuous operation, acting as safeguards against the effects of unexpected power failures. The strength of these technologies lies in their ability to quickly and effectively respond to sudden power shortages, preventing severe disruptions to crucial operations and services. This rapid deployment of backup power helps maintain not only operational integrity but also the essential societal functions these facilities support. Furthermore, the importance of these technologies extends to disaster resilience, preserving data integrity, and, importantly, ensuring patient safety in healthcare environments. In healthcare, where the well-being of patients and the smooth functioning of medical processes are paramount, integrating these technologies is critical. They not only keep medical equipment and data systems running without interruption but also ensure that patient care continues unhindered during external challenges. Therefore, deploying these technologies in healthcare settings is not just about operational efficiency; it is a vital strategy for comprehensive patient care and organizational resilience. Overall, introducing these technologies in critical facilities is a complex task that goes beyond technical needs, addressing broader issues of operational stability, disaster readiness, and societal well-being. Adopting these solutions indicates a proactive approach to minimizing potential disruptions and strengthening the core aspects of societal health and safety [54]. In conclusion, EES systems are reshaping the energy sector, offering innovative solutions to the challenges of grid stability, renewable energy integration, electric mobility, and beyond. Their ability to deliver high-performance energy storage and rapid discharge capabilities across a variety of applications highlights the critical role of EESS in facilitating a sustainable, efficient, and resilient energy future. As technology advances and adoption increases, the potential of EESS to drive the global energy transition will undoubtedly continue to expand, heralding a new era of energy innovation.

8.3.4 Chemical Energy Storage Systems

Chemical ESSs represent vital technologies that involve the storage of energy in chemical form and its subsequent release when needed. These systems are utilized for purposes such as balancing fluctuations in renewable energy sources, managing power imbalances within electrical grids, and efficiently storing energy. Among the various technologies offered to meet energy storage capacity, conversion efficiency, and long-term storage requirements are batteries, hydrogen fuel cells, and redox accumulators. The widespread adoption of these systems has the potential to enhance the security, efficiency, and sustainability of energy systems, thus making significant contributions to energy justice, equality, security, and sustainability in the future.

8.3.4.1 Sodium Sulfur

The sodium sulfur (NaS) battery, a groundbreaking energy storage solution, operates by utilizing sodium and sulfur as electrode materials, with β-alumina ceramics serving as the electrolyte. Collaborative efforts between NGK INSULATORS, LTD. (NGK) and the Tokyo Electric Power Co. (TEPCO) since the mid-1980s have spearheaded the development of NaS battery technology in Japan, culminating in NGK's commencement of large-scale commercial production in April 2003. Originally conceived to address the fluctuations in electricity demand within Japan, NaS batteries have evolved to cater to a myriad of applications, including emergency power supply, combined emergency power and load balancing, and bolstering electric transmission and distribution systems. Persistent endeavors for project development and deployment are now underway in both North America and Europe. Within each NaS battery cell, sodium and sulfur serve as the active components for the negative and positive terminals, respectively, while the electrolyte consists of sodium-ion-conductive β-alumina ceramic. Operating at temperatures hovering around 300°C, these batteries maintain the electrode materials in a molten state, thus mitigating resistance to the flow of sodium ions through the solid β-alumina electrolyte. An illustrative depiction of the operational principle of an NaS battery cell is provided in Figure 8.2, showcasing the reversible electrochemical reaction that occurs during both discharge and charge cycles. The nominal EMF of an NaS cell is approximately 2 V. During the discharge phase, sodium at the negative electrode liberates electrons to the external circuit, while sodium ions traverse the β-alumina electrolyte to the positive electrode, where they combine with sulfur to generate sodium polysulfide. Conversely, during the charging phase, this reaction undergoes reversal: the sodium polysulfide present at the positive electrode decomposes, facilitating the return of sodium ions to the positive electrode [58].

8.3.4.2 Lead Acid

Lead–acid (LA) batteries typically consist of two plates: one made of a lead alloy and the other with a PbO_2 base. When the batteries discharge, $PbSO_4$ is produced. The PbO_2 base plate primarily comprises a lead alloy grate infused with a PbO_2 paste. These grates are typically made of alloys containing small quantities of

Figure 8.2 Lead–acid battery.

calcium, antimony, and tin. Sulfuric acid solution serves as the electrolyte in use. Usually, a polypropylene container holds the electrolyte along with a set of six cells. LA batteries come in two main types: sealed and flooded. Sealed lead–acid (SLA) batteries require minimal maintenance, while flooded lead–acid (FLA) batteries need more frequent electrolyte top-ups.

This chapter will primarily delve into SLA batteries due to their superior characteristics in almost every aspect except cost. FLA batteries are cheaper but necessitate user maintenance. Deep-cycle LA batteries are equipped with thicker plates, resulting in higher active material density and energy density. Although this allows for deeper discharge potential, it does not necessarily extend the cycle life of the battery. Absorbent glass mat (AGM) LA batteries utilize a glass fiber separator soaked in just enough electrolyte to saturate the mat. This design prevents gas buildup, leakage, and the need for electrolyte top-ups, making it maintenance-free and suitable for various orientations. AGM batteries also referred to as valve-regulated lead-acid (VRLA) batteries, boast increased energy density due to significant compression. Unlike gel variations, AGM batteries allow for efficient ion movement, although they may have reduced surge current capability [59].

8.3.4.3 Flow Batteries

Flow batteries, heralded for their prowess in storing electrical energy, offer a multitude of benefits. These encompass the capacity for scalability with utmost ease, enabling seamless adjustment to varying energy demands and facilitating the integration of renewable energy sources into the grid infrastructure. Moreover, their rapid response time ensures swift adaptation to fluctuations in energy demand, thereby enhancing grid stability and reliability. The high power density inherent in flow batteries allows them to accommodate quick load changes, ensuring smooth operation even during periods of peak demand or sudden spikes in consumption. Additionally, the efficient round-trip energy conversion capability of flow batteries optimizes the utilization of stored energy, maximizing overall system efficiency and minimizing energy losses. Furthermore, their extended service life, characterized by the ability to endure numerous charge and discharge cycles without significant degradation, contributes to their cost-effectiveness and sustainability as long-term energy storage solutions. It is important to recognize that one of the most significant advantages offered by redox flow batteries (RFBs) lies in their distinctive capability to disentangle rated power from energy storage capacity. This feature sets them apart from traditional battery technologies, wherein these critical parameters are often intricately linked, resulting in limitations in adaptability and flexibility. Through the decoupling of power and capacity, RFBs provide a level of versatility that enables finer control over energy management and utilization, accommodating diverse operational demands and variable grid conditions with greater precision. This inherent flexibility not only enhances the overall performance and efficiency of ESSs but also unlocks new opportunities for optimizing grid operations, integrating renewable energy sources, and addressing dynamic energy demand patterns more effectively. Thus, the capacity of RFBs to separate rated power from energy storage capacity represents a notable advancement in energy storage technology, with far-reaching implications for the future sustainability and resilience of global energy systems. The separation of rated power from energy storage capacity in RFBs provides an unparalleled level of flexibility in the design and implementation of ESSs, allowing for highly customized solutions tailored to specific application requirements and operational demands. Unlike conventional battery technologies, where power output and energy capacity are tightly intertwined, the distinct architecture of flow batteries enables independent optimization of these parameters. In flow batteries, the power output is primarily determined by the size and configuration of the cell stack, while the energy capacity is defined by the dimensions and capacity of the storage tanks housing the electrolyte solutions used in the redox reactions. This decoupling of power and energy capacity not only enhances the versatility and adaptability of flow battery systems but also facilitates finer control over their performance characteristics, enabling them to effectively address a wide range of grid-scale stationary energy storage applications in the electricity generation and transmission industries. As a result, flow batteries are increasingly recognized as a promising solution for meeting the growing demand for reliable, scalable, and sustainable energy storage solutions in modern power systems. Despite the array of configurations and chemical compositions available for flow batteries, they can generally be classified into two main categories: RFBs and hybrid flow batteries (HFBs). Within the realm of RFBs, the Vanadium Redox Battery (VRB) holds a prominent position, widely acknowledged as one of the most meticulously researched and promising advancements in the field. Originating in the early 1980s, the VRB has garnered significant attention due to its unique characteristics and potential applications in energy storage. However, it is important to note that the development and implementation of VRB technology have not been without challenges. Issues pertaining to materials selection, technological limitations, and operational considerations have been subjects of extensive analysis and discussion in various scholarly works and industry publications. These investigations have aimed to identify and address key barriers to the widespread adoption and commercialization of VRB systems, thereby advancing the field of flow battery technology as a whole. The all-VRB represents a significant breakthrough in the evolution of energy storage technologies, particularly in addressing a persistent challenge faced by earlier iterations of RFBs. This challenge revolves around the inherent tendency for chemical interaction to occur across the separator between different redox species, which often leads to cross-contamination between the anolyte and catholyte solutions. This phenomenon not only compromises the efficiency and reliability of battery systems but also shortens their operational lifespan. In response to this challenge, researchers have turned to the unique properties of vanadium, a versatile element capable of existing in four distinct oxidation states. The innovative concept underlying the VRB involves harnessing vanadium's versatility to utilize the same element in both the anolyte and catholyte solutions. By doing so, the VRB effectively mitigates the risk of chemical interaction and cross-contamination, thereby significantly enhancing the overall performance and durability of the battery system. Furthermore, the strategic selection of soluble V + 2/V + 3 species for the anolyte and V + 4/V + 5 species (in the form of VO + 2/VO2+) for the catholyte further optimizes the electrochemical reactions within the battery, ensuring efficient energy storage and retrieval processes. Beyond these technical advancements, the VRB offers several additional advantages that position it as a highly promising solution for grid-scale and stationary energy storage applications. These include enhanced scalability to accommodate varying energy demands, improved cycle life to prolong battery longevity, and reduced environmental impact due to the recyclability of vanadium-based electrolytes. As a result, the VRB has garnered considerable attention and interest within the energy storage industry, with ongoing research and development efforts focused on further refining and optimizing its performance characteristics. As these advancements continue to unfold, the VRB is poised to play a pivotal role in shaping the future landscape of sustainable energy storage solutions, driving innovation and progress in the transition toward a cleaner and more resilient energy infrastructure. These electrolytes are both comprised of an aqueous 5 M H_2SO_4 solution, separated by an ionically conducting membrane. The Zn–Br_2 flow battery belongs to the category of HFB, which differs from the conventional definition of an RDF system. Similar to the Zn–air battery, the Zn–Br_2 flow battery utilizes Zn metal as the negative electrode. During discharge and charge cycles, Zn is dissolved and electrodeposited at the negative electrode, while Br_2 gas, which is soluble in the electrolyte rather than being in the ionized redox couple form, is present at the positive electrode, where the conversion of Br-/Br2 occurs. Zn-Br2 flow batteries typically offer higher energy densities (65–75 Wh/kg) compared to traditional RFBs. This is partly attributed to the use of Zn metal anode, which eliminates the need for large volumes of anolyte required in conventional RFBs.

In addition to their notable features, it is important to delve deeper into the characteristics and challenges associated with Zn–Br_2 flow batteries. These batteries exhibit an open-circuit voltage (OCV) that surpasses that of traditional RFBs, marking a significant advantage in terms of their overall performance and suitability for energy storage applications. Their impressive storage efficiencies, often exceeding 70%, coupled with high coulombic and voltage efficiencies reaching 90% and 85%, respectively, underscore their potential to meet the demands of grid-scale energy storage and other stationary applications effectively. However, despite these promising attributes, Zn–Br_2 flow batteries present certain inherent challenges that need to be addressed to ensure their practical viability and widespread adoption. Chief among these challenges is the corrosive and hazardous nature of Br_2 gas, which poses significant safety and environmental concerns. Leakage of Br_2 gas from the battery system can lead to adverse health effects and environmental pollution, necessitating stringent safety protocols and preventive measures to mitigate risks effectively. Moreover, the tendency of Br_2 gas to permeate through the membrane separator due to its inadequate selectivity for Br ions presents a critical technical challenge. This phenomenon results in a range of detrimental effects on battery performance, including a decrease in the working cell potential, an increase in polarization losses, and a reduction in the utilization of reactants. These issues collectively undermine the efficiency, reliability, and longevity of Zn–Br_2 flow battery systems, highlighting the importance of ongoing research and development efforts to overcome these challenges and optimize their performance. Despite these challenges, the unique advantages offered by Zn–Br_2 flow batteries, such as their high voltage and efficiency, make them a compelling option for grid-scale energy storage and other stationary applications. By addressing the technical and safety challenges associated with Br_2 gas and membrane permeation, these batteries have the potential to play a pivotal role in facilitating the transition toward a more sustainable and resilient energy infrastructure. Continued innovation and investment in Zn–Br_2 flow battery technology are essential to unlock their full potential and accelerate their deployment in real-world ESSs [60].

8.3.4.4 Lithium-Ion (Li-Ion)

A mere 25 years ago, in 1991, Sony Corporation made a significant breakthrough in battery technology with the introduction of a groundbreaking product known as the lithium-ion (Li-ion) battery. This innovation was spurred by a critical incident involving the recall of phones utilizing Moli Energy lithium/MoS_2 batteries, which led to a fire incident causing harm to users. In response to this incident and the subsequent need for safer and more reliable energy storage solutions, Sony Corporation embarked on a journey to develop and commercialize the Li-ion battery. This marked a pivotal moment in the history of battery technology, as the Li-ion battery emerged as a safer and more efficient alternative to existing battery technologies. Since its introduction, the Li-ion battery has revolutionized numerous industries, including consumer electronics, EVs, and renewable energy storage, playing a crucial role in powering modern-day devices and facilitating the transition toward a more sustainable energy future. Sony, alongside a cohort of other pioneering companies, embarked on an ambitious journey to pioneer the development of lithium metal batteries featuring a positive electrode constructed from manganese dioxide. However, this pursuit encountered a series of hurdles, particularly revolving around safety concerns that emerged throughout the developmental phase. Drawing from personal experience, the author, who was actively involved in researching Li/MnO_2 systems at the Energizer Research Laboratory during this epoch, possesses an intimate familiarity with the intricacies of these safety incidents. Comprehensive testing of the cells illuminated troubling signs, notably dendritic lithium short circuits, which manifested as fluctuations in negative voltage during the charging process, alongside sporadic and unpredictable instances of cell explosions. Despite the setbacks posed by safety concerns, the allure of lithium metal technology persisted, owing to its remarkable specific energy and energy density characteristics. These tantalizing attributes presented a significant leap forward compared to previous aqueous systems, instigating considerable excitement within the battery industry. Central to the appeal of lithium metal batteries was their unparalleled performance potential, underpinned by lithium's exceptional cell potential and remarkably low atomic weight relative to alternative negative electrode materials. This unique combination endowed lithium metal batteries with the promise of achieving unprecedented levels of energy storage efficiency and density, thereby propelling them to the forefront of battery innovation endeavors. Thus, while the road to realizing the full potential of lithium metal batteries was fraught with safety challenges, the prospect of unlocking unparalleled energy storage capabilities served as a powerful catalyst for continued research and development efforts. Armed with invaluable insights gained from firsthand experiences and empirical observations, researchers remained undeterred in their quest to surmount the obstacles posed by safety concerns and unlock the transformative potential of lithium metal technology. In doing so, they aimed not only to revolutionize the landscape of energy storage solutions but also to usher in a new era of sustainable and efficient battery technologies that could power the world of tomorrow. In addition to the research conducted by Sony, exploration into lithium battery technology extended to include endeavors such as Exxon's Li/TiS_2 system and Bell Laboratories' Li/$NbSe_3$ system. Despite the promising potential of these systems, their advancement was hindered by various safety challenges that emerged during the investigative process. For instance, the Exxon system encountered significant setbacks in addressing safety concerns, primarily stemming from the vulnerability of both the lithium negative electrode and the composition of $LiClO_4$ salt and ether (typically dioxolane) used as a solvent. These components proved to be shock-sensitive and susceptible to explosion when subjected to strong shock conditions, highlighting critical safety vulnerabilities inherent in the system design. These safety concerns underscored the intricate balance between performance optimization and safety considerations inherent in the development of lithium battery technologies. While the pursuit of enhanced energy storage capabilities remained a driving force behind these endeavors, the imperative to ensure the safety and reliability of battery systems emerged as a paramount concern. The failure of the Exxon system to adequately address safety issues served as a sobering reminder of the complex interplay between technological innovation and risk mitigation within the realm of battery research and development. Despite the setbacks encountered by the Exxon system and other lithium battery technologies, these challenges catalyzed further exploration and innovation within the field. Researchers and engineers redoubled their efforts to devise novel approaches and design strategies aimed at mitigating safety risks while maximizing the performance potential of lithium battery systems. By addressing these critical concerns head-on, they sought to unlock the full spectrum of benefits offered by lithium battery technologies, paving the way for their widespread adoption and integration into a diverse array of applications spanning consumer electronics, transportation, and renewable energy storage. The constrained cycle life of lithium batteries was further linked to the suboptimal recycling efficiency of the lithium electrode. Many investigations resorted to incorporating excessive quantities of lithium metal within the cell, ostensibly in pursuit of heightened efficiency. However, upon the depletion of this surplus lithium, a rapid decline in capacity ensued with each subsequent cycle. This phenomenon underscored the intricate dynamics governing the interplay between electrode composition, cycling performance, and overall battery longevity. The propensity to augment lithium quantities within the cell configuration reflected a strategic attempt to bolster efficiency metrics. However, this approach proved to be counterproductive in the long run, as the excessive lithium supply was depleted over successive cycles, leading to a precipitous drop in capacity. This highlights the critical need for a nuanced understanding of the intricate balance between electrode composition and cycling behavior in lithium batteries. Moreover, the observed decline in capacity following the depletion of excess lithium underscored the complexity of optimizing battery performance while maintaining sustainability and longevity. These findings underscored the imperative for researchers and engineers to adopt more holistic approaches to battery design and optimization. By integrating insights gleaned from fundamental studies on electrode composition and cycling behavior, future advancements in lithium battery technology can strive to achieve a harmonious balance between efficiency, longevity, and sustainability. These unsuccessful attempts provided valuable insights into process and material requirements. It was discovered that secondary cells were much more sensitive to impurities such as water in electrolyte and electrode materials. Furthermore, it was found that processing methods differed from those of primary cells. This investigation contributed to the current outstanding state of Li-ion batteries. The current market for Li-ion batteries has evolved far beyond its initial scope, which primarily focused on small electronic devices for the 3C market. Nowadays, numerous additional markets have emerged, including toys, various types of lighting (such as liquid crystal display [LCD] and fluorescent lights), e-cigarettes, medical devices, and more. Furthermore, the realization that Li-ion battery packs using 18,650, 26,700, and 26,650 sizes can operate at significantly higher power levels than previously believed has led to the expansion of markets for portable electric tools, garden equipment, e-bikes, and various other products. While some high-energy 18,650 cells now boast capacities of up to 3.4 Ah, high-power cells have had to compromise on capacity in order to achieve continuous discharge capabilities of 20A or higher within the same cell size. Although certain cells claim capacities as high as 2.5 Ah, maintaining such high capacity levels during cycling proves to be challenging [61].

The fabrication of a Li-ion battery entails the integration of individual Li-ion cells, which can be interconnected in various configurations to optimize performance characteristics. These cells are commonly arranged in parallel to augment current delivery or in series to elevate voltage output, with the possibility of employing combined configurations to achieve tailored outcomes. The grouping of these cells into modules enables further scalability and customization, allowing for the assembly of multiple modules into a cohesive battery pack. A notable example is found in the battery pack of a typical Tesla vehicle, which boasts an energy capacity of 85 kW-hours and comprises an impressive array of 7104 individual cells. At the heart of a basic Li-ion cell lies a cathode, serving as the positive electrode, and an anode, serving as the negative electrode, both immersed within an electrolyte solution housing lithium ions. This fundamental configuration facilitates the essential electrochemical reactions underlying battery operation, wherein lithium ions shuttle between the cathode and anode during charge and discharge cycles. This intricate interplay between electrode materials and electrolyte composition forms the cornerstone of Li-ion battery technology, enabling the efficient storage and release of electrical energy to power a diverse range of applications, from portable electronics to EVs and renewable ESSs. Within a Li-ion battery, the critical components, namely the electrodes, are delineated by a separator, typically crafted from a microporous polymer membrane. This separator plays a pivotal role in enabling the seamless exchange of Li-ions between the electrodes while simultaneously impeding the flow of electrons, thereby facilitating the battery's electrochemical processes. It is worth noting that alongside conventional liquid electrolytes, researchers have explored alternative materials such as polymers, gels, and ceramics for use in Li-ion batteries, aiming to enhance performance, safety, and longevity. Despite significant strides in the development of electrode materials, electrolytes, and separators, the fundamental architecture of Li-ion cells has remained relatively unchanged since their commercialization by Sony over two decades ago. This steadfast adherence to the core principles of Li-ion battery design underscores the enduring reliability and efficacy of these energy storage devices. Despite ongoing innovations and advancements in materials science and battery technology, the basic framework established by early Li-ion batteries continues to serve as the foundation for contemporary battery designs. Indeed, the quest for improved performance and sustainability has spurred continual research and development efforts aimed at refining the constituent components of Li-ion batteries. Yet, the overarching objective remains consistent: to optimize energy storage capacity, efficiency, and safety while minimizing environmental impact. As such, while the specific materials and manufacturing processes may have evolved over time, the fundamental principles governing Li-ion battery design endure as a testament to their enduring relevance and utility in powering a myriad of modern-day applications. Addressing safety concerns is a crucial challenge highlighted by recent incidents of Li-ion battery fires involving Boeing 787 aircraft and Tesla Model S cars. This underscores the paramount importance of battery safety, particularly for ensuring the widespread acceptance of EVs and the continued growth of the Li-ion battery-powered vehicle market. Automakers need to make significant investments in battery management systems to enhance the safety of large battery packs in vehicles. Alternatively, efforts should be directed toward developing nonflammable Li-ion batteries, including those based on aqueous or ceramic electrolytes, as well as all-solid-state batteries. Future Li-ion batteries are likely to incorporate high-voltage cathodes and high-capacity anodes, necessitating intensive research into electrode materials to better understand their stability and interaction with electrolytes. Emphasis on battery safety should be increased alongside efforts to achieve high energy density. Standardized battery safety testing procedures should be widely adopted. The endeavor to create electrode materials with minimal environmental impact, often referred to as "green batteries," presents a multifaceted challenge in the realm of battery development. In an ideal scenario, forthcoming iterations of Li-ion batteries would integrate biologically derived organic or inorganic electrodes paired with aqueous electrolytes, thus paving the way for sustainable energy storage solutions. Key components such as carbon and silicon could be sourced from renewable biomass, aligning with the overarching goal of reducing carbon emissions and minimizing ecological footprint. Recent strides in research have sparked optimism surrounding the utilization of virus-assisted synthesis techniques for crafting electrode materials, offering a novel avenue for exploration in the pursuit of green battery technology. The prospect of scaling up these synthesis methods to achieve large-scale production at room temperature using biological templates, including genetically modified viruses, holds considerable promise for the future of sustainable battery manufacturing. Moreover, the development of organic electrodes engineered to withstand dissolution by electrolytes represents another promising avenue for enhancing the sustainability of Li-ion batteries. By leveraging organic materials with inherent resistance to degradation, researchers can mitigate the environmental impact associated with battery disposal and recycling processes, thereby fostering a more sustainable approach to energy storage technology. Looking ahead, future research endeavors should prioritize the advancement of sustainable and eco-friendly Li-ion batteries, with a concerted focus on optimizing electrode materials, electrolyte compositions, and manufacturing processes. By embracing a holistic approach to battery design that prioritizes environmental stewardship and resource efficiency, the transition toward green battery technology can be accelerated, ushering in a new era of energy storage solutions that are both environmentally sustainable and technologically advanced. Additionally, it is essential for future battery research to consider LCAs to evaluate the environmental impact of technologies. Little is known about the environmental implications of next-generation Li-ion batteries' production, use, and disposal. Mass production of Li-ion batteries for EVs may result in significant volumes of contaminated waste and environmental degradation. LCAs can provide a comprehensive understanding of environmental aspects and potential impacts throughout a battery's life cycle, from raw materials acquisition to disposal. Techno-economic analysis (TEA) of Li-ion batteries is also crucial to determine their technical and economic feasibility for large-scale production. Integrating LCAs and TEAs will help develop next-generation Li-ion batteries with optimized social, environmental, and economic impacts (Figure 8.3) [62].

Li-ion battery technologies have been utilized for several decades in sectors like portable electronics and power tools. However, their utilization in large-scale grid applications has only emerged recently, with significant research efforts aimed at enhancing their energy density, reducing costs, improving charging speed, and ensuring safety. Furthermore, ongoing efforts in the realm of large-scale battery systems are dedicated to advancing efficiency and optimizing performance across various components, including control systems, power electronics, and heating, ventilation, and air conditioning (HVAC) systems. At the heart of these systems lie multiple interconnected subsystems meticulously designed to facilitate operation, control, thermal regulation, and seamless integration with the grid. Employing a modular setup, these subsystems are intricately interdependent, with proper sizing playing a critical role in maximizing overall performance and financial viability. Given the inherent degradation of Li-ion cells over time, the design of large-scale battery energy storage systems (BESSs) often necessitates oversizing or allowance for reconfiguration to mitigate the impacts of cell degradation and ensure long-term operational resilience. This underscores the indispensable role of component modularity in accommodating evolving operational requirements and optimizing system longevity. It is imperative to acknowledge that the specific implementation scope of each subsystem may vary significantly depending on a

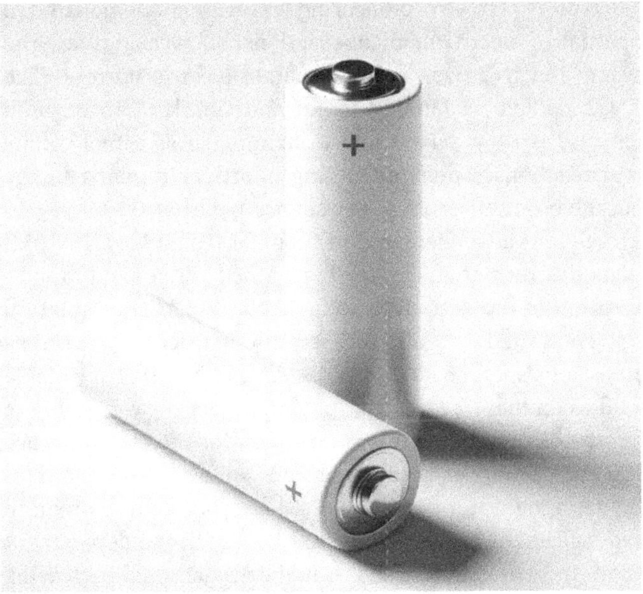

Figure 8.3 The Li-ion batteries.

multitude of factors, including the BESS manufacturer, the engineering, procurement, and construction (EPC) contractor, as well as the prevailing environmental conditions of the project site. This variability underscores the importance of adaptability and customization in tailoring large-scale battery systems to meet the unique needs and challenges of each individual project. [63].

8.3.4.5 Fuel Cell

In the annals of scientific history, the year 1801 marked a pivotal moment when Humphry Davy, a distinguished British chemist, embarked on groundbreaking investigations into electrolysis utilizing the voltaic pile. His pioneering experiments led to the revelation of several hitherto unknown metals, including sodium, potassium, and alkali metals, thereby reshaping our understanding of chemical elements and compounds. Davy's seminal research laid the groundwork for subsequent developments in electrochemical technology, including the emergence of fuel cells. Fast forward to 1838, when Christian Friedrich Schönbein, a German-Swiss chemist, made significant strides in the field by inventing fuel cells. Building upon Davy's work, Schönbein's innovative contributions marked a significant leap forward in harnessing electrochemical processes for practical applications. Just a year later, in 1839, William Robert Grove, a distinguished Welsh chemical physicist, achieved a major breakthrough with the creation of the first gas voltaic battery. Grove's pioneering experiment demonstrated the transformative potential of electrochemical reactions, showcasing how the interplay between oxygen and hydrogen could generate electric current. The term "fuel cell" itself was coined by Ludwig Mond and Charles Langer, both prominent British chemists, as they delved into utilizing coal as a fuel source to drive electrochemical reactions. Their pioneering efforts culminated in the achievement of a remarkable output of 20 A per square meter at 0.73 V, laying the groundwork for subsequent advancements in fuel cell technology. In 1932, Francis Bacon, an esteemed British engineer, further refined Langer and Mond's work, leading to the creation of the first alkaline fuel cell (AFC). The evolution of fuel cell technology continued to gain momentum, particularly with its integration into space missions spearheaded by NASA in the late 1950s. Notable milestones include the development of the first polymer electrolyte membrane fuel cell by Willard Thomas Grubb and Leonard Niedrach for the Gemini space program, as well as the utilization of alkali fuel cells in Apollo missions to provide vital power and drinking water to astronauts. The space shuttle program further propelled the advancement of fuel cell technology with the deployment of a 12 kW alkali fuel cell. In subsequent decades, fuel cells garnered increasing interest as potential power sources for EVs, culminating in their commercial availability since 2007. This ongoing trajectory of innovation and application underscores the enduring significance of fuel cells as a versatile and sustainable energy solution with far-reaching implications for various industries and sectors (Figure 8.4).

Fuel cells operate on a technology called proton exchange membrane fuel cell (PEMFC), which stands for proton exchange membrane or polymer electrolyte membrane fuel cell. These innovative cells incorporate a water-based and acidic polymer membrane serving as the electrolyte, coupled with electrodes catalyzed by platinum. Functioning primarily on pure hydrogen, they also have the capability to utilize reformed natural gas post-carbon monoxide removal. One of the distinguishing features of PEMFC is their operational temperature, which remains below 100 °C. This technology represents a significant advancement in sustainable energy solutions, offering a versatile and efficient means of generating electricity while minimizing environmental impact. Through the precise control of proton exchange across the polymer membrane, PEMFCs enable the seamless conversion of chemical energy into electrical power, making them ideal for a wide range of applications across industries, from transportation to stationary power generation. Another type, called high-temperature PEMFC (HT–PEMFC), is essentially a PEMFC, but it uses a mineral acid-based system as its electrolyte instead of a water-based one. Operating within a temperature range of up to 200°C, a variety of fuel cell technologies offer distinct advantages and applications. One such technology, direct methanol fuel cell (DMFC), features a polymer

Figure 8.4 A hydrogen vehicle refueling.

membrane as its electrolyte and leverages a platinum–ruthenium catalyst on its anode to directly utilize hydrogen sourced from liquid methanol. This innovative approach allows for efficient energy conversion while providing flexibility in fuel sources. Another noteworthy fuel cell variant is molten carbonate fuel cell (MCFC), which operates at substantially higher temperatures, typically around 650°C. This technology relies on molten carbonate salt contained within a porous ceramic matrix as its electrolyte. By functioning at elevated temperatures, MCFCs unlock unique benefits such as enhanced reaction kinetics and increased tolerance to impurities, paving the way for applications in power generation and industrial settings. The diverse array of fuel cell technologies underscores the versatility and potential of this clean energy solution across various sectors. From the efficient utilization of liquid methanol in DMFCs to the robust performance of molten carbonate salt in MCFCs, each technology offers distinct advantages tailored to specific operational requirements and environmental conditions. As research and development continue to drive innovation in fuel cell technology, these systems hold immense promise for powering a sustainable energy future. Phosphoric acid fuel cell (PAFC) employs an intricate silicon carbide framework adorned with finely dispersed platinum catalyst on both its anode and cathode. Within this sophisticated structure, phosphoric acid serves as the electrolyte, enabling efficient electrochemical reactions. Operating within a temperature range of up to 200°C, PAFC offers a compelling solution for various applications requiring reliable and sustainable energy generation. In contrast, solid oxide fuel cell (SOFC) relies on a robust solid ceramic electrolyte to facilitate proton conduction and oxygen ion migration across its electrodes. This innovative design ensures optimal efficiency and durability, positioning SOFC as a promising technology for stationary power generation and other demanding industrial applications. On the other hand, AFC adopts an alkaline electrolyte, which, when combined with pure hydrogen and oxygen, drives electrochemical reactions to produce electricity. This distinctive configuration offers advantages such as high efficiency and low emissions, making AFC a compelling choice for aerospace and other specialized applications where stringent performance requirements must be met. By harnessing the unique properties of different electrolytes and catalyst materials, these diverse fuel cell technologies collectively contribute to the advancement of clean and sustainable energy solutions. As research and development efforts continue to drive innovation in fuel cell technology, these systems hold immense potential to address pressing energy challenges and accelerate the transition toward a greener and more sustainable future. [64].

8.3.5 Mechanical Energy Storage Systems (MESSs)

Mechanical energy storage systems (MESSs) have gained substantial recognition as a cornerstone within contemporary energy management paradigms, particularly amidst escalating concerns surrounding sustainability and the imperative integration of renewable energy sources. MESS embodies a spectrum of meticulously engineered technologies aimed at the storage of energy in mechanical modalities. This encompasses the capture of potential energy, exemplified by methodologies like pumped hydro storage, alongside the retention of kinetic energy through mechanisms such as flywheel systems. These reservoirs of stored energy serve as pivotal assets, poised for efficient conversion back into electrical power upon demand, thereby assuming a critical role in harmonizing the delicate equilibrium between energy supply and demand dynamics [60].

The incorporation of MESS solutions contributes indispensably to the preservation of grid reliability and stability, constituting a linchpin in ensuring the seamless functionality and resilience of the electrical grid infrastructure. By virtue of their ability to facilitate the judicious allocation of energy resources, MESS technologies bolster the grid's capacity to accommodate the intermittency inherent in renewable energy sources, mitigating potential disruptions and enhancing overall system robustness. Moreover, the strategic deployment of MESS configurations engenders multifaceted benefits across diverse sectors and applications, transcending the mere provision of electrical power. Through their adept manipulation of mechanical energy storage principles, these systems afford opportunities for enhanced grid flexibility, demand response management, and optimization of energy utilization patterns. Such capabilities render MESS not only instrumental in addressing contemporary energy challenges but also instrumental in fostering a sustainable energy landscape conducive to long-term environmental preservation and socioeconomic advancement. The significance of multi-energy storage systems transcends its conventional role as a mere energy storage solution. Its impact is profound in bolstering the viability and efficiency of renewable energy sources, notably wind and solar power, which although sustainable, are beset by their inherent intermittency and unpredictability. The intermittent nature of these renewable sources poses formidable challenges to the management of energy grids, primarily attributable to their variable and non-dispatchable output. In this context, MESS emerges as a pivotal technology offering a multifaceted solution. At its core, MESS functions as a crucial buffer, adeptly mitigating the variability in energy production through the judicious storage of surplus energy during periods of high production and low demand. This stored energy reservoir is strategically tapped into during periods characterized by elevated demand or diminished production, thereby effectually smoothing out the oscillations in energy availability. Consequently, MESS serves as a linchpin in the endeavor to harmonize the intermittency of renewable energy sources with the requisite stability and reliability of modern energy grids. Moreover, from a theoretical perspective, the integration of MESS into the energy ecosystem necessitates a comprehensive understanding of complex interdisciplinary dynamics encompassing engineering, economics, and environmental sustainability. Its deployment entails intricate considerations ranging from technological compatibility and grid infrastructure optimization to economic feasibility and environmental impact assessment. In practical terms, the deployment of MESS promises to revolutionize the energy landscape, fostering a paradigm shift toward a more resilient, sustainable, and responsive energy infrastructure. Through its adept modulation of energy supply and demand dynamics, MESS not only augments the integration of renewable energy sources but also enhances grid resilience, mitigates energy market volatility, and facilitates the transition toward a low-carbon future [65].

Furthermore, the incorporation of renewable energy sources into the grid through the utilization of modern energy storage systems technologies offers multiple benefits. This integration not only decreases reliance on fossil fuels but also bolsters grid stability. By reducing the need for expensive peak-time energy generation methods, which frequently rely on nonrenewable energy sources notorious for environmental pollution, MESS facilitates a shift toward cleaner energy production. Consequently, this integration holds the promise of reducing energy expenditures, curbing greenhouse gas emissions, and fostering the development of a more sustainable and robust energy infrastructure, thereby contributing positively to environmental preservation and economic efficiency [66]. Moreover, beyond the aforementioned advantages, MESSs play a pivotal role in facilitating the transition toward a sustainable energy landscape by fostering enhanced efficiency and reliability in energy provision. A prime illustration of this is in the realm of pumped hydro storage, a prevalent form of MESS technology. This method involves utilizing surplus electricity to pump water to higher altitudes, thus storing potential energy. Subsequently, when demand surges, the stored water is released to flow downhill, powering turbines and generating electricity. This dual-purpose mechanism not only serves as a means of energy storage but also fulfills critical roles in peak load management and the provision of ancillary grid services such as frequency regulation and load following. In essence, MESS technologies like pumped hydro storage represent a multifaceted solution that not only addresses energy storage challenges but also contributes significantly to the overall efficiency and sustainability of energy systems [67]. Moreover, in the realm of cutting-edge mechanical storage technologies, options such as flywheels and CAES present compelling opportunities for bolstering the adaptability and agility of the grid in response to varying energy requirements. Flywheel systems function by harnessing kinetic energy through the continuous rotation of a rotor within a vacuum environment, enabling swift deployment of stored energy to provide grid support as needed. Conversely, CAES systems operate by storing energy in the form of compressed air, typically within subterranean caverns or specialized vessels, which can subsequently be released to power turbines for electricity generation during periods of peak demand. These innovative approaches hold considerable potential for optimizing grid efficiency and resilience in the face of dynamic energy dynamics [68]. In summary, the utilization of MESSs plays a crucial role in enabling the seamless incorporation of renewable energy sources into the existing power grid infrastructure. This integration is vital for fostering a more sustainable, efficient, and dependable energy ecosystem. As the world transitions toward a heightened dependence on renewable energy sources, the significance of MESS in upholding grid stability and sustainability escalates, solidifying its status as an indispensable component of the evolving global energy landscape. These systems not only offer a means of storing excess renewable energy generated during periods of peak production but also provide a reliable mechanism for supplying power during times of high demand or when renewable sources are not actively generating electricity. Their multifaceted functionality contributes to enhancing the overall resilience and reliability of the grid, thereby mitigating potential disruptions and ensuring uninterrupted energy supply to consumers. Moreover, the deployment of MESS facilitates the optimization of renewable energy utilization, enabling grid operators to balance supply and demand dynamics more effectively while minimizing wastage and maximizing efficiency. In essence, the continued advancement and widespread adoption of MESS are pivotal in driving the transition toward a more sustainable, resilient, and environmentally conscious energy infrastructure on a global scale [69].

8.3.5.1 Types of Mechanical Energy Storage Systems

Pumped Hydroelectric Storage (PHS) PHS emerges as a cornerstone within the array of large-scale energy storage solutions, playing a pivotal role in enhancing the dependability and adaptability of global electricity grids. Grounded in a straightforward yet profound principle, this technology leverages surplus electricity to tackle one of the most pressing challenges in contemporary energy landscapes: the effective integration of intermittent renewable energy sources such as wind and solar into the power grid. By harnessing excess electricity to propel water from a lower to a higher elevation reservoir, PHS transcends its basic operational tenets, evolving into a sophisticated energy storage mechanism. This strategy not only optimizes the utilization of surplus renewable energy but also establishes a reservoir of potential energy primed for conversion back into electrical power during peak demand periods or diminished renewable energy output. At the heart of PHS technology lies its dual-reservoir architecture, intricately connected by water conduits and outfitted with cutting-edge pump–turbine apparatus. During periods of low electricity demand or when renewable energy production surpasses current consumption levels, this surplus energy is channeled into pumps responsible for raising water to the upper reservoir, thereby converting and storing electrical energy as potential energy. The strategic selection of geographical locations for these reservoirs holds paramount importance, as the elevation disparity between them directly correlates with the potential energy storage capacity. A greater disparity in elevation denotes a larger reservoir of potential energy that can be tapped into and utilized later when needed [70]. During periods of peak electricity demand, pumped hydro storage (systems facilitate the release of stored water from elevated reservoirs, channeling it through turbines to regenerate electrical energy for distribution across the grid. Renowned for its adaptability, PHS exhibits remarkable responsiveness to fluctuations in both demand and supply, thus serving as a pivotal asset in maintaining grid stability. This versatility holds particular significance in the context of integrating variable renewable energy sources into the grid, as it offers a robust solution to address their inherent intermittency. Notably, PHS systems boast commendable operational efficiency, typically achieving round-trip energy efficiency rates ranging from 70% to 80%. Furthermore, their scalability and enduring service lifespan render them capable of expanding to meet evolving energy storage demands while reliably serving the grid for extended durations [71].

Despite necessitating significant environmental alterations during installation, such as the creation of reservoirs and modifications to water flows, contemporary PHS initiatives are increasingly characterized by efforts to mitigate ecological impacts. These endeavors encompass the integration of environmental safeguards, the preservation of local ecosystems, and the promotion of sustainable water management practices. Such initiatives underscore a paradigmatic shift within technology toward more environmentally conscientious implementations.

In summary, pumped hydro storage systems represent a versatile and efficient means of energy storage and grid stabilization, particularly well-suited to complementing the integration of renewable energy sources. Moreover, contemporary efforts prioritize environmental considerations, thereby reflecting a commitment to sustainable development practices within the realm of energy infrastructure.

The economic outlook for pumped hydro storage is increasingly promising, marked by its substantial operational advantages and its pivotal function in facilitating the smooth assimilation of renewable energy sources. The evolving regulatory incentives and financial structures aimed at bolstering energy storage technologies serve to augment the appeal of PHS, owing to its capacity to bolster energy security, reinforce grid stability, and advance objectives related to emissions mitigation [72]. Ongoing advancements within PHS technology persist in expanding its practical utility while mitigating associated constraints. Novel methodologies, including the implementation of artificial reservoirs, subterranean storage facilities, and utilization of seawater, are under examination to surmount geographical and ecological barriers, thereby enhancing the feasibility of PHS deployment across diverse landscapes. However, persistent challenges, notably substantial initial investment prerequisites, negotiation of intricate regulatory frameworks, and harmonization with overarching global sustainability objectives, endure [73].

In summary, PHS is emerging as a key technology in the development of a sustainable and resilient electricity infrastructure. Its unmatched capability to store and release energy on demand, especially against the backdrop of growing reliance on renewable energy, establishes PHS as a fundamental element of future energy systems. Through continuous innovation and strategic implementation, PHS not only addresses the immediate challenges of energy storage and grid stability but also plays a significant role in advancing sustainable energy transitions (Figure 8.5).

Flywheel Energy Storage (FES) Flywheel energy storage (FES) systems epitomize a cutting-edge technological paradigm designed for the efficient storage of electrical energy. These systems harness the inherent properties of rotating flywheels to store energy in the form of kinetic motion. The innovative methodology underlying FES involves the initial conversion of electrical energy into kinetic energy by imparting high velocities to the flywheel. Subsequently, when energy demand arises, this stored kinetic energy is reclaimed as electricity by decelerating the flywheel, thereby activating a generator that facilitates the reconversion of kinetic energy into electrical form. This intricate process is rooted in the fundamental

Figure 8.5 Hydroelectric storage.

principle of angular momentum conservation, underpinned by the seamless transmutation between kinetic and electrical energy states. Such sophisticated energy storage mechanisms hold considerable promise for addressing the escalating demands of modern energy grids while advancing the frontiers of sustainable energy technologies (Figure 8.6) [74].

The design and construction of FES systems are meticulously tailored around the central component, the flywheel, which holds pivotal significance. Contemporary FES implementations harness state-of-the-art materials such as carbon fiber composites in flywheel fabrication to attain an optimal amalgamation of robustness and lightweight characteristics. This delicate equilibrium serves to amplify the energy density of the flywheel while concurrently fortifying it against the formidable centrifugal forces encountered at heightened rotational velocities. Moreover, the integration of sophisticated bearing technologies, encompassing magnetic and air bearings, assumes a pivotal role in this paradigm. These bearings serve as instrumental elements in mitigating frictional losses by facilitating nearly friction-free rotation of the flywheel, thereby substantially augmenting the overall efficiency and durability of the system [75]. Additionally, the utilization of a flywheel within a controlled vacuum environment serves to significantly mitigate air resistance, thereby contributing to a notable reduction in energy dissipation

Figure 8.6 An illustration of flywheel energy storage system.

throughout the operational lifespan of the system. This particular facet underscores the paramount significance of curtailing operational energy consumption as a means to bolster the comprehensive efficiency of the system. Consequently, optimizing the conditions surrounding the flywheel's operation within a vacuum chamber represents a critical strategy in augmenting overall system performance and sustainability [76]. The role of power electronics in FES systems cannot be overstated. These components meticulously manage the electrical energy's flow into and out of the flywheel, crucial for optimizing the charging and discharging cycles. Power electronics ensure the system's operational efficiency is maximized and that it can swiftly respond to fluctuating energy demands. One of the most lauded benefits of FES technology is its exceptional energy efficiency, with systems often achieving round-trip efficiencies exceeding 90%. This efficiency, combined with the system's robustness and low maintenance needs, renders FES a cost-effective solution over its operational lifespan. Furthermore, FES systems offer environmental advantages by not depending on chemical processes or containing hazardous substances, starkly differentiating them from many conventional energy storage technologies [77]. The capability of FES systems to instantaneously release energy renders them invaluable for applications requiring immediate power delivery. This attribute is particularly crucial for stabilizing electrical grids, where FES systems can immediately counterbalance demand or supply fluctuations, thus bolstering the reliability and stability of power networks. Beyond grid stabilization, FES technology facilitates the integration of renewable energy by storing surplus energy from sources like wind and solar for subsequent utilization. This function enables a smoother transition of renewable energies into the grid, fostering a more sustainable and resilient energy framework. Furthermore, the application of FES systems extends to the transportation sector, notably in regenerative braking systems for trains and buses, enhancing energy conservation by recuperating and reutilizing kinetic energy during braking [78]. Despite the myriad advantages offered by FES systems, challenges such as elevated initial costs and the necessity for advancements in energy density and scalability remain. These challenges are central to ongoing research efforts aimed at developing more economical and efficient FES solutions. Future innovations in materials science could yield lighter, more durable flywheels capable of higher-speed operations, thus augmenting the energy storage capabilities of FES systems. Similarly, progress in bearing technology and power electronics is vital for boosting overall system performance and reducing operational expenses [79]. In conclusion, FES systems present a forward-looking and efficacious approach to energy storage, characterized by high efficiency, swift response capabilities, and environmental friendliness. As technological advancements continue to surmount current limitations, FES technology is set to play an increasingly pivotal role in global energy management strategies, significantly contributing to the shift toward more sustainable and reliable energy systems.

Compressed Air Energy Storage (CAES) CAES emerges as a critical technology in the renewable energy storage sector, ingeniously leveraging a simple yet profoundly efficient principle: utilizing surplus energy to compress air into a contained space, which can then be released to generate electricity through the expansion process when needed. This technology symbolizes a significant shift toward sustainable energy solutions, serving as a beacon of hope for addressing the intermittency challenges associated with renewable energy sources such as wind and solar power and it is an innovative method for storing energy, leveraging the simple concept of using compressed air as a medium to store and release energy. This method effectively tackles the intermittent nature inherent in renewable energy sources like solar and wind power. It achieves this by providing a means to store surplus energy generated during peak production periods, thereby circumventing wastage and subsequently releasing this stored energy during periods of heightened demand or reduced production capacity. By seamlessly balancing supply and demand dynamics, this approach not only enhances the reliability and stability of renewable energy systems but also maximizes their overall efficiency and utility, paving the way for a more sustainable and resilient energy landscape. The process of CAES involves two main stages: the storage of energy and the release of energy. Initially, electrical energy, which can be sourced from the grid or directly from renewable energy installations, is used to operate a compressor. This compressor sucks in atmospheric air, compresses it to high pressure, and stores it in a containment area, which could be an underground cavern or above-ground tanks designed to handle the high pressure. The process of compressing air within CAES systems initiates a notable heat generation phenomenon, particularly evident in advanced CAES configurations. In these sophisticated setups, the generated heat can be harnessed and preserved utilizing thermal storage technologies, thereby enhancing the system's overall efficiency. By capturing and storing this thermal energy, CAES systems optimize energy utilization and minimize wastage, contributing to a more sustainable energy ecosystem. When the stored energy is required, the compressed air is meticulously released from its storage reservoir and directed through a turbine. This carefully orchestrated process facilitates the conversion of stored pneumatic energy into mechanical energy, which is then further transformed into electrical energy through the operation of a generator. By seamlessly transitioning between energy storage and release phases, CAES systems offer a reliable and flexible solution for balancing supply and demand dynamics within the power grid. Furthermore, the integration of thermal storage technologies into CAES systems represents a significant advancement in energy storage technology. By effectively capturing and utilizing excess heat generated during the compression process, these innovative systems maximize energy efficiency and minimize environmental impact. This holistic approach to energy storage not only enhances the overall performance of CAES systems but also underscores their potential as a key enabler of renewable energy integration and grid stability initiatives. Incorporating thermal storage technologies into CAES systems marks a significant stride toward achieving greater energy efficiency and sustainability. This integration not only optimizes the utilization of excess heat but also enhances the overall performance and reliability of CAES systems. By harnessing the synergies between compressed air storage and thermal energy capture, these integrated systems offer a promising solution for addressing the evolving challenges of energy storage and grid management in an increasingly renewable-centric energy landscape. Before or during expansion through the turbine, the air is heated, either by burning a fuel, such as natural gas or by using the previously captured heat from the compression stage. This heating process increases the energy content of the expanding air, thereby enhancing the efficiency of electricity generation. The turbine is connected to a generator, which converts the mechanical energy of the expanding air into electrical energy that can be fed back into the grid. The efficiency of CAES systems is influenced by the efficiency of the compressors and turbines used, as well as by the effectiveness of the heat management system. Advanced CAES systems aim to maximize efficiency by improving these components and processes, potentially reducing or eliminating the need for fossil fuels by effectively utilizing captured heat [80]. At its essence, CAES capitalizes on the excess electricity available during periods of low demand or high renewable production. This surplus electricity powers an air compressor that forces air into an underground cavern or a specially designed storage vessel, thus storing the energy in the form of high-pressure compressed air. When the demand for electricity experiences a sudden surge or when there's a dip in the availability of renewable energy sources, the stored compressed air within the system comes into action. This stored energy is released from its containment and undergoes a carefully orchestrated process. First, it is subjected to a heating phase, where the thermal energy captured during the compression stage is utilized to further elevate its temperature. Subsequently, the heated compressed air is directed through a turbine, where it undergoes expansion. This expansion process drives the turbine's rotation, effectively converting the stored pneumatic energy into mechanical energy. Finally, this mechanical energy is harnessed to generate electricity through the operation of a generator, completing the transformation process. The electricity generated in this manner is seamlessly integrated back into the grid, where it serves to meet the immediate demand. By effectively bridging the gap between fluctuating energy supply and demand, this process plays a pivotal role in maintaining grid stability and reliability. Moreover, it facilitates the optimal utilization of renewable energy sources by providing a means to store excess energy during periods of abundance and release it when needed, thereby maximizing overall energy efficiency and sustainability. In essence, the release, heating, and expansion of stored compressed air through a turbine represent a critical mechanism for balancing the dynamics of energy supply and demand, ensuring a consistent and reliable electricity supply for consumers while advancing the transition toward a greener and more resilient energy infrastructure [81, 82]. The versatility of CAES is one of its most striking features. It can be implemented in a variety of geological formations, including salt caverns, aquifers, and depleted natural gas fields, each providing unique advantages in terms of scalability, storage capacity, and geographical distribution [67]. This flexibility allows CAES to be a highly adaptable solution, overcoming the geographic and infrastructural limitations of different regions and facilitating localized optimization of energy storage and distribution networks. Moreover, CAES systems are noted for their longevity and durability, offering sustainable energy storage solutions that can last for decades, thereby complementing the transient nature of some renewable energy sources [83]. A pivotal attribute of CAES is its potential to enhance grid stability and reliability. By acting as a buffer against the variability of renewable energy sources, CAES plays a crucial role in maintaining a steady electricity supply, even amidst fluctuating supply levels. This capability is essential for integrating higher percentages of renewable energy into the grid, a vital step toward reducing carbon emissions and combating climate change [84]. Additionally, CAES systems can be quickly mobilized to address sudden spikes in energy demand, providing grid operators with a crucial tool for managing load and ensuring uninterrupted power supply [80]. Despite its numerous benefits, CAES technology encounters challenges and limitations. The efficiency of CAES systems, particularly concerning the energy expended to compress air versus the energy generated during expansion, remains a focal point of ongoing research and development. Technological advancements, including the integration of TES to minimize energy losses during the compression and expansion phases, are expected to improve the efficiency and economic feasibility of CAES systems [81]. Furthermore, the substantial initial capital investment and the necessity for suitable geological sites for air storage are significant barriers to the widespread adoption of CAES. Nevertheless, with continuous innovation and technology scaling, these obstacles are progressively being surmounted, clearing the path for CAES to play an indispensable role in the future energy landscape [85]. In conclusion, CAES stands as a testament to human ingenuity in the pursuit of sustainable energy solutions. By providing a reliable and adaptable method for storing surplus renewable energy, CAES not only addresses the inherent challenges of energy intermittency but also contributes to the stability and resilience of the electrical grid. As the world advances toward a future dominated by renewable energy sources, technologies like CAES will be essential in ensuring that this transition is both sustainable and seamless, marking a significant step in the journey toward a greener, more sustainable future.

Gravity Energy Storage Gravity energy storage is predicated on the utilization of the omnipresent force of gravity to effectuate the conversion of potential energy into electrical energy.

The emergent methodology of power storage, characterized by its sustainability and operational efficiency, is increasingly captivating both scholarly and practical interest, particularly when integrated with renewable energy sources such as wind and solar power. This approach presents a promising avenue to address the pressing need for effective energy storage solutions in the context of the transition toward renewable energy. Its capacity to store excess energy generated during peak production periods of intermittent renewables enhances grid stability and reliability while maximizing overall energy efficiency. Moreover, its scalability and versatility render it suitable for a wide range of applications, spanning from grid-scale storage to decentralized MG solutions, thereby contributing to the global efforts toward a low-carbon future. The growing attention toward this innovative storage methodology reflects a significant shift in energy paradigms, emphasizing sustainability and resilience in the pursuit of a more environmentally friendly and sustainable energy landscape. The foundational principle underpinning gravity energy storage is elegantly simplistic yet ingeniously efficacious: it revolves around the elevation of substantial masses during periods characterized by energy surplus, subsequently orchestrating their deliberate descent during instances of heightened energy demand. This orchestrated descent facilitates the seamless reconversion of potential energy into kinetic energy, thereby engendering a conducive milieu for the subsequent transformation of kinetic energy into electrical energy through adeptly integrated generators. Such a conceptual framework not only underscores the feasibility of gravity energy storage as a viable alternative within the contemporary energy landscape but also underscores its potential to address the exigencies of sustainable power provision in an increasingly dynamic energy ecosystem (Figure 8.7) [86].

Gravity energy storage operates on a fundamental yet robust principle: it leverages surplus electricity to elevate a substantial mass to a heightened position, thereby storing energy within the gravitational potential framework. Upon demand for electricity, this process undergoes reversal wherein the elevated mass is permitted to descend, driving a generator using its kinetic energy, thereby transforming it back into electrical energy for incorporation into the power grid. This approach capitalizes on the innate laws of gravity, wherein an object's elevation imparts potential energy. The initial phase entails utilizing excess electricity, frequently sourced from renewable outlets such as wind or solar,

Figure 8.7 Energy storage with gravity law state-of-the-art.

particularly during their peak production periods, to energize electric motors. These motors undertake the task of lifting a mass, which may encompass solid entities like rocks or concrete blocks or alternatively involve the elevation of water via pumping mechanisms. As long as the mass remains positioned at an elevated state, energy continues to be stored within the system.

Subsequently, when electricity demand surpasses supply levels, the stored potential energy undergoes conversion back into electricity. This is executed by initiating the descent of the mass, thereby allowing gravity to facilitate its downward trajectory. As the mass descends, it propels a generator, thus effectuating the conversion of motion into electrical energy.

The efficiency of this mechanism is contingent upon the efficacy of the employed motors and generators, alongside the elevation from which the mass is released. While acknowledging that no ESS is devoid of losses attributable to factors such as friction and air resistance, it is noteworthy that gravity ESSs can attain commendable efficiency rates. Furthermore, they exhibit potential for extended operational lifespans with minimal degradation in storage capacity. Gravity energy storage demonstrates scalability and adaptability across diverse contexts and environments, ranging from modest-scale installations to expansive endeavors utilizing natural topographical features. Importantly, its environmental footprint remains relatively diminutive, particularly in contrast with chemical battery alternatives, given its reliance on mechanical energy storage sans the necessity for deleterious chemicals or emissions [87].

This innovative approach primarily hinges on the manipulation of substantial masses to harness energy. In times of surplus energy, these masses are elevated to higher altitudes, effectively storing potential energy. Conversely, during periods of heightened demand, these masses are gradually lowered, thereby converting potential energy into kinetic energy. Subsequently, sophisticated generators and mechanical systems facilitate the conversion of kinetic energy into electrical energy. The elegance of gravity energy storage lies in its reliance on gravity, a force characterized by its constancy and predictability. The predictability inherent in gravity's operation renders gravity energy storage an appealing solution for bolstering the resilience of the electrical grid. This is particularly relevant with the increasing integration of renewable energy sources, which inherently exhibit intermittency in their energy output due to weather patterns and diurnal variations. The capacity to store excess energy becomes imperative in managing the fluctuations in energy production associated with renewable sources. The adoption of gravity energy storage introduces a compelling opportunity to augment the dependability and robustness of energy systems. By leveraging the fundamental principles of gravity, this approach provides a reliable mechanism to tackle the intermittent nature of renewable energy sources, thereby marking a significant stride in the pursuit of sustainable and resilient energy solutions.

This innovative method holds promise in mitigating the variability inherent in renewable energy generation, offering a dependable means to store and release energy as needed. Consequently, it plays a pivotal role in enhancing the overall stability and reliability of energy systems, contributing to the ongoing efforts to transition toward a more sustainable and environmentally conscious energy landscape [88]. Recent technological advancements have significantly improved the viability and efficiency of gravity energy storage. While traditional forms of gravitational storage, such as PHS, have been utilized for decades, newer iterations aim to overcome some of the limitations associated with these traditional methods. For example, pumped hydro requires large bodies of water and is constrained by geographical and environmental factors. In contrast, modern gravity ESSs employ solid weights, ranging from massive concrete blocks to specially designed composite materials. These can be elevated and lowered using cranes, rails, or other mechanical means, thus obviating the need for large water bodies and expanding the potential for wider geographical implementation [86]. Gravity energy storage presents a conspicuous advantage owing to its inherent capacity for prolonged-duration energy retention. Unlike conventional battery technologies, which are susceptible to degradation over time and possess finite limitations concerning both storage capacity and duration, gravity-based storage systems exhibit the capability to retain energy over extended periods with minimal loss. This enduring resilience renders gravity energy storage systems particularly adept at fulfilling the exigencies of grid stabilization across protracted temporal cycles, spanning days to weeks, rather than merely accommodating short-term fluctuations on an hourly basis. Moreover, the requisite physical infrastructure integral to gravity energy storage, encompassing weight masses and intricate mechanical apparatus, manifests a protracted operational lifespan. This characteristic engenders the potential for reduced lifecycle costs and a mitigated environmental footprint over the system's operational tenure, particularly when juxtaposed with the environmental impact associated with chemical battery technologies. The enduring robustness of gravity-based storage mechanisms thus underscores their viability as sustainable and economically advantageous alternatives for addressing the evolving demands of contemporary energy grid management [89]. Despite its considerable potential, the development and deployment of gravity ESSs face several challenges. The initial investment required for setting up such systems can be substantial, given the need for robust infrastructure capable of lifting and supporting massive weights. Additionally, there are engineering challenges related to optimizing the efficiency of energy conversion and minimizing energy losses during the storage cycle. Furthermore, while the environmental impact of gravity energy storage is generally lower than fossil-fuel-based storage methods, the construction and operation of these systems can still have significant environmental and ecological implications.

These impacts vary depending on the scale and location of the gravity ESSs [90]. As the global community continues to pivot toward renewable energy sources, the demand for innovative and efficient energy storage solutions will undoubtedly escalate. Gravity energy storage stands out as a promising option, offering a blend of scalability, sustainability, and efficiency. The successful development and integration of gravity energy storage into the energy system will necessitate collaborative efforts among engineers, scientists, policymakers, and industry stakeholders. Overcoming the present challenges will be crucial to unlocking the full potential of gravity energy storage. In achieving this, gravity energy storage could play a pivotal role in realizing a sustainable energy future. It offers a reliable means of storing and delivering renewable energy, thereby helping to stabilize the electrical grid amidst increasing variability and demand [86].

Spring Energy Storage Spring energy storage has emerged as a pioneering innovation within the realm of renewable energy, heralding a novel approach for the effective retention and discharge of energy via the utilization of mechanical springs. This cutting-edge technological advancement is rooted in the fundamental principle of elasticity, whereby materials possess an intrinsic capability to amass potential energy when subjected to compressive or tensile forces. In contrast to conventional energy storage methodologies such as batteries or PHS systems, spring energy storage systems (SESSs) distinguish themselves by their simplicity, scalability, and limited environmental impact. This innovative paradigm in energy storage is predicated upon the utilization of mechanical springs as a means to capture and subsequently release energy. Unlike traditional methods that rely on chemical reactions or hydraulic mechanisms, SESS harnesses the inherent properties of elastic materials to store and transmit energy efficiently. Through the application of compressive or tensile forces, these springs accumulate potential energy, which can be readily accessed and utilized when required. This departure from reliance on complex and resource-intensive systems underscores the elegance and ingenuity of spring energy storage.

Moreover, the scalability of SESS offers significant advantages in adapting to varying energy demands and infrastructural requirements. By modularizing the storage units and adjusting the compression or tension levels of the springs, the storage capacity of SESS can be tailored to meet specific needs, whether at the individual or grid level. This flexibility not only enhances the reliability and resilience of energy storage solutions but also facilitates their integration into diverse settings and applications. Furthermore, the environmental benefits of SESS cannot be overstated. Unlike some conventional storage methods that rely on hazardous materials or involve extensive land and water usage, SESS offer a notably reduced environmental footprint. The absence of toxic chemicals or reliance on large-scale infrastructure minimizes adverse ecological impacts, aligning with the imperative of sustainable energy development [91]. At the core of the stored energy spring system lies its fundamental constituents: a mechanical spring and an apparatus designed for the compression or elongation of said spring. These springs are meticulously fashioned from a diverse spectrum of materials, including but not limited to steel, composite materials, and state-of-the-art polymers, each meticulously chosen for their exceptional elasticity and resilience. This selection process ensures the robustness and longevity necessary for the system's optimal functionality.

During periods characterized by surplus energy supply surpassing demand, typically observed during intervals of heightened renewable energy generation or reduced consumption, the SESS capitalizes on this abundance through the employment of external forces. These forces, often in the form of electric motors or hydraulic pistons, are strategically utilized to compress the spring, thereby imbuing it with potential energy, which becomes intricately stored within its molecular structure. This sophisticated mechanism underscores the intricate interplay between material science, mechanical engineering, and energy storage principles within the framework of SESS. By harnessing cutting-edge materials and sophisticated compression techniques, SESS exemplifies a paradigm shift in energy storage methodologies, offering a promising avenue toward sustainable energy utilization and grid stability optimization [92]. The latent potential energy contained within the spring is preserved in a quiescent state until a demand arises for its liberation. During such instances, a mechanism is activated within the system, inducing a rapid expansion or contraction of the spring. This mechanical manipulation engenders the conversion of potential energy into kinetic energy, subsequently transmuted into electrical energy via generators or analogous conversion apparatus. This intricate process facilitates a seamless interplay between kinetic and potential energy reservoirs, thereby enabling the efficient storage and retrieval of energy. Such functionality assumes paramount importance in reconciling the intermittent output of renewable energy sources with the variable exigencies of the energy grid, thereby enhancing the overall stability and reliability of the energy infrastructure [93]. One of the foremost advantages of SESS is its inherent scalability and modular design. Contrary to the requirements of large-scale PHS, which necessitates specific geographical conditions and substantial infrastructure, SESS can be implemented in a wide variety of settings, ranging from densely populated urban areas to isolated rural locations. Furthermore, the system's design allows for easy adjustment in scale to meet different energy storage needs, making it applicable for both large-scale utility applications and smaller, distributed energy storage scenarios [94]. Moreover, SESS presents significant environmental benefits over traditional energy storage technologies. By relying on mechanical rather than chemical processes, it circumvents the environmental hazards associated with the production and disposal of batteries, such as the extraction of resources, the handling of toxic chemicals, and recycling challenges. Operationally, SESS has a negligible environmental impact, emitting no greenhouse gases or harmful by-products [95]. In addition to environmental merits, SESS showcases exceptional efficiency and longevity. Unlike battery systems, which suffer from degradation and a finite number of charge–discharge cycles, mechanical springs are capable of enduring millions of cycles with minimal wear and tear, ensuring long-lasting and reliable performance. This resilience leads to lower maintenance needs and reduced lifecycle costs, further bolstering the economic appeal of spring energy storage solutions [96]. In summation, spring energy storage exemplifies a pioneering approach to energy storage, offering a viable, scalable, and efficient alternative for addressing the integration challenges of renewable energy sources into the electrical grid. By exploiting the principle of elasticity, SESS holds the promise of revolutionizing energy storage and utilization, setting the stage for a more sustainable and resilient energy future.

8.3.5.2 A Comprehensive Comparative Analysis of Mechanical Energy Storage Systems

The ongoing pursuit of energy storage solutions that are not only efficient and cost-effective but also environmentally sustainable is of paramount importance as societies worldwide endeavor to transition toward resilient, low-carbon energy systems. Amidst this imperative, MESS has emerged as pivotal components, offering a diverse range of options to contend with the intermittent nature of renewable energy sources and bolster grid stability. This expanded analysis seeks to delve comprehensively into the intricate landscape of several key MESS technologies, notably including PHS, FES, CAES, gravity energy storage (GES), and spring energy storage (SES). By focusing on four critical dimensions—namely, efficiency, cost, scalability, and environmental impact—this examination aims to provide a thorough understanding of the capabilities and limitations of each technology within the broader context of energy storage. Within the academic discourse surrounding energy systems and sustainability, such an investigation serves to illuminate the nuanced interplay between technological innovation, economic viability, and environmental stewardship, thereby contributing to informed decision-making processes and the advancement of energy transition agendas on a global scale. The assessment of ESSs encompasses multifaceted considerations ranging from efficiency and cost-effectiveness to scalability and environmental impact. Within this purview, MESS hold prominence for their capacity to address the intermittency inherent in renewable energy sources and enhance grid stability. Diving into the intricacies of various MESS technologies reveals a nuanced landscape ripe for analysis. Efficiency serves as a fundamental metric in evaluating the performance of MESS technologies. PHS systems, renowned for their high round-trip efficiency, typically ranging between 70% and 80%, capitalize on the mature and simplistic conversion of gravitational potential energy into electricity through water displacement. In contrast, FES technologies boast efficiencies exceeding 85%, owing to minimal energy dissipation during both storage and retrieval phases. CAES systems, however, exhibit variable efficiencies influenced by factors such as thermodynamic processes and geological characteristics. GES and SES systems showcase efficiency profiles contingent upon nuanced design features and operational parameters, adding complexity to their evaluation [67]. The economic viability of MESS technologies hinges on considerations of initial capital investments, lifecycle costs, and operational expenses. Despite substantial upfront costs associated with infrastructure development, PHS systems often demonstrate favorable lifecycle costs owing to their longevity and low maintenance requirements. FES systems, while initially capital-intensive due to precise engineering needs, offer competitive lifecycle costs driven by durability and minimal maintenance demands. Conversely, CAES encounters challenges related to cavern development and equipment expenses, potentially impacting its economic feasibility. GES and SES exhibit cost advantages at smaller scales, yet material constraints and scalability challenges may restrict their applicability for larger-scale implementations [60]. Scalability emerges as a critical aspect in adapting MESS technologies to evolving energy storage demands. PHS systems exhibit remarkable scalability, ranging from small-scale installations to massive multi-gigawatt systems capable of regional support. FES systems offer modularity, enabling capacity adjustments through the addition of flywheels, thereby facilitating scalability. CAES scalability is contingent upon geological and infrastructural factors, which may constrain deployment possibilities. Conversely, GES and SES face inherent scalability limits due to geographical and material constraints, necessitating careful consideration during deployment planning [42]. Environmental impact assessment is indispensable in evaluating the sustainability of MESS technologies. PHS systems typically incur low environmental footprints by leveraging existing water bodies with minimal emissions. FES systems operate with environmental benignity, devoid of emissions or hazardous materials. However, CAES may necessitate rigorous environmental assessments due to concerns surrounding underground storage and energy-intensive compression/expansion cycles. GES and SES, while potentially low-impact, require meticulous site selection and material considerations to mitigate environmental risks effectively [67]. In conclusion, the comprehensive analysis of MESS technologies underscores the complex interplay between technological innovation, economic feasibility, and environmental sustainability. By evaluating efficiency, cost-effectiveness, scalability, and environmental impact, stakeholders can make informed decisions aligned with their specific needs and sustainability goals. Furthermore, ongoing advancements in materials science and engineering hold promise for enhancing the performance and viability of MESS technologies in fostering a resilient and low-carbon energy landscape.

8.3.5.3 Applications and Case Studies of Mechanical Energy Storage Systems (MESSs)

Utility-Scale Energy Storage The Bath County Pumped Storage Station, nestled in the verdant landscapes of Virginia, United States, represents a monumental achievement in the realm of MESS. Boasting an impressive capacity exceeding 3 gigawatts, this facility not only stands as a testament to the vast potential of PHS technology but also delineates the critical role such systems play in addressing the complex energy storage needs of expansive utility grids. The operational dynamics of the Bath County station are predicated on the strategic utilization of two reservoirs positioned at disparate elevations. This ingenious configuration facilitates the storage of surplus energy during periods of diminished demand by pumping water to the elevated reservoir. Conversely, during peak demand intervals, this stored energy is unleashed by allowing the water to flow back down, thereby generating electricity through hydroelectric turbines. This cyclical process is instrumental in fostering equilibrium between energy supply and demand, thereby ensuring a stable and reliable grid operation. Furthermore, the ancillary services provided by such facilities, encompassing frequency regulation, load management, and emergency power supply, are indispensable for the seamless integration of renewable energy sources. These services significantly bolster the grid's resilience, facilitating a transition toward a more sustainable and renewable energy landscape [97].

Grid Stabilization In the arena of grid stabilization, FES systems emerge as paragons of innovation, offering unparalleled rapid response capabilities and precision control. The Beacon Power Flywheel Energy Storage Plant, situated in Stephentown, New York, epitomizes the application of this technology in augmenting grid stability. FES systems distinguish themselves through their ability to either absorb or dispatch energy within the span of milliseconds, a feature that is paramount in maintaining the grid frequency within the stipulated norms. This rapid modulation capability ensures the continuity of a stable and reliable electricity supply, a critical requirement in today's energy landscape. The flywheel technology, characterized by its high efficiency, expeditious response times, and a virtually unlimited life cycle, stands as a formidable solution to the challenges of grid stabilization. By complementing existing power generation and transmission infrastructures, FES systems like those deployed at the Beacon Power Plant significantly enhance the grid's overall resilience, ensuring its capability to withstand and adapt to the evolving demands of energy consumption [78].

Renewable Energy Integration The incorporation of renewable energy sources into existing grid infrastructure is a multifaceted challenge, one that demands innovative solutions capable of navigating the complexities of energy storage and management. In this arena, MESS emerges as a compelling option, offering a diverse array of capabilities to address the dynamic nature of renewable energy generation. The Hornsdale Power Reserve, located in South Australia, stands as a shining example of ingenuity in this field. Through the fusion of Li-ion battery storage with FES solutions, this facility showcases the transformative potential of MESS technologies in facilitating the seamless integration of renewable energy into the grid. More than just a storage facility, the Hornsdale Power Reserve serves as a versatile hub for a myriad of grid services essential for maintaining stability and reliability. From frequency regulation to voltage support, the facility's capabilities span a spectrum of vital functions crucial for accommodating the intermittent nature of renewable energy sources like wind and solar. By providing rapid response capabilities to fluctuations in energy production and demand, the Hornsdale Power Reserve exemplifies the agility and resilience inherent in MESS technologies. This versatility and responsiveness not only ensure a harmonious integration of renewable energy into the grid but also lay the groundwork for a sustainable energy future. As societies worldwide strive to reduce reliance on fossil fuels and transition toward cleaner, more sustainable alternatives, innovative storage solutions like MESS play a pivotal role in enabling this transition. By harnessing the power of mechanical energy storage, facilities like the Hornsdale Power Reserve pave the way for a future characterized by enhanced grid stability, reduced carbon emissions, and a greater reliance on renewable energy sources [67].

Cross-Sector Applications Beyond their pivotal role in the energy sector, MESS have demonstrated remarkable versatility and adaptability, finding applications across a diverse array of industries. In the realm of transportation, for instance, FES systems have been increasingly adopted in the design of hybrid vehicles, including buses and trains. These systems are ingeniously designed to capture and store kinetic energy during braking phases, which is subsequently utilized for propulsion. This innovative approach not only enhances the overall energy efficiency of these vehicles but also contributes to a significant reduction in greenhouse gas emissions, aligning with global sustainability goals. Furthermore, the application of MESS technologies extends into the industrial manufacturing sector, where they serve as reliable backup power sources. This application is crucial in ensuring the continuity of operations during power outages, thereby mitigating the risk of operational disruptions. The broad applicability and adaptability of MESS technologies across these sectors underscore their potential to drive innovation and sustainability, heralding a new era of energy efficiency and environmental stewardship [98].

8.3.6 Limitations

Following the examination of these storage systems, our intention in this section is to discuss their favorable and unfavorable aspects. The drawbacks of ESSs typically revolve around the inefficient performance of the storage device in static situations, resulting in excessive self-discharge. Furthermore, they are not ideal for extended storage due to friction issues. In recent times, more sophisticated bedding materials like high-temperature superconductors (HTS) have been employed to minimize idle losses and alleviate self-discharge in storage devices. However, utilizing HTS necessitates liquid nitrogen for cooling at lower temperatures, thereby raising the initial investment required for the ESS. Li-ion batteries are susceptible to environmental influences, which can impact their cyclic depth of discharge and longevity, necessitating the implementation of internal overcharge protection mechanisms. Consequently, they may not be deemed suitable for applications mandating complete discharge. Furthermore, there exists a concern regarding the potential release of oxygen from metal oxide electrodes due to undesirable side reactions at elevated temperatures. Hence, Li-ion batteries commonly incorporate a temperature control apparatus to uphold operational parameters within optimal ranges. Present research endeavors predominantly center on the incorporation of nanostructured materials and graphite within Li-ion batteries to augment their charging/discharging kinetics, extend cycle life, and enhance overall performance. Additionally, emphasis is placed on the utilization of advanced electrode materials and electrolyte formulations to bolster specific energy density. The drawbacks of LA batteries include their shorter lifespan in electrochemical cells, their adverse environmental effects, and their lower energy density. Current research endeavors typically concentrate on mitigating these drawbacks by employing novel materials. For instance, the development of alternative materials like lead–antimony batteries, valve-regulated sealed LA batteries, and starting, lighting, and ignition batteries offers the potential to extend cycle life and improve discharge capacity. Moreover, investigations into integrating LA batteries with renewable energy sources represent a significant advancement in meeting the demands of intermittent wind and photovoltaic energy generation. Among the disadvantages of NaS batteries are commonly high corrosion, high internal resistance, and high annual operating costs. These types of batteries need to be maintained in a molten state of approximately

300°C to function effectively, necessitating the use of a temperature control device. Additionally, various serious safety issues such as explosions and fires may arise in the event of molten electrode contact. Current research efforts generally focus on improving the performance of NaS batteries and reducing the requirement for high operating temperatures. In this context, studies on different low-temperature NaS batteries are extensively examined in the literature. Furthermore, nickel–cadmium batteries are susceptible to memory effects, which means their rated capacity may decrease significantly due to improper discharging and recharging processes. On the other hand, Ni–MH batteries have a lower memory effect, allowing them to disregard environmental impact and operate within a broader temperature range, resulting in superior performance compared to Ni–Cr batteries. Despite this, Ni–MH batteries have lower energy efficiency, ranging from 65 to 70%. Ni – H2 batteries offer benefits such as increased rated capacity, extended lifespan, and resistance to overcharging. However, their limited adoption is due to their high cost, elevated self-discharge rate, and reduced volumetric energy density. RFB boasts several advantages, including heightened efficiency, extensive discharge capabilities, remarkable stability in cycle life, and a prolonged lifespan spanning 15 to 20 years. Electrodes remain stable without undergoing any physical or chemical alterations. Encapsulation of the electrolyte within a storage vessel ensures RFB prevents excessive self-discharge, catering to the needs of long-term storage or large-scale applications. Conversely, VRB presents challenges such as elevated energy capital costs, diminished energy density, and substantial spatial requirements in fixed applications. Furthermore, VRB may experience V2O5 precipitation under high-temperature conditions, thus constraining its energy density. Current research endeavors in VRB primarily concentrate on two fronts: augmenting energy density by enhancing the stability and solubility of electrolytes and electrodes. For instance, additives like oxalic acid are introduced to stabilize the cathode, while the utilization of graphite felt or pliable carbon nanofibers facilitates electrode fabrication. Additionally, efforts are directed toward mitigating the relatively steep operating expenses. A reduction to $300/kWh in operating costs would render VRB technology markedly cost-competitive. ZnBr batteries offer several benefits, including a high energy density ranging from 30 to 85 Wh/L, favorable reversibility in redox reactions, the ability to discharge deeply, a long lifespan of 10 to 20 years, and an extended discharge duration of around 10 hours. Conversely, they have drawbacks such as the utilization of corrosive materials, relatively low cycle efficiency of 65 to 70%, and strict operational requirements, notably the need to operate within a narrow temperature range. SCs present numerous benefits, including elevated power density ranging from 500 to 5000 W/kg, prolonged cycle life exceeding 105 cycles, high operational efficiency of 85–97%, swift response time of less than 5 milliseconds, an extended operational lifespan of roughly 40 years, and the capability for rapid charging. However, they also exhibit drawbacks such as a notable self-discharge rate ranging from 5 to 40% per day, substantial initial investment costs, and the limitation of stored energy to supplying power for only a brief duration to the load. Present research endeavors in this domain entail the exploration of novel electrode materials like polyethylene infused with silicon macromolecules, graphite, and carbon nanotubes alongside initiatives aimed at minimizing initial capital outlays. SMES systems offer numerous advantages such as quick response times, high power densities (up to 4000 W/L), outstanding efficiency (95–98%), minimal energy loss, short discharge periods (less than a minute), unlimited charge and discharge cycles, and long operational lifespans (30 years). Conversely, disadvantages include the significant initial investment needed, a high rate of self-discharge (10–15% per day), susceptibility of coils to temperature changes, and the environmental impact of strong magnetic fields. Currently, research efforts in SMES mainly focus on reducing costs associated with superconducting coils and liquid nitrogen cooling systems, as well as improving materials for HTS coils with reduced temperature sensitivity and lower initial expenses [99].

8.4 Utilization of Data Science Methods

Electric power distribution constitutes a pivotal phase in the energy delivery process, facilitating the transmission of electricity from production facilities to end users. Typically overseen by distribution utilities, this process operates through intricate electrical grids. Within the realm of electric distribution, an array of data is systematically gathered and leveraged. This includes core electrical parameters such as voltage magnitudes and current intensities, consumer load profiles, fault occurrence and outage statistics, maintenance chronicles of distribution assets, and the geometric configuration of the grid. The monitoring and regulation of voltage levels and current magnitudes are fundamental to overseeing grid integrity, preempting potential overloads, and ensuring operational safety. Analysis of consumer load profiles is instrumental in forecasting demand dynamics and optimizing grid capacity utilization. Strategizing grid maintenance initiatives is contingent upon comprehensive insight into fault occurrence and outage data, enabling timely interventions and fostering grid resilience. The maintenance history of distribution assets serves as a cornerstone for devising structured maintenance schedules, thereby minimizing service disruptions and bolstering system reliability. Moreover, the delineation of grid topology and connectivity informs infrastructure expansion endeavors, encompassing tasks such as network augmentation and subscriber integration. The accurate compilation, processing, and interpretation of these data sets are pivotal to the effective governance and safeguarding of electric distribution infrastructure. Furthermore, the integration of advanced data analytics methodologies and smart grid technologies affords continuous surveillance and enhancement of grid performance metrics. Ultimately, these endeavors converge toward the delivery of enhanced reliability and uninterrupted electric services to end-users, underscoring the significance of data-driven management paradigms in the domain of electric power distribution. Alongside these, the processing of these data yields invaluable insights for energy storage as well. Particularly, it contributes significantly to the accurate utilization of concepts such as energy equity and justice.

8.4.1 Big Data Platforms

8.4.1.1 Cloud-Based Big Data Platforms

Cloud-based big data is a method that enables the storage, processing, and analysis of large and complex datasets, utilizing cloud computing infrastructure and services. In comparison to traditional on-premises data centers, cloud-based big data solutions leverage the scalability, flexibility, and cost-effectiveness of cloud platforms. Cloud platforms provide flexibility in storing and processing large volumes of data for organizations without any infrastructure constraints. Additionally, cloud-based big data solutions offer flexibility through various storage options, processing frameworks, and analytical tools. The cloud computing model follows a pay-as-you-go approach, where organizations pay for the resources they consume, potentially offering a more cost-effective solution compared to building and maintaining on-premises infrastructure. Deployment and distribution of cloud-based big data solutions can be faster and simpler than establishing on-premises infrastructure. Cloud providers offer managed services and preconfigured environments for various big data technologies. Cloud-based big data solutions can seamlessly integrate with other cloud services such as data storage, business intelligence, and IoT platforms, enabling organizations to develop comprehensive data-driven applications. Cloud providers invest in security measures and compliance certifications to ensure the security of data stored and processed in their infrastructure. However, it is important for organizations to implement appropriate security measures and adhere to regulatory requirements. Overall, cloud-based big data solutions enable organizations to leverage the power of data analytics and focus on deriving insights and driving innovation from their data, freeing them from the complexities of managing infrastructure.

The analysis plays a crucial role in effectively managing the generation and distribution of power within a smart grid context. This framework offers a plethora of advantages, leveraging cloud computing and big data analytics to execute a wide array of functions within the smart grid ecosystem. For instance, the utilization of historical data analysis enables accurate forecasting of energy production, while the analysis of consumer behavior facilitates demand forecasting. Moreover, the framework enables the identification of high and low-priority demands, thereby optimizing resource allocation. Built upon industry standards and cloud APIs like Amazon Web Services (AWS), the platform also supports virtualization technologies such as vSphere, Xen, virtual machine ware (VMWare), and kernel-based virtual machine (KVM), making it compatible with major operating systems like Ubuntu and Debian. Structured into distinct layers, the framework provides access to a diverse range of cloud computing services tailored to different needs. At the infrastructure as a service (IaaS) layer, core tasks such as data gathering, storage, and processing are handled, catering to the resource requirements of various services and applications within the smart grid system. Moving up the hierarchy, the software as a service (SaaS) level houses all system services, offering user-friendly interfaces for easy access and utilization. For instance, Google PowerMeter offers real-time monitoring of electricity usage, empowering users to optimize energy consumption efficiently. Meanwhile, the platform as a service (PaaS) layer serves as a stage for developing applications and services using cloud computing tools and libraries like Salesforce. This facilitates tailored application development while integrating specific security measures and legal regulations for effective data management. Additionally, it inherits cloud capabilities from applications, enhancing scalability and adaptability. The data as a service (DaaS) level provides analytical insights from extensive smart grid data files, accessible to both consumers and power providers. However, it operates solely on a read-only basis, preventing data downloads to maintain security and integrity. Overall, the framework underscores a robust security approach, leveraging hierarchical identity-based cryptography to ensure not only data security but also scalability and adaptability, thereby laying a strong foundation for efficient management and operation within the smart grid ecosystem [100].

8.4.1.2 Scaling

Scaling refers to a system's capacity to handle growing demands for data processing. Various big data platforms employ different scaling methods. Generally, these platforms can be classified into two types of scaling. Horizontal scaling, also termed as scale out, distributes the workload across multiple servers, potentially including commercial machines. This strategy entails adding numerous independent machines to boost processing power. Typically, each machine operates distinct instances of the operating system. Vertical scaling, or scale-up, entails augmenting a single server with additional processors, more memory, and faster hardware. This approach typically concentrates on enhancing the capabilities of a single server hosting a sole instance of an operating system [101, 102].

Horizontal Scaling Platforms Horizontal scaling platforms, also referred to as scale-out architectures, are engineered to manage escalating workloads by augmenting hardware resources horizontally, specifically by incorporating additional machines or nodes into the infrastructure. Notable instances of horizontal scaling platforms encompass P2P networks and Apache Hadoop. P2P networks facilitate decentralized processing and storage by dispersing tasks and data across interconnected nodes. This decentralized approach is frequently employed in scenarios where individual nodes contribute resources to the collective network, as observed in applications like file sharing or distributed computing. Apache Hadoop, an open-source framework, is tailored for the distributed storage and processing of extensive datasets across clusters comprising commodity hardware. It comprises the Hadoop Distributed File System (HDFS) for storage functionality, coupled with the MapReduce programming model, which orchestrates parallel data processing across multiple nodes.

Continual scholarly endeavors are dedicated to the refinement of horizontal scaling tools, epitomized by Apache Spark, aimed at transcending the constraints of extant platforms. Apache Spark, recognized for its expeditious and versatile distributed computing capabilities, boasts an in-memory data processing engine, heralding superior performance and user-friendly attributes vis-à-vis conventional MapReduce-based systems such as Apache Hadoop. Noteworthy features of Spark encompass support for multiple programming languages and an array of libraries catering to diverse tasks, including SQL processing, machine learning, graph analytics, and real-time stream processing.

In essence, horizontal scaling platforms empower systems to accommodate heightened workloads through the dispersion of tasks and data across manifold nodes. P2P networks and Apache Hadoop exemplify such platforms, while the ongoing scholarly endeavors exemplified by Apache Spark strive to augment scalability, performance, and functionality within this paradigm [101, 102].

Vertical Scaling Platforms Vertical scaling platforms typically encompass various technologies commonly used in applications requiring extensive data processing and intensive computing high-performance computing clusters (HPC), multicore processors, graphics processing units (GPUs), and field programmable gate arrays (FPGAs) stand out as some of the widely recognized vertical scaling approaches. HPC are commonly employed in scientific computations and other applications demanding substantial data processing. These clusters, formed by combining multiple high-performance servers, provide parallel computing power. Multicore processors enable a single processor to concurrently serve multiple threads, enhancing performance and efficiency, particularly for multi-threaded applications. GPUs are specialized hardware primarily used in graphics-intensive applications such as gaming and graphic design. However, they have also been increasingly utilized for general-purpose computing. GPUs process large amounts of data in parallel, facilitating rapid and efficient computation. FPGAs are customizable and reprogrammable hardware components. Their customization capability allows for hardware acceleration tailored to specific applications. FPGAs are renowned for their high-performance solutions characterized by low latency and high parallel processing capacity [101, 102].

8.4.2 Most Used Big Data Analytics Algorithms

Big data analytics is a discipline aimed at extracting meaningful insights and understanding from the increasingly voluminous and complex datasets prevalent in today's world. It is supported by various data mining and machine learning techniques. The primary objective of big data analytics is to uncover patterns, relationships, and insights from large-scale datasets. The algorithms employed in big data analytics are typically designed to process datasets, understand their features, uncover latent information, and predict future trends. For instance, clustering algorithms are utilized to group data points with similar characteristics and transform the dataset into a more understandable structure. These algorithms encompass methods such as K-means and hierarchical clustering. Moreover, classification algorithms are used to assign data points to specific categories or classes. This is particularly important in supervised learning problems, where algorithms such as support vector machines (SVMs), decision trees, and K-nearest neighbors (KNN) are commonly favored. Another area of focus is regression analysis, which is employed to determine the relationship between a dependent variable and one or more independent variables. Techniques like Linear Regression and Logistic Regression fall within this category. Additionally, association rules analysis focuses on discovering patterns and relationships within datasets. This is particularly useful in areas such as marketing strategies and product placement. The Apriori algorithm is a commonly used example in this type of association rules analysis. Finally, validation and anomaly detection algorithms are used to identify abnormal or unexpected behaviors within a dataset. This is critical for ensuring data integrity and preventing potential errors or fraud. All of these algorithms form a comprehensive toolkit covering various aspects of big data analytics. These tools facilitate businesses in making data-driven decisions, thereby enabling them to gain a competitive advantage. However, effective utilization of these algorithms requires expertise and knowledge in processing and analyzing complex datasets.

8.4.2.1 K-Means Clustering

K-means clustering stands as a widely utilized technique in data analysis, aiming primarily to segment a dataset based on specific attributes. Its core function involves breaking down the dataset into homogeneous subsets, ensuring that items within the same cluster exhibit similarity while those in separate clusters demonstrate dissimilarity. Consequently, each subset encompasses data points sharing akin features, while a clear distinction persists between diverse subsets. K-means clustering operates as an unsupervised learning approach devoid of reliance on labels or pre-established information to unveil underlying data structure. Through clustering, the aim is to grasp the dataset's arrangement and uncover inherent patterns. Hence, clustering algorithms serve as pivotal tools for exploring and comprehending data. In the realm of big data, methods like K-means clustering gain significance due to the inherent complexity and substantial size of large datasets, which poses challenges for manual analysis. By segmenting extensive datasets into more manageable fragments, clustering facilitates enhanced comprehension of the dataset. Additionally, many formulations of these methods hinge on representatives for each cluster, aiding in summarizing the dataset. These representatives offer a concise overview of the original dataset's structure, thereby simplifying analysis. Commonly employed clustering methods such as K-means find application across diverse domains including data mining, pattern recognition, image processing, and bioinformatics. Each method boasts distinct advantages and drawbacks, with the choice of the most suitable method contingent upon the dataset's characteristics and analysis objectives [103].

8.4.2.2 Hierarchical Clustering

Hierarchical clustering algorithms are methods that partition data points into groups within a hierarchical structure, defining relationships between these groups. These algorithms are typically categorized into two main types: agglomerative and divisive. Agglomerative methods start by considering each data point as an individual cluster and progressively merge these clusters to expand the clustering structure. In contrast, divisive methods begin with all data points in a single cluster and iteratively partition this large cluster into smaller subsets. Hierarchical clustering algorithms generally operate based on a distance or similarity metric. This metric determines the level of distance or similarity between data points, guiding the clustering process. The algorithm constructs clustering structures using a specific criterion (such as shortest distance or maximum similarity) and often presents the results using visual aids like dendrograms. These structures are used to understand patterns and relationships within the dataset, providing a valuable tool for interpreting clustering outcomes [104].

8.4.2.3 Support Vector Machine

SVM, known as support vector machine, is a supervised learning algorithm primarily employed for classification and regression tasks. Its primary objective is to establish a decision boundary to segregate data points into multiple classes or predict output values. The fundamental principle of SVM involves identifying a hyperplane that optimally segregates data points. This hyperplane is designed to maximize the margin between different classes, ensuring the greatest separation between the hyperplane and the nearest data point. This results in optimal class separation and often leads to enhanced generalization. An appealing aspect of SVM is its ability to handle intricate scenarios, such as datasets that are not linearly separable. In such cases, SVM utilizes kernel methods to transform data into higher-dimensional feature spaces, aiming to find linearly separable hyperplanes. Another advantage of SVM is its incorporation of regularization parameters (denoted by C), which mitigate the risk of overfitting by balancing data fitting with model generalization. Despite its strengths, SVM may suffer from sluggishness when processing extensive datasets, as analyzing all data points can be time-consuming. Moreover, it does not directly accommodate multi-class problems, necessitating the use of one-vs-one or one-vs-all classifiers. Nonetheless, with proper tuning and parameter selection, SVM demonstrates remarkable performance in classification tasks and finds applications across diverse domains.

The SVM technique offers an efficient method for supervised classification tasks on hyperspectral images. SVMs have frequently demonstrated superior performance metrics, including classification accuracies, computational efficiency, and robustness to parameter settings, when compared to other prevalent classifiers such as maximum likelihood, KNN, and radial basis function (RBF) neural networks. Particularly, SVMs are notably effective in scenarios featuring heterogeneous classes with limited training samples. A notable aspect of SVM-based supervised classification lies in its ability to handle high-dimensional data without requiring the traditional feature selection process, as it integrates feature extraction and regularization elements within its learning process, typically separately required in alternative algorithms. However, despite its commendable attributes, traditional SVM classifiers solely use spectral information from hyperspectral images and overlook spatial information, including the interdependencies among adjacent pixels. This limitation is noteworthy given the potential benefits of spatial information, such as insights into the shapes and sizes of various structures, which often leads to improved classification accuracy and more precise classification maps. Moreover, dealing with hyperspectral data presents challenges due to its voluminous nature, comprising numerous bands that may contain redundancy and noise, alongside the increasing demand for labeled samples for effective training. A particular concern is the scalability issue faced by SVMs with a growing number of samples, which can often hinder their effectiveness in such scenarios [105].

8.4.2.4 Decision Trees

Decision trees are frequently employed in machine learning and are adept at handling classification and regression tasks alike. This approach involves dividing the data by constructing a tree-like structure based on the dataset's features. Each internal node of the tree represents a decision point regarding a feature, while each leaf node indicates a class label or regression value. These trees identify decision points to effectively split the dataset and maximize information gain. Consequently, they can either classify the dataset using straightforward decisions or conduct regression analysis. Their popularity stems from their straightforward interpretation and easy visualization, aiding in understanding the relationships and significance of features. Nonetheless, decision trees are susceptible to overfitting and can be sensitive to the data. Thus, they often require careful adjustment and oversight to enhance the model's generalization capability and overall performance. The decision tree classification problem can be described as follows N represents a collection of instances comprised of the X vector and the discrete class label y. In this context, Xi denotes the i-th attribute within the X vector, and it is allocated one of the values $Xi1; Xi2; \ldots; Xij$, where $1 \leq i; j \leq d$. The objective of classification is to construct a decision tree model capable of accurately predicting the y classes of forthcoming instances. In the realm of data stream mining, the magnitude of instances is boundless, indicating $N \rightarrow \infty$ [106].

8.4.2.5 K-Nearest Neighbors (KNN) Algorithm

The KNN algorithm is a machine learning approach utilized for solving classification and regression tasks. Its fundamental principle is to utilize the labels or values of the nearest K neighbors to determine the class or value of an instance to be predicted. The operation of this algorithm is as follows: First, the KNN algorithm is fed with a training dataset. This dataset consists of the attributes (independent variables) of examples and their labels (class labels). Then, when a prediction is required, the KNN algorithm computes the target variable or class of each training example. The similarity or distance between the example to be predicted and other examples is calculated, usually using metrics like Euclidean distance. Subsequently, based on the computed distances, the KNNs to the predicted example are selected. A class prediction is made based on the class labels of the selected K neighbors (for classification), for instance, the majority vote of the labels of the nearest K neighbors can be taken. For regression problems, a prediction can be made, such as the mean or weighted mean of the target variables of the K neighbors. As a result, the predicted class or value is reported. KNN's advantages include its simple implementation, lack of training time, and ability to discover nonlinear relationships. However, it can be computationally expensive for large datasets, and a balanced choice of neighborhood is required [107, 108].

8.5 Energy Storage and Data Analysis Applications in Energy Justice and Equality

Today, the concepts of energy justice and equity are gaining significance as key components of sustainable development goals. Energy justice concerns the fair distribution of access to and utilization of energy resources, emphasizing equal accessibility to energy services for all. This entails reducing energy poverty and expanding opportunities for benefiting from energy services. On the other hand, energy equity aims for equal access to energy services for all segments of society and the fair distribution of these services across all layers of society. In this context, the use of ESSs based on renewable energy sources plays a significant role in efforts to achieve energy justice and equity. These systems can increase energy access through their capacity for storing and distributing energy, ensuring the safe and stable delivery of energy services, thus improving society's access to energy. Therefore, the adoption of ESSs based on renewable energy sources is recognized as an important tool in achieving energy justice and equity goals.

ESSs hold critical importance in balancing the fluctuating nature of renewable energy sources, regulating energy supply, and enhancing energy security. These systems enable the storage of energy generated from solar, wind, hydroelectric, and other renewable sources for later use when needed. Energy storage technologies come in various forms, including battery storage systems, hydroelectric pumped storage systems, TES systems, and CAES systems. Each storage technology carries its own advantages and disadvantages. For instance, while battery storage systems provide rapid response and flexibility, hydroelectric pumped storage systems offer large-scale energy storage capacity and are effective in seasonal energy storage. TES systems are utilized for storing thermal energy at high temperatures, whereas CAES systems offer flexible site options and relatively low costs. Therefore, when selecting an ESS, the project's requirements, local conditions, and specific needs of the application area should be taken into account.

The deployment and operation of ESSs increasingly rely on the use of big data analytics and AI technologies. Big data analytics enables the analysis of large amounts of data collected in energy storage projects to derive meaningful insights. These data can be obtained in various areas, such as energy production, consumption, storage, pricing, and demand forecasts. Machine learning and deep learning algorithms can leverage these data to be used in various fields, such as energy demand forecasting, renewable energy source forecasting, and grid management. For example, energy demand forecasting is a critical factor in ensuring the optimal use of ESSs. These forecasts can help predict future demand by analyzing existing energy consumption trends and variables. Similarly, forecasting renewable energy sources' production is crucial in determining when and how much energy storage is needed. AI and big data analytics enable more accurate and efficient predictions, enhancing the efficiency and effectiveness of ESSs. Therefore, the use of big data analytics and AI technologies in the energy sector can contribute to the smarter and more sustainable management of ESSs.

Machine learning algorithms can be effectively utilized in energy demand forecasting. These algorithms can predict future energy demand based on past energy consumption data, making them a critical tool for planning ESS capacity. Particularly, techniques like time series analysis and regression models can be used to forecast future demand based on past consumption data. These predictions are crucial in determining when and how much energy storage is needed. Deep learning algorithms can be applied to forecast renewable energy sources. Especially, the nature of renewable sources such as solar and wind can be challenging to predict due to their dependency on various factors and variability over time. However, deep learning algorithms can learn from large amounts of data to make these predictions more accurately. Deep learning models capable of handling the complexity of variables like sunlight and wind speed can forecast future renewable energy production more accurately. These predictions can enhance the efficiency of ESSs, optimize integration with renewable energy sources, increase energy supply capacity, and maintain grid balance.

In the installation of ESSs, considering the participation and needs of the community is crucial for ensuring energy justice and equity. The placement and operation of these systems should aim to increase access to energy services for various segments of society and ensure equal access to energy for everyone. This is particularly vital for people living in rural areas, low-income communities, or marginalized groups whose access to energy services is often limited. Properly locating and effectively operating ESSs can improve their quality of life and strengthen social justice. As a result, energy justice and equity are fundamental requirements of modern energy systems, and ESSs play a significant role in achieving these goals. These systems can facilitate a fairer distribution of energy derived from renewable sources and make energy services accessible to everyone. However, achieving these objectives requires not only technological solutions but also a focus on community participation and needs. Technologies like big data analytics and AI can enable more effective use of ESSs and contribute to a fairer distribution of energy services. Nevertheless, it's critical to use these technologies correctly and in conjunction with community involvement [109]

In recent years, the widespread use of smart systems, especially with the rise of new technologies like AI, has gained momentum. Smart systems have been integrated into many aspects of our lives and have become the subject of intensive research in the literature. Reference [110] addresses the use of information and communication technologies to optimize the production, distribution, and consumption of electricity in smart grids. The opportunities, concepts, and challenges brought by these technologies play a crucial role in the effective implementation and management of smart grids. Particularly, big data technologies can be used to address requirements such as processing, storing, and visualizing smart grid data. The mentioned article extensively examines the steps, tools, and technical requirements for implementing and distributing these technologies. Such research provides a vital foundation for understanding the impact of technological advancements in the energy sector and building more sustainable and efficient energy systems.

The importance of using big data analytics and machine learning in the electricity industry was highlighted in a study published in 2020 [111]. This study examines the use of these techniques for various applications in smart grids and provides a comparative analysis to determine the most suitable methods and algorithms. Technological advancements in the electricity industry offer solutions to important goals such as increasing energy efficiency, ensuring reliability, and facilitating the integration of renewable energy sources. The mentioned study aims to understand the potential of big data analytics and machine learning techniques in the electricity industry and contribute to the more effective and efficient management of smart grids. Such research can contribute to the development of knowledge-based decision-making processes in the energy sector and the creation of smarter and more sustainable energy systems in the future. This topic has become a valuable field of study for energy production and distribution companies. Transforming a large amount of data in the production and distribution processes of electricity companies into useful information for real-time decision-making has become a priority. Technologies like blockchain have been observed to provide reliability and security in big data processing in renewable energy systems and have been successfully applied in smart grid systems [112]. Evaluating the real-world applicability of energy storage technologies underscores the importance of focusing on real-world problems beyond theoretical foundations and understanding the challenges encountered in the commercialization of these technologies. Reference [113] aims to fill a gap in the literature by providing a detailed analysis of real-world applications of energy storage technologies. This study examines the concept of energy storage, different technologies, and the performance of projects worldwide. Additionally, it highlights the challenges to the commercial implementation of ESSs. These challenges include factors such as cost, technical compatibility, regulatory barriers, and market maturity. The research aims to evaluate the potential of energy storage technologies under real-world conditions, contribute to strategies for their widespread adoption and commercialization. Such detailed analyses are seen as an important step toward the wider adoption of ESSs and have the potential to contribute to planning and managing future energy transformation processes.

The reference [114] titled "Energy justice beyond the wire: Exploring the multidimensional inequities of the electrical power grid in the United States" examines the multiple dimensions of inequality within the US electricity grid. It discusses concepts of energy justice and climate justice, focusing on technical aspects such as high-voltage direct-current transmission. The article presents a theoretical framework, noting that no specific data was used in the research described.

The article in reference [115], emphasizes how big data structures and relevant metrics can play a critical role in understanding and addressing energy justice and energy poverty issues. It discusses how data policies and metrics handle information regarding energy access and usage and how this information can be evaluated from an energy justice perspective. However, selective and biased use of data and metrics can adversely affect vulnerable groups and compromise justice. Therefore, it highlights the need for developing and effectively using data infrastructures to enhance justice in the energy sector.

The reference [116] addresses the importance and challenges of big data usage in the distributed generation segment of smart grids. It discusses how big data standards and maturity models can improve data management and analysis processes in smart grid operations. The article emphasizes that data collection, storage, and analysis processes are critical for the integration and management of distributed energy resources. Additionally, it examines how big data analytics can play a strategic role in optimizing energy production and distribution.

The article [117] emphasizes the importance of big data analytics and management in smart grids. It provides a detailed view from data sources to integration, storage, and analysis processes. Big data analytics plays a critical role in making the smart grid smarter, more efficient, and profitable. These processes enable more effective use of data in smart grids, offering significant improvements in energy management and distribution.

In reference [118], it discusses the use of AI and machine learning in the installation of photovoltaic systems and BESS in homes. AI and machine learning technologies are utilized to predict energy production and consumption, optimize system performance, and enhance energy management. During the installation process, these technologies help manage uncertainties such as changes in sunlight, weather conditions, and seasonal variations. Additionally, AI is used to improve the efficiency of energy storage and distribution, thereby enhancing the overall efficiency and reliability of smart home systems.

8.6 Conclusion

In this era of rapid technological advancement and growing global energy demands, the strategic development and integration of ESSs stand out as critical enablers of sustainability and equity in energy distribution. Our discussion has highlighted the multifaceted role of these systems, not only in meeting increasing energy needs but also in promoting fair access and efficient utilization of resources. As we delve deeper into the intricacies of energy storage, it becomes clear that the use of AI and big data analytics offers transformative potential to optimize these systems for better performance and inclusivity.

Looking forward, it is essential to continue advancing the technology and infrastructure of ESSs. This involves enhancing the scalability, efficiency, and sophistication of various storage methods—mechanical, electrochemical, and electromagnetic. By improving these technologies, we can ensure a more resilient energy network that can withstand the challenges posed by fluctuating supply and demand dynamics. Moreover, the regulatory frameworks governing energy storage must be robust and adaptive, designed to support equitable energy distribution and protect the right of every individual to access energy. These policies should encourage innovation and guide the deployment of storage solutions that are not only technologically advanced but also socially and economically inclusive. The potential of decentralized ESSs also warrants further exploration. These systems could democratize energy access, allowing for more localized and community-driven approaches to energy management. This shift toward decentralization could significantly impact how energy equity is achieved, particularly in underserved regions. Additionally, as energy storage technologies become more complex, the importance of interdisciplinary approaches becomes increasingly apparent. Collaboration across various scientific, technological, and regulatory domains will be vital in addressing emerging challenges and ensuring that the benefits of energy advancements are shared across all sectors of society.

In conclusion, the journey toward a sustainable and equitable energy future is complex and multifaceted. By continuing to innovate and adapt in the field of energy storage and by ensuring that these innovations are guided by the principles of fairness and sustainability, we can look forward to a future where energy is not only abundant and clean but also distributed on a just and equitable basis. The path forward will require concerted efforts and a commitment to integrating these advanced energy solutions into the broader energy system, ensuring they contribute effectively to global energy security and sustainability.

References

1 Al-Ali, A.R., Zualkernan, I.A., Rashid, M. et al. (2017). A smart home energy management system using IoT and big data analytics approach. *IEEE Transactions on Consumer Electronics* 63 (4): 426–434.

2 Sidortsov, R. and McCauley, D. (2023). Energy justice. In: *Theorising Justice: A Primer for Social Scientists* (ed. J. Ohlsson and S. Przybylinski), 171–190. Bristol: Bristol University Press.

3 Chang, C. and Farnum, B. (2014). *Artificial Photosynthesis for Solar Energy Storage: Toward a Sustainable and Equitable Future*, 10–11. Communications: Cambridge University Science and Policy Exchange.

4 Papapostolou, C., Kondili, E.M., and Kaldellis, J.K. (2014). Energy supply chain optimisation: special considerations for the solution of the energy planning problem. *Computer Aided Chemical Engineering* 33: 1525–1530.

5 Sovacool, B.K. and Blyth, P.L. (2015). Energy and environmental attitudes in the green state of Denmark: implications for energy democracy, low carbon transitions, and energy literacy. *Environmental Science & Policy* 54: 304–315.

6 Von Oertzen, D. (2015). *REEE-Powering Namibia*. Konrad Adenauer Stiftung.

7 Athari, M.H. and Ardehali, M.M. (2016). Operational performance of energy storage as function of electricity prices for on-grid hybrid renewable energy system by optimized fuzzy logic controller. *Renewable Energy* 85: 890–902.

8 Prehoda, E.W. (2016). Energy justice and US energy policy: case study applications exploring US energy policy through an energy justice framework. Doctoral dissertation. Michigan Technological University.

9 Becker, S. and Naumann, M. (2017). Energy democracy: mapping the debate on energy alternatives. *Geography Compass* 11 (8): e12321.

10 Chalvatzis, K.J. and Ioannidis, A. (2017). Energy supply security in southern Europe and Ireland. *Energy Procedia* 105: 2916–2922.

11 Merzic, A., Music, M., and Haznadar, Z. (2017). Conceptualizing sustainable development of conventional power systems in developing countries–a contribution towards low carbon future. *Energy* 126: 112–123.

12 Sareen, S. and Haarstad, H. (2018). Bridging socio-technical and justice aspects of sustainable energy transitions. *Applied Energy* 228: 624–632.

13 Goddard, G. and Farrelly, M.A. (2018). Just transition management: balancing just outcomes with just processes in Australian renewable energy transitions. *Applied Energy* 225: 110–123.

14 Milčiuvienė, S., Kiršienė, J., Doheijo, E. et al. (2019). The role of renewable energy prosumers in implementing energy justice theory. *Sustainability* 11 (19): 5286.

15 Fortier, M.O.P., Teron, L., Reames, T.G. et al. (2019). Introduction to evaluating energy justice across the life cycle: a social life cycle assessment approach. *Applied Energy* 236: 211–219.
16 Hajiaghasi, S., Salemnia, A., and Hamzeh, M. (2019). Hybrid ESS for microgrids applications: a review. *Journal of Energy Storage* 21: 543–570.
17 Demski, C., Thomas, G., Becker, S. et al. (2019). Acceptance of energy transitions and policies: public conceptualisations of energy as a need and basic right in the United Kingdom. *Energy Research & Social Science* 48: 33–45.
18 Adam Darmawan, A. (2019). Designing a Just Peer-to-Peer Energy Trading System: an exploration of actions to be performed by the Municipality of Rotterdam to approach a just peer-to-peer energy trading system for the community of Bospolder-Tussendijken.
19 Thomas, G., Demski, C., and Pidgeon, N. (2020). Energy justice discourses in citizen deliberations on systems flexibility in the United Kingdom: vulnerability, compensation and empowerment. *Energy Research & Social Science* 66: 101494.
20 Smale, R. and Kloppenburg, S. (2020). Platforms in power: householder perspectives on the social, environmental and economic challenges of energy platforms. *Sustainability* 12 (2): 692.
21 Koirala, B.P., van Oost, E.C., van der Waal, E.C., and van der Windt, H.J. (2021). New pathways for community energy and storage. *Energies* 14 (2): 286.
22 Tiwari, S., Schelly, C., and Sidortsov, R. (2021). Developing a legal framework for energy storage technologies in the US: the case of pumped underground storage hydro. *The Electricity Journal* 34 (10): 107048.
23 Tarekegne, B., Kerby, J., Bharati, A.K., and O'Neil, R. (2022). Assessing the energy equity benefits of energy storage solutions. *2022 IEEE Electrical Energy Storage Application and Technologies Conference (EESAT)*, 1–5. IEEE.
24 Kerby, J., Bharati, A.K., and Tarekegne, B. (2022). Assessing the energy equity benefits of mobile energy storage solutions. *6th E-Mobility Power System Integration Symposium (EMOB 2022)*, volume 2022, 100–106. IET.
25 Zengin, Z.O. (2023). The energy justice imperative for clean energy storage alternatives. In: *The Power of Energy Justice & the Social Contract* (ed. R.J. Heffron and L. de Fontenelle), 217–225. Cham: Springer Nature Switzerland.
26 Noorman, M., Espinosa Apráez, B., and Lavrijssen, S. (2023). AI and energy justice. *Energies* 16 (5): 2110.
27 Mohammadi, S., Eliassen, F., and Jacobsen, H.A. (2023). Applying energy justice principles to renewable energy trading and allocation in multi-unit buildings. *Energies* 16 (3): 1150.

28 Shan, R. and Kittner, N. (2024). Allocation of policy resources for energy storage development considering the Inflation Reduction Act. *Energy Policy* 184: 113861.

29 Haugen, M., Blaisdell-Pijuan, P.L., Botterud, A. et al. (2024). Power market models for the clean energy transition: state of the art and future research needs. *Applied Energy* 357: 122495.

30 Gee, A.M., Robinson, F.V., and Dunn, R.W. (2013). Analysis of battery lifetime extension in a small-scale wind-energy system using supercapacitors. *IEEE Transactions on Energy Conversion* 28 (1): 24–33.

31 Kollimalla, S.K., Mishra, M.K., Ukil, A., and Gooi, H.B. (2016). DC grid voltage regulation using new HESS control strategy. *IEEE Transactions on Sustainable Energy* 8 (2): 772–781.

32 Sikkabut, S., Mungporn, P., Ekkaravarodome, C. et al. (2016). Control of high-energy high-power densities storage devices by Li-ion battery and supercapacitor for fuel cell/photovoltaic hybrid power plant for autonomous system applications. *IEEE Transactions on Industry Applications* 52 (5): 4395–4407.

33 Song, Y.D., Cao, Q., Du, X., and Karimi, H.R. (2013). Control strategy based on wavelet transform and neural network for hybrid power system. *Journal of Applied Mathematics* 2013 (1): 375840.

34 Hredzak, B., Agelidis, V.G., and Jang, M. (2013). A model predictive control system for a hybrid battery-ultracapacitor power source. *IEEE Transactions on Power Electronics* 29 (3): 1469–1479.

35 Alnejaili, T., Drid, S., Mehdi, D. et al. (2015). Dynamic control and advanced load management of a stand-alone hybrid renewable power system for remote housing. *Energy Conversion and Management* 105: 377–392.

36 Acar, C. (2018). A comprehensive evaluation of energy storage options for better sustainability. *International Journal of Energy Research* 42 (12): 3732–3746.

37 Krishan, O. and Suhag, S. (2019). An updated review of energy storage systems: classification and applications in distributed generation power systems incorporating renewable energy resources. *International Journal of Energy Research* 43 (12): 6171–6210.

38 Fang, R. (2019). Life cycle cost assessment of wind power–hydrogen coupled integrated energy system. *International Journal of Hydrogen Energy* 44 (56): 29399–29408.

39 Wang, K., Wang, W., Wang, L., and Li, L. (2020). An improved SOC control strategy for electric vehicle hybrid energy storage systems. *Energies* 13 (20): 5297.

40 Ferreira, J.A. and Van Wyk, J.D. (2001). Electromagnetic energy propagation in power electronic converters: toward future electromagnetic integration. *Proceedings of the IEEE* 89 (6): 876–889.

41 Zhao, J.M. and Zhang, Z.M. (2015). Electromagnetic energy storage and power dissipation in nanostructures. *Journal of Quantitative Spectroscopy and Radiative Transfer* 151: 49–57.

42 Nikolaidis, P. and Poullikkas, A. (2018). Cost metrics of electrical energy storage technologies in potential power system operations. *Sustainable Energy Technologies and Assessments* 25: 43–59.

43 Wagner, L. (2014). Overview of energy storage technologies. In: *Future Energy* (ed. T.M. Letcher), 613–631. Elsevier.

44 Buckles, W. and Hassenzahl, W.V. (2000). Superconducting magnetic energy storage. *IEEE Power Engineering Review* 20 (5): 16–20.

45 Vatamanu, J. and Bedrov, D. (2015). Capacitive energy storage: current and future challenges. *The Journal of Physical Chemistry Letters* 6 (18): 3594–3609.

46 Schoenbach, K.H., Kristiansen, M., and Schaefer, G. (1984). A review of opening switch technology for inductive energy storage. *Proceedings of the IEEE* 72 (8): 1019–1040.

47 Tan, K.M., Babu, T.S., Ramachandaramurthy, V.K. et al. (2021). Empowering smart grid: a comprehensive review of energy storage technology and application with renewable energy integration. *Journal of Energy Storage* 39: 102591.

48 Amrouche, S.O., Rekioua, D., Rekioua, T., and Bacha, S. (2016). Overview of energy storage in renewable energy systems. *International Journal of Hydrogen Energy* 41 (45): 20914–20927.

49 Behabtu, H.A., Messagie, M., Coosemans, T. et al. (2020). A review of energy storage technologies' application potentials in renewable energy sources grid integration. *Sustainability* 12 (24): 10511.

50 Zhou, X., Lin, Y., and Ma, Y. (2015). The overview of energy storage technology. *2015 IEEE International Conference on Mechatronics and Automation (ICMA)*, 43–48. IEEE.

51 Atcitty, S., Neely, J., Ingersoll, D. et al. (2013). Battery energy storage system. In: *Power Electronics for Renewable and Distributed Energy Systems: A Sourcebook of Topologies, Control and Integration* (ed. S. Chakraborty, M.G. Simões, and W.E. Kramer), 333–366. London: Springer-Verlag.

52 Showers, S.O. and Raji, A.K. (2019). Benefits and challenges of energy storage technologies in high penetration renewable energy power systems. *2019 IEEE PES/IAS PowerAfrica*, 209–214. IEEE.

53 Liu, C., Li, F., Ma, L.P., and Cheng, H.M. (2010). Advanced materials for energy storage. *Advanced Materials* 22 (8): E28–E62.

54 Yao, L., Yang, B., Cui, H. et al. (2016). Challenges and progresses of energy storage technology and its application in power systems. *Journal of Modern Power Systems and Clean Energy* 4 (4): 519–528.

55 Sun, Y., Zhao, Z., Yang, M. et al. (2019). Overview of energy storage in renewable energy power fluctuation mitigation. *CSEE Journal of Power and Energy Systems* 6 (1): 160–173.

56 Kumar, K. and Jaipal, B. (2022). The role of energy storage with renewable electricity generation. In: *Electric Grid Modernization* (ed. M. Ghofrani). IntechOpen.

57 Han, L., Chen, W., Zhuang, B., and Shen, H. (2017). A review on development practice of smart grid technology in China. *IOP Conference Series: Materials Science and Engineering* 199 (1): 012062.

58 Oshima, T., Kajita, M., and Okuno, A. (2004). Development of sodium-sulfur batteries. *International Journal of Applied Ceramic Technology* 1 (3): 269–276.

59 Townsend, A. and Gouws, R. (2022). A comparative review of lead-acid, lithium-ion and ultra-capacitor technologies and their degradation mechanisms. *Energies* 15 (13): 4930.

60 Gür, T.M. (2018). Review of electrical energy storage technologies, materials and systems: challenges and prospects for large-scale grid storage. *Energy & Environmental Science* 11 (10): 2696–2767.

61 Blomgren, G.E. (2016). The development and future of lithium ion batteries. *Journal of the Electrochemical Society* 164 (1): A5019.

62 Deng, D. (2015). Li-ion batteries: basics, progress, and challenges. *Energy Science & Engineering* 3 (5): 385–418.

63 Killer, M., Farrokhseresht, M., and Paterakis, N.G. (2020). Implementation of large-scale Li-ion battery energy storage systems within the EMEA region. *Applied Energy* 260: 114166.

64 Lucia, U. (2014). Overview on fuel cells. *Renewable and Sustainable Energy Reviews* 30: 164–169.

65 Ali, M. (2023). A role of mechanical energy storage systems in grid application. *Brilliance: Research of Artificial Intelligence* 3 (1): 1–8.

66 Alami, A.H. (2020). *Mechanical Energy Storage for Renewable and Sustainable Energy Resources*, 71. Springer.

67 Mahmoud, M., Ramadan, M., Olabi, A.G. et al. (2020). A review of mechanical energy storage systems combined with wind and solar applications. *Energy Conversion and Management* 210: 112670.

68 Peña-Alzola, R., Sebastián, R., Quesada, J., and Colmenar, A. (2011). Review of flywheel based energy storage systems. *2011 International Conference on Power Engineering, Energy and Electrical Drives*, 1–6. IEEE.

69 Amirante, R., Cassone, E., Distaso, E., and Tamburrano, P. (2017). Overview on recent developments in energy storage: mechanical, electrochemical and hydrogen technologies. *Energy Conversion and Management* 132: 372–387.

70 Kougias, I. and Szabó, S. (2017). Pumped hydroelectric storage utilization assessment: forerunner of renewable energy integration or Trojan horse? *Energy* 140: 318–329.

71 Rehman, S., Al-Hadhrami, L.M., and Alam, M.M. (2015). Pumped hydro energy storage system: a technological review. *Renewable and Sustainable Energy Reviews* 44: 586–598.

72 Pinnangudi, B., Kuykendal, M., and Bhadra, S. (2017). Smart grid energy storage. In: *The Power Grid* (ed. B. D'Andrade), 93–135. Academic Press.

73 Ma, T., Yang, H., and Lu, L. (2014). Feasibility study and economic analysis of pumped hydro storage and battery storage for a renewable energy powered island. *Energy Conversion and Management* 79: 387–397.

74 Genta, G. (2014). *Kinetic Energy Storage: Theory and Practice of Advanced Flywheel Systems*. Butterworth-Heinemann.

75 Faraji, F., Majazi, A., and Al-Haddad, K. (2017). A comprehensive review of flywheel energy storage system technology. *Renewable and Sustainable Energy Reviews* 67: 477–490.

76 Olabi, A.G., Wilberforce, T., Abdelkareem, M.A., and Ramadan, M. (2021). Critical review of flywheel energy storage system. *Energies* 14 (8): 2159.

77 Amiryar, M.E. and Pullen, K.R. (2017). A review of flywheel energy storage system technologies and their applications. *Applied Sciences* 7 (3): 286.

78 Espinar, B. and Mayer, D. (2011). The role of energy storage for mini-grid stabilization.

79 Jin, C., Jiang, X., Zhong, G., and Li, X. (2017). Research on coordinated control strategy of flywheel energy storage array for island microgrid. *2017 IEEE Conference on Energy Internet and Energy System Integration (EI2)*, 1–6. IEEE.

80 Wang, J., Lu, K., Ma, L. et al. (2017). Overview of compressed air energy storage and technology development. *Energies* 10 (7): 991.

81 Lund, H. and Salgi, G. (2009). The role of compressed air energy storage (CAES) in future sustainable energy systems. *Energy Conversion and Management* 50: 1172–1179. https://doi.org/10.1016/j.enconman.2009.01.032.

82 Boicea, V.A. (2014). Energy storage technologies: the past and the present. *Proceedings of the IEEE* 102 (11): 1777–1794.

83 Cavallo, A. (2007). Controllable and affordable utility-scale electricity from intermittent wind resources and compressed air energy storage (CAES). *Energy* 32 (2): 120–127.

84 Duhan, J. (2018). Compressed air energy storage in salt caverns: geomechanical design workflow, CAES siting study from a geomechanics perspective, and deep brine disposal. Master's thesis. University of Waterloo.

85 Wang, H., Wang, Z., Liang, C. et al. (2022). Underwater compressed gas energy storage (UWCGES): current status, challenges, and future perspectives. *Applied Sciences* 12 (18): 9361.

86 Tong, W., Lu, Z., Chen, W. et al. (2022). Solid gravity energy storage: a review. *Journal of Energy Storage* 53: 105226.

87 Berrada, A., Loudiyi, K., and Zorkani, I. (2017). System design and economic performance of gravity energy storage. *Journal of Cleaner Production* 156: 317–326.

88 Hunt, J.D., Zakeri, B., Falchetta, G. et al. (2020). Mountain Gravity Energy Storage: a new solution for closing the gap between existing short-and long-term storage technologies. *Energy* 190: 116419.

89 Bowen, T., Chernyakhovskiy, I., Xu, K. et al. (2021). USAID grid-scale energy storage technologies primer (No. NREL/TP-6A20-76097). Golden, CO: National Renewable Energy Lab.(NREL).

90 Berrada, A., Emrani, A., and Ameur, A. (2021). Life-cycle assessment of gravity energy storage systems for large-scale application. *Journal of Energy Storage* 40: 102825.

91 Saxena, V., Kumar, N., and Nangia, U. (2021). Smart grid: a sustainable smart approach. *Journal of Physics: Conference Series* 2007 (1): 012042.

92 Zhou, X., Liu, G., Han, B. et al. (2021). Design of a human lower limbs exoskeleton for biomechanical energy harvesting and assist walking. *Energy Technology* 9 (1): 2000726.

93 Hamada, A.T. and Orhan, M.F. (2022). An overview of regenerative braking systems. *Journal of Energy Storage* 52: 105033.

94 Dan, G., Zailin, P., Lidi, W. et al. (2016). The research on the integration of energy storage and generating wind turbine system model. *2016 China International Conference on Electricity Distribution (CICED)*, 1–4. IEEE.

95 Hosseini, S.M., Soleymani, M., Kelouwani, S., and Amamou, A. (2023). Energy recovery and energy harvesting in electric and fuel cell vehicles, a review of recent advances. *IEEE Access* 11: 83107–83135.

96 Hill, F.A., Havel, T.F., and Livermore, C. (2009). Modeling mechanical energy storage in springs based on carbon nanotubes. *Nanotechnology* 20 (25): 255704.

97 Carnegie, R., Gotham, D., Nderitu, D., and Preckel, P.V. (2013). *Utility Scale Energy Storage Systems*, 1. State Utility Forecasting Group (SUFG) – Purdue University.

98 Olympios, A.V., McTigue, J.D., Farres-Antunez, P. et al. (2021). Progress and prospects of thermo-mechanical energy storage—a critical review. *Progress in Energy* 3 (2): 022001.

99 Zhang, Z., Ding, T., Zhou, Q. et al. (2021). A review of technologies and applications on versatile energy storage systems. *Renewable and Sustainable Energy Reviews* 148: 111263.

100 Zainab, A., Ghrayeb, A., Syed, D. et al. (2021). Big data management in smart grids: technologies and challenges. *IEEE Access* 9: 73046–73059.

101 Singh, D. and Reddy, C.K. (2015). A survey on platforms for big data analytics. *Journal of Big Data* 2 (1): 1–20.

102 Dafir, Z., Lamari, Y., and Slaoui, S.C. (2021). A survey on parallel clustering algorithms for big data. *Artificial Intelligence Review* 54: 2411–2443.

103 Feldman, D., Schmidt, M., and Sohler, C. (2020). Turning big data into tiny data: constant-size coresets for k-means, PCA, and projective clustering. *SIAM Journal on Computing* 49 (3): 601–657.

104 Xie, W.B., Lee, Y.L., Wang, C. et al. (2020). Hierarchical clustering supported by reciprocal nearest neighbors. *Information Sciences* 527: 279–292.

105 Cavallaro, G., Riedel, M., Richerzhagen, M. et al. (2015). On understanding big data impacts in remotely sensed image classification using support vector machine methods. *IEEE Journal of Selected Topics in Applied Earth Observations and Remote Sensing* 8 (10): 4634–4646.

106 Yang, H. and Fong, S. (2013). Improving adaptability of decision tree for mining big data. *New Mathematics and Natural Computation* 9 (01): 77–95.

107 Zhang, S., Li, X., Zong, M. et al. (2017). Learning k for kNN classification. *ACM Transactions on Intelligent Systems and Technology (TIST)* 8 (3): 1–19.

108 Guo, G., Wang, H., Bell, D. et al. (2003). KNN model-based approach in classification. *On The Move to Meaningful Internet Systems 2003: CoopIS, DOA, and ODBASE: OTM Confederated International Conferences, CoopIS, DOA, and ODBASE 2003*, Catania, Sicily, Italy (3–7 November 2003). Proceedings, 986–996. Berlin, Heidelberg: Springer-Verlag.

109 Hossain, E., Khan, I., Un-Noor, F. et al. (2019). Application of big data and machine learning in smart grid, and associated security concerns: a review. *IEEE Access* 7: 13960–13988.

110 Daki, H., El Hannani, A., Aqqal, A. et al. (2017). Big Data management in smart grid: concepts, requirements and implementation. *Journal of Big Data* 4: 1–19.

111 Dhupia, B., Usha Rani, M., and Alameen, A. (2020). The role of big data analytics in smart grid management. *Emerging Research in Data Engineering Systems and Computer Communications: Proceedings of CCODE* 2019: 403–412.

112 Colmenares-Quintero, R.F., Quiroga-Parra, D.J., Rojas, N. et al. (2021). Big Data analytics in Smart Grids for renewable energy networks: systematic review of information and communication technology tools. *Cogent Engineering* 8 (1): 1935410.
113 Aneke, M. and Wang, M. (2016). Energy storage technologies and real life applications–a state of the art review. *Applied Energy* 179: 350–377.
114 Sovacool, B.K., Carley, S., and Kiesling, L. (2024). Energy justice beyond the wire: exploring the multidimensional inequities of the electrical power grid in the United States. *Energy Research & Social Science* 111: 103474.
115 DellaValle, N. and Sareen, S. (2020). Nudging and boosting for equity? Towards a behavioural economics of energy justice. *Energy Research & Social Science* 68: 101589.
116 Sundararajan, A., Hernandez, A.S., and Sarwat, A.I. (2020). Adapting big data standards, maturity models to smart grid distributed generation: critical review. *IET Smart Grid* 3 (4): 508–519.
117 Sovacool, B.K., Martiskainen, M., Hook, A., and Baker, L. (2019). Decarbonization and its discontents: a critical energy justice perspective on four low-carbon transitions. *Climatic Change* 155: 581–619.
118 Rojek, I., Mikołajewski, D., Mroziński, A., and Macko, M. (2023). Machine learning-and artificial intelligence-derived prediction for home smart energy systems with PV installation and battery energy storage. *Energies* 16 (18): 6613.

9

Energy Trading Mechanisms for Energy Justice and Equity

Iman Rahmati[1] *and Asghar Akbari Foroud*[2]

[1] *Electricity Market Monitoring & Studies, Iran Grid Management Company, Tehran, Iran*
[2] *Electrical and Computer Engineering Faculty, Semnan University, Semnan, Iran*

Acronyms

AEMO	Australian Electricity Market Operator
CNG	compressed natural gas
ETS	emission trading systems
FiT	feed-in-tariff
GGE	gasoline gallon equivalent
GHG	green-house-gas
GNP	gross national product
IEA	International Energy Agency
IMF	International Monetary Fund
NEM	National Electricity Market
WHO	World Health Organization

9.1 Introduction

The global energy sector has undergone a profound transformation over the past four decades, transitioning from a predominantly public service model to a market-driven commodity system. This shift necessitates robust regulatory frameworks that explicitly incorporate trading mechanisms to facilitate the democratization of energy access. Historical evidence demonstrates that unfettered market forces are insufficient to ensure equitable social and economic outcomes. Furthermore, the history of renewable energy technologies in the current century shows that the sustainability of the energy provision was not

Communities for Clean Energy Justice and Equity in Grid Modernization, First Edition.
Edited by Mohammadreza Daneshvar, Behnam Mohammadi-Ivatloo, and Amjad Anvari-Moghaddam.

left to the competition and market hands but greatly relied on the regulator's intervention within the market design. This chapter seeks to have a look at the current energy markets and the trading mechanisms through the lens of energy justice. We highlight the nexus between energy markets and energy justice to understand how the market and trading mechanisms can promote or hinder achieving energy justice. The most common trading mechanisms regarding energy justice, and especially energy sustainability, are briefly reviewed. Also, we will point out some of the shortcomings in the energy markets and trading mechanisms and review the amendments proposed in the literature.

9.2 Energy Market and Justice

Markets play a noticeable role in human lives globally and market forces define how commodities and services are exchanged among suppliers and consumers. Moreover, markets are mainstreams of economic growth and development. Markets determine the price of the goods and services that people need, and hence, the affordability of these goods and services is affected. On the other hand, energy has become one of the most essential commodities in daily human life. Almost every aspect of our activities is directly connected to energy. Many goods we consume daily are the direct and indirect products of energy use, and most services we use would be undeliverable without energy consumption. That's why solid access to energy has been one of the main indicators of the quality of human life and energy poverty has been greatly credited as a threat to social justice [1].

Historically, human dependence on energy has rendered the energy issue a multifaceted problem, encompassing both energy availability and affordability. Consequently, energy markets have traditionally been expected to ensure a continuous supply of energy at a price deemed reasonable, or at least acceptable, by consumers. However, later and as the concerns regarding climate change arose, the sustainability of energy provision became more and more essential. Hence, the already challenging problem of energy became more complex which is also titled as energy trilemma: affordability, availability, and sustainability. Due to this fact, almost all the governments in the world have put a great deal of effort and resources into ensuring energy availability at an acceptable price to their citizens while directing the transition to a sustainable future [1, 2].

9.2.1 Energy as a Tradable Commodity

There is a huge literature discussing the nexus between trade and development, and a wide range of people, including economists, governmental bodies and policymakers, and academic writers, have enlightened this matter. The history approves that expanding the trade can contribute to higher employment factors, poverty eradication, and better nutrition levels. It has been shown that when a commodity is tradeable, people try to make a profit out of more optimized production and trade and this leads to knowledge integration and diffusion of ideas [3]. The resulting innovation due to the trade of a commodity not only happens in that industry but also spills over the related industries. Such finding is compatible with the outcome of the analysis performed by Wacziarg and Welch that showed liberalization of trade generally has a positive impact on economic growth [4]. This conclusion is also applicable to the energy industry. Throughout the history of trading energy carriers have expanded their usage, increased demand, and resulted in technological improvement. Energy tradability refers to the features of energy carriers that make them easy to buy, sell, and exchange across different locations and actors. Such features encompass the level of technology that is needed to capture and use an energy carrier and also the accessibility, storability, and transportability of that energy source. However, as we will see in the next section, the trade and mechanisms for trading have not been the same for all the energy carriers and not through the history.

From the viewpoint of tradability, the energy sources can be categorized as:

a) Sorts of energy carriers that are no longer considered as energy sources. Such as whale oil that once was traded and used for illumination, especially in the 18th and 19th centuries.
b) The fossil energy sources, such as coal, oil, and gas, have been widely used and traded by humans to this day.
c) Nuclear energy has recently been located within the reach of human knowledge and technology, and its application as a peaceful energy source became operational in less than 70 years. The first nuclear power plant commenced its operation in June 1954 in Obninsk, a town located 100 km from Moscow [5].
d) Furthermore, a distinct category of energy sources merits attention. These energies have been known and utilized for millennia; however, their tradability has only become feasible in recent decades. The archeological findings suggest that the application of sun, water, and wind energy goes back to ancient civilizations when people in ancient Greece and Roman Empire, Persia and China incorporated wind energy to empower water pumps in the cooling systems and created water wheels and windmills to grind grains [6, 7]. Although these energy sources had to wait a very long time for the invention of commercial electricity and become tradable.

9.2.2 Energy Justice Versus Affordability and Access to Energy

Energy is not only the driving force for development but also an essential commodity in day-to-day life and plays a key role in the well-being of the people in every society [8]. Air-conditioning, transportation, home appliances, telecommunication and information technology, etc., all depend on access to energy. Hence, the first element of energy justice is energy accessibility. Energy poverty refers to a situation that part of some people do not have access to a minimum level of modern energy that supports their basic household needs [9]. Such a situation may stem from the insufficient energy supply, or in most cases, the financial inability of households to meet their basic domestic energy needs [10].

The impact of markets and trading mechanisms on energy access is twofold. As Yong Yan and colleagues argue, through absorbing foreign and domestic investments, trade activities can provide the necessary technologies and expertise to abate energy poverty. On the other hand, without proper regulation, increased trade activities can go south and lead to environmental degradation, which worsens energy poverty issues in the long term [11].

Based on what the World Bank reports, 1.1 billion people didn't have access to electricity in 2010 that accounted for 16% of the global population. In 11 years, this declined to 675 million persons equals 9% of the world's population [12]. The International Energy Agency (IEA) has defined "Modern Energy Access" as a basic lower limit for annual electricity consumption which is 250 kWh in rural households and 500 kWh in urban areas [13]. Although, due to the differences in culture, lifestyle, and environmental conditions, defining such a strict line for energy poverty cannot rationally relate to different parts of the world [14].

Apart from electricity, around 2.3 billion people around the world (around 30% of the global population) do not have access to clean fuels and use polluting fuels such as solid fuels and kerosene [12]. The intense negative effects of the pollution released by such inefficient household cooking endanger women and children (especially girls) more severely and this is a manifestation of energy injustice and gender inequality. World Health Organization (WHO) claims that annually 3.2 million deaths have roots in lack of access to clean cooking fuels and technology, including 240,000 deaths in children under the age of five [15].

A great deal of studies has been done around the measurement techniques and indicators of energy poverty, and many of them propose indicators that are based on the comparison between household income and expenditures [16]. However, in some cases, the lack of access to modern energy is not directly caused by the financial inability of the people but the lack of proper infrastructures to support and provide the needed amount of energy, which of course can be due to the financial insufficiency of the state.

As stated before, market approach and trading mechanisms have had an undeniable role in providing goods and services to people and by a proper design can be means to promote universal access. For example, in several academic studies, peer-to-peer energy trading has been considered and proposed as an efficient tool to promote electricity access and eradicate energy poverty [17–19].

As another instance, the European Clean Energy Package establishes energy-saving targets through a long-term renovation strategy and by enhancing the functioning of retail electricity markets in order to protect vulnerable consumers. Similarly, the European Commission's March 14, 2023, proposal to reform the EU's electricity market design includes a new provision on energy sharing. This provision mandates that Member States implement fair and equitable measures to ensure vulnerable and energy-poor households can participate in energy-sharing initiatives [20].

Moreover, as is elaborated in [21], particularly financial liberalization and market-based approaches, has a fostering effect on renewable energy R&D investment and combating energy poverty. As another example, Liaison Entre Actions de Développement de l'Économie Rurale (LEADER) program in European countries, which started in the 1990s, was a set of actions to address inequalities against marginalized communities in rural areas which in many cases had limited or no access to electricity. Later in 2014, this program was expanded to community-led local development (CLLD) in order to address the need of marginalized communities in urban areas too [22, 23]. For example, more than 37 marginalized urban areas in Romania having energy access problems were approved to be covered by CLLD [24]. LEADER and CLLD, in a multitude of cases, include joint cooperation of public and private entities in which private entities have more than 51% of the votes in decision-making.

A study on the energy poverty problem in 27 European countries from 2000 to 2019 revealed that an increase in bilateral trades within the energy sector positively expedited economic activities, resulting in higher accessibility of energy in these countries. The study reveals that greater service openness and higher trade levels improve energy affordability within the EU. The results indicate that a 1% increase in service openness resulted in short-run and long-run energy cost reductions of 5.3% and 7.9%, respectively [25].

Yet as stated earlier, the impact of increased trade activities on energy accessibility can be both positive and negative. For example, in a study performed by Zhao et al. using a 15-year data panel of 64 countries, the impacts of several factors including:

- renewable energy consumption
- energy efficiency
- urbanization level
- trade openness
- economic growth

on energy poverty have been estimated globally [26] It is interesting that they found a negative correlation between energy poverty and trade openness in the Asian region, which means increasing trade openness would alleviate energy poverty in this region. Nevertheless, the authors suggest a positive correlation exists between the prevalence of energy poverty and the degree of trade openness regarding Africa.

While trade openness may exacerbate energy poverty in Africa, Amoah and colleagues believe it can increase the renewable share in total energy consumption in this region [27]. In their analysis on a 22-year set of data (1996–2017) for 32 African countries, they found that renewable energy integration and open market components are positively correlated. In other words, although trade openness may worsen the energy poverty issue in Africa, it improves the sustainability aspect of energy justice in this region.

Such bidirectional effects of energy trade on different aspects of energy justice and equity definitely have a complicating effect and make policymaking in this area a challenging task that should be done with delicacy.

9.2.3 How Tradability Contributes to Energy Justice?

As stated in the opening of this chapter, trade is the mainstream of getting the goods and services we need. Our history on earth shows that when people start to trade something, a line of business opens up, and as the volume of trade goes up, more resources are allocated to the chain of activities related to that business. This chain of activities may include providing land, input raw materials, banking and financial services, human resources, machinery, energy, shipment and transportation, marketing and advertisement, storage, retailing, etc. This increasing allocation of resources leads to a higher concentration on research and development to find more efficient ways to deliver those services. This way, the whole process entails lower production costs and it means that more persons can afford acquiring that specific good or service which elevate the social justice and equity.

The identical principle can be observed in the context of energy. Initially, commercial electricity was perceived as a luxury commodity accessible only to affluent households for the purpose of illumination. However, the tradability of electricity facilitated the expansion of the electricity market. Consequently, the utilization of economies of scale and the implementation of innovative technologies aimed at enhancing efficiency led to a reduction in electricity costs. As a result, access to electricity was democratized, thereby enhancing social equity in terms of electricity accessibility.

On the other hand, it has been witnessed in the history that even valuable things may lose their value, or at least being undervalued because of their lack of tradability. As an example for this case, let us have a look at the history of modern oil discovery which took place in the 1850s in Pennsylvania, United States. Since petroleum extraction is mainly due to the gas pressure within the inner layers of the earth, a considerable amount of gas comes out along with the oil. Nevertheless, due to the non-tradability of natural gas during that era, it was initially regarded as a by-product waste in the early stages of the oil industry and was often disposed of through burning. It was not until the 1920s that natural gas evolved into a tradable fuel, thereby gaining acceptance in industrial complexes, residential settings, and power plants [28].

Furthermore, natural gas is considered a better fossil fuel compared to gasoline and diesel fuel. Natural gas not only is less polluting but also more technically favorable to power-generating units.

However, unexpectedly the price of natural gas is generally lower than diesel fuel. For instance, the comparison between the average price for gasoline, compressed natural gas (CNG), and diesel fuel in the United States is depicted in Figure 9.1, which is based on the data published by the US Department of Energy [29]. Such a price difference is, inter alia, due to the fact that diesel fuel is liquid, more storable, has a larger market, and thence higher tradability.

We want to conclude here that if the trade mechanisms can be aligned toward the key elements of energy justice, then it would be more easily achievable. If the energy markets fail to valorize energy justice and equity or continue to undervalue this goal, reliance solely on conventional wisdom and moral principles will likely fall short of meeting societal expectations. For instance, as stated in Section 9.2.2, access to energy for all and energy affordability represent crucial components in

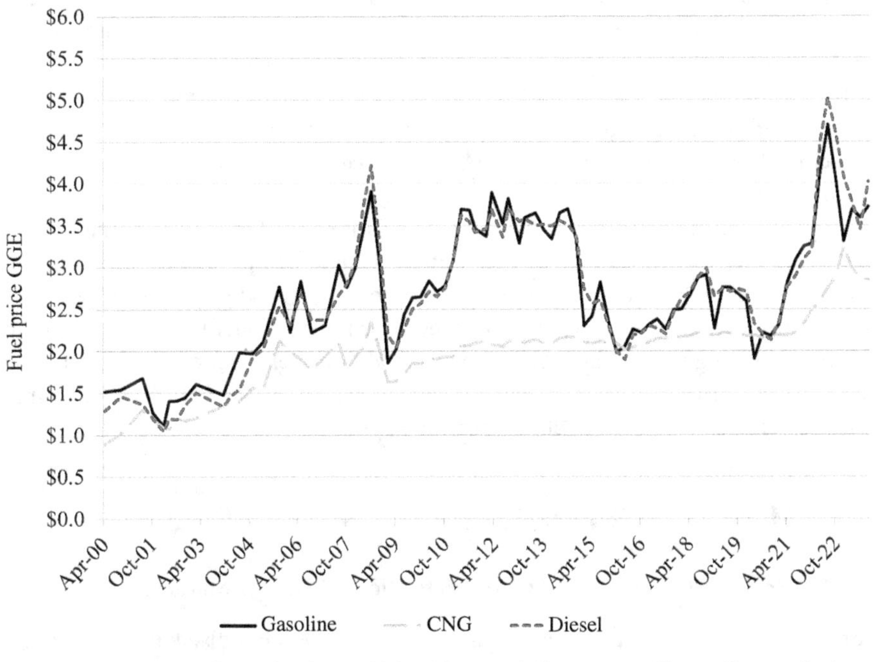

Figure 9.1 Average fuel price in the United States, dollar per gasoline gallon equivalent ($/GGE).

the pursuit of energy justice. The tradability of energy has the potential to harness market dynamics, thereby enhancing efficiency within the energy supply chain and reducing costs. As another example, renewable energies have been known to humans for millennia, however, their massive integration into the modern world did not happen until they found their path to the interconnected power grids and new mechanisms were established in the electricity markets to make these energies tradeable. Hence, improving the tradability of renewable energies can be an effective catalyst for sustainability, which is another key element of energy justice. Along with such manifestation, there have been great efforts to introduce and establish market mechanisms to reduce carbon footprint in the energy sector and ameliorate intra generation energy justice which will be discussed in the further sections.

9.2.4 How Tradability Threatens Energy Justice?

In contrast with what is described in the previous section, market mechanisms can act as a diminishing factor for energy justice and equity. The reason behind this possibility lies within one of human instincts: selfishness. Humans in general, care about the common wealth and social welfare much less than they do for their benefit. Microeconomic principles posit that market structures should be designed and monitored to ensure that individual actors' pursuit of self-interest ultimately results in greater societal benefit. However, the pursuit of individual profit, coupled with factors such as racism, discrimination, and political conflicts between nations, can lead to market distortions and undermine energy justice and equity. Additionally, errors in market design and shortsighted decision-making can contribute to these negative outcomes, exacerbating inequalities within the energy sector.

Furthermore, even in very sophisticated designs of markets, it is very likely that the social welfare of the living people attracts higher importance, compared to the right and welfare of the generations to come. Normally, people respond to short-term threats more radically than they do to remote benefits. Hence, it would not be surprising to witness a situation in which long-term benefits are sacrificed in favor of short-term ones. This has been one of the biggest barriers to establishing and implementing righteous climate change policies.

9.3 Markets for Energy Carriers

9.3.1 Brief History and the Current Situation of Energy Markets

Among the energy sources that are still largely merchandised in the energy market, coal is the oldest and also the most polluting one. It has been known to humans since ancient times in China, and its usage was also witnessed by Marco Polo in the 13th century. The first documented use of the term "coal" for this substance is seen in Aristotle's writings. Although there are evidences for using coal in the Roman Empire, due to the abundance of wood, the use of coal was very limited. Later, around the early 13th century, coal was found as a suitable heat source for metalworking. The extraction and utilization of coal experienced significant growth during the 18th century, coinciding with the onset of the Industrial Revolution, driven by a steadily increasing demand for coal. Subsequently, the discovery of coal gas as a combustible fuel introduced another dimension to the realm of coal energy. Coal maintained its prominence as a primary energy source until the conclusion of World War II, at which point oil-based fuels came to dominate the transportation sector, followed by the integration of natural gas for residential heating purposes. Despite these developments, the coal industry persisted, albeit with a primary focus on power generation [30].

Nowadays, around 27% of the global energy provision and 40% of worldwide electricity generation are based on coal, which is a considerable share. In addition, coal accounts for 40% of energy-related CO_2 emissions. However, according to the y and as depicted in Figure 9.2 [31], more than 70% of coal is used in China and India. In fact, around 55.5% of total primary energy consumption in China and 55.1% of India came from coal in 2022 [32].

China and India are the biggest coal producers, and 52.8% and 8.6% of the global coal in 2022 [33].

Variations in material specifications, particularly in coal extracted from diverse geographic locations, result in disparities in heating values, which in turn

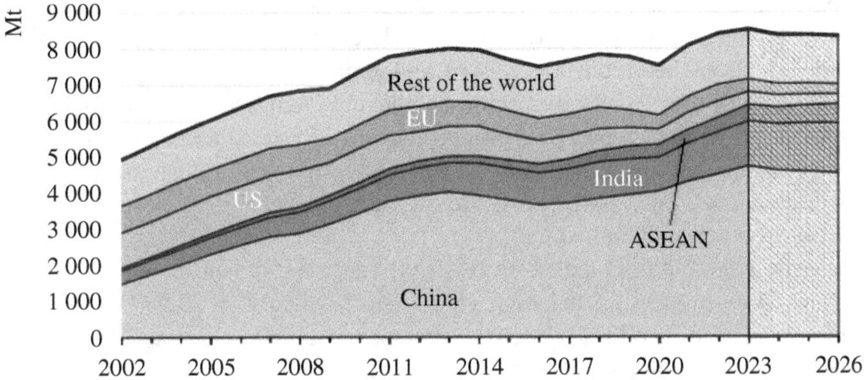

Figure 9.2 Use of coal around the world. Source: [31]/International Energy Agency / CC BY 4.0. (IEA, Global coal consumption, 2002–2026, IEA, Paris, https://iea.blob.core.windows.net/assets/a72a7ffa-c5f2-4ed8-a2bf-eb035931d95c/Coal_2023.pdf, IEA. License: CC BY 4.0IEA, IEA.)

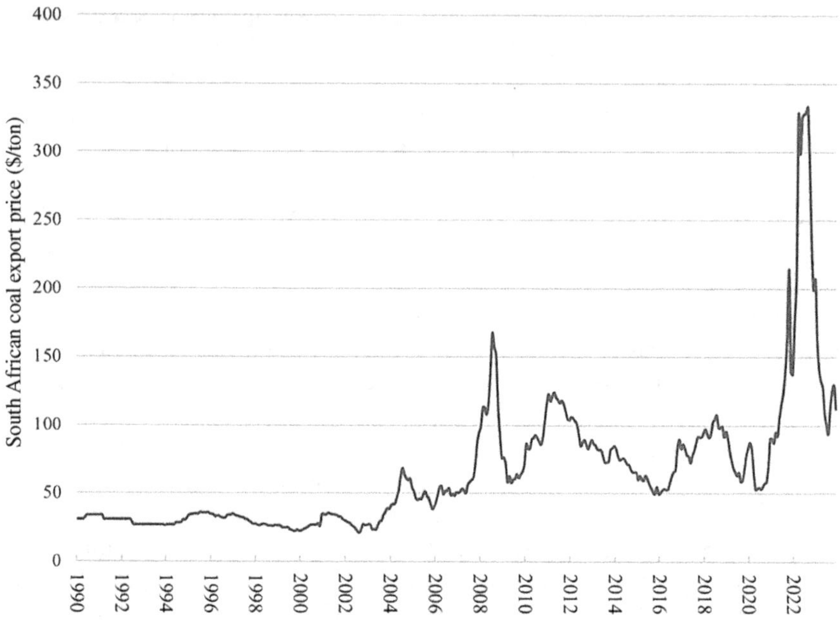

Figure 9.3 Monthly average price of South African coal export.

influence their commercial value. To get a general picture regarding the variation of coal prices through time, the average price index for South African coal is depicted in Figure 9.3 which is based on data available in [34].

As seen in Figure 9.3, following the war in Ukraine, coal prices sharply responded to the European energy crisis in 2022 and again fell back down.

Since commercial oil production in the middle of the 19th century, the demand for fossil fuels has been constantly rising, and soon after World War II, coal yielded its place to oil-based fuels, and these combustible materials became dominant energy sources that empowered human life in every imaginable aspect. These energy sources have been favorable for being easy to use, storable, transportable, and thence tradeable. Specifically, the oil market has been one of the biggest markets ever recorded in history.

Based on the data published by IEA [35] and as is shown in Figure 9.4, oil covered around 29.5% of the total global energy supply in 2021. Fossil fuels, containing gas, oil, and coal, together have more than 80% share and this shows a high level of dependence on these energy sources.

The invention of electricity and interconnected power grids changed the shape of human life greatly. Electricity created new forms of energy demand, and almost all the types of machinery that once were powered by fossil fuels became considerably electrified. However, it may sound unexpected that only

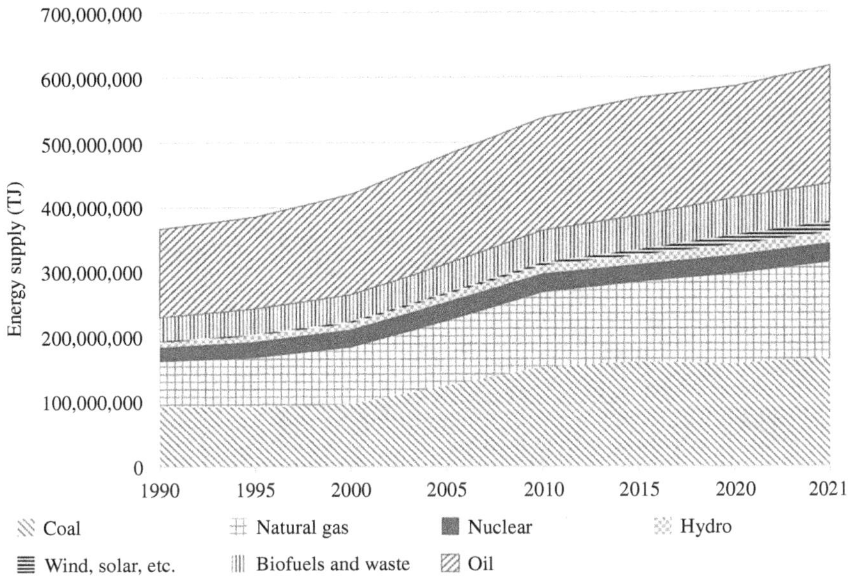

Figure 9.4 Total energy supply based on the primary energy.

less than 20% of the primary energy sources are consumed to generate electricity, and the rest is used in other applications like heating, transportation, industrial manufacturing, etc. Figure 9.5 shows the share of electricity generation in the primary energy use through time [36]. The power sector in its early stages was mainly unregulated, privately owned, and geographically fragmented. Electricity providers owning the whole supply chain, from generation to retailing, normally in a vertically integrated structure started to sell electricity which was normally used for illumination. However, as the electricity demand rose, the public sector intervention in the power industry intensified and state-owned monopolies appeared [37]. In contrast with oil and other fossil energy sources, electricity has lower tradability. This is mainly due to the following characteristics of electricity:

- Limited potential for storage and its cost.
- Necessity of an instantaneous balance between electricity generation and demand.
- Being almost unreplaceable.
- Inelasticity of electricity demand.

Restricted by these limitations, in most parts of the world electricity was managed centrally and via vertically integrated monopolies for decades. Such a structure left very limited space for competition and market, and in most cases, electricity was treated as a public service that was delivered by state-owned

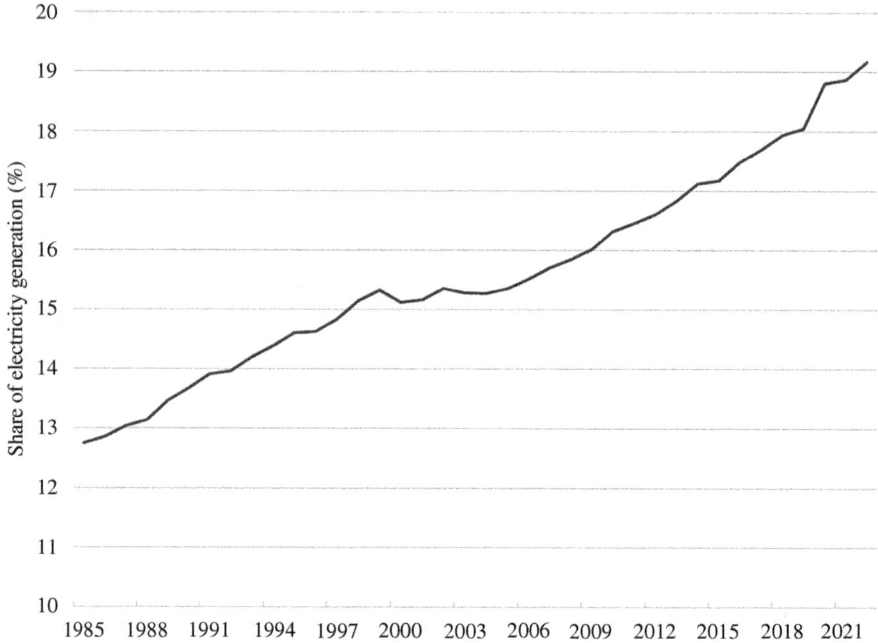

Figure 9.5 Share of electricity generation in primary energy use.

utilities. In fact, many countries came to this conclusion that the electricity sector inherently exhibits characteristics of a natural monopoly, wherein centralized management results in lower overall costs due to enhanced coordination. Although, in some parts of the world, private utilities persisted to exist and gained a geographical monopolistic advantage [38]. The first steps to introduce electricity as a tradable commodity were taken in Chile (1982), where Augusto Pinochet followed the ideas of Chicago school economics to achieve a "Revolutionary market society" [39]. The same idea was followed by England and Wales at the time when Margaret Thatcher was in power (1990). The power industry reform process is generally referred to as restructuring and includes privatization, opening of wholesale competition, and unbundling of the generation, transmission, distribution, and retail sectors [40]. The restructuring of the power industry soon became popular and many countries have been taking this path toward injecting competition in this sector and making room for private sector participation. This process has transformed electricity from a public service to a tradable commodity, and competition, mainly in the generation sector and retailing, has become a driving force for higher technical and financial efficiency.

Despite the diversity of electricity market designs, almost all of them include some structures that can contribute to energy justice and equity, especially in the case of the sustainability of electricity provision to the consumers. Most electricity markets in the world are equipped with the elements to support renewable electricity, and there are mechanisms in place to make renewable electricity tradable [41, 42]. Such mechanisms will be further reviewed in this chapter. Furthermore, electricity markets have fostered innovations such as microgrids and energy hubs, which hold significant potential for mitigating electricity access deficits, a critical aspect of energy justice [12].

9.3.2 Challenges of Energy Markets to Promote Energy Justice

As stated in Sections 9.2.3 and 9.2.4, energy markets can have both positive and negative impacts on energy justice, depending on how well the market is designed and executed. From the viewpoint of energy justice and equity, there are barriers in the current energy markets that can be hindering. As mentioned before in this chapter, eliminating energy poverty and providing access to clean affordable energy is a prominent item in energy justice. Energy markets, in general, pave the way for profitable energy-related investments, and in most cases, those who suffer from energy poverty are not eligible customers for such a profit-oriented environment. Hence, market forces may become futile to bring modern energy to those people. It should be noted that energy justice, particularly access to modern energy, is of higher importance when it comes to residential energy needs.

Moreover, energies in general, and especially electricity, are perceived as commodities with very low levels of demand elasticity. The inelastic behavior of electricity demand is more visible in residential electricity demand [43]. It means that the price increase would have a minor effect on lowering the demand. Labandeira et al. performed a meta-analysis to probe the short-term and long-term elasticity regarding energy carriers, including electricity, gasoline, diesel, natural gas, and heating oil. They came to this conclusion that the average values of demand elasticity of these products categorize them as price inelastic [44]. In such cases, the market would react drastically to major changes in demand or supply and sudden rises in demand, or any sudden decrease in the supply side, may lead to intense price spikes. Such situations have been observed and documented frequently in the history of energy markets. For example, after the Islamic revolution in Iran (1979) and later the Iran–Iraq war (1980), due to the decrease in oil supply, the oil price from \$14.95 per barrel in 1978 rose to \$37.42 per barrel in 1980 which showed more than 150% increase [45].

Another key point that's worth discussing here is that energy has emerged to become a vital element for industrial and economic development, and historically, there is an intensive correlation between the amount of energy use and gross national product (GNP) [8, 46, 47]. This reliance on energy carriers has been an effective driver for the government authorities to intervene in the energy sector.

There is a wide spectrum regarding government intervention and in some parts of the world, state-owned companies are directly involved in the energy supply chain and the government takes an active role in determining final usage tariffs, injecting subsidies and cross-subsidies, etc. But, in some other countries government intervention is limited to some regulatory measures. Government intervention in markets is one of the most polemic issues in politics and economics and in the case of energy markets government intervention has a strong historic footprint. The energy sector in many countries includes a high level of subsidy and in most cases, it's perceived to serve higher affordability and social equity. While there are continuous disputes on the negative effect of energy subsidy on consumer's behavior and energy efficiency on the consumption side and consequently higher carbon emissions which is a threat to energy justice and sustainable development [48]. This issue will be further discussed in Section 9.3.3.

Putting altogether, due to the intrinsic low elasticity of their demand, energy markets are vulnerable to market power which endanger the market performance and competition fairness. On the other hand, since the energy carriers are strategic commodities, generally there are a lot of government/state interventions and subsidies within the energy sector and this can divert the energy markets from their true objective and deteriorate energy justice.

9.3.3 Affordability, Subsidy, and Market

As described in Section 9.2.2, access to modern energy sources is one of the most fundamental plausible goals in the context of energy justice. However, the main challenge is that market structures may entail situations in which the energy price becomes too high to be accessible to people. The negative impact of the failure to provide people with modern forms of energy carriers is twofold. First, this failure substantially contributes to deteriorating living conditions and social inequality that leads to persisting poverty, social disputes, and in extreme cases rebels and social disorder.

As a mean to tackle energy vulnerability and eliminate energy poverty, many countries have designed subsidy structure in the end-user energy tariffs. According to the World Bank, energy subsidy is "*a deliberate policy action by the government that specifically targets electricity, fuels, or district heating and that results in one or more of the following effects: 1) It reduces the net cost of energy purchased, 2) It reduces the cost of energy production or delivery, and 3) It increases the revenues retained by those engaged in energy production and delivery (energy suppliers)*" [49]. The amount and share of subsidy can be quite different from one country to another. The annual amount and share of subsidy within the energy sector for 25 countries with the highest energy subsidy are shown in Figure 9.6. As is seen in this figure, Russia annually spends around 63 billion US$

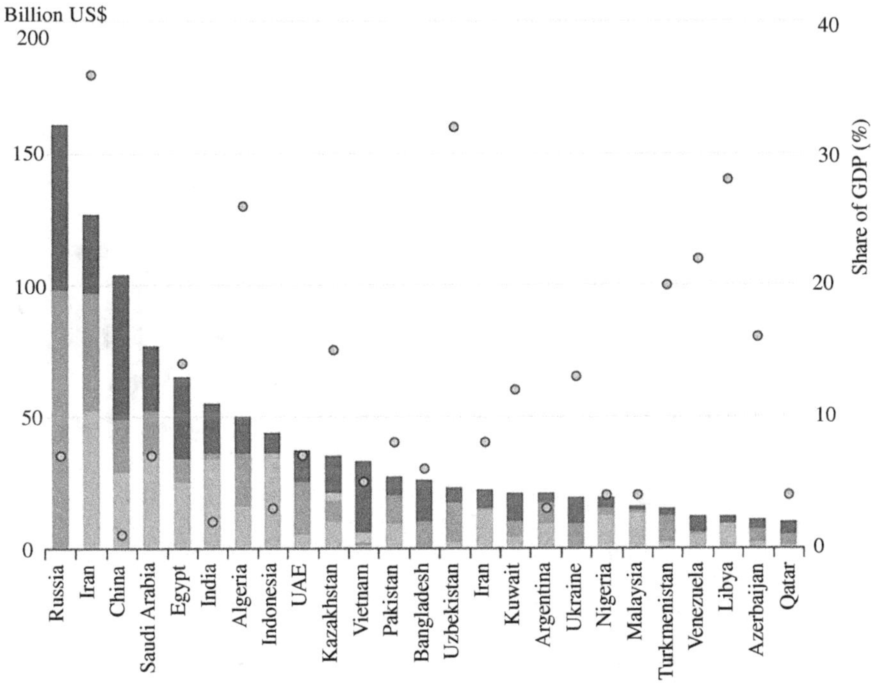

Figure 9.6 Amount energy subsidy (bars), and share of fuel subsidy (gray dots). Source: [50]/International Energy Agency / CC BY 4.0. (IEA, Value of fossil-fuel subsidies by fuel in the top 25 countries, 2022, IEA, Paris, https://www.iea.org/data-and-statistics/charts/value-of-fossil-fuel-subsidies-by-fuel-in-the-top-25-countries-2022, IEA. License: CC BY 4.0.)

for energy subsidy which is the highest in the world. However, the subsidy in Russia covers only 7% of the end user's energy cost. On the other hand, Iran has the highest subsidy share in the energy price, and about 36% of end user's energy cost is covered through subsidies. The scattered round dots in Figure 9.6 confirm that different states around the world apply different policies regarding energy economy and there's no global consensus in this matter.

In addition to Figure 9.6 and in order to demonstrate the wide spectrum of energy product pricing in the world and as an indicator for energy prices, the data related to the gasoline price prepared by the World Bank [51], is depicted in Figure 9.7. As can be seen in this figure, there is a noticeable gap among the prices, and this is not limited to the gasoline and the same situation exists for other energy products as well.

In many cases subsidies proved to be a plausible instrument to improve social equity. For instance, "Warm Front" Scheme was an energy cost assisting program that was launched in 2000 in England, and it helped 2.3 million

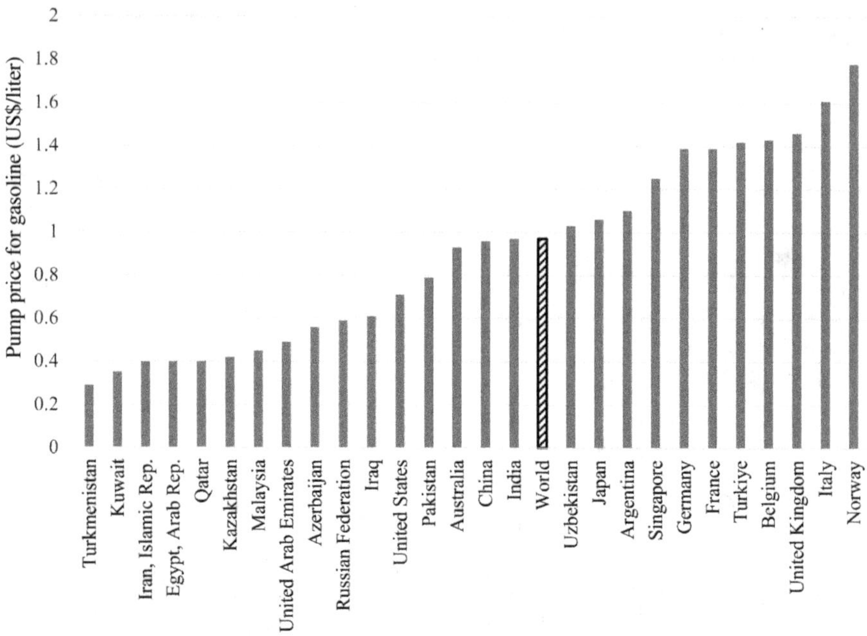

Figure 9.7 Pump price for gasoline in some selected countries in 2016.

households to reduce their energy cost. Since the scheme mainly focused on insulation improvement, not only helped the end users to lower their energy bill but also decreased their carbon emission and made their energy usage more sustainable [52]. The subsidy scheme conducted in South Africa to make liquified petroleum gas affordable for household cooking is another successful example that entailed less carbon emission, lower household cost, and higher living conditions [48].

Notwithstanding the successful practices and prevalence of energy subsidy schemes around the globe, many energy analysts believe that intensively subsidized energy structure ultimately entrenches social benefits. They highlight the contradiction between subsidized energy and energy efficiency, and this has become a polemic and ongoing debate in energy policy. It's worth emphasizing that the consequence of energy subsidy generally far exceeds the scope of the energy sector. Energy subsidy differs energy valuation from market price and its real economic value and the consequent undervaluation of energy changes the consumer's behavior and aggravates energy intensity which consequently leads to higher amount of carbon emission, unhealthy air quality, and related health concerns. Moreover, intensive energy subsidy exhausts the national budget and incapacitates the government to serve people properly in other sectors such as public transportation, medical support and healthcare, and education.

International Monetary Fund (IMF) after a vast investigation on the global experience on energy subsidy suggests six fundamental factors in energy subsidy reform to achieve higher productivity and social justice [53]:

- Pursuing clear long-term objectives through a comprehensive plan.
- Having vast and inclusive communication with citizen to justify the necessity of reform and its positive impact on their quality of life.
- Having a step-by-step plan for subsidy reduction considering a rational sequence across energy carriers.
- Decreasing production costs via making the state-owned companies more efficient.
- Targeting financially vulnerable families and maintaining financial and instrumental support for them.
- Depoliticize energy pricing via institutional reform and limiting government intervention in the energy market, especially at the end user frontier.

Establishment of a market-based environment with active private participation can be aligned with the latter item in the above list and exonerate the government for the increasing prices and pave the way for a comprehensive subsidy reform that benefits a broader range of people while protecting the vulnerable groups.

9.3.4 The Interplay Between Efficiency and Justice

The rising awareness of global warming and negative impacts of fossil fuels emissions continuously attract more attention to the policies regarding sustainable development and international forces are in play to develop and implement such policies. Generally, sustainable development has two wings:

- Less reliance on fossil-based energy sources and expanding renewable energies
- Attaining a higher level of energy efficiency in energy production/consumption

A general interpretation of energy efficiency is "to achieve more with less energy consumption." Improving energy efficiency is perceived as a passage toward a more sustainable economy. As IEA describes, energy efficiency has the potential to be seen as "the surest energy supply" [54] and since it is a cost-effective way of supply and its vast unrealized potential it can be considered as "the first fuel" [55]. Energy efficiency can lead to lower primary energy usage and the corollary is lower carbon emission which serves to improve energy justice and sustainability indicators. The European Commission has recognized energy efficiency as "the EU's biggest energy source" and a major means to achieve energy security and attain climate change goals [55]. Moreover, implementing energy efficiency programs contributes to investment savings in the energy sectors which can be then utilized in other sectors and gain higher economic value and improve quality of life. The evidence is evident and pervasive that energy saving can offer a wide range of significant and valuable advantages to multiple socioeconomic aspects of people's lives around the world, and some of the most prominent benefits due to implementing energy efficiency programs are as follows [54]:

- Releasing public budget.
- Less local air pollution.
- Less greenhouse gas (GHG) emissions.
- Creation of job opportunities.
- Improve family affordability.
- Lower energy prices.
- Higher level of energy security.

It's quite evident that all of the abovementioned benefits, in one way or another, contribute positively to a better situation regarding energy justice and because of that energy efficiency is one of the most viable options to achieve sustainability and equity.

Although surprisingly, energy efficiency can affect energy usage in the opposite direction too. There are evidences that energy efficiency in the first run leads to less energy consumption and thus less demand for energy with entails lower energy price. Lower energy costs, coupled with the income effect, can potentially stimulate increased energy consumption, this phenomenon may partially or entirely offset the energy savings achieved through improved efficiency measures. [56]. In other words, when the "rebound value" is higher than zero and less than 100%, part of the energy savings is lost due to higher demand for energy-related goods and services. For example, a 10% rebound effect due to 30% increase in energy efficiency, eliminates 3% of the energy gain of higher efficiency and total energy use would decrease by only 27%. Because of the rebound effect, the ultimate energy saving may fall short of what has been predicted to gain from implementing energy efficiency programs. On the other hand, rebound values higher than 100% mean that due to backfire effect, the higher demand for energy outvalue the energy savings out of higher efficiency and the total energy use would increase [57].

9.3.5 Market Mechanisms to Promote Energy Justice

As explained in the previous sections, energy markets and energy tradability can have both positive and negative effect on energy justice level, and there are two reasons behind that. First, while opening the energy sector to market competition and increase in trade activities may promote some justice aspects such as energy accessibility, can also exacerbate some other aspects such as energy sustainability.

Second, surprisingly, liberalization energy markets and promoting market-based competition in the energy sector can have both positive and negative effect on a single aspect of energy justice at the same time. As mentioned previously, in Section 9.2.2, liberalizing energy industry and increase in market-based activities promote investment in the energy infrastructure and expand energy accessibility. But at the same time, since there is less financial benefit in delivering energy to poor marginalized people, leaving the energy services to the hands of private companies without proper adjustments, can lead to a situation in which energy companies are likely to leave these people behind worsening the energy poverty in such regions worse.

In order to address these issues and based on accumulated experiences around the world, different states and countries have tried to incorporate proper mechanisms and adjustments within the energy markets and related trading mechanisms. For instance, a set of mechanisms has been proposed and implemented in order to address energy poverty issues and/or lack of energy affordability. Some of the most common ones are as follows:

Subsidy Schemes: Many countries have set energy subsidies within their energy trades. These subsidies are widely used in the residential sector to ensure that financially vulnerable families can afford to use a minimum level of energy [58].

Local Mini-grids: In many cases and specially in the countries where the distance between rural areas is long, the electrification of the villages with small population may not be commercially viable. In such circumstances, local mini-grids may be considered as a solution to provide far-reach communities with electricity. Because this way, the high upfront cost of high-voltage transmission lines and substations is saved. Mini-grids significantly contribute to the UN SE4All Initiative's efforts to achieve universal electricity access by 2030 [59, 60].

Energy Efficiency Improvement Low-interest Loans: This mechanism has been introduced as a tool to help low-income families to decrease their energy costs. Efficiency improvement solutions may not a be easily affordable to people, and especially low-income household who need it the most. Such economic loans not only help them to decrease their monthly payment for energy but also help the overall energy system to achieve higher performance [61].

Community-owned Renewable Facilities: Community–private partnerships (CPPs) in renewable energy emerge when neither local communities nor the private sector alone possess the resources, skills, and capabilities for successful project implementation. CPPs typically combine private sector finance, technology, and expertise with community-owned land (especially in the case of biomass or biofuels) [62].

Also, some other mechanisms and adjustments have been used to improve energy justice through sustainability improvement. A short review on these mechanisms is presented in Section 9.4.3.

9.3.6 Energy Access During Crisis

One of the most challenging situations regarding energy justice and equity is when there is an extreme scarcity in the supply of at least one form of energy. Depending on the magnitude of the shortage, such condition can affect social stability in zonal, country-wide, regional, and even international level. In such circumstances, the concerns regarding reliable energy access rise and both secure energy provision of the households and economic condition of commercial firms will be at risk. At first glance, one may construe that such situation would be at the benefit of energy providers and give them the opportunity to enjoy soaring prices. However, as the history shows this would be a false assumption. High energy prices would act as a barrier to expansion in energy-related industries, and this is a menace to the business continuity of energy-providing firms.

While short-term energy supply shortages may be manageable, prolonged periods of scarcity typically necessitate government intervention to mitigate the social consequences. Such interventions often involve price suppression or, in more extreme cases, the suspension of market mechanisms altogether.

As an example, for extreme state intervention in common execution of the market, we can refer to the situation that took place in Australia in June 2022. Australia is a major energy exporting country, and yet for the first time of its history, National Electricity Market (NEM) experienced 9 days suspension during June 15 till June 23, 2022. The primary cause of this unusual event was the capacity shortage due to international high fuel cost and insufficiency of demand elasticity which together led to the formation of high electricity prices in that period. Frequent price spikes in June 2022 triggered a course of actions that finally convinced and authorized Australian Electricity Market Operator (AEMO) to intervene and suspend competition in NEM and replace it with the Market Suspension Pricing Schedule [63].

In most cases, government intervention during energy crisis is to shield consumers from the exorbitant energy prices. Although government support is more common to help financially vulnerable citizens and families, depending on the magnitude of the price shock, it could include the vulnerable firms too. For example after the energy shock due to Russia's invasion of Ukraine, several European states commenced support schemes totaled over 650 billion euros to protect households as well as commercial firms from energy price spikes [64].

As Sgaravatti et al. report, following the war in Ukraine, most EU countries dedicated a sizeable supporting fund to hedge energy consumers from the high energy prices. As can be observed in Figure 9.8, Malta, Bulgaria, Austria, Italy, Greece, and Lithuania have dedicated more than 5% of their national gross domestic product (GDP) to such supporting schemes, and in Germany alone the total budget for this purpose exceeds 150 billion euros [65, 66].

While claiming to protect the citizens and economy, state intervention and support schemes normally have extensive negative side-effects. Government

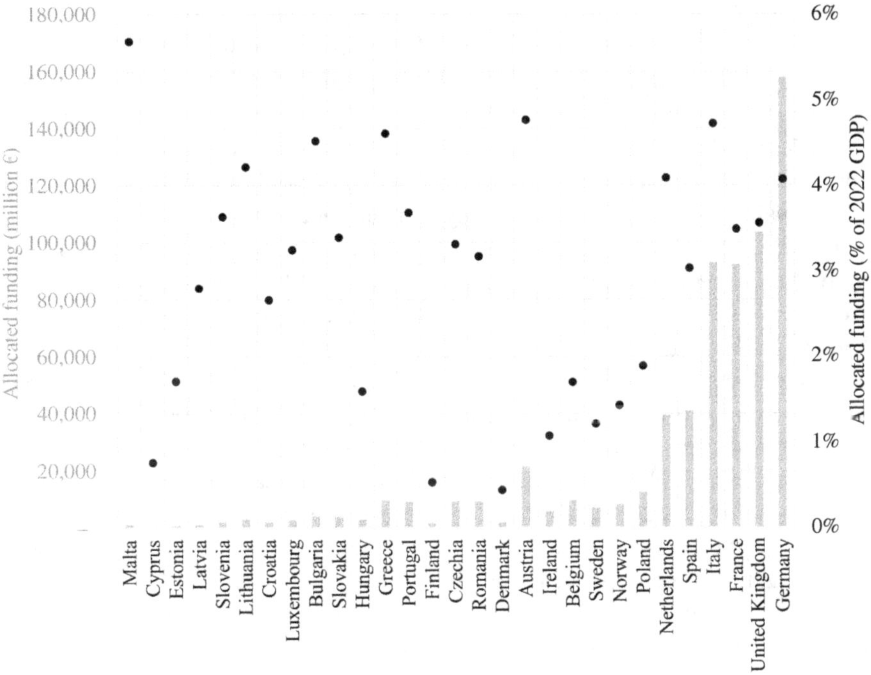

Figure 9.8 Allocated funding in EU countries to shield consumers from high energy prices after Russia–Ukraine conflict.

support mechanisms during energy crisis can have a weakening effect on long-term objectives such as sustainability goals, demand-side management, and energy security plan. Moreover, due to the flaws in design and prediction, the financial burden of the support mechanisms can overweigh the economic benefits. The authorities should keep it in mind that intervention in the normal course of markets is a serious decision to make and once it's done it will put the government at the front and makes the government liable to react to similar events too. Another challenge regarding the implementation of support mechanisms is the intrinsic political inertia of such schemes, once they are defined, it takes a lot of political burden to eliminate them [64, 67].

9.4 Trading Mechanisms to Support Energy Justice

9.4.1 Energy Justice and Sustainability

As explained in Sections 9.2.3 and 9.2.4, the ultimate impact of market mechanisms on energy justice indicators can be positive and negative, depending on the market performance and structure. Among the key elements of energy justice, sustainability is the one that is still far from the target levels, even in modern economies. That's why in the context of energy justice there is a noticeable emphasis on the mechanisms and policies to push the sustainability indicators up. In fact, notwithstanding general expectations, if the share of renewable energies in final energy consumption is chosen as an indicator, the low-income countries would be ranked much higher than the developed countries. Among 30 countries with the highest GDP per capita in 2020, in only 11 countries, the ratio of renewable energy to the final energy consumption is higher than 20%. At the same time, among 30 countries with the lowest GDP per capita, in 23 countries more than 50% of total consumed energy is provided by renewable sources [67]. However, it should be noted that most of these low-income countries suffer from energy poverty and lack of sufficient access to modern energy forms. Yet this is a plausible notice that in order to achieve sustainability goals and a clean balanced life on mother earth, the role of leading economies is influential and there is a long way ahead of them to fulfill their responsibility. The significance of sustainability as a core component of energy justice has become increasingly recognized. This shift has spurred a concerted effort in numerous regions to restructure energy markets, particularly electricity markets, and integrate mechanisms that promote sustainability. Based on the data published by the World Bank [68], the ratio of renewable energy in final energy consumption for the world, low-income, and high-income country groups during three decades is illustrated in Figure 9.9.

9.4.2 The Case for and Against the Coexistence of Market and Sustainability

In a very basic term, markets are places where people, as buyers and sellers, interact to gain profit. In other words, markets are intrinsically based on economic profit. Where sustainability restrictions decrease the profitability of a considerably large group of firms, it's unrealistic to expect free market forces to drive the sustainability indicators upward. That's the common situation in the energy, and particularly electricity markets. Back to what is explained in Section 9.2.1, energy markets have a noticeable long history, and suffices it to have a look at environmental and sustainability indicators to realize what has been the contribution of the energy market, inter alia, on the environmental destruction of mother nature around the world.

Moreover, the nexus between energy poverty eradication and sustainability, which are both important aspects of energy justice, adds to the complication of the problem. Normally, the people who suffer from a lack of access to sufficient modern energy forms cannot afford to harness renewable and sustainable forms of energy, and energy markets in these regions describe renewable energy more as a luxury than a profitable solution for economic firms. And because of that, most approaches toward taking these people out of energy poverty results in higher

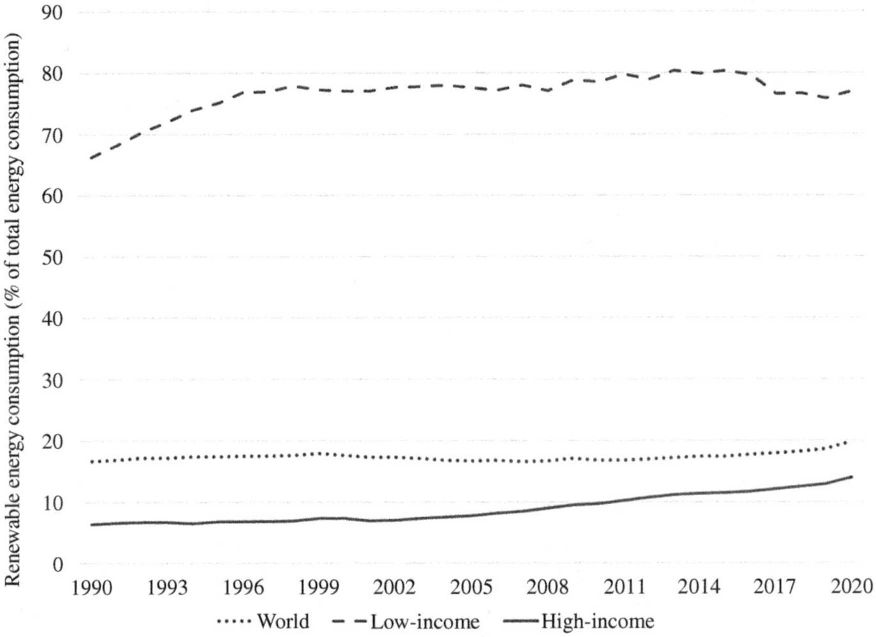

Figure 9.9 Share of renewable energy in total energy consumption.

carbon emissions. In other words, fulfilling one prominent goal of energy justice entrenches another major goal.

However, what was mentioned before is not the whole story. As Steve Jobs once said "*humans are basically tool builders*," and creative humans in the field of energy markets have come up with plans and arrangements that transform electricity markets into tools to boost the sustainability of energy sector, and we are going to shortly review some major examples of these tools in Section 9.4.3.

9.4.3 Short Review on Sustainability-driven Trading Mechanisms

In general, we categorize the mechanisms that in principle are design to support sustainability in three different forms:

- **Renewable Energy Support**

This category encompasses all sorts of support schemes that are designed to encourage investment in renewable energy production. Despite the rising general awareness of the environmental cost of fossil fuels combustions, as well as continuing progress in the technology and R&D efforts, the high investment cost of harnessing renewable energies has been a major barrier and as mentioned before in this chapter still a considerable share of worldwide final energy use depends on fossil energy sources. In order to surmount this challenge, several supporting mechanisms have been practiced in the world to augment the economic feasibility of investment in renewable energy production [69]. However, as the competition in renewable technologies is increasing, government intervention is declining [70].

In most countries, kind of financial support is applied to encourage private investors toward renewable technologies. Some of the most common support schemes in renewable electricity include tax benefits, quota systems, feed-in tariff (FiT), premium tariff, etc. Table 9.1 shows the major European supporting mechanisms for renewable electricity [71].

- **Carbon Taxation**

This category includes financial and taxation mechanisms to make economic incentives for less carbon emission. Such instruments that may put an extra tax burden or provide tax partial relief (due to excess emission or emission reduction) can be applied to both the energy-producing and energy-consuming sectors.

Carbon taxation is a surcharge that is normally defined per ton CO_2 equivalent and levied by the government to make buying and using fossil fuels financially less desirable to individual people, business firms, and organizations. This gives a premium advantage to sustainable approaches such as energy saving, energy efficiency, and renewable energy usage. Furthermore, the revenue out of carbon taxation enables government authorities to invest and financially support projects to boost sustainability. However, as Nadiri et al. describe, carbon taxation alone does not produce sufficient driving force for boosting energy sustainability and should be backed by local and international initiatives to promote commercial leverage of pro-environmental technologies [72].

Carbon taxation was first proposed by David Gordon Wilson in the 1970s. He offered this mechanism as a price instrument instead of emission limits which had been already introduced in some of the states in the United States. Despite the origin of the idea, carbon taxation was for the first time implemented in Finland in 1990. The carbon taxation mechanism is easy to administer and incurs low administration costs. As the Swedish Tax Administration reports, the administrative cost is about 0.1% of the government's revenue from energy and carbon tax [73]. Thanks to this implementation simplicity, many countries and states around the world have already adopted carbon taxation as a means toward a more sustainable economy. Based on the data published in [74], the magnitude of the carbon levy and the date of promulgation for some European countries are depicted in Figure 9.10. For the more complete list of countries around the world that have already implemented carbon taxation, please refer to [75].

Table 9.1 Renewable electricity support mechanisms is the EU countries.

	Feed-in tariff	Premium tariff	Tax benefit	Quota system	Contract for difference	Investment grant	Low interest loan	Subsidy	Tendering system	Net metering
France	•	•	•							
Norway				•						
Denmark		•					•			•
Belgium				•						
Netherlands		•	•			•				•
Germany	•	•					•			
Sweden			•	•				•		
Spain	•	•								
Italy	•	•							•	•
United Kingdom	•		•	•	•					
Switzerland	•									
Poland	•	•	•				•	•	•	
Hungary	•	•						•		
Austria	•					•		•		

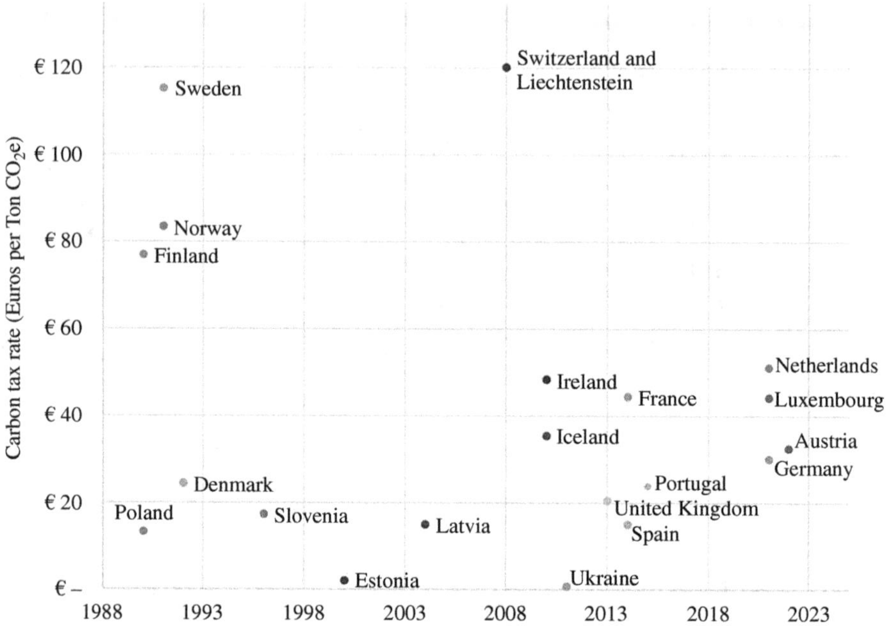

Figure 9.10 Carbon tax rate and date of initiation in the European countries.

- **Carbon Trading**

Carbon trading are interesting mechanisms that provide a competitive platform for giving incentives for sustainable approaches to energy production and consumption. The unique feature of these mechanisms is that compared to the aforementioned categories, they are fairly compatible with the general idea of free markets and the level of government intervention in them is much less. Carbon trading mechanisms are globally known as emission trading systems (ETS) are places where carbon credits are traded. Carbon credits and carbon offsets are generally priced per ton of CO_2 equivalent. It's worth mentioning that there are two types of carbon markets: Voluntary markets and Compliance markets. The difference is simple: in compliance markets, firms buy carbon offsets to compensate their excess emission, compared to their emission allowance, which is predetermined by the governmental authorities. While, in voluntary markets, the players' incentives to buy carbon offsets do not stem from governmental limitations or penalties.

Where the government authorities define the permissible emission of the firms, emission allowance, on the one hand, firms with excess emissions can escape from the incompliance penalties by purchasing carbon credits from ETS. On the other hand, firms that have established emission offsets can participate in

ETS to make money. Established emission offsets should be validated by the authorities before being tradable in ETS. Emission offsets can be earned through implementing energy efficiency programs, energy savings projects, and renewable energy production.

The first influential theoretical efforts that paved the way for the establishment of pollution markets, go back to 1960 when Ronald Coase argued in his article that in comparison with central regulations for fining polluting activities, it's better to let the participants (both polluting ones, and the ones who suffer the pollution) negotiate and reach a solution. Coase's work, which later brought him the 1991 Noble Prize in economics, was later supported and refined by numerous academicians, policy analysts, and regulators and these synergic efforts led to the United Nations' Kyoto Protocol on climate change in 1997, where more than 150 countries founded a basis for international carbon trading. However, each country has a unique position and conditions, and without considering these dissimilarities among the countries, the carbon-credit market would not be able to motivate just transition [76]. Later, during the first decade of the 21st century, the first two environmental market exchanges, China Beijing Environment Exchange (CBEEX) in Beijing and Shanghai Environment and Energy Exchange (SEEE) in Shanghai, were launched in China. In 2005, the Regional Greenhouse Gas Initiative (RGGI) emerged which was the first mandatory carbon market within 10 states of the United States. During 2005–2007, the first European ETS was successfully implemented, which experienced major expansions in the following years. In 2006, the state of California established its state-wide cap and trade mechanism. Starting in 2013, seven Chinese provinces launched their carbon trading mechanisms until in 2017 the Chinese national carbon markets commenced their operation. Currently, ETS programs are widespread in Europe and Canada. Also as mentioned before, several states in the United States have already implemented ETS. Moreover, ETSs are active in some Asian and Pacific states, such as China, Kazakhstan, Japan, Australia, and New Zealand [75, 77]. The state of Carbon Tax and ETS implementation worldwide as of 2023 is depicted in Figure 9.11.

9.4.4 How Trading Mechanisms Affect Energy Justice and Equity?

In all sectors, markets serve as a pivotal component of the economy, exerting a profound influence not only within the specific industry itself but also across interconnected industries that supply essential inputs or consume outputs. Consequently, they significantly impact the livelihoods of individuals employed in these sectors and their families. This assertion is particularly pertinent in the context of energy markets, given that energy stands as a fundamental primary input for nearly all industries. As a result, energy markets wield substantial influence over the economic conditions of societies, potentially altering the landscape

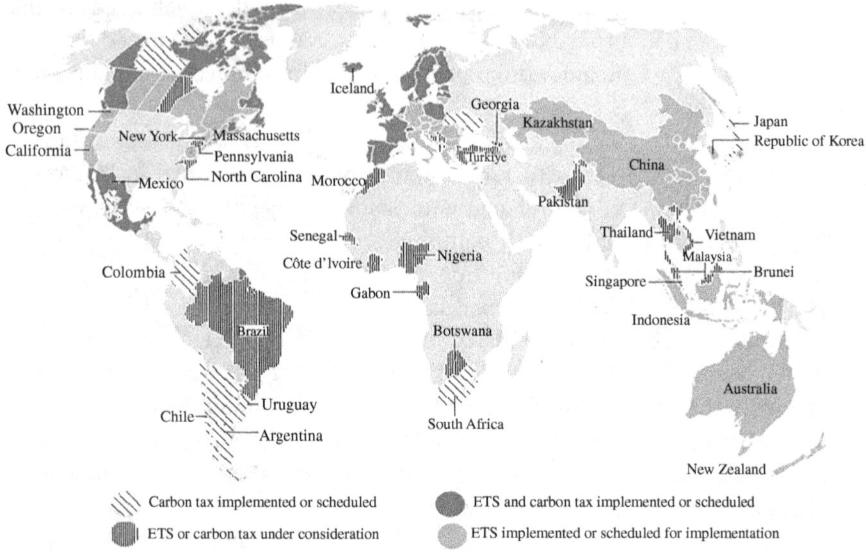

Figure 9.11 ETS and carbon tax mechanisms in the world. Source: [75]/The World Bank / CC by 3.0. (The original picture is published in "State and Trends of Carbon Pricing" prepared by World Bank in 2023 under License: Creative Commons Attribution CC BY 3.0 IGO. Some minor modifications have been made to increase readability.)

of economic justice in a positive or negative manner. From this perspective, the structures of energy markets can have a significant impact on the economic dimensions of energy justice, encompassing aspects like energy affordability and accessibility. However, the impact of energy markets extends beyond economic considerations. As previously outlined in this chapter, the prevailing challenges of global warming and greenhouse gas emissions are, to a significant extent, a consequence of the interactions within energy markets over time. In this section, we intend to discuss how the different energy-related trading mechanisms affect the transition toward energy justice.

Traditional trading mechanisms in energy sectors are generally based on microeconomic theories, which define how the economic and industrial firms respond to the supply and demand forces. In such trading structures, it is the financial profit of energy companies that motivates them to take decisions both in the investment and operation phase. Economic profit in this context normally refers to short-term benefits achievable for the company's managers as well as shareholders. Of course, since as a vital commodity, energy demand suffers from low price-elasticity, especially in the case of electricity, unregulated profiteering behavior of energy companies may lead to severe impacts on both economic and noneconomic dimensions of energy justice. Due to capital-intensiveness of energy industry, mere competition in energy markets can lead to large oligopolies and endanger fairness, equity, participation and inclusiveness, and accountability of energy industry.

Moreover, because of the firms' tendencies toward short-time and exclusive benefits, it's very likely that energy companies fail to respond effectively to some long-run and remote threats such as climate change.

At the first glance, one may hastily conclude that traditional energy markets and trading mechanisms can be aligned with energy justice concerns through some minor adjustments in their designs. In fact, what's said above is just one side of the coin as there exist intricate dilemmas associated with harmonizing trading mechanisms within the energy sectors. These dilemmas stem from the inherent complexity of climate change and its interplay with the social and economic framework of societies. Trading mechanisms have potential to motivate commercial entities to enhance economic performance and lower the energy cost and consequently boost the energy justice via improving energy affordability. Nevertheless, this approach may inadvertently favor fossil-fuel-based energy technologies over renewable alternatives due to their higher fixed costs. Conversely, supporting initiatives for the expansion of renewable energy sources could lead to increased energy costs, potentially compromising energy affordability. Introducing subsidies to reduce the cost of renewable energy not only disrupts market equilibrium but also places a financial strain on the government and diminishes individuals' motivation for energy conservation. On the other hand, market mechanisms have led to heightened R&D efforts in constructing renewable equipment and facilities resulted in lower fixed costs making market instruments more compatible with sustainability-driven approaches [78].

Furthermore, competitive trading mechanisms within the energy markets prove to be better solutions to attract investment for justice-aligned strategies such as renewable expansion and enhancing energy efficiency, compared to more central mechanisms such as FiT and guaranteed purchase. Quota system, Carbon trading, and renewable energy certificate (REC) trading are examples of such competitive trading mechanisms [79, 80].

9.4.5 The Need for Amendments

Implementing ETS, carbon taxation, and renewable energy support have not been without hardships. Since the first application of carbon taxation, many debates have existed regarding the optimum amount of carbon tax. The centrality of deciding the carbon tax level by authority entities makes it a political decision that can lead to unending disputes between parties.

On the other hand, the determination of the carbon footprint of different commercial and industrial activities is a highly technical burden. Considering the diversity in production technology, efficiency, and fuel type, having 100% transparency in the determination of emission allowance is unlikely. For example, let's assume two companies that buy exactly the same amount of fossil fuels. However, due to the different performance of the combustion process and/or emission filtering, the final amounts of emission would be substantially different. In such conditions, on the one hand, applying the same carbon tax on buying fossil fuels would be inefficient. It's worth mentioning that, misallocation of emission allowance or carbon offsets not only threaten the stability of the carbon market but also can pose negative impacts on the market of the products whose production leads to carbon emission [81].

The same concern is valid for the determination of emission offsets due to different projects in the field such as energy efficiency, energy saving, and environmentally friendly projects. Designing sound accrediting procedures to deal with registered projects and verification of the carbon offsets while eliminating potential conflicts of interest can be a challenge. Having all these together, there is still a need to elaborate accurate, technically defendable, and robust estimation processes to assess the carbon footprint or carbon savings of different activities [75].

Furthermore, the energy crisis in Europe following the Russian invasion of Ukraine underscored the potential threat posed by energy crisis scenarios to the realization of sustainability objectives and the fragility of international agreements in this domain. For instance, due to high energy prices in the European territory Slovenia suspended its carbon taxation program, which had been launched in 1996. For the same reason, Austria put a delay in its program for the commencement of its national ETS until October 2022. Germany decided to put the expansion of its national ETS on hold. Moreover, due to the global stress on energy prices, Indonesia decided to postpone the opening of its national ETS, which was supposed to commence its operation in 2022.

Such examples demonstrate the fragile nature of sustainability boosting programs, which without them the future of energy justice would fade into shadows and there is a need to make these arrangements less vulnerable.

9.5 Summary

This chapter explores the intersection of energy justice realization and established trading mechanisms within the energy sector. It emphasizes the critical importance of energy as a fundamental commodity for daily human life and development, highlighting the necessity of achieving a prosperous future that ensures justice for both current and future generations. The chapter also addresses the adverse environmental impacts of human activities related to energy production and consumption, such as air pollution, greenhouse gas emissions, and global climate change, which pose threats to justice and equity.

Various aspects of the energy justice concept, including energy affordability, access, and sustainability, are significantly influenced by energy market structures and trading mechanisms. Traditional arrangements in energy markets have proven insufficient in addressing energy justice issues, leading to the proposal and implementation of auxiliary mechanisms on a global scale. While progress has been made in certain areas, such as improving access to modern energy sources, significant challenges remain in other aspects.

Furthermore, given the interdependency between market structures and socioeconomic infrastructures, there is no single solution that can be applied universally. Thus, the need for innovation and amendments in designing market structures is evident to effectively address these challenges.

References

1 International Energy Agency, I.E.A. and W. Bank (2014). *Sustainable Energy for All 2013–2014: Global Tracking Framework Report*. The World Bank.

2 Jones, E.C. and Reyes, A. (2023). Identifying themes in energy poverty research: energy justice implications for policy, programs, and the clean energy transition. *Energies* 16 (18): 6698.

3 Buera, F.J. and Oberfield, E. (2020). The global diffusion of ideas. *Econometrica* 88 (1): 83–114.

4 Wacziarg, R. and Welch, K.H. (2008). Trade liberalization and growth: new evidence. *The World Bank Economic Review* 22 (2): 187–231.

5 Fischer, D. (1997). *History of the International Atomic Energy Agency: The First Forty Years*. International Atomic Energy Agency (IAEA).

6 Belessiotis, V. and Delyannis, E. (2000). The history of renewable energies for water desalination. *Desalination* 128 (2): 147–159.

7 Smil, V. (2004). World history and energy. *Encyclopedia of Energy* 6: 549–561.

8 Brand-Correa, L.I. and Steinberger, J.K. (2017). A framework for decoupling human need satisfaction from energy use. *Ecological Economics* 141: 43–52.

9 González-Eguino, M. (2015). Energy poverty: an overview. *Renewable and Sustainable Energy Reviews* 47: 377–385.

10 Scarpellini, S., Rivera-Torres, P., Suárez-Perales, I. et al. (2015). Analysis of energy poverty intensity from the perspective of the regional administration: empirical evidence from households in Southern Europe. *Energy Policy* 86: 729–738.

11 Yan, Y., Li, L., Madureira, L. et al. (2023). Tackling energy poverty through trade activities: analyzing social well-being in China. *Energy Policy* 293: 113176.

12 IEA, IRENA, UNSD, World Bank, WHO (2023). Tracking SDG 7: The Energy Progress Report. World Bank, Washington DC, License: Creative Commons Attribution—NonCommercial 3.0 IGO (CC BY-NC 3.0 IGO).

13 IEA (2023). World Energy Outlook, International Energy Agency.

14 Samarakoon, S. (2019). A justice and wellbeing centered framework for analysing energy poverty in the Global South. *Ecological Economics* 165: 106385.

15 World Health Organization (2021). *Measuring Energy Access: A guide to Collecting Data Using "The Core Questions on Household Energy Use"*. International Bank for Reconstruction and Development.

16 Herrero, S.T., Ahmad, F., and Umer, K. (2017). Energy poverty indicators: a critical review of methods. *Indoor and Built Environment* 26 (7): 1018–1031.

17 Aoun, A., Boemi, S.N., Corovessi, A. et al. (2021). *Fighting Energy Poverty through Innovation: Challenges, Opportunities and Solutions*. Athens, Greece: INZEB. https://inzeb.org/wp-content/uploads/2021/05.

18 Amin, W., Ahmad, F., Umer, K. et al. (2022). P2p energy trading: an optimal solution for energy shortage in Pakistan. *IEEE Access* 10: 89614–89633.

19 Khan, I. (2019). Drivers, enablers, and barriers to prosumerism in Bangladesh: a sustainable solution to energy poverty? *Energy Research and Social Science* 55: 82–92.

20 Koukoufikis, G., Schockaert, H., Paci, D. et al. (2023). *Energy Communities and Energy Poverty*. Publications Office of the European Union.

21 Gozgor, G., Mahalik, M.K., Demir, E. et al. (2020). The impact of economic globalization on renewable energy in the OECD countries. *Energy Policy* 139: 111365.

22 Kah, S., Martinos, H., and Budzich-Tabor, U. (2023). CLLD in the 2014–2020 EU programming period: an innovative framework for local development. *World* 4 (1): 122–139.

23 Servillo, L. and De Bruijn, M. (2018). From LEADER to CLLD: the adoption of the new fund opportunities and of their local development options. *European Structural and Investment Funds Journal* 6 (3): 223–233.

24 Gaman, F., Iacoboaea, C., Aldea, M. et al. (2022). Energy transition in marginalized urban areas: the case of Romania. *Sustainability* 14 (11): 6855.

25 Zhao, X., Ramzan, M., Sengupta, T. et al. (2022). Impacts of bilateral trade on energy affordability and accessibility across Europe: does economic globalization reduce energy poverty? *Energy and Buildings* 262: 112023.

26 Zhao, J., Dong, K., Dong, X. et al. (2022). How renewable energy alleviate energy poverty? A global analysis. *Renewable Energy* 186: 299–311.

27 Amoah, A., Kwablah, E., Korle, K. et al. (2020). Renewable energy consumption in Africa: the role of economic well-being and economic freedom. *Energy, Sustainability and Society* 10: 1–17.

28 Yergin, D. (2011). *The Prize: The Epic Quest For Oil, Money and Power*. Simon and Schuster.

29 DOE (2023). Alternative Fuels Data Center. U.S. Department of Energy.

30 Miller, B.G. (2005). *Coal Energy Systems*, 31–74. Academic Press.

31 IEA (2023). Coal 2023, Analysis and Forecast to 2026. International Energy Agency.

32 Ritchie, H. and Rosado, P. (2017). Fossil fuels. *Our World in Data* OurWorldInData.org.

33 Jorli, M., Van Passel, S., Sadeghi, H. et al. (2017). Estimating human health impacts and costs due to Iranian fossil fuel power plant emissions through the impact pathway approach. *Energies* 10 (12): 21–36.

34 IMF IMF Primary Commodity Prices. International Monetary Fund.

35 International Energy Agency. Total energy supply (TES). https://www.iea.org/data-and-statistics/ (accessed 25 January 2024).

36 Ritchie, H., Rosado, P., and Roser, M. (2023). Total electricity generation as share of primary energy. part of the following publication: - Energy. Data adapted from Ember, Energy Institute. Retrieved from https://ourworldindata.org/grapher/electricity-as-a-share-of-primary-energy.

37 Foster, V. and Rana, A. (2019). *Rethinking Power Sector Reform in the Developing World*. World Bank Publications.

38 Kirschen, D.S. and Strbac, G. (2018). *Fundamentals of Power System Economics*. Wiley.

39 Silva, E. (1993). Capitalist coalitions, the state, and neoliberal economic restructuring: Chile, 1973–88. *World Politics* 45 (4): 526–559.

40 Huneault, M., Galiana, F., and Gross, G. (1999). A review of restructuring in the electricity business. In: *Proceedings of 13th Power Systems Computation Conference*, 19–31.

41 Namir, Y.E., Li, Z., and Hamdi, M. (2019). Restructuring in the power systems for promotion of renewable energy: an updated review. *Journal of Circuits, Systems and Computers* 1–14.

42 Abbruzzese, M., Infante, D., and Smirnova, J. (2024). The diffusion of renewable energy production in European countries: the role of incentives. *Economics of Innovation and New Technology* 1–24.

43 Zhu, X., Li, L., Zhou, K. et al. (2018). A meta-analysis on the price elasticity and income elasticity of residential electricity demand. *Journal of Cleaner Production* 201: 169–177.

44 Labandeira, X., Labeaga, J.M., and López-Otero, X. (2017). A meta-analysis on the price elasticity of energy demand. *Energy Policy* 102: 549–568.

45 Estrada, M.A.R., Park, D., Tahir, M. et al. (2020). Simulations of US-Iran war and its impact on global oil price behavior. *Borsa Istanbul Review* 20 (1): 1–12.

46 Goldemberg, J., Johansson, T.B., Reddy, A.K. et al. (1985). Basic needs and much more with one kilowatt per capita. *Ambio* 14: 190–200.

47 Okushima, S. (2021). Energy poor need more energy, but do they need more carbon? Evaluation of people's basic carbon needs. *Ecological Economics* 187: 107081.

48 Sovacool, B.K. (2017). Reviewing, reforming, and rethinking global energy subsidies: towards a political economy research agenda. *Ecological Economics* 135: 150–163.

49 Kojima, M. (2017). *Identifying and Quantifying Energy Subsidies: Energy Subsidies*. The World Bank.

50 IEA (2022). Value of fossil-fuel subsidies by fuel in the top 25 countries. https://www.iea.org/data-and-statistics/charts/value-of-fossil-fuel-subsidies-by-fuel-in-the-top-25-countries-2022 (accessed January 2024). IEA. Licence: CC BY 4.0.

51 The World Bank (2016). Pump price for gasoline (US$ per liter). https://data.worldbank.org/indicator/EP.PMP.SGAS.CD?end=2016&name_desc=false&start=1991&view=chart (accessed 11 February 2024).

52 Watson, C. and Bolton, P. (2013). Warm front scheme. In: *Science and Environment*. UK Parliament: House of Commons Library https://commonslibrary.parliament.uk/research-briefings/sn06231/.

53 Clements, M.B.J., Coady, M.D., Fabrizio, M.S. et al. (2013). *Energy Subsidy Reform: Lessons and Implications*. International Monetary Fund.

54 Campbell, N., Ryan, L., Rozite, V. et al. (2014). *Capturing the multiple benefits of energy efficiency*. Paris, France: IEA.

55 Strategy, C. (2015). Energy Efficiency–The First Fuel for The EU Economy-Final Report.

56 Herring, H. and Sorrell, S. (ed.) (2009). Energy efficiency and sustainable consumption. In: *The Rebound Effect*, 1–266. Hampshire: Palgrave Macmillan London.

57 Hanley, N., PG, M.G., Swales, J.K. et al. (2009). Do increases in energy efficiency improve environmental quality and sustainability? *Ecological Economics* 68 (3): 692–709.

58 Bonan, J., Pareglio, S., and Tavoni, M. (2017). Access to modern energy: a review of barriers, drivers and impacts. *Environment and Development Economics* 22 (5): 491–516.

59 Peters, J., Sievert, M., and Toman, M.A. (2019). Rural electrification through mini-grids: challenges ahead. *Energy Policy* 132: 27–31.

60 Antonanzas-Torres, F., Antonanzas, J., and Blanco-Fernandez, J. (2021). State-of-the-art of mini grids for rural electrification in West Africa. *Energies* 14 (4): 990.

61 Walls, M. (2014). Comparing subsidies, loans, and standards for improving home energy efficiency. *City* 16 (1): 253–278.

62 Eitan, A., Herman, L., Fischhendler, I. et al. (2019). Community–private sector partnerships in renewable energy. *Renewable and Sustainable Energy Reviews* 105: 95–104.

63 Tan, Z., Geng, H., Xu, X. et al. (2022). Suspension of Australian National Electricity Market in 2022 necessitates mechanism evolution ensuring power supply security. *Journal of Modern Power Systems and Clean Energy* 11 (2): 674–679.

64 Ari, M.A., Engler, P., Li, G. et al. (2023). *Energy Support for Firms in Europe: Best Practice Considerations and Recent Experience.* International Monetary Fund.

65 Sgaravatti, G., Tagliapietra, S., Zachmann, G. et al. (2021). National policies to shield consumers from rising energy prices. *Bruegel Datasets*. https://www.bruegel.org/dataset/national-policies-shield-consumers-rising-energy-prices (accessed February 2024).

66 Eurostat Data Browser (2022). GDP and main components (output, expenditure and income). https://ec.europa.eu/eurostat/databrowser/bookmark/4ad27e6f-358a-4a3d-82a0-587d69a833eb?lang=en (accessed 22 April 2024).

67 Samadi, A.H. and Afroozeh, S. (2024). Legitimacy, institutional inertia, and climate change. In: *Institutional Inertia* (ed. N. Faghih and A.H. Samadi), 275–302. Springer.

68 The World Bank (2020). GDP per capita, and renewable energy consumption (% of total final energy consumption). https://databank.worldbank.org/ (accessed 13 February 2024).

69 Held, A., Ragwitz, M., Del Río, P. et al. (2019). Do almost mature renewable energy technologies still need dedicated support towards 2030? *Economics of Energy and Environmental Policy* 8 (2): 81–98.

70 Shah, I.H., Hiles, C., and Morley, B. (2018). How do oil prices, macroeconomic factors and policies affect the market for renewable energy? *Applied Energy* 215: 87–97.

71 European Comission (2018). Legal sources on renewable energy: renewable energy policy database and support. http://www.res-legal.eu/ (accessed 22 October 2021).

72 Nadiri, A., Gündüz, V., and Adebayo, T.S. (2024). The role of financial and trade globalization in enhancing environmental sustainability: evaluating the effectiveness of carbon taxation and renewable energy in EU member countries. *Borsa Istanbul Review* 24 (2): 235–247.

73 Åkerfeldt, S. (2017). The Benefits of a Carbon Tax–Swedish Experiences and a Focus on Developing Countries. Domestic Resource Mobilization and Tax Base

Protection. In UN Workshop on Practical Issues in Protecting the Tax Base of Developing Countries.

74 Mengden, A. Carbon Taxes in Europe (2023). Tax foundation. https://taxfoundation.org/data/all/eu/carbon-taxes-in-europe-2023/#:~:text=In%201990%2C%20Finland%20was%20the,gas%20emissions%2C%20such%20as%20methane (accessed 15 February 2024).

75 State and Trends of Carbon Pricing (2023). *World Bank, License: Creative Commons Attribution CC BY 3.0 IGO.* https://doi.org/10.1596/978-1-4648-2006-9. Washington, DC: World Bank.

76 Bodnar, Z., Lanzarini, C.A., and de Mesquita, T.B. (2024). Transnational climate justice and the carbon credit market. *Beijing Law Review* 15: 828–842.

77 Gorman, H.S. and Solomon, B.D. (2002). The origins and practice of emissions trading. *Journal of Policy History* 14 (3): 293–320.

78 Barnea, G., Hagemann, C., and Wurster, S. (2022). Policy instruments matter: support schemes for renewable energy capacity in worldwide comparison. *Energy Policy* 168: 113093.

79 Herbes, C., Rilling, B., MacDonald, S. et al. (2020). Are voluntary markets effective in replacing state-led support for the expansion of renewables?–a comparative analysis of voluntary green electricity markets in the UK, Germany, France and Italy. *Energy Policy* 141: 111473.

80 Hulshof, D., Jepma, C., and Mulder, M. (2019). Performance of markets for European renewable energy certificates. *Energy Policy* 128: 697–710.

81 Betz, R., Michaelowa, A., Castro, P. et al. (2022). *The Carbon Market Challence: Preventing Abuse Through Effective Governance.* Cambridge University Press https://doi.org/10.1017/9781009216500.

10

Peer-to-Peer Energy Communities: An Equitable Approach to Expanding Social Welfare and Energy Justice

Mohammad Hasan Ghodusinejad, Setareh Peirov, and Hossein Yousefi

School of Renewable Energies and Environmental Engineering, College of Interdisciplinary Science and Technologies, University of Tehran, Tehran, Iran

Nomenclature

Abbreviations

DC	direct current
DES	distributed energy systems
FIT	feed-in-tariff
GAMS	general algebraic modeling system
LINDO	linear, interactive, and discrete optimizer
PtX	power to X
PV	photovoltaic
P2P	peer-to-peer
RES	renewable energy sources
ToU	time-of-use

Sets

t	index for time (hour)
i	index for prosumers

Parameters

$C_t^{G,imp}$	energy cost from the grid at time t (\$/kWh)
$C_t^{G,exp}$	energy cost to the grid at time t (\$/kWh)
C_t^{p2p}	P2P energy trade cost at time t (\$/kWh)

Communities for Clean Energy Justice and Equity in Grid Modernization, First Edition.
Edited by Mohammadreza Daneshvar, Behnam Mohammadi-Ivatloo, and Amjad Anvari-Moghaddam.

$P^{L}_{i,t}$	load of prosumer i at time t (kWh)
EF^{G}	emission factor of the grid (kg/kWh)
C^{CO_2}	emission cost of the grid ($/kg)
$P^{PV}_{i,t}$	PV (photovoltaic) power of prosumer i in time t (kWh)
$P^{G,imp}_{\max}$	maximum power import from grid (kWh)
$P^{G,exp}_{\max}$	maximum power export to grid (kWh)
$P^{p2p,buy}_{\max}$	maximum P2P power buy (kWh)
$P^{p2p,sell}_{\max}$	maximum P2P power sell (kWh)
SDR_t	supply to demand ratio at time t

Variables

$P^{G,imp}_{i,t}$	energy import from grid, prosumer i at time t (kWh)
$P^{G,exp}_{i,t}$	energy export to grid, prosumer i at time t (kWh)
$Z^{G,imp}_{i,t}$	binary variable for grid import, prosumer i at time t
$Z^{G,exp}_{i,t}$	binary variable for grid export, prosumer i at time t
$P^{p2p,buy}_{i,t}$	P2P energy buy of prosumer i at time t (kWh)
$P^{p2p,sell}_{i,t}$	P2P energy sell of prosumer i at time t (kWh)
$Z^{p2p,buy}_{i,t}$	binary variable for P2P power buy, prosumer i at time t
$Z^{p2p,sell}_{i,t}$	binary variable for P2P power sell, prosumer i at time t

10.1 Energy Transition and Democratization

Meeting the requirements of modern human life has caused extensive changes in energy systems in the past decades. Traditional energy systems no longer fulfill the energy needs of today's modern life due to the imbalance between energy supply and demand [1]. As a result, the modernization and transformation of these systems according to the needs of the communities has been placed on the agenda of the experts. In the meantime, due to the higher consumption of electricity compared to other forms of energy and its prominent role in the modern world, it is of great significance to examine the evolution of electricity networks.

Energy scientists believe that power grids should be transformed in a way that transfers energy flow in a more democratic, equitable, and sustainable way [2]. On the other hand, negative socio-ecological effects of energy production systems need to be minimized as well as environmental costs have to be considered in the development of energy systems. Consequently, renovating the electricity networks should be followed up based on the concept of energy democratization, which addresses the above concerns.

Energy democracy represents a shift from a centralized economy based on fossil fuels to environmentally friendly energy systems based on community participation [3]. Therefore, this concept links the transition to renewable energies with the process of democratization, focusing on social benefits and equitable distribution of energy in societies. In order to attain the goals of energy democratization, expanding the share of clean energies in power production and emphasizing the role of the final consumer in energy supply are two vital issues that should be addressed in modern electricity supply structures.

In the past, energy supply systems generally operated centrally wherein the energy demand had been provided centrally through national distribution networks. But in recent years, the implementation of local energy systems like microgrids has increased in line with the strategy of decentralization [4]. One of the important benefits of these systems is the possibility of increasing the share of renewable energy in distributed production systems. Moreover, the role of end users in energy generation has been noticed in these schemes. In other words, unlike in the past, energy consumers are not passive actors in the energy systems but play an active role in the energy production process. So, these systems provide the possibility of turning energy consumers into prosumers in line with the realization of energy democratization.

Examining the relationship between grid modernization and grid transition with energy democracy is indispensable for a better analysis of the factors affecting the reconstruction of energy networks, especially electricity networks. Figure 10.1 demonstrates a qualitative trend relating to the changes in the structure of electricity networks in recent decades. According to the figure, the process of modernization of electricity networks and moving from traditional grids to modern networks has been accompanied by increasing the role of consumers and changing them from passive peers to active actors in the energy system. Meanwhile, several factors are involved in the evolution of energy networks, which are mentioned inside the bubbles of the figure.

As mentioned in Figure 10.1, traditional grids produce and distribute energy based on different strategies. In the beginning, constant rates were considered for energy trade, which was a primitive, undeveloped, and inflexible method. Then, ToU pricing was introduced, in which the price of electricity depends on the time of use and alters during the day and night and also through the whole year, according to the electricity demand [5]. In the next stage, an electricity pricing system was proposed based on real time or dynamic pricing, which was more developed than the previous rating methods. In this scheme, affordable prices for

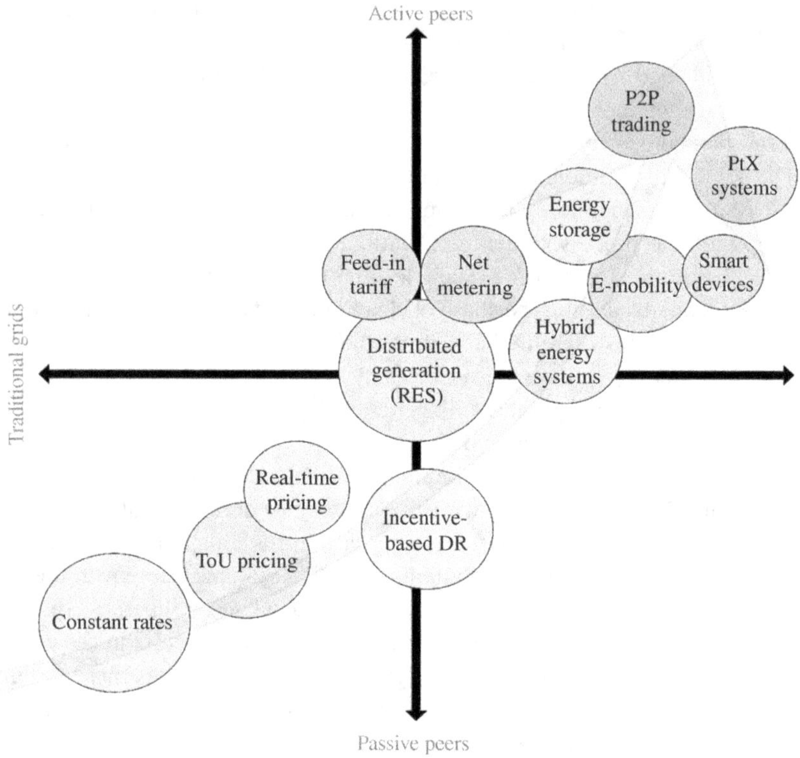

Figure 10.1 The transition trend from traditional grids/passive peers toward modern grids/active peers.

electricity were determined based on market research so that the interests of both consumers and distributors were taken into account [6].

End users do not play an active role in traditional systems. However, their role has been passively defined in incentive-based demand response. Based on this strategy, consumers are encouraged to reduce energy intake during peak demand hours and receive incentives from service providers, in turn [7]. This strategy was established in order to balance the grid by shifting power demand to times when energy is more abundant, normally through charges or financial incentives.

The transition from traditional networks to novel energy generation units was realized through the development of distributed generation networks and the wide application of renewable energies. In these systems, energy is generated by several sources based on the obtainable potential of renewable energies such as sun, wind, and biomass, along with nonrenewable sources like diesel generators. They perform in the form of hybrid energy systems or microgrids.

Authorities support network modernization and the implementation of renewable energies in energy supply by various means. In this regard, one of the most widely used strategies is the feed-in-tariff (FIT) policy. In FIT, the government guarantees to buy the generated renewable electricity from the producers at a price above the market. FITs frequently include long-term contracts, from 15 to 20 years [8]. In addition to FIT, the net metering system was also considered to support solar power producers. According to this system, local producers will be given credit if they produce excess solar power and inject it into the grid [9].

E-mobility is another energy modernization approach in terms of transportation which includes different types of electric vehicles, electric micromobility equipment, and transit automobiles. The electric light-duty vehicle business is progressing quickly, and possibly, the future belongs to electromobility [10]. In line with the modernization of energy systems, PtX platforms are developed to convert power into other forms of energy. In this system, a set of exciting technologies are developed to convert generated power from renewable sources into heat, hydrogen, methanol, or other energy carriers [11]. Due to the development of small-scale distributed energy systems (DES), energy distribution systems have been converted in a way that enables prosumers to exchange electricity. Additionally, the techniques of production and consumption of electricity are quickly varying in a way that electricity consumers are progressively becoming prosumers.

Finally, thanks to the expansion of small-scale DES, energy administration systems have been renovated to allow prosumers to commerce electricity and gain profit. Therefore, peer-to-peer (P2P) trading, known as P2P trading has been introduced as the most developed and progressive method of energy exchange in societies as the next generation of smart grids [12], while the term "peers" in the energy market specifies the grid-connected structures, which could be two or more [13]. In this system, energy consumers and producers exchange energy directly, while it is possible for end users to be energy prosumers [14]. In other words, this concept offers new markets for prosumers and causes opportunities for innovative services in the electricity distribution network [15]. This context has been developed in the direction of decentralization of energy systems and increasing the consumers' role from passive users to active prosumers. This platform supports the central electricity service to diminish reserve requirements, reduce investment costs and operational charges, moderate peak demand, and promote reliability.

P2P technology offers flexible options to customers in line with the development of clean energy applications and the transition to low-carbon systems. In this regard, producers can trade the surplus energy to their counterparts and increase the revenue of the system. Thus, P2P energy trading can take place via different marketplace configurations, such as coordinated markets, decentralized markets, or community market [16].

Local distributed electricity generators and consumers commonly pay a monthly subscription fee to become participants of an energy trading platform. Similar to an open marketplace, prosumers attempt to set the highest possible rate, considering their expenses and turnover, while clients seek the lowest charge, based on their personal preferences and requirements [17]. Indeed, solar energy trading, for example, is possible to take place not only among neighbors but between larger communities as well. Subsequently, P2P energy trading is a modern facility that empowers both end users and prosumers to trade energy without reliance on an intermediary platform.

In this chapter of the book, considering the concept of grid modernization, two important issues of "*creating energy communities*" and "*P2P energy trading*" are examined, and achieving energy justice by following these approaches is investigated. According to the stated introduction, it can be claimed that the two aforementioned issues are the chief and fundamental aspects of grid modernization and energy transition. In this regard, this chapter represents evidence and detailed analysis to prove that these items are two prominent paths to achieve energy justice.

10.2 Peer-to-Peer Energy Communities

The energy transition in terms of moving toward the modernization of energy systems can be implemented and explored in the two frameworks, including the emergence of energy communities and P2P energy trading. These concepts are structural and technical aspects of modern energy systems that have become the main trigger for the development of energy communities. On the other hand, the economic and exchange concerns of the modern energy market have been the main factors in P2P energy market emergence.

10.2.1 Advent of Energy Communities

One of the consequences of energy democratization is the emergence of energy communities, which refers to plans where citizens, small businesses, and local authorities possess or contribute to the production and/or consumption of sustainable energy [18]. The link between cities and energy is compound and requires a worldwide approach that incorporates advanced technologies, sustainable urban planning policies, and the conscious participation of the community [19]. Based on this concept, citizens are enabled to take control of their energy production and requirements and decrease their dependence on the main grid. Hence, decentralizing energy services and moving toward local energy societies is one of the means of energy democratization that will reduce energy scarcity and save unprotected populations.

The policy of raising the application of renewable energies has a very noticeable role in the development of energy communities. This approach can specifically reduce the environmental impact of the energy sector on societies and help moving toward a low-carbon economy [20]. Based on this issue, the European Union (EU) motivates its member states to raise the share of renewable energy sources (RES) in their energy basket [21]. One of the basic aims of this policy is to moderate climate change effects, commonly originating from the emission of greenhouse gases (GHGs) into the atmosphere once energy is traditionally produced via fossil fuels.

Facilitating the access of consumers to energy and their self-reliance on energy supply is one of the main characteristics of evolving energy societies. Additionally, the possibility of energy storage and the affordability of energy prices are other pros of energy communities. Furthermore, this platform provides a flexible energy exchange via balancing power supply and demand at the local magnitude [22]. Therefore, by considering four major aspects of energy security, including availability, affordability, accessibility, and acceptability, evolving energy communities have a direct effect on the promotion of energy security among societies [23].

10.2.2 Peer-to-Peer Energy Trading

In line with the emergence of distributed energy societies, new mechanisms governing energy markets should also be discussed, which are briefly mentioned in this section.

The modernization of energy systems not only introduces technological revolutions but also affects structural transformation. One of the examples of such mechanical advances is P2P trades employed in prosumer energy interchange.

The widespread application of information technology in societies and the expansion of smart methods of trading has made it possible to organize scheduled market exchanges openly between consumers [12]. Therefore, the digitization of communities, has significantly accelerated the rate of development of such markets and enabled energy transactions on a universal scale without mediums or through arranged principles functioning electronically. P2P platform establishes superior flexibility and potentially higher financial profits for both prosumers and consumers by supporting energy exchange with real-time market conditions [24].

Generally, energy sharing in societies can be implemented through three different market structures, as presented in Figure 10.2. The first structure used in traditional energy systems is the centralized energy market so that energy services and pricing are vertically organized with communications and trade controlled

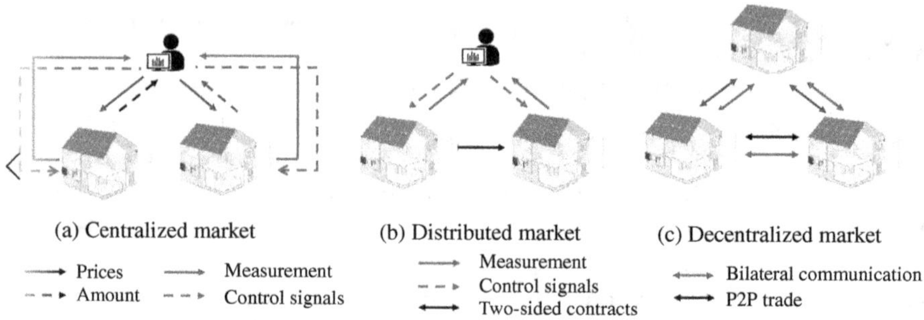

Figure 10.2 Classic market arrangements for energy sharing: (a) centralized, (b) distributed, (c) decentralized. Source: [25]/John Wiley & Sons.

centrally by the operator [26]. The next structure is the distributed energy market, where the operator's role in controlling the market is reduced. In this approach, agents exchange energy horizontally under the supervision of an operator instead of directly purchasing energy from a central system. In this method, consumers experience more autonomy in the energy market, while the operator indirectly affects the energy market by providing incentives and legal signals [27]. The third structure is a decentralized energy market wherein energy exchange is done directly by agents without the intervention of the operator. This structure offers the highest level of flexibility and privacy to users and is considered the most modern means of energy exchange in communities [25].

The development of P2P energy transactions has several benefits for both prosumers and customers. For example, this innovative exchange method considers the benefits of buyers and producers in a flexible and dynamic manner and covers the weaknesses of traditional energy markets to a large extent. Moreover, this market provides the possibility of meeting a wide range of energy demands at an acceptable charge. It also represents the applicability of up-to-date technologies in the procedure of advancing the operation of the modern market, particularly e-commerce [28].

P2P energy trading is facilitated by several key technologies, primarily blockchain, Internet of Things (IoT), and advanced algorithms. These technologies enhance energy management, promote sustainability, and reduce costs for users. For instance, blockchain ensures secure and transparent transactions among prosumers, minimizing fraud and enhancing trust [29], while real-time monitoring with the aid of IoT devices enables real-time tracking of energy production and consumption, facilitating informed trading decisions [30]. IoT integration also leads to better management of energy resources, contributing to lower carbon footprints [8]. Real-time pricing (RTP) strategies allow consumers to respond to market fluctuations, optimizing energy use and costs [31].

Moreover, there are advanced algorithms like the multi-agent peer-to-peer Dairy Farm Energy Simulator (MAPDES) that demonstrate significant cost savings and increased energy sales in agricultural settings [32]. While these technologies present numerous advantages, challenges such as market stability, privacy concerns, and the need for regulatory frameworks remain critical for the successful implementation of P2P energy trading systems.

To sum up, dissimilar to the traditional centralized network of the electricity service, P2P trading permits direct energy exchange between nearby prosumers with a particular amount of electricity production capacity. By this mechanism, energy consumers are encouraged to convert to prosumers and take part in energy trading in order to gain revenue and promote energy markets. Hence, this platform provides a sustainable and consistent equilibrium between district energy generation and energy demand [33]. It also supports the grid by decreasing peak demand, reserve load, and network losses [14]. So, P2P transaction tools are one of the upcoming exploration agendas for modern communities in terms of contributing to DES.

10.2.3 Regulatory of P2P Markets

P2P energy trading is governed by various regulatory frameworks and policies that adapt to the unique challenges of different regions. These frameworks are essential for facilitating decentralized energy transactions and ensuring market integrity. A framework for P2P trading includes mechanisms for participant credit evaluation and bidding strategies to maintain market order and enhance transaction feasibility [34]. While these frameworks aim to promote sustainable energy practices, they also highlight the need for ongoing adaptation to address emerging challenges in the energy landscape.

The implementation of P2P energy communities in Europe faces complex legal challenges that require comprehensive regulatory solutions beyond sector-specific regulations [35]. These challenges stem from complex legal frameworks, tariff structures, and socioeconomic conditions across different regions. The EU is revising its legal frameworks to accommodate P2P trading, addressing consumer protection, contract law, and data law to facilitate a more inclusive energy market [36]. Each community experiences unique regulatory barriers based on its socioeconomic context, necessitating tailored solutions [35]. Inefficient retail tariffs can exacerbate the challenges faced by energy communities, as they prevent effective pricing of local network congestion and hinder the financial viability of P2P. Tariff reform is essential to facilitate the establishment of energy communities and optimize their operations [37]. In areas with limited information communication technology (ICT) access, establishing trust among participants is crucial for successful P2P energy exchange [38]. The integration of advanced technologies, such as the blockchain, can help address these trust issues by ensuring transparent transaction tracing [38].

While these challenges are significant, they also present opportunities for innovation in regulatory frameworks and market designs, potentially leading to more resilient and efficient energy systems.

10.3 Energy Justice and P2P Energy Communities

Energy-related subjects have recently become a novel focus of gravity for justice scholars. Energy justice is an emerging concept that addresses equity and morality in modern energy systems [39]. Energy justice, in terms of social sustainability, on the other hand, focuses on the well-being of people and communities [40]. Energy inequality is not merely a theoretical concept; it is a tangible issue with significant repercussions. It is a substantial impediment to the well-being and sustainability of future generations and undermines the quality of life in contemporary societies [41]. In other words, energy justice regards to the target of equity achievement in both social and economic contribution to the energy scheme, whereas also modifying social, economic, and health burdens on those excessively affected by the energy structure.

The concept of energy justice has a vital role in different aspects of energy organization, such as conversion, distribution, market, and accessibility. In addition, considering the challenges of the system from the perspective of energy justice in the decision-making process promotes overall energy equity and enhances the awareness and responsibility of energy politicians [42].

Energy justice is one of the fundamental concerns that should be given special attention in the development and modernization of energy systems. Keeping in mind that energy communities and P2P energy trading, which are two important symbols of energy modernization, are examined in this study in terms of energy justice (Figure 10.3).

One of the issues of concern from the perspective of energy justice is the availability and accessibility of energy for all society members [43]. The formation of energy communities has significantly improved the level of energy access in communities by enabling and facilitating the direct exchange of energy between agents. Consequently, it has generally improved equality and energy justice in modern societies. On the other hand, local energy generation and decentralization of energy services allow the participation of different classes of society in the energy production process [44]. As a result, by increasing the role of agents in energy systems, the benefit of all members of the community is considered in the process of energy production and conversion according to their needs, and energy justice is established at a higher level in society.

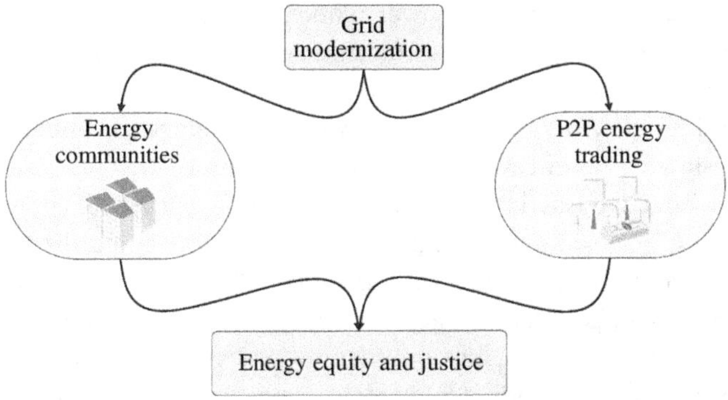

Figure 10.3 P2P energy communities: a key to energy equity and justice.

Affordability of energy for all energy community members is another significant aspect of energy justice, which can be attainable through P2P trading [45]. P2P transactions provide a dynamic and flexible platform for energy trading in which the benefits of agents are considered appropriately. In fact, P2P trading has reduced the cost of energy access for communities by eliminating intermediaries and in this regard, has promoted energy justice and equity. Different trading mechanisms govern diverse types of P2P markets. During these various mechanisms, the price of energy is determined through various methods, such as auctions, pricing based on supply–demand, etc. Most of these pricing schemes lead to providing energy at a lower price than the retail price of electricity distribution companies. Therefore, in P2P markets, mainly the costs of prosumers are reduced. This basically could be one of the main axes of energy justice. On the other hand, most of the regulations governing P2P energy markets are formulated locally and based on the interests of market participants. This leads to the deregulation of the energy network, increasing energy democratization and justice, and ensuring the maximum benefits of market members.

With a closer look at the concept of energy justice, the ethical consequences of our collective energy decisions may become clearer, as they indirectly and even directly affect the future of the next generation, mostly in terms of environmental issues like climate change [42]. On the other hand, the protecting of resources against depletion is one of the moral responsibilities of the current generation toward future peers. A possible potential to fight against global warming, climate change, and resource depletion is to modify policies in the field of energy and proliferate the share of renewable energies in meeting the human energy demand [46]. This could be conceivable in the context of DES and the establishment of energy communities in which agents are able to generate energy mainly with renewable resources such as PV panels, wind turbines, and fuel cells [47].

This will considerably reduce the load of power plants, which are commonly based on fossil fuels, and consequently, control climate change and preserve exhaustible energy resources.

P2P energy communities are emerging as innovative solutions for local energy management, showcasing successful real-world implementations. Several case studies illustrate their effectiveness in enhancing energy efficiency and community engagement [48]. For example, Taouil et al. tested a distributed direct current (DC) microgrid architecture in rural areas, enabling P2P energy exchange through a blockchain marketplace. This system achieved 11% of total demand from community resources while reducing production losses by 16% [38]. In a Norwegian neighborhood, P2P trading platforms were assessed, revealing significant cost savings for end users and minimal grid impact when integrating PV systems and batteries. This study highlighted the transition from consumers to prosumers, enhancing local energy distribution [49]. An Australian case study demonstrated a blockchain-integrated local energy market, allowing prosumers and consumers to trade electricity efficiently. The model outperformed traditional methods, showcasing the potential of decentralized energy trading [50]. P2P trading allows low-income households to negotiate electricity prices and quantities, enabling them to access sufficient energy at lower costs. For instance, households in Sub-Saharan Africa can trade excess solar energy, achieving prices as low as £0.35/kWh [51]. A multi-community trading mechanism can reduce average daily electricity costs by 64%, making energy more affordable for marginalized groups [52]. While these examples highlight the benefits of P2P energy communities, challenges such as regulatory frameworks and technology adoption remain critical for broader implementation.

In terms of reinforcement of energy justice, P2P energy trading presents a transformative approach to alleviating energy poverty in marginalized communities by facilitating direct electricity exchanges among local producers and consumers, as well as enhanced access to renewable energy, particularly in regions with limited infrastructure. P2P trading allows low-income households to negotiate and access excess electricity generated from renewable sources, such as solar home systems, at reduced prices [53]. For instance, simulations indicate that households can secure sufficient energy (1–1.1 kWh/day) at competitive rates, enabling them to transition to higher tiers of electricity access [51]. A multi-community trading mechanism can further optimize energy distribution, reducing average daily electricity costs by 64%, thus making energy more affordable for marginalized groups [52]. P2P markets also promote economic sustainability and efficiency by minimizing energy waste and lowering costs for consumers. Experimental frameworks demonstrate significant cost reductions while aligning energy supply with community demand [54]. The integration of P2P trading within local energy markets fosters the adoption of RES, addressing barriers to implementation and enhancing community resilience [55].

In addition to reducing financial costs, P2P energy communities provide significant environmental and social advantages. One of the principal advantages of P2P systems is the augmented utilization of renewable energy, such as solar and wind power. This particularly helps in mitigating GHG emissions and negative environmental effects. Consequently, these systems can assume a pivotal role in alleviating climate change and enhancing air quality within communities [56]. Besides, as the social issues are regarded, another beneficial facet of these systems is the enhancement of energy equity. With the democratization of energy generation and distribution, a greater number of individuals within the community can attain access to sustainable and dependable energy. This matter is particularly salient for communities that are marginalized in terms of energy accessibility. Furthermore, the reduction of energy costs through P2P systems can contribute to the improvement of individuals' quality of life and mitigate energy poverty. However, despite the aforementioned advantages, there also exist some challenges. One of the issues raised is the necessity for new infrastructure for these systems, which may be linked with environmental costs in the short term. Additionally, P2P platforms necessitate social and cultural coordination, which may engender resistance in diverse communities.

Along with all the benefits that P2P energy communities offer to raise the level of energy justice, these markets also come with a number of challenges. P2P energy communities are new markets under existing electricity market structures, providing new business based on the community's needs or interests. Such business model designs require precise testing to ensure that it can benefit all market participants. As these models are particularly expected to equitably distribute costs and benefits among all of market participants, this can lead to some difficulties in implementing the model [57, 58]. Apart from this, P2P energy communities are implemented to promote local energy trade. Hence, the overall framework has to be designed based on not only market-driven trading but also the values, pain points, and social justice issues of the community [57, 58]. In other words, social issues and human emotions can sometimes have a negative impact on reaching consensus at the market level.

Additionally, local P2P energy markets face a number of implementation challenges that can hinder the successful implementation of such markets. The most basic of these challenges include legal and regulatory barriers. Many countries still do not have clear legal frameworks for P2P platforms. Regulating such systems requires legislative changes that may be time-consuming and complex. Some existing regulations may also hinder the innovation and development of the P2P market. On the other hand, infrastructure costs are worth noting. Developing and maintaining the necessary infrastructure for P2P energy transactions, such as digital monitoring systems and accurate measurement equipment, requires significant investments. These costs can be challenging, especially in the early stages of implementation. It is also important to prepare the market and pave the way to accept this new type of energy market. For P2P systems to succeed, the market must be technically and culturally ready. Wide acceptance of this type of transaction requires public awareness and education so that energy consumers are aware of their role as prosumers and have the ability to use these systems.

10.4 A Confirmatory Analytical Study

In line with the contents stated in the previous sections and in order to clarify the represented concepts, in this section, a practical example is characterized and resolved in order to explain the relationship between energy communities and P2P trading with the concept of energy equity. In Section 10.4.1, the generalities of the problem are stated and in Section 10.4.2 the outcomes of the case study are described.

10.4.1 Study Framework

A simple microgrid consisting of four units with a specific electrical load for a 24-hour day is considered as the case study. This microgrid is being investigated under three different scenarios. In these scenarios, the evolution and transition of energy systems from traditional arrangements to modern structures and the formation of energy communities have been investigated.

In the first scenario, all the load required by the units is traditionally provided through the central grid. This scenario describes a traditional energy system where the agents are passive energy consumers and the national grid is the only source of power supply that performs in a conventional way.

In the second scenario, each unit is equipped with a photovoltaic panel to generate power and has the possibility of exchanging energy with the grid. In this scenario, agents are transformed into active energy prosumers rather than passive consumers and a primary energy community is shaped through the energy trade between the prosumers and the grid.

The third scenario represents a situation where in addition to energy exchange with the grid, agents have the possibility of P2P energy transactions through a local market under the supervision of a community organizer. This scenario offers the most modern energy system, among other ones. It describes a contemporary energy community where agents benefit from the advantages of P2P energy transactions as well as power trade with the grid.

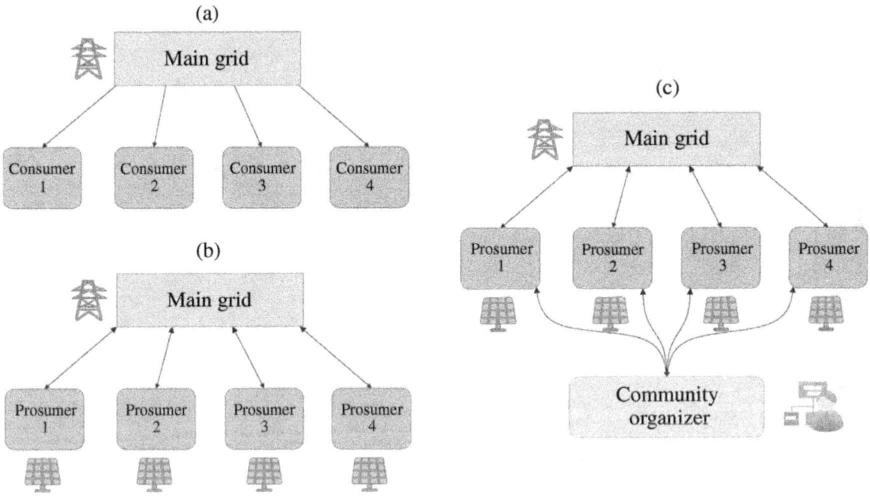

Figure 10.4 Schematic diagram of the system: (a) first scenario: supply from the grid, (b) second scenario: energy community, (c) third scenario: P2P energy sharing community.

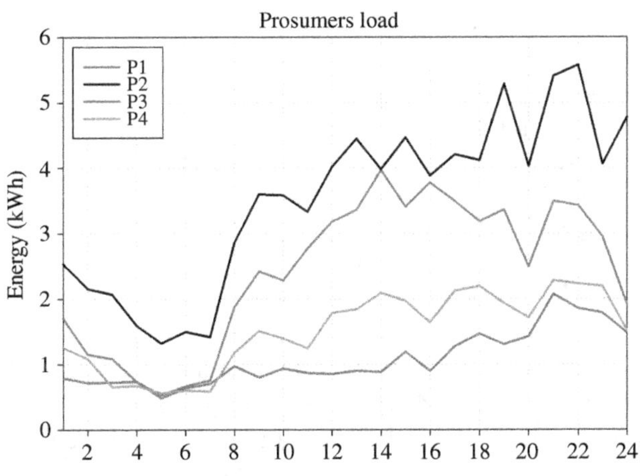

Figure 10.5 Prosumers daily load.

The three abovementioned scenarios are represented in Figure 10.4

Figure 10.5 displays the electrical load of each unit during a 24-hour day and Figure 10.6 shows the generated power by the photovoltaic panels of each prosumer. It is clear from this picture that P2 and P4 have the same capacity for PV installation.

The basic specifications and constraints of the present model are defined in Table 10.1. Based on the represented data, the maximum power that can be

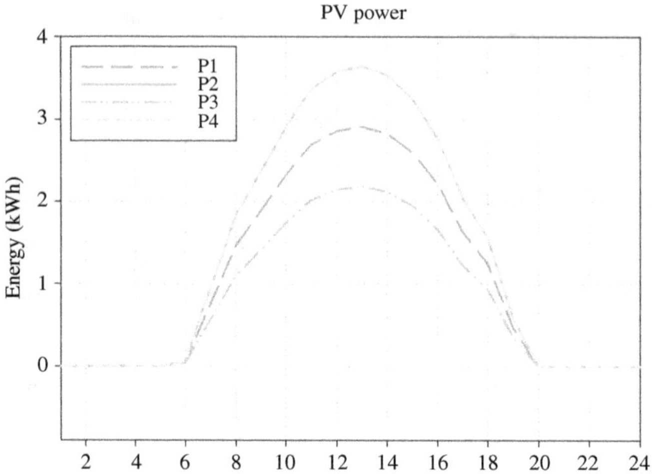

Figure 10.6 Prosumers PV power.

Table 10.1 Techno-economic parameters of the model.

Parameter	Value
Maximum import/export from/to the grid (kWh)	6
Maximum P2P buy/sell energy trading (kWh)	6
Grid CO_2 emission cost ($/kg)	0.0005
Grid CO_2 emission factor (kg/kWh)	0.97

exchanged between the network and each prosumer and the maximum energy that can be traded between agents are considered to be 6 kWh per each time step. The environmental cost of electricity generation from the conventional power plant and the grid CO_2 emission factor are also specified in the table.

To optimize the P2P energy system in this chapter of the book, four main assumptions are considered in the model:

1) **Energy Balance**: At any time step, the total energy produced and entered into the system is equal to the total energy consumed and exported. This helps to maintain the stability of the system.
2) **Maximum Power of Energy Exchange**: For each of the prosumers, there are limits on the amount of energy imported or exported. These limits are determined based on the physical capacity of the equipment and network.

3) **CO_2 Emission Rate**: The amount of environmental cost for each unit of energy entered into the system through the grid is calculated, taking into account the amount of carbon dioxide emission and its environmental effects.
4) **P2P Pricing**: In the model, it is assumed that the price of energy exchanged between prosumers is determined based on the ratio of local supply and demand, and this price is between the price of energy import and export from the grid.

10.4.1.1 First Scenario: Supply From the Grid

In the first scenario, the total load of four units is supplied through the central network. The total cost (TC) of this system includes the price of purchasing electricity from the grid and the environmental cost of electricity generation by the conventional, which is defined in Eq. (10.1).

$$TC = \sum_t \sum_i C_t^{G,imp} P_{i,t}^L + \sum_t \sum_i EF^G C^{CO_2} P_{i,t}^L \tag{10.1}$$

10.4.1.2 Second Scenario: Energy Community

In the second scenario, the energy exchange between the prosumers and the grid is possible. So, the case is modeled as an optimization problem. The goal of this optimization problem is to minimize the total cost of the system, which is defined by Eq. (10.2). The first term of this equation defines the cost of energy exchange with the grid, which includes the cost of buying electricity from the grid minus the profit from retailing energy with it. The second term expresses the environmental costs of power supply by the conventional grid.

$$\min\ TC = \sum_t \sum_i \left(C_t^{G,imp} P_{i,t}^{G,imp} - C_t^{G,exp} P_{i,t}^{G,exp} \right) + \sum_t \sum_i EF^G C^{CO_2} P_{i,t}^{G,imp} \text{s.t.}(3) - (6) \tag{10.2}$$

Equations (10.3)–(10.6) state the model constraints of the optimization problem. Equation (10.3) represents the energy balance in the system. According to this equation, the total power input and electricity generated in each unit is equal to the total electric load and energy exported from that unit.

$$P_{i,t}^L + P_{i,t}^{G,exp} = P_{i,t}^{PV} + P_{i,t}^{G,imp} \tag{10.3}$$

Equations (10.4) and (10.5) express the upper limit of energy export and import for each unit. Z in these relations is a binary variable and is defined to guarantee that import and export do not happen simultaneously, which is stated in Eq. (10.6).

$$P_{i,t}^{G,imp} \leq P_{\max}^{G,imp} Z_{i,t}^{G,imp} \tag{10.4}$$

$$P_{i,t}^{G,exp} \leq P_{\max}^{G,exp} Z_{i,t}^{G,exp} \tag{10.5}$$

$$Z_{i,t}^{G,imp} + Z_{i,t}^{G,exp} \leq 1 \tag{10.6}$$

10.4.1.3 Third Scenario: P2P Energy Sharing Community

This scenario is quite similar to the second one. But the only difference is that in this scenario P2P energy exchange between agents is possible in a local market. The objective function is to minimize the total cost of the system, which contains three terms (Eq. (10.7)). The first term is the cost of energy exchange with the grid, the second term is the environmental cost of electricity production on the grid, and the third term is the cost of P2P energy trading.

$$\min TC = \sum_t \sum_i \left(C_t^{G,imp} P_{i,t}^{G,imp} - C_t^{G,exp} P_{i,t}^{G,exp} \right) + \sum_t \sum_i EF^G C^{CO_2} P_{i,t}^{G,imp} + \sum_t \sum_i \left(C_t^{p2p} P_{i,t}^{p2p,buy} - C_t^{p2p} P_{i,t}^{p2p,sell} \right) \text{ s.t.}(8) - (19) \tag{10.7}$$

Equations (10.8)–(10.19) denote the optimization constraints. Equation (10.8) illustrates the overall energy balance for each unit. Whereas, Eq. (10.9) shows the energy balance at the community organizer level. Based on this relationship, the total energy purchased by the units is equal to the total energy sold to them, which means that the energy organizer is not allowed to store, produce, and sell energy.

$$P_{i,t}^L + P_{i,t}^{G,exp} + P_{i,t}^{p2p,sell} = P_{i,t}^{PV} + P_{i,t}^{G,imp} + P_{i,t}^{p2p,buy} \tag{10.8}$$

$$\sum_t \sum_i P_{i,t}^{p2p,sell} = \sum_t \sum_i P_{i,t}^{p2p,buy} \tag{10.9}$$

Equations (10.10)–(10.12) express the limitation of energy exchange with the network and are the same as relations Eqs. (10.4)–(10.6) in the previous section.

$$P_{i,t}^{G,imp} \leq P_{\max}^{G,imp} Z_{i,t}^{G,imp} \tag{10.10}$$

$$P_{i,t}^{G,exp} \leq P_{\max}^{G,exp} Z_{i,t}^{G,exp} \tag{10.11}$$

$$Z_{i,t}^{G,imp} + Z_{i,t}^{G,exp} \leq 1 \tag{10.12}$$

Equations (10.13) and (10.14) determine the upper limit of buying and selling energy between agents, and again, Eq. (10.15) guarantees that buying and selling energy in the local market does not take place at the same time.

$$P_{i,t}^{p2p,buy} \leq P_{\max}^{p2p,buy} Z_{i,t}^{p2p,buy} \tag{10.13}$$

$$P_{i,t}^{p2p,sell} \leq P_{\max}^{p2p,sell} Z_{i,t}^{p2p,sell} \tag{10.14}$$

$$Z_{i,t}^{p2p,buy} + Z_{i,t}^{p2p,sell} \leq 1 \tag{10.15}$$

Equations (10.16)–(10.18) are defined to determine the price of energy in P2P trading. The supply demand ratio (SDR) index, which stands for SDR, is explained in Eq. (10.16) as the ratio of the sum of total photovoltaic power produced in each unit to the total load of all units.

$$SDR_t = \frac{\sum_i P_{i,t}^{PV}}{\sum_i P_{i,t}^{L}} \tag{10.16}$$

Equations (10.17) and (10.18) introduce two parameters called balance cost and control cost. The first is defined as the average price of energy import and export to the grid, and the second is the difference in the cost of export and import divided by two [59].

$$C_t^{bal} = \frac{\left(C_t^{G,imp} + C_t^{G,exp}\right)}{2} \tag{10.17}$$

$$C_t^{con} = \frac{\left(C_t^{G,imp} - C_t^{G,exp}\right)}{2} \tag{10.18}$$

Using these two parameters, the price of energy for sale in the local market between peers is obtained using Eq. (10.19) [59]. This price is necessarily between the cost of selling electricity to the grid and purchasing it.

$$C_t^{p2p} = -\frac{2}{\pi} C_t^{con} \tan^{-1}\left((\ln SDR_t)^3\right) + C_t^{bal} \tag{10.19}$$

10.4.2 Results and Discussion

The optimization problem has been solved by employing the general algebraic modeling system (GAMS) software and via the linear, interactive, and discrete optimizer (LINDO) solver. The objective function was minimizing the total cost of the system in each scenario with regard to the constraints mentioned in the above equations. GAMS is a high-level modeling system for mathematical programming and optimization problems. This software is widely used in complex problems that require accurate mathematical modeling. GAMS was chosen for this study due to its high flexibility and ability to solve large models with a large number of variables. One of the advantages of this software is its ability to solve nonlinear and mixed models (including continuous and discrete variables), which is important in energy models such as P2P. The LINDO solver used in this study is one of the powerful solvers in GAMS, which is used to solve linear and nonlinear programming problems. This solver was chosen due to its high speed and appropriate accuracy in solving optimization problems, especially in energy networks and economic analysis problems.

Table 10.2 represents the total cost of the system under three defined scenarios. As is clear from the results, the total cost has been reduced significantly under

Table 10.2 Total costs of the prosumers in all scenarios.

Scenario	Total cost ($)
First scenario: supply from the grid	19.467
Second scenario: energy community	9.706
Third scenario: P2P energy sharing community	9.249

the second and third scenarios, which means that moving toward the energy communities and establishment of modern energy structures has an obvious positive economic influence on the model. In fact, the transition from the traditional structure and the formation of an energy community (scenarios 1 to 2) have reduced the daily electricity cost of these four prosumers by nearly half. In other words, just turning consumers into prosumers and providing the possibility of exchanging energy with the grid will decrease energy costs to a very high extent. Also, it can be seen that by converting the energy system into a P2P energy community, nearly 5% of the total energy costs are reduced. In fact, with the transition from scenario 2 to scenario 3, one of the main manifestations of energy justice, i.e. reducing the cost of energy for everyone, has been realized.

The daily price of energy input from the grid, exported to the grid, and P2P sales during 24 hours are represented in Figure 10.7. As specified in the previous section, the charge for energy in P2P transactions is always between the cost of export and the price of purchasing electricity from the grid, which can also be realized in the figure. This clearly leads to the promotion of public interest in the energy sector. Since energy sellers can sell their excess energy at a price higher than the

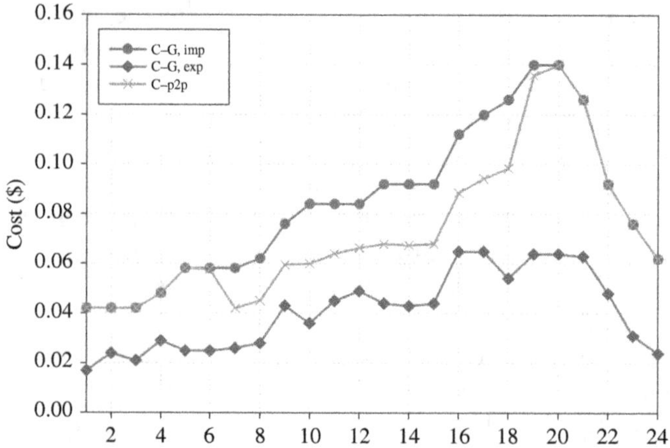

Figure 10.7 Daily energy prices (grid import/export and P2P local price).

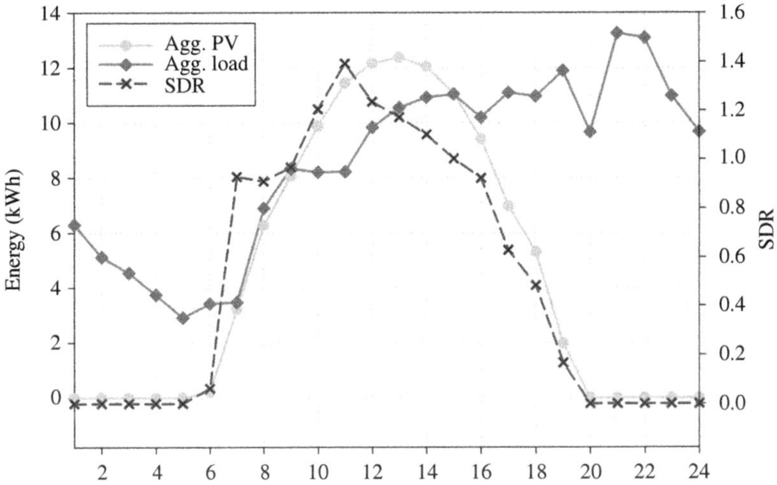

Figure 10.8 Aggregated PV power and load versus supply–demand ratio.

network rate. On the other hand, energy buyers can also meet their needs at a lower price than the network's retail rate. This means that this new structure can provide energy to consumers more easily and at a lower price.

Figure 10.8 depicts the total hourly renewable power produced in four units, the accumulative load of all units per hour, and the ratio of the previous two parameters to each other, which is defined as SDR, during one day.

Figures 10.9 and 10.10 show the daily energy profiles of prosumers in the second and third scenarios, which include power exchange with the grid in the second scenario and both energy exchange with the grid and P2P energy trading in the third one.

In the second scenario, the solar power generation profile in all four units is almost similar. In the middle hours of the day, available solar power reaches its highest value. Therefore, the amount of electricity imported from the grid has touched its lowest value in those times in P1 and P2, while in P3 and P4, this amount is zero. This means that all the load in these units is completely supplied from the photovoltaic panels, and the rest is injected into the grid. The electricity export to the grid is also zero in P1 and P2, and all the generated power is spent on their own loads, but in P3 and P4 throughout the middle hours of the day, when electricity production is maximum, energy export to the grid is observed.

Figure 10.10 describes the active role of agents in a modern energy community as they are capable of energy trading with each other as well as exchanging power with the grid. The energy import from the grid is usually followed during the hours of the day when PV generation is low or near zero. Electricity export to the grid was detected from P1 and P2 at some hours of the day, which indicates the presence of

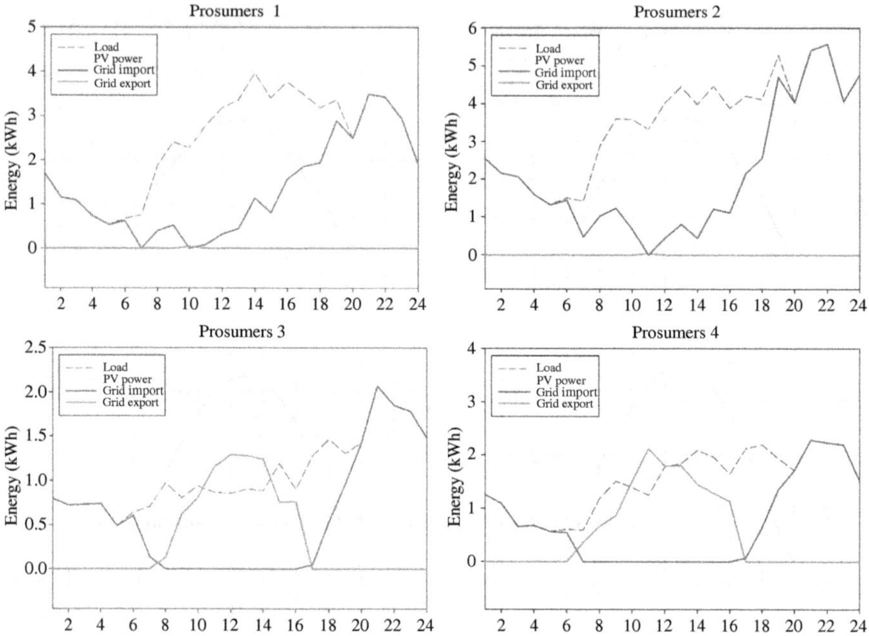

Figure 10.9 Daily energy profiles of prosumers in the second scenario.

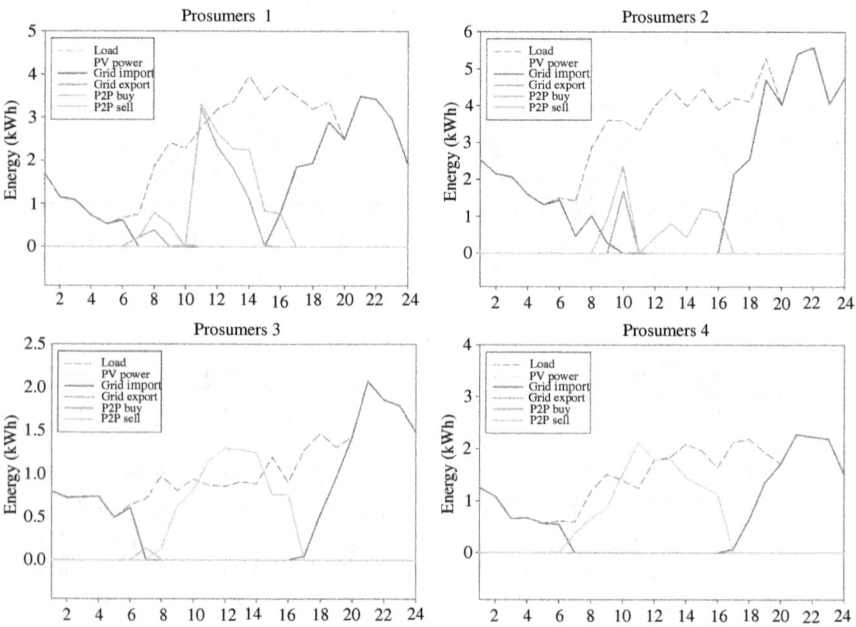

Figure 10.10 Daily energy profiles of prosumers in the third scenario.

excess solar power in these units, while in P3 and P4, no power has been injected into the network. Since the overall load of P3 and P4 is lower than P1 and P2, in this microgrid, P3 and P4 were energy sellers to P1 and P2, and this energy exchange occurred mainly in the middle hours of the day.

10.5 Conclusion

In this chapter, the obligation of transitioning energy systems from traditional structures to modern schemes was investigated. In this regard, the weaknesses and drawbacks of the traditional energy systems were expressed, and the necessity of transforming these systems into modern structures with the aim of increasing energy justice in societies was brought up. Then, the characteristics of modern energy systems were examined in the two contexts of energy communities and P2P energy trading. Then, the impact of these concepts on the development of energy integrity in societies was analyzed. Based on the stated contents, decentralization of the energy systems and giving an active role to the agents are considered the core requirements of modern energy societies, which led to the establishment of energy democracy in countries.

A practical example of energy transition in societies was presented in the form of three scenarios to deepen and conceptualize the topics discussed earlier. In this example, a microgrid consisting of four units was considered as a case study. This microgrid acts as a passive agent in the first scenario; it can supply its required load only from the central grid. But in the second scenario, the units are equipped with photovoltaic panels and are capable of exchanging power with the grid and creating an energy community. Finally, in the third scenario, in addition to exchanging power with the network, prosumers are able to trade energy with each other directly in the form of P2P energy trading. By comparing the total cost of these three scenarios, it has become clear that the more modern systems have a much lower cost than the traditional energy arrangement. As a consequence, the benefits for society members and social welfare in energy communities are considered fairer in a more democratized way compared with traditional systems.

As a brief summary of this chapter, the following can be mentioned:

- Energy democratization in terms of increasing the role of different members in the energy market, is progressing in different ways through applying different concepts. This movement is clearly in line with the modernization of energy systems.
- Two important aspects of the modernization of energy systems are the emergence of *energy communities* and the advent of *P2P energy exchange*

markets. With the reviews and explorations presented in this chapter, it was found that the development of these concepts in energy systems will ultimately facilitate the achievement of energy justice and ensure the benefit of all parties in an energy system.

References

1 Fan, J., Zhang, J., Yuan, L. et al. (2024). Deep low-carbon economic optimization using CCUS and two-stage P2G with multiple hydrogen utilizations for an integrated energy system with a high penetration level of renewables. *Sustainability* 16 (13): 1–20. 10.3390/su16135722.

2 Pesch, U. (2019). Elusive publics in energy projects: the politics of localness and energy democracy. *Energy Research and Social Science* 56: 101225. 10.1016/j.erss.2019.101225.

3 Szulecki, K. and Overland, I. (2020). Energy democracy as a process, an outcome and a goal: a conceptual review. *Energy Research and Social Science* 69: 101768. 10.1016/j.erss.2020.101768.

4 Zhang, S. and Hou, C. (2021). Model of decentralized cross-chain energy trading for power systems. *Global Energy Interconnection* 4 (3): 324–334. 10.1016/j.gloei.2021.07.006.

5 Mbungu, N.T., Naidoo, R.M., and Bansal, R.C. (2017). Real-time electricity pricing: TOU-MPC based energy management for commercial buildings. *Energy Procedia* 105: 3419–3424. 10.1016/j.egypro.2017.03.781.

6 Guo, X., Zhao, Q., Zheng, D. et al. (2020). A short-term load forecasting model of multi-scale CNN-LSTM hybrid neural network considering the real-time electricity price. *Energy Reports* 6: 1046–1053. 10.1016/j.egyr.2020.11.078.

7 Wang, L., Han, L., Tang, L. et al. (2024). Incentive strategies for small and medium-sized customers to participate in demand response based on customer directrix load. *International Journal of Electrical Power & Energy Systems* 155: 109618. 10.1016/j.ijepes.2023.109618.

8 Zhang, Z. and Sun, Y. (2024). Smart tariff calculator for feed-in tariff scheme of renewable resources. *Measurement: Sensors* 31: 100920. 10.1016/j.measen.2023.100920.

9 Marańda, W., Tylman, W., Kotas, R. et al. (2023). Parametric study on utilization of PV-energy in residential microgrids with hot water heating. *Energy Reports* 10: 1555–1564. 10.1016/j.egyr.2023.08.025.

10 Wang, K., Wang, H., Yang, J. et al. (2022). Electric vehicle clusters scheduling strategy considering real-time electricity prices based on deep reinforcement learning. *Energy Reports* 8: 695–703. 10.1016/j.egyr.2022.01.233.

11 Nady, S., El Fadil, H., Koundi, M. et al. (2022). Power to X systems: state-of-the-art (PTX). *IFAC-Papers OnLine* 55 (12): 300–305. 10.1016/j.ifacol.2022.07.328.

12 Das, A., Peu, S.D., Akanda, M.A.M., and Islam, A.R.M.T. (2023). Peer-to-peer energy trading pricing mechanisms: towards a comprehensive analysis of energy and network service pricing (NSP) mechanisms to get sustainable enviro-economical energy sector. *Energies* 16 (5): 10.3390/en16052198.

13 Muhsen, H., Allahham, A., Al-halhouli, A. et al. (2022). Business model of peer-to-peer energy trading: a review of literature. *Sustainability* 14 (3): 10.3390/su14031616.

14 Chen, Y., Lei, X., Yang, J. et al. (2022). Decentralized P2P power trading mechanism for dynamic multi-energy microgrid groups based on priority matching. *Energy Reports* 8: 388–397. 10.1016/j.egyr.2022.08.109.

15 Islam, S.N. (2024). A review of peer-to-peer energy trading markets: enabling models and technologies. *Energies* 17 (7): 1702. 10.3390/en17071702.

16 Larbwisuthisaroj, S. and Chaitusaney, S. (2024). A method for determining the maximum P2P penetration considering stakeholder benefits. *IET Generation, Transmission and Distribution* 18 (1): 145–158. 10.1049/gtd2.13078.

17 Cárdenas-Álvarez, J.P., España, J.M., and Ortega, S. (2022). What is the value of peer-to-peer energy trading? A discrete choice experiment with residential electricity users in Colombia. *Energy Research and Social Science* 91: 10.1016/j.erss.2022.102737.

18 Gomes, L. and Vale, Z. (2024). Costless renewable energy distribution model based on cooperative game theory for energy communities considering its members' active contributions. *Sustainable Cities and Society* 101: 105060. 10.1016/j.scs.2023.105060.

19 Aruta, G., Ascione, F., Fistola, R., and Iovane, T. (2024). The city as a power hub for boosting renewable energy communities: a case study in Naples. *Sustainability* 16 (18): 7988. 10.3390/su16187988.

20 Standal, K., Leiren, M.D., Alonso, I. et al. (2023). Can renewable energy communities enable a just energy transition? Exploring alignment between stakeholder motivations and needs and EU policy in Latvia, Norway, Portugal and Spain. *Energy Research and Social Science* 106: 103326. 10.1016/j.erss.2023.103326.

21 Queiroz, H., Lopes, R.A., Martins, J. et al. (2023). Assessment of energy sharing coefficients under the new Portuguese renewable energy communities regulation. *Heliyon* 9 (10): 10.1016/j.heliyon.2023.e20599.

22 Shahinzadeh, H., Nasrabadi, M. N., Nafisi, H. et al. (2024). Enhancing transparency in the peer-to-peer transactive energy market: smart contract monitoring in bilateral negotiations with the ethereum platform. *In 2024 9th*

International Conference on Technology and Energy Management (ICTEM), pp. 1–8, 10.1109/ICTEM60690.2024.10631949.

23 Zahedi, R., Ghodusinejad, M.H., and Gitifar, S. (2023). Threats evaluation of border power plants from the perspective of fuel type and providing solutions to deal with them: a case study of Iran. *Transactions of Indian National Academy of Engineering* 8 (1): 55–67. 10.1007/s41403-022-00377-5.

24 Aoun, A., Adda, M., Ilinca, A. et al. (2024). Comparison between blockchain P2P energy trading and conventional incentive mechanisms for distributed energy resources—a rural microgrid use case study. *Applied Sciences* 14 (17): 7618. 10.3390/app14177618.

25 Chen, Y. and Zhao, C. (2022). Review of energy sharing: business models, mechanisms, and prospects. *IET Renewable Power Generation* 16 (12): 2468–2480. 10.1049/rpg2.12438.

26 Wang, Z., Liu, F., Ma, Z. et al. (2021). Distributed generalized nash equilibrium seeking for energy sharing games in prosumers. *IEEE Transactions on Power Apparatus and Systems* 36 (5): 3973–3986. 10.1109/TPWRS.2021.3058675.

27 Chen, Y., Wei, W., Liu, F. et al. (2018). Optimal contracts of energy mix in a retail market under asymmetric information. *Energy* 165: 634–650. 10.1016/j.energy.2018.09.139.

28 Jonek-Kowalska, I. (2023). P2P transactions as a market innovation from a prosumer perspective: level of potential involvement and behavioral incentives and disincentives of development. *Journal of Open Innovation: Technology, Market, and Complexity* 9 (3): 100140. 10.1016/j.joitmc.2023.100140.

29 Widayanti, R., Aini, Q., Haryani, H. et al. (2021). Decentralized electronic vote based on blockchain P2P. *In 2021 9th International Conference on Cyber and IT Service Management (CITSM)*, pp. 1–7, 10.1109/CITSM52892.2021.9588851.

30 Baig, M.J.A., Iqbal, M.T., Jamil, M., and Khan, J. (2021). Peer-to-peer energy trading in a micro-grid using internet of things and blockchain. *Electronics* 25 (2): 39–49. 10.53314/ELS2125039B.

31 Mahmoud, M. and Ben Slama, S. (2024). Driving innovation: prosumer incentives in peer-to-peer energy trading. *Engigneering Technology and Applied Science Research* 14 (3): 14600–14611. 10.48084/etasr.7367.

32 M. I. Shah, E. Barrett, and K. Mason, "*Reinforcement Learning Enabled Peer-to-Peer Energy Trading for Dairy Farms*," 2024.

33 Tao, J., Ting, H., Hao, X. et al. (2023). P2P transaction method for distributed energy prosumers based on reputation value. *Global Energy Interconnection* 6 (3): 308–323. 10.1016/j.gloei.2023.06.005.

34 Wu, Q., Song, Q., He, X. et al. (2024). Distributed peer-to-peer energy trading framework with manufacturing assembly process and uncertain renewable energy plants in multi-industrial micro-grids. *Energy* 302: 131876. 10.1016/j.energy.2024.131876.

35 Pigliautile, I. et al. (2022). Peer-to-peer energy communities: regulatory barriers in the EU context. *Open Research Europe* 2: 147. 10.12688/openreseurope.15282.1.

36 Lavrijssen, S., Reins, L., and ten Caten, T. (2023). Realizing peer-to-peer trading in the electricity market in the EU and its member states. In: *Handbook of Energy Law in the Low-Carbon Transition*, 513–530. 10.1515/9783110752403-039.

37 Biggar, D. and Hesamzadeh, M.R. (2022). Energy communities: challenges for regulators and policymakers. In: *Energy Communities* (ed. S. Löbbe, F. Sioshansi, S. Löbbe, et al.), 131–149. Academic Press.

38 Taouil, K., Aloulou, R., Bradai, S. et al. (2024). P2P energy exchange architecture for swarm electrification-driven PV communities. *Energies* 17 (15): 10.3390/en17153680.

39 Jenkins, K., McCauley, D., Heffron, R. et al. (2016). Energy justice: a conceptual review. *Energy Research and Social Science* 11: 174–182. 10.1016/j.erss.2015.10.004.

40 Ozkan, C. and Okay, N. (2024). Renewable energy, economic policy uncertainty and climate policy uncertainty: new evidence for environmental kuznets curve from emerging and developed countries. *Sustainability* 16 (14): 6049. 10.3390/su16146049.

41 Volodzkiene, L. and Streimikiene, D. (2024). Towards energy equity: understanding and addressing multifaceted energy inequality. *Energies* 17 (17): 4500. 10.3390/en17174500.

42 Sovacool, B.K. and Dworkin, M.H. (2015). Energy justice: conceptual insights sand practical applications. *Applied Energy* 142: 435–444. 10.1016/j.apenergy.2015.01.002.

43 Campos, I. and Marín-González, E. (2020). People in transitions: energy citizenship, prosumerism and social movements in Europe. *Energy Research and Social Science* 69: 101718. 10.1016/j.erss.2020.101718.

44 Wu, Y., Wu, Y., Cimen, H. et al. (2022). Towards collective energy Community: potential roles of microgrid and blockchain to go beyond P2P energy trading. *Applied Energy* 314: 119003. 10.1016/j.apenergy.2022.119003.

45 Park, C. and Yong, T. (2017). Comparative review and discussion on P2P electricity trading. *Energy Procedia* 128: 3–9. 10.1016/j.egypro.2017.09.003.

46 Peirow, S., Razi Astaraei, F., and Saifoddin Asl, A. (2023). Techno-economic and environmental assessment of a hybrid renewable energy system for a hospital using multi-criteria decision-making method. *Energies* 16 (4): 1916. 10.3390/en16041916.

47 Sattich, T., Freeman, D., Scholten, D., and Yan, S. (2021). Renewable energy in EU-China relations: policy interdependence and its geopolitical implications. *Energy Policy* 156: 112456. 10.1016/j.enpol.2021.112456.

48 Ghodusinejad, M.H., Peirov, S., Yousefi, H., and Astaraei, F.R. (2024). Multi-objective optimal design and performance analysis of a residential microgrid. *Journal of Energy Storage* 92: 112201. 10.1016/j.est.2024.112201.

49 Alzubaidi, L. H., Thakur, G., Vijayalakshmi, G.2024). Peer-to-peer energy trading platforms in local energy markets. *In E3S Web of Conferences*, vol. 540, 10.1051/e3sconf/202454007003.

50 Ali, L, Azim, M.I., Peters, J. et al.. (2023). P2P trading-enabled local energy market supplemented with blockchain technology: an Australian case study. *In 2023 IEEE Green Technologies Conference (GreenTech)*, pp. 51–55, 10.1109/GreenTech56823.2023.10173850.

51 Etukudor, C., Couraud, B., Robu, V. et al. (2020). Automated negotiation for peer-to-peer electricity trading in local energy markets. *Energies* 13 (4): 10.3390/en13040920.

52 Shafiekhani, M., Qadrdan, M., Zhou, Y., and Wu, J. (2024). Development of a multi-community peer-to-peer electricity trading mechanism. *Cardiff University Engineering Research Conference 2023,* Cardiff, UK (12-14 July 2023). pp. 148–152, 10.18573/conf1.ah.

53 Ghodusinejad, M.H. and Yousefi, H. (2023). Peer-to-peer energy trading in multi-carrier energy systems. In: *Demand-Side Peer-to-Peer Energy Trading*, 183–201. Cham: Springer International Publishing 10.1007/978-3-031-35233-1_9.

54 Bokkisam, H.R. and Selvan, M.P. (2021). Effective community energy management through transactive energy marketplace. *Computers and Electrical Engineering* 93: 107312. 10.1016/j.compeleceng.2021.107312.

55 Ghaemi, S. and Anvari-Moghaddam, A. (2023). Local energy communities with strategic behavior of multi-energy players for peer-to-peer trading: a techno-economic assessment. *Sustainable Energy, Grids and Networks* 34: 101059. 10.1016/j.segan.2023.101059.

56 Ghodusinejad, M.H., Lavasani, Z., and Yousefi, H. (2023). A combined decision-making framework for techno-enviro-economic assessment of a commercial CCHP system. *Energy* 276: 127609. 10.1016/j.energy.2023.127609.

57 Wilkinson, S., Hojckova, K., Eon, C. et al. (2020). Is peer-to-peer electricity trading empowering users? Evidence on motivations and roles in a prosumer business model trial in Australia. *Energy Research and Social Science* 66: 10.1016/j.erss.2020.101500.

58 Junlakarn, S., Kokchang, P., and Audomvongseree, K. (2022). Drivers and challenges of peer-to-peer energy trading development in Thailand. *Energies* 15 (3): 1–25. 10.3390/en15031229.

59 Song, J.G., Kang, E.S., Shin, H.W., and Jang, J.W. (2021). A smart contract-based P2P energy trading system with dynamic pricing on Ethereum Blockchain. *Sensors* 21 (6): 1–27. 10.3390/s21061985.

11

Sustainable Development of Energy Communities for Energy Justice and Equity

Arash Rajaei[1], *Sobhan Dorahaki*[2], *Masoud Rashidinejad*[1], *Peyman Afzali*[3], *and Mahdi Bedrood*[1]

[1] *Department of Electrical Engineering, Shahid Bahonar University of Kerman, Kerman, Iran*
[2] *Department of Electrical Engineering, Qatar University, Doha, Qatar*
[3] *Department of Energy (AAU Energy), Aalborg University, Aalborg, Denmark*

11.1 Introduction

In our pursuit of a sustainable and equitable future, integrating renewable energy sources (RESs), energy efficiency measures, and community engagement has become pivotal. This chapter explores the intricate aspects of sustainable development within energy communities (ECs), where clean energy, social participation, and environmental responsibility come together to build resilient, locally driven energy systems. Focusing on energy justice and equity, this chapter uncovers the connections between sustainable energy practices and social justice.

Sustainable ECs prioritize using renewable resources like solar and wind power while promoting energy conservation. The active participation of residents and businesses in decision-making reflects a commitment to balancing social, economic, and environmental goals. As we explore sustainable development, the importance of linking energy practices with social justice becomes clearer.

This chapter focuses on the connection between sustainable energy and social justice. By examining case studies, policy frameworks, and technological interventions, we reveal the complexities of energy justice and equity. Empowering communities to shape their energy systems not only fosters environmental sustainability but also sparks societal change. This exploration offers a comprehensive view of how ECs drive a just and equitable energy future.

Taking an interdisciplinary approach, this chapter examines how environmental sustainability, social inclusivity, and economic empowerment intersect. By analyzing successes and challenges, we provide actionable insights for future sustainable energy strategies. This outlook emphasizes the key role ECs play in

Communities for Clean Energy Justice and Equity in Grid Modernization, First Edition.
Edited by Mohammadreza Daneshvar, Behnam Mohammadi-Ivatloo, and Amjad Anvari-Moghaddam.

driving positive change, creating a future where clean, affordable energy coexists with a strong commitment to justice and equity. Addressing both environmental sustainability and social justice is urgent, positioning sustainable energy as a foundation for building a more equitable world.

We achieve sustainable development in ECs by implementing energy-saving initiatives, such as energy efficiency programs (EEPs), and promoting clean energy use. Figure 11.1 presents a multi-energy community structure based on sustainability, considering social, environmental, economic, and technical factors. Each energy community has a manager who actively engages in EEPs to enhance sustainability. The manager of the multi-energy communities works to optimize both energy efficiency and overall sustainability.

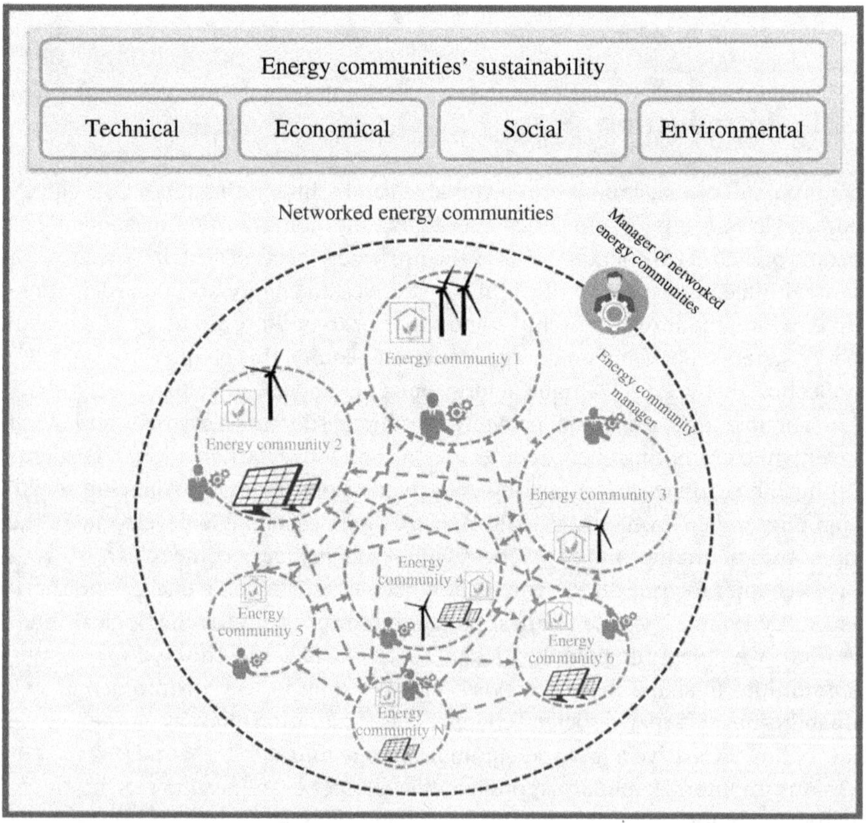

Figure 11.1 The overview of a multi-energy communities structure based on the sustainability development concept.

In this chapter, we discuss how ECs play a vital role in aligning sustainable electricity practices with social justice and equity. By integrating RESs, such as solar and wind power, and promoting EEPs, this study shows how local energy systems can be both resilient and socially equitable.

This chapter also explores innovative strategies and technological advances designed to empower communities, helping them shape their energy landscapes and make strides toward environmental sustainability and economic empowerment. The overarching goal is to demonstrate how sustainable ECs can act as catalysts for broader societal change, ensuring a future where energy systems are inclusive, just, and aligned with the urgent needs of social fairness and environmental stewardship.

11.2 Foundations of Energy Justice and Equity

11.2.1 The Conceptual Framework of Energy Justice

The conceptual framework of energy justice lays the foundation for understanding and addressing the complex dynamics between energy systems and societal equity [1]. At its core, energy justice encompasses the fair distribution of the benefits and burdens tied to energy production, distribution, and consumption. This subsection explores the foundational principles shaping the conceptual landscape of energy justice, providing a nuanced understanding of its various dimensions.

The distributional aspect of energy justice emphasizes the fair sharing of both the benefits and burdens related to energy systems. It examines how the advantages of clean energy, like reduced air pollution and economic growth, are distributed across social groups while also addressing the unequal impacts, such as environmental harm or health risks, often borne by marginalized communities. By prioritizing fairness in this distribution, this dimension of energy justice seeks to correct historical and systemic inequalities in energy practices.

Next, the procedural dimension focuses on ensuring fairness in decision-making processes related to energy policies and projects. Energy justice requires the meaningful and inclusive participation of all stakeholders, especially those historically marginalized or excluded. This means dismantling power imbalances, fostering community engagement, and promoting transparent and accessible information. A just energy transition demands not only the equitable sharing of benefits but also a democratic approach to shaping the energy landscape [2].

The third dimension, interactional justice, explores the fairness of relationships and communication within energy practices. It addresses the social and cultural dimensions of energy decisions, recognizing diverse perspectives and fostering inclusive dialogue. Understanding how energy choices affect communities on a personal and cultural level is crucial for cultivating a sense of justice that extends beyond just distributive and procedural aspects [3].

In summary, the conceptual framework of energy justice integrates distributional, procedural, and interactional dimensions. It requires a holistic understanding of how energy systems intersect with social structures, acknowledging the historical inequities that shape these interactions. As we transition from the conceptual framework to real-world implications, it becomes essential to examine how energy justice impacts society. By analyzing the fairness in distributing benefits and burdens, ensuring inclusive decision-making, and fostering equitable interpersonal relationships, this framework helps us understand the complexities of energy justice in the context of sustainable development within ECs.

The conceptual framework of energy justice lays the foundation for understanding and addressing the complex dynamics within the intersection of energy systems and societal equity [1]. At its core, energy justice is a multidimensional concept that encompasses the fair distribution of benefits and burdens associated with energy production, distribution, and consumption. This subsection navigates through the foundational principles that shape the conceptual landscape of energy justice, aiming to provide a nuanced understanding of its various dimensions.

The distributional aspect of energy justice emphasizes the fair distribution of both the benefits and burdens related to energy systems. It explores how the advantages of clean energy, such as lower air pollution and economic growth, are shared among various social groups while also addressing the unequal distribution of negative effects, like environmental damage or health risks, often borne by marginalized communities. By prioritizing fairness in this distribution, this dimension of energy justice seeks to correct historical and systemic inequalities in energy practices.

Secondly, the procedural dimension focuses on the fairness of decision-making processes related to energy policies and projects. In the pursuit of energy justice, it is imperative to ensure that all stakeholders, especially those historically marginalized or excluded, have meaningful and inclusive participation in decision-making. This involves dismantling power imbalances, fostering community engagement, and promoting transparent and accessible information. A just energy transition demands not only the equitable distribution of benefits but also a democratic approach to shaping the energy landscape [2].

The third dimension, interactional justice, explores the fairness of relationships and communication within energy practices. This includes addressing the social and cultural dimensions of energy decisions, recognizing diverse perspectives, and fostering inclusive dialogues. Understanding how energy choices affect communities on a personal and cultural level is crucial for cultivating a sense of justice that extends beyond just distributive and procedural aspects [3].

In summary, the conceptual framework of energy justice integrates distributional, procedural, and interactional dimensions. It necessitates a holistic understanding of how energy systems intersect with social structures, acknowledging the historical inequities that have often defined these interactions. As we transition from the conceptual framework to real-world implications, it becomes essential to examine how energy justice impacts society.

As we transition from the conceptual framework to real-world implications, it becomes essential to examine how energy justice impacts society. By thoroughly analyzing the fairness in distributing benefits and burdens, the inclusiveness of decision-making processes, and the interpersonal relationships involved in energy decisions, this framework lays the foundation for understanding the complexities of energy justice in the context of sustainable development within ECs.

11.2.2 Social Implications of Energy Practices

As we shift focus from the broader theoretical framework, exploring the social implications of energy practices becomes pivotal in the discourse on energy justice and equity. Energy choices are not merely technical decisions but have profound and differential impacts on diverse communities, shaping and reflecting existing social structures. This subsection aims to provide a comprehensive exploration of the intricate web of social consequences associated with energy practices, shedding light on the challenges and opportunities that arise within the realm of sustainable development [4].

One primary consideration is energy poverty, which disproportionately affects marginalized communities and deepens existing socioeconomic disparities existing socioeconomic disparities. Examining how energy choices impact vulnerable populations, this section addresses the need to ensure that the transition to sustainable energy does not inadvertently deepen inequalities. Strategies such as targeted policy interventions and community-driven initiatives are explored to promote a more inclusive and just energy transition.

Additionally, energy justice recognizes that energy choices intersect with broader societal issues like displacement and housing disparities. Large-scale energy projects, such as the development of renewable energy infrastructure, can sometimes lead to the displacement of communities, particularly those already marginalized. By analyzing case studies and policy frameworks, this section endeavors to unpack the social dimensions of energy-related displacement and emphasizes the need for approaches that prioritize social equity and community well-being.

Social determinants of energy consumption are also crucial in understanding how energy practices affect different demographic groups. This subsection delves into how factors such as income, race, and education influence energy access and usage patterns. By analyzing these determinants, we can identify and address disparities in energy consumption, ensuring policies and initiatives meet the needs of diverse communities. With these social implications in mind, the focus shifts naturally to the economic aspects of energy justice, as the distribution of energy-related benefits and costs also plays a crucial role in fostering equity.

In conclusion, the social implications of energy practices go beyond environmental considerations, encompassing a wide range of social dynamics and disparities. By examining issues such as energy poverty, displacement, and the social determinants of energy consumption, this subsection contributes to the foundation for an inclusive and just energy transition. It highlights the importance of considering social dimensions alongside technical and economic aspects, emphasizing the interconnectedness of energy choices with broader societal structures and inequalities.

11.2.3 Economic Dimensions of Energy Equity

Understanding the economic dimensions of energy equity is crucial for developing strategies that ensure sustainable energy development is shared equitably among stakeholders. This subsection delves into the multifaceted economic considerations surrounding energy practices, emphasizing the transformative potential of sustainable energy in fostering economic equity within communities [5].

A key economic aspect of energy equity is the potential for job creation and local economic empowerment. The transition to sustainable energy sources, such as solar and wind power, not only offers environmentally friendly alternatives but also serves as a catalyst for employment opportunities. This section examines case studies and initiatives that highlight the positive economic impact of clean energy investments, focusing on how these initiatives can stimulate local economies. By fostering a decentralized energy infrastructure, sustainable energy practices can stimulate job growth, reduce dependency on centralized power systems, and empower local communities economically [6].

Another critical factor is the fair distribution of costs and benefits in the energy transition. While sustainable energy development brings about numerous benefits, it is crucial to address potential disparities in how these advantages are distributed among different socioeconomic groups. This involves examining the financial implications for both consumers and producers, ensuring that the costs of transitioning to sustainable energy are not disproportionately borne by vulnerable communities do not bear a disproportionate share of the costs. Addressing these economic dimensions naturally leads to an exploration of how energy affordability and access can be maintained for all segments of society.

Additionally, energy affordability is a vital part of economic equity. Sustainable energy practices should not only contribute to economic empowerment but also ensure that clean energy remains accessible to all, regardless of socioeconomic status. This section analyzes affordability challenges and explores innovative financing models to craft policies that promote inclusive access to clean energy. By understanding the economic barriers to adopting sustainable energy, this subsection seeks to identify pathways for making clean energy technologies more financially accessible and beneficial to a broader spectrum of society.

In summary, the economic dimensions of energy equity highlight sustainable energy development as a catalyst for positive economic transformation. By examining job creation, equitable cost and benefit distribution, and approaches to improve energy affordability, this section supports the foundation for a fair and inclusive energy transition. As we consider economic factors, it becomes clear that environmental sustainability must also be addressed to reinforcing the interconnected nature of energy decisions with broader economic systems and opportunities.

11.2.4 Environmental Sustainability as a Pillar of Energy Justice

The pursuit of energy justice is closely linked to environmental sustainability, emphasizing that a just and equitable energy future must prioritize the planet's health and well-being. In continuing from the social and economic dimensions, this subsection delves into the intricate relationship between sustainable energy development and environmental justice, emphasizing the imperative of fostering ecological harmony within the framework of energy justice [7].

At its core, environmental sustainability in the context of energy justice involves a critical examination of the ecological impacts of different energy sources and technologies [8]. While the previous sections focused on social and economic equity, this exploration extends beyond the reduction of carbon emissions to address broader environmental considerations, including land use, water consumption, and biodiversity. By analyzing the ecological footprint of various energy practices, this section establishes a foundation for sustainable energy development that considers the overall health of ecosystems, not just climate change.

Environmental justice within the energy justice framework also requires examining how energy practices disproportionately affect vulnerable and marginalized communities. Certain populations, often socioeconomically disadvantaged, bear a higher burden of environmental degradation and pollution resulting from energy production. Just as social and economic inequities demand attention, this subsection explores case studies and policies that highlight environmental injustices and suggests strategies for transitioning to cleaner energy while addressing historical disparities in exposure to environmental harm.

A critical aspect of environmental sustainability in energy justice is the shift to RESs like solar and wind power. Tying into the economic opportunities of renewable energy discussed earlier, this section highlights how cleaner alternatives can reduce environmental harm and foster energy independence. By understanding the technological advancements and policy incentives driving the shift toward renewables, it contributes to a vision of an energy landscape that is not only just and equitable but also environmentally regenerative.

In sum, environmental sustainability is a cornerstone of energy justice, highlighting the deep connection between social equity and ecological well-being. By scrutinizing the ecological impacts, addressing environmental injustices, and promoting the transition to RESs, this subsection supports the foundation for an inclusive and just energy transition. As emphasized throughout the chapter, it stresses the importance of considering environmental dimensions alongside social and economic factors, urging sustainable energy practices that benefit both people and the planet.

11.3 Energy Communities: Conceptual Framework

11.3.1 Defining Energy Communities

Defining ECs requires navigating the complex interactions of groups, entities, and individuals actively engaged involved in energy decision-making, generation, and consumption within a shared framework [9]. As we explore the core aspects of ECs, it becomes clear that they are not confined solely to geographical boundaries but are rather characterized by a shared commitment to sustainable energy practices and collective participation in shaping the local energy landscape. These communities manifest in various forms, ranging from neighborhoods and municipalities to cooperatives, indigenous groups, and even virtual communities connected by a common energy-related goal.

At the core of ECs is the concept of shared ownership and participation. Building on this notion, ECs represent a departure from traditional, centralized energy models, emphasizing the democratization of energy decision-making [10]. Whether through community-owned renewable energy projects, cooperatives, or participatory governance structures, they aim to distribute benefits and responsibilities more equitably across the energy sector. This definition acknowledges the diversity of structures and scales at which ECs operate, encapsulating the richness of community-driven energy initiatives globally.

Moreover, further expanding the definition, ECs are characterized by a collective vision for sustainable energy development. This vision transcends mere energy production and consumption; it incorporates broader goals such as environmental sustainability, social equity, and economic empowerment. Community members actively shape this vision to reflect their unique needs, values, and aspirations. Thus, defining ECs goes beyond a technical understanding of energy systems to encompass the social, cultural, and economic dimensions that shape these communities.

In addition to these defining features, understanding ECs also means recognizing the interconnectedness of stakeholders both within and outside the community. Residents, businesses, local governments, and organizations become integral components of the energy ecosystem, collaborating to foster resilience, self-sufficiency, and a shared sense of responsibility [11]. This interconnectedness highlights the importance of cooperation in achieving the collective goals of ECs.

In summary, ECs are defined by their fluidity and adaptability, reflecting a paradigm shift toward more inclusive, participatory, and locally driven approaches to energy decision-making and development.

11.3.2 Principles of Community Engagement in Energy Development

Fostering meaningful and inclusive community engagement is crucial to the success of ECs within sustainable energy development [12]. Building on the conceptual framework of energy justice, this section explores the principles that guide the collaboration between communities and energy development projects, recognizing that effective engagement is essential for achieving energy justice and equity.

At the core of community engagement is the active participation of members in decision-making, from project conception to implementation and beyond. A key element of this engagement is transparency, which serves as a foundational principle, ensuring that relevant information is readily accessible to all stakeholders. This openness builds trust and empowers community members to make informed decisions about energy projects that shape their lives and surroundings.

Examples of engagement strategies include involving communities in planning, promoting transparent communication, and ensuring economic benefits like job creation and infrastructure improvements. Furthermore, these principles contribute to the success and acceptance of renewable energy projects, as seen in studies of ECs across Europe and the United States [13]. The approach goes beyond consultation by actively including the voices of marginalized groups, ensuring that the benefits and burdens of sustainable energy development are distributed equitably.

Participatory decision-making is another cornerstone, giving communities a genuine role in shaping their energy landscapes [14]. This involves not only consulting community members but also involving them in substantive ways, allowing their input to influence the trajectory of energy projects. Mechanisms like workshops, citizen panels, and forums ensure that community concerns and aspirations drive energy development initiatives.

Equity is a guiding principle that seeks to address historical disparities and rectify past injustices within the energy sector [15]. This involves acknowledging and actively mitigating any negative impacts that sustainable energy projects might have on vulnerable or marginalized communities. Equity aligns with the broader goals of energy justice, ensuring a fair distribution of both benefits and burdens.

In conclusion, these principles of community engagement lay the foundation for a collaborative and equitable relationship between energy initiatives and the communities they impact. By prioritizing transparency, inclusivity, participatory decision-making, and equity, these principles offer a clear path for navigating the complexities of sustainable energy development. Recognizing the integral role of communities, these principles contribute to the overarching objective of fostering energy justice and equity within the conceptual framework of ECs.

11.3.3 Local Governance and Decision-Making Structures

The structure of local governance and decision-making within ECs plays a crucial role in guiding the path toward sustainable energy development. As we build on the social and economic dimensions of energy justice, a case study from Sweden [16] illustrates how different governance structures influence the success of decentralized energy systems. The contrast between a distributed energy system (DES) and a citizen energy community (CEC) highlights how differing stakeholder priorities can shape the transition toward more community-oriented energy models. This section delves into the diverse models and mechanisms employed by ECs, emphasizing the importance of tailoring governance structures to the unique characteristics and needs of different communities.

Decentralized governance structures enable more inclusive and responsive decision-making processes, ensuring that community members have a direct influence on the planning, implementation, and management of energy projects. In exploring these governance models, from community cooperatives to participatory councils, valuable insights are gained into the effectiveness of decentralization in promoting resilience and sustainability.

Furthermore, in line with the need for inclusive governance, the establishment of clear and transparent decision-making processes is paramount in fostering trust and accountability within ECs. This requires defining roles, responsibilities, and decision-making authority at the local level. By examining case studies that exemplify transparent decision-making structures, this subsection aims to illuminate how clear governance frameworks contribute to the effective functioning of ECs. Transparent decision-making not only enhances community participation but also serves as a mechanism for ensuring that energy projects align with the values and aspirations of the community [17].

In addition to transparency, local governance within ECs is also closely tied to issues of social equity and justice [18]. Recognizing the historical disparities in decision-making power, successful governance structures prioritize inclusivity, striving to amplify the voices of marginalized groups. This subsection explores initiatives that actively engage with diverse community members, promoting representation and ensuring that decision-making structures address existing power imbalances. By doing so, it contributes to the creation of governance models that genuinely reflect the needs and aspirations of the entire community.

Moreover, collaborative partnerships between ECs and external stakeholders play a crucial role in governance structures [19]. This involves creating mechanisms for dialogue and collaboration with local governments, utilities, and other relevant entities. Analyzing successful collaborations, this section sheds light on how ECs can navigate external partnerships to maximize their influence in decision-making processes, fostering a sense of shared responsibility for the local energy landscape.

To conclude, local governance and decision-making structures are integral components that shape the resilience and sustainability of ECs. By embracing decentralization, transparency, inclusivity, and collaborative partnerships, these structures empower communities to actively participate in and influence the trajectory of their energy systems. Ultimately, recognizing the diverse ways governance can be structured provides valuable insights into tailoring approaches that best suit the unique characteristics of each energy community, contributing to the overarching goal of fostering energy justice and equity.

11.3.4 Collaborative Energy Planning and Implementation

Collaborative energy planning and implementation are transformative processes within ECs, emphasizing collective decision-making, shared visioning, and inclusive participation. This paper provides a comprehensive analysis of ECs as a strategy for addressing climate change, emphasizing the importance of bottom-up involvement and the challenges posed by local policies, community readiness, and technological availability. Through a threefold methodology, the paper reviews European and Italian policies, Horizon 2020 projects, and successful European ECs, offering insights and lessons for future implementation pathways [20].

Collaborative Visioning and Goal Setting: Central to collaborative energy planning is the process of community visioning and goal setting. The U.S. Department of Energy outlines a structured approach to community energy strategic planning (CESP), which includes steps for developing a shared energy vision and setting goals based on the community's values and needs [21]. This process engages community members in a participatory manner, allowing them to collectively articulate their aspirations and priorities related to energy development. Through facilitated workshops, town hall meetings, or online platforms, ECs can collaboratively envision their ideal energy future. This shared vision forms the foundation for setting goals and objectives that align with the community, ensuring energy projects resonate with the broader aspirations of members.

Technological Integration and Innovation: Technology and innovation play crucial roles in collaborative energy planning [22]. This subsection explores how ECs can leverage technological advancements to enhance energy efficiency, optimize renewable energy generation, and facilitate community-wide monitoring and management. By examining case studies, communities can adopt cutting-edge solutions while ensuring these benefits are accessible to all members, promoting sustainability and inclusivity.

Policy Frameworks and Regulatory Advocacy: Collaborative energy planning extends beyond the community level, engaging broader policy frameworks and regulatory landscapes. This involves advocating for supportive policies at local, regional, and national levels to create an enabling environment for sustainable energy initiatives [23]. By analyzing successful advocacy strategies, this section highlights how ECs can influence the regulatory landscape, ensuring they have a voice in shaping the policies that govern their energy future.

Community-Owned Energy Projects: A hallmark of collaborative energy planning is the development of community-owned energy projects [24]. These initiatives, ranging from community solar installations to microgrids, empower communities to control their energy production and consumption. Examining case studies of successful community-owned projects shows how collaborative planning drives the realization of locally driven, sustainable energy solutions. Additionally, this subsection explores financing models and support mechanisms that help communities overcome financial barriers and ensure the economic viability of community-owned projects.

Education and Capacity Building: Effective collaboration in energy planning requires ongoing education and capacity building within ECs. This involves fostering a collective understanding of energy systems, renewable technologies, and the broader implications of sustainable energy choices. By examining successful educational initiatives, this section demonstrates how capacity building strengthens informed decision-making and enhances community participation in energy planning and implementation.

In conclusion, collaborative energy planning and implementation represent a holistic approach to sustainable energy development within ECs. By fostering collaborative visioning, integrating technology, advocating for supportive policies, promoting community-owned projects, and investing in education, ECs can navigate the complexities of planning and implementing projects that align with their values and priorities. This comprehensive approach ensures that the benefits of sustainable energy development are equitably distributed, contributing to the overarching goals of energy justice and equity.

11.4 Policy Interventions for Energy Justice

11.4.1 Frameworks for Inclusive Policy Development

Inclusive policy development stands as a cornerstone for advancing energy justice, recognizing the need to involve diverse stakeholders in shaping the trajectory of sustainable energy initiatives. As we move forward from broader discussions of energy justice, this subsection explores frameworks that prioritize inclusivity, community engagement, and justice within the policy-making process. These frameworks aim to create an enabling environment for equitable access to clean energy [25].

Central to these frameworks is the belief that energy policies should be crafted through collaboration, involving a wide range of voices. An essential approach is participatory decision-making, where community members, particularly those historically marginalized, actively help shape energy policies. This can be achieved through community consultations, advisory councils, or other platforms that facilitate meaningful engagement. By incorporating diverse perspectives, these frameworks ensure that policies align with the specific needs, values, and goals of the communities they affect.

In addition to participatory mechanisms, these frameworks stress the importance of transparency and accessibility. Clear, open communication **empowers communities to engage** in decision-making. This involves providing accessible educational resources, transparent policy documentation, and clear channels for community feedback. An informed community can better advocate for its interests and contribute to the development of just and equitable energy policies.

Addressing historical power imbalances is another critical element of inclusive policy frameworks. These frameworks aim to correct past exclusions by recognizing that certain communities have historically been excluded from decision-making processes. Targeted outreach, capacity-building initiatives, and affirmative action ensure the active participation of marginalized groups. By centering equity, these frameworks create a more level playing field, giving all communities an equal say in shaping their energy futures.

Expanding beyond local engagement, inclusive policy frameworks also acknowledge the interconnectedness of regional, national, and global energy systems. They recognize that the benefits and impacts of energy policies extend beyond the immediate community. Engaging with a wide range of stakeholders, including environmental advocates, indigenous groups, and regional authorities, ensures a holistic approach that addresses the multifaceted dimensions of energy justice.

In conclusion, inclusive policy development forms the foundation for advancing energy justice. By prioritizing community engagement, transparency, addressing historical inequities, and recognizing the interconnected nature of energy systems, these frameworks guide policymakers in crafting policies that not only advance sustainable energy goals but also contribute to a more just and equitable society.

11.4.2 Financial Mechanisms for Equitable Access

The pursuit of energy justice requires dismantling financial barriers and ensuring that sustainable energy benefits are accessible to all, regardless of socioeconomic status. As we shift from broader discussions on energy justice to its economic dimensions, this subsection examines financial mechanisms that promote equitable access to clean energy technologies and explores strategies that foster economic empowerment in marginalized communities.

A key starting point in this effort is the establishing subsidies and grants for vulnerable populations [26]. These financial tools reduce the upfront costs of adopting clean energy technologies, making them more accessible to low-income households. By examining successful subsidy programs and grant initiatives, this subsection identifies approaches that effectively prioritize and benefit historically disadvantaged communities, fostering an inclusive transition to sustainable energy.

Low-interest financing models also play a crucial role in expanding access to clean energy. Favorable financing terms allow individuals and communities with limited resources to invest in renewable energy projects. Policymakers and financial institutions can collaborate to design innovative financing solutions that address the economic challenges faced by marginalized groups, thereby democratizing access to clean energy and promoting economic resilience.

Community-based financial models offer another avenue for promoting financial inclusion. Mechanisms such as community investment funds and cooperative financing empower communities to invest in and own clean energy projects, keeping the economic benefits within the community. This subsection examines successful community-based initiatives to provide insights into how these models can foster local economic development and social cohesion, aligning with the principles of energy justice.

Furthermore, it is essential to recognize that financial mechanisms for equitable access must address the intersectionality of social and economic disparities. Tailoring financial instruments to account for the diverse needs of different communities – considering factors like race, ethnicity, and gender – is essential for avoiding unintended consequences and ensuring an inclusive transition to sustainable energy. Analyzing case studies that highlight the successes and challenges of intersectional financial approaches, this section contributes to the development of strategies that prioritize equity.

In conclusion, financial mechanisms for equitable access are critical for advancing energy justice. By examining and implementing various tools such as subsidies, grants, low-interest financing, and community-based financial models, policymakers can dismantle economic barriers and ensure that sustainable energy benefits are shared equitably. These financial strategies enhance access to clean energy technologies while contributing to broader economic empowerment and social equity within communities.

11.4.3 Regulatory Frameworks for Community-Owned Energy Projects

Regulatory frameworks play a pivotal role in shaping community-owned energy projects, providing the guidelines and support that communities need to take control [27]. As we move from financial mechanisms to the policy environment, this subsection examines how regulatory frameworks encourage and facilitate the development of community-owned energy projects, fostering local resilience, economic empowerment, and sustainable energy solutions.

Central to these frameworks is the recognition that traditional energy regulations often fail to align with the unique needs of community-based initiatives. To address this, regulatory frameworks strive to create an enabling environment by streamlining permitting processes, simplifying administrative hurdles, and tailoring regulations to suit the scale and nature of community-owned projects. By examining successful cases, policymakers can identify strategies that help communities to navigate the complexities of project development efficiently.

Building on this enabling environment, by examining successful cases, policymakers can identify strategies community ownership structures within the energy sector. This includes legal provisions that acknowledge community cooperatives, energy collectives, and other ownership models, ensuring they have clear legal status and can access the necessary support mechanisms. By establishing these legal pathways, regulatory frameworks contribute to the creation of a conducive environment for grassroots energy initiatives.

In addition to ownership structures, successful regulatory frameworks address fair compensation for energy contributions from community-owned projects. This involves implementing mechanisms such as feed-in tariffs or power purchase agreements that ensure communities receive just compensation for the energy they generate. By exploring examples of effective compensation models, policymakers can design regulations that promote economic sustainability and create a level playing field for community-owned projects in the broader energy market.

Grid integration is another critical element of these frameworks, which is essential for the scalability and viability of community-owned energy projects. Policies that facilitate the connection of these projects to existing grids support their success. By examining regulatory approaches that prioritize grid access and integration, communities can seamlessly become active participants in the larger energy ecosystem.

Additionally, it is important to note that these frameworks also balance the need for local autonomy with overarching regulatory goals. Striking a balance between community empowerment and adherence to broader energy policies ensures that community-owned projects meet sustainability objectives while meeting the unique needs of the communities they serve. By incorporating flexibility, policymakers can create an adaptive and supportive environment that encourages innovation and diversity in community-owned energy projects.

In conclusion, regulatory frameworks for community-owned energy projects are essential in fostering a transition to a more decentralized and community-driven energy landscape. By streamlining processes, recognizing ownership structures, ensuring fair compensation, and facilitating grid integration, these frameworks lay the foundation for communities to actively participate in and benefit from sustainable energy initiatives.

11.4.4 Energy Affordability and Inclusivity Policies

Ensuring energy affordability and inclusivity sits at the heart of energy justice, recognizing that the benefits of sustainable energy must be accessible to all members of society [28]. As we transition from broader themes of equity to specific policy measures, this subsection focuses on policies that address economic challenges, foster inclusivity, and contribute to dismantling barriers to clean energy access.

A fundamental component of these policies is energy affordability policies. Central to energy affordability policies is the recognition that the cost of energy must not disproportionately burden vulnerable and low-income communities. To achieve this, measures such as income-based energy assistance programs, utility bill assistance, and targeted subsidies for energy-efficient appliances aim to reduce the financial strain. By examining successful affordability policies, we see how governments and utilities can collaborate to ease the economic strain on marginalized households, ensuring clean energy remains affordable and accessible to those who need it most.

Building on affordability, inclusive rate structures offer another crucial approach to promoting energy justice. These structures go beyond traditional flat-rate billing and consider socioeconomic factors when determining energy costs. Time-of-use pricing, tiered rates, and income-based billing are examples of inclusive rate structures that can be tailored to the economic circumstances of different communities. Through analyzing case studies that illustrate the implementation and impact of inclusive rate structures, policymakers can discover ways to prioritize fairness and equity in energy pricing.

Complementing these rate structures are energy efficiency incentives that play a key role in energy affordability policies. Recognizing that reducing overall energy consumption can alleviate financial burdens, these incentives involve offering financial incentives, rebates, or tax credits for energy-efficient upgrades in homes and businesses. By exploring successful energy efficiency incentive programs helps identify strategies that not only promote affordability but also contribute to a more sustainable and resilient energy future.

Beyond financial strategies, community engagement in policy decision-making is an essential aspect of energy affordability and inclusivity policies. Engaging communities in the decision-making processes that shape these policies ensures that they align with the realities and priorities of the people they impact. Creating opportunities for inclusive dialogue, such as participatory planning and community-driven decision-making, helps policymakers design solutions that resonate with diverse communities, fostering ownership and fairness in the policy development process.

Furthering this inclusivity requires intersectional approaches, which must be adopted in energy affordability and inclusivity policies to consider the diverse needs and circumstances of different communities. These policies must account for race, ethnicity, gender, and geographic location, acknowledging that specific populations encounter unique barriers to accessing affordable and clean energy. By applying an intersectional approach to policy design, governments, and stakeholders can work to dismantle systemic inequalities and push forward energy justice for everyone.

In conclusion, energy affordability and inclusivity policies form essential parts of the energy justice framework. By addressing economic challenges, implementing inclusive rate structures, incentivizing energy efficiency, engaging communities in decision-making, and adopting intersectional approaches, policymakers help create a more just and equitable energy landscape. These policies not only improve access to clean energy but also empower marginalized communities economically, fostering a transition that prioritizes the well-being of all citizens.

11.4.5 Community Engagement in Policy Decision-Making

Community engagement in policy decision-making plays a crucial role in fostering energy justice, recognizing that the active participation of communities is essential for crafting policies that align with their values, needs, and aspirations [29]. Transitioning from broader energy justice principles, this subsection delves into the multifaceted aspects of community engagement, exploring mechanisms that empower communities to have a meaningful impact on the policies that shape their energy futures.

A key aspect of this process is inclusive decision-making. This involves creating spaces for dialogue, deliberation, and collaboration where community members can contribute their insights, concerns, and aspirations. By providing forums such as town hall meetings, community workshops, and participatory planning sessions, opportunities are created for diverse voices to be heard, ensuring that policy decisions are informed by a broad spectrum of perspectives.

Community engagement extends beyond consultation to genuine partnerships between policymakers and communities. These collaborative approaches involve codesigning policies with community members, recognizing their expertise and lived experiences. Policymakers can use community knowledge to develop more nuanced, context-specific policies, fostering a sense of ownership and shared responsibility for the outcomes.

Transparency is another critical element of community engagement. Frameworks emphasizing the importance of clear communication and information accessibility. By making policies clear and understandable to all community members and ensuring transparency in the decision-making process, communities can track policy progress and understand the factors shaping decisions. This transparency builds trust and strengthens the legitimacy of the process.

Moreover, addressing power imbalances within decision-making is essential. Policymakers must actively work to remove barriers that historically marginalized communities face. This may involve targeted outreach, capacity-building initiatives, and the establishment of support structures to ensure that all community members have an equal opportunity to participate and influence policy decisions.

In addition to addressing power dynamics, ongoing and iterative engagement is also crucial for adapting policies to evolving community needs. Regular feedback loops and mechanisms for continuous dialogue ensure that policies remain responsive to the dynamic nature of community priorities and challenges. Policymakers should view community engagement as an ongoing, collaborative process that evolves alongside the community.

In summary, community engagement in policy decision-making actively empowers communities to shape their energy futures. By fostering inclusive decision-making, building transparent communication channels, addressing power imbalances, and establishing ongoing engagement mechanisms, policymakers can ensure that energy policies reflect the diverse needs and aspirations of the communities they impact. This collaboration directly supports energy justice by positioning communities as active contributors and beneficiaries in the sustainable energy transition.

11.4.6 Addressing Environmental Justice Through Policies

Addressing environmental justice through policies is essential for ensuring that the benefits and burdens of sustainable energy development are equitably distributed, particularly considering historical disparities in environmental impacts [30]. As we transition from the broader framework of energy justice to its environmental dimension, this subsection explores policy interventions that aim to mitigate and rectify environmental injustices, promoting a sustainable and just energy transition that prioritizes both communities and the environment.

A critical starting point in this effort is implementing robust environmental impact assessments (EIAs) as part of the decision-making process for energy projects. Policymakers should use these assessments to evaluate the environmental, social, and health impacts of projects, particularly on vulnerable and communities. By integrating a thorough EIA process, policymakers can identify and address potential disproportionate environmental harms, ensuring that communities are not unfairly burdened by the negative consequences of energy projects.

In addition to EIAs, establishing emissions reduction targets is another critical strategy for addressing environmental justice. Policymakers must set ambitious goals for reducing greenhouse gas emissions and other pollutants, policymakers can contribute to mitigating climate change and improving air quality, with a specific focus on areas that have historically experienced higher pollution levels. This approach mitigates environmental harm and aligns with environmental justice principles, addressing past disparities in pollution exposure.

Another critical policy mechanism is zoning regulations that consider environmental justice concerns, which play a critical role in shaping the spatial distribution of energy projects. Policymakers can design zoning policies that prioritize the well-being of vulnerable communities, preventing the concentration of polluting industries and infrastructure in these areas. Careful zoning practices help create healthier environments, addressing historical patterns of environmental injustice.

Furthermore, community engagement should be integral to policies addressing environmental justice. Policymakers must actively involve communities in decision-making processes related to energy projects, ensuring that residents can voice their concerns and contribute local knowledge. Meaningful engagement leads to policies that better address specific environmental justice concerns.

In addition to addressing individual environmental concerns, policies should consider the cumulative impacts of multiple environmental stressors on communities. Environmental justice is not just about addressing individual sources of pollution but understanding the combined effects of various stressors on vulnerable populations. By adopting a holistic, intersectional approach enables policymakers to address the compounding impacts of environmental, social, and economic factors, leading to more comprehensive and effective interventions.

In conclusion, policies addressing environmental justice are crucial for creating a sustainable and just energy future. By integrating robust EIAs, setting emissions reduction targets, implementing equitable zoning regulations, prioritizing community engagement, and adopting an intersectional approach, policymakers can contribute to rectifying historical environmental injustices. These policies foster a more inclusive and environmentally sustainable energy transition.

11.5 Technological Innovations for Inclusive Energy

11.5.1 Accessible Clean Energy Technologies

Ensuring access to clean energy technologies for all communities, regardless of their economic or geographic circumstances, is a fundamental goal within the realm of technological innovations for inclusive energy [31]. As we shift from broader discussions of energy justice to the role of technology, this subsection explores strategies and advancements aimed at making clean energy technologies more accessible, thereby contributing to the overarching objectives of energy justice and equity.

One pivotal aspect of this approach involves the development and deployment of decentralized energy solutions. These technologies, such as off-grid solar systems and small-scale wind turbines, are designed to provide reliable and affordable energy to communities that may be off the traditional grid. By prioritizing decentralized solutions, technological innovations aim to bridge the energy access gap and empower communities to generate their own sustainable power, reducing dependence on centralized energy infrastructure.

Building on this, advancements in energy storage technologies further enhance access to clean energy. Improved energy storage solutions, such as advanced batteries and innovative storage materials, enable communities to store excess energy generated during peak production periods for use during times of low energy production. This mitigates the intermittency of renewable sources like solar and wind, ensuring a more consistent and reliable energy supply. Access to reliable energy storage is particularly beneficial for off-grid and remote communities, offering a stable power source independent of external factors.

In addition to storage solutions, smart grid technologies represent another avenue for enhancing the accessibility of clean energy. By integrating advanced communication and control systems, smart grids enable more efficient and responsive energy distribution. This ensures that clean energy is distributed where it is needed most, optimizing the use of renewable resources and reducing transmission losses. For underserved communities, smart grids improve energy reliability and affordability by optimizing renewable resource use and reducing transmission losses.

Moreover, financial innovations play a key role in making clean energy technologies more accessible. Community-based financing models, pay-as-you-go systems, and innovative microfinance initiatives help overcome financial barriers, allowing individuals and communities to adopt clean energy technologies without the need for significant upfront investments. This democratization of financing ensures that the economic benefits of clean energy are more widely distributed, aligning with the principles of energy justice.

User-friendly designs and adaptability are also key to accessibility. Technological innovations that prioritize simplicity, ease of maintenance, and compatibility with local contexts contribute to the widespread adoption of clean energy solutions. By acknowledging the diverse needs of different communities, these technologies ensure that clean energy solutions can be effectively utilized and maintained by a broad spectrum of users.

In summary, pursuing accessible clean energy technologies requires a multifaceted approach. This includes decentralized solutions, energy storage advancements, smart grid technologies, innovative financing models, and user-friendly designs. By addressing barriers related to both infrastructure and affordability, technological innovations aim to create an inclusive energy landscape where all communities can harness the benefits of clean and sustainable power.

11.5.2 Smart Grids and Energy Management Systems

Smart grids and energy management systems offer cutting-edge innovations that can significantly transform energy distribution and consumption [32]. As we shift from discussions on access to clean energy technologies to the role of infrastructure, this subsection delves into the comprehensive implications of these technologies, emphasizing their role in enhancing efficiency, reliability, and inclusivity within the energy sector.

At the core of these innovations is the fact that smart grids fundamentally redefine traditional energy distribution by integrating advanced communication and control systems [33]. These systems enable real-time monitoring, analysis, and optimization of energy usage, creating a more responsive and adaptable energy infrastructure. With the deployment of sensors, smart meters, and intelligent devices, the grid can collect granular data, offering insights into energy consumption patterns. This data-driven approach identifies inefficiencies and opportunities for optimization, contributing to a more efficient and resilient energy grid.

Smart grids also accommodate decentralized energy sources, seamlessly integrating renewable systems like solar panels and wind turbines. This flexibility enhances the reliability of the energy supply by reducing dependence on centralized power plants and mitigating disruptions. By supporting diverse energy sources, smart grids contribute to a more inclusive energy landscape.

In addition to grid flexibility, smart grids empower consumers by providing real-time information about their energy consumption. This transparency allows individuals and communities to make informed decisions about their energy usage, identify opportunities for energy conservation, and ultimately reduce their electricity bills. For vulnerable or low-income communities, this aspect of smart grids enhances energy affordability and supports the principles of energy justice by putting more control over energy consumption in the hands of consumers.

Energy management systems complement smart grids by using algorithms and artificial intelligence to optimize energy use, predict demand, and manage resources more effectively. These systems reduce waste and enhance grid stability, ensuring reliable energy access, especially in underserved or remote areas.

Furthermore, another critical component is the integration of demand–response mechanisms within smart grids and energy management systems, which fosters a more interactive and participatory energy landscape. Demand–response allows consumers to actively engage in shaping their energy consumption patterns by responding to price signals or grid conditions. This level of participation not only promotes energy efficiency but also empowers communities to play a more active role in the overall functioning of the energy system, aligning with the principles of community engagement and energy justice.

In conclusion, smart grids and energy management systems represent transformative technologies that have the potential to revolutionize the energy sector. By enhancing efficiency, accommodating decentralized energy sources, empowering consumers, and promoting community engagement, these innovations contribute to a more inclusive and resilient energy landscape. Policymakers and industry stakeholders must advance these technologies to address diverse communities' needs, ensuring that the benefits of smart grids and energy management systems are accessible to all.

11.5.3 Decentralized Energy Storage Solutions

Decentralized energy storage solutions emerge as pivotal innovations in creating a more accessible, reliable, and inclusive energy landscape [34]. As we transition from the broader discussion of smart grids and energy systems, this subsection explores the multifaceted dimensions of decentralized energy storage, elucidating its role in mitigating the intermittency of renewable sources, enhancing energy resilience, and fostering community empowerment.

One of the primary challenges in integrating RESs such as solar and wind into the energy grid is their inherent intermittency. Decentralized energy storage solutions overcome this by capturing excess energy during periods of high production and storing it for later use. With technologies like advanced batteries, flywheels, and compressed air storage systems, communities can store renewable energy locally, ensuring a continuous and reliable power supply even when the primary energy sources are not actively generating.

Decentralized storage also addresses energy access disparities, especially in remote or underserved areas. These solutions reduce dependence on centralized grids, enabling off-grid regions to benefit from localized energy storage. This fosters greater energy independence, ensuring a more equitable distribution of energy resources and supporting the resilience of isolated or marginalized areas.

Furthermore, beyond addressing energy access challenges, decentralized energy storage solutions contribute to the democratization of energy generation and consumption. Community-owned and community-managed storage systems empower residents to actively participate in shaping their energy landscape. This can manifest in the form of local energy cooperatives or shared energy storage facilities, where the benefits of stored energy are collectively managed and distributed among community members. By promoting local ownership and control, decentralized energy storage aligns with the principles of community engagement and economic empowerment.

Decentralized storage also mitigates the impacts of grid failures and natural disasters. Microgrids with storage can operate independently during emergencies, ensuring essential services remain operational. **This resilience is crucial for vulnerable areas**, supporting reliable energy access during crises.

As costs decrease, decentralized storage becomes more affordable and scalable. These advancements make storage solutions accessible to more communities, allowing for tailored systems that meet specific needs. This flexible approach ensures energy access adapts to each community's unique situation.

In conclusion, decentralized energy storage solutions play a transformative role in advancing an inclusive and resilient energy future. By addressing intermittency challenges, promoting energy independence, democratizing energy management, enhancing resilience during emergencies, and ensuring affordability and scalability, these technologies contribute to a more equitable distribution of energy benefits. Policymakers and industry stakeholders must continue to prioritize and invest in decentralized energy storage as a key enabler of a sustainable, just, and accessible energy transition.

11.5.4 Inclusive Technological Design and User-Centered Innovation

Inclusive technological design and user-centered innovation prioritize developing technologies tailored to meet the diverse needs of communities. As we move from a broader discussion of decentralized energy systems, this subsection explores the comprehensive implications of these strategies, emphasizing their role in creating technologies that are accessible, user-friendly, and aligned with the principles of energy justice.

At the core of these approaches is the recognition that diverse communities have varying requirements, capacities, and contextual considerations. This approach involves actively involving end users, including community members, in the design and development process of clean energy technologies. By engaging users throughout the process, innovators ensure that technologies are culturally sensitive, adaptable to local contexts, and effectively address the specific challenges faced by different communities. This collaborative design process democratizes technology, ensuring that the benefits of clean energy are accessible to all, regardless of socioeconomic status or geographic location.

Building on this engagement, user-centered innovation extends beyond initial design phases to embrace ongoing feedback loops and iterative improvements [35]. By establishing mechanisms for continuous engagement with end users, innovators can gather insights into the performance, usability, and impact of technologies in real-world scenarios. This iterative process refines technologies, evolving them to meet communities' changing needs. A user-centric mindset is crucial for promoting energy justice, empowering communities to shape technological advancements.

Inclusive design also emphasizes considering the full technology life cycle, from production to disposal. Sustainable and equitable technologies focus on environmental responsibility, ethical sourcing of materials, and proper waste management practices. By adopting a holistic approach, innovators minimize environmental impacts, contributing to a more just energy transition.

Moreover, fostering inclusivity in design also means addressing digital divides and ensuring that technologies are accessible to diverse user groups. This includes catering to individuals with varying levels of digital literacy, different language preferences, and physical or cognitive disabilities. User interfaces, manuals, and support systems must prioritize inclusivity, ensuring all users engage with clean energy technologies.

Furthermore, another crucial element is the recognition of affordability and economic accessibility in inclusive technological design. Clean energy technologies should be designed with cost considerations that align with the financial capacities of different communities. This involves exploring innovative financing models, such as pay-as-you-go systems or community-based financing, to ensure that the economic benefits of clean energy are equitably distributed.

In conclusion, inclusive technological design and user-centered innovation represent transformative approaches that prioritize the active involvement of communities in the development, improvement, and deployment of clean energy technologies. By considering cultural context, ongoing user feedback, environmental sustainability, accessibility, and economic affordability, innovators can create technologies that contribute to the principles of energy justice. This user-centric mindset ensures that clean energy technologies are not just accessible but actively shaped by and for the communities they serve, fostering a more equitable and inclusive energy future.

11.5.5 Mobile Technologies for Energy Access

Mobile technologies have emerged as powerful tools in bridging the energy access gap, offering innovative solutions that empower communities and individuals to engage with and benefit from clean energy resources. As we move from a discussion of broader technological solutions, this subsection explores the comprehensive impact of mobile technologies on energy access, emphasizing their role in democratizing information, facilitating energy transactions, and enhancing the efficiency of energy systems.

A key contribution of mobile technologies is the democratization of information. Mobile devices, particularly smartphones, provide individuals in both urban and remote areas with access to real-time information about energy usage, pricing, and availability. Through mobile applications, users can gain insights into their consumption patterns, empowering them to make informed decisions about their usage and adopt more sustainable practices. This access is critical for promoting energy literacy and enabling users to actively engage in shaping their energy consumption behaviors [36].

Beyond information, mobile technologies facilitate energy transactions, especially in regions with limited traditional banking services. Mobile payment platforms allow users to pay for energy services, such as electricity or cooking fuel, using their mobile devices. This innovation reduces the reliance on cash transactions – which can be challenging in remote or underserved areas – and enhances the affordability and accessibility of clean energy resources. In this way, mobile payments promote financial inclusion, ensuring participation in the clean energy economy even without traditional banking.

In addition to facilitating transactions, mobile technologies support the development and management of decentralized energy systems. For off-grid or rural areas where traditional energy infrastructure may be lacking, mobile-enabled microgrids or standalone solar systems can provide a reliable source of electricity. These systems can be monitored, controlled, and maintained remotely through mobile applications, minimizing downtime and ensuring efficient operation. The integration of mobile technologies in decentralized energy solutions enhances their scalability, adaptability, and ease of management, contributing to their widespread adoption.

Furthermore, mobile technologies enable the implementation of demand–response strategies within energy systems by sending users alerts or notifications about peak demand periods. This allows users to adjust their energy consumption in response to price signals or grid conditions, promoting more efficient energy use. This level of engagement empowers consumers to actively participate in the management of energy demand, contributing to overall grid efficiency and sustainability.

Inclusive technological design is a critical factor in leveraging mobile technologies for energy access. Ensuring that mobile applications and platforms are user-friendly, culturally relevant, and accessible to diverse populations is essential for maximizing their impact. By tailoring these technologies to the unique needs and preferences of different communities, mobile innovations foster greater adoption and active engagement, aligning with the principles of energy justice.

In conclusion, mobile technologies have become instrumental in advancing energy access by providing information, facilitating transactions, supporting decentralized energy systems, and enabling user participation in demand-response initiatives. Their ubiquity and versatility make them powerful tools for empowering individuals and communities to actively engage with and benefit from clean energy resources, contributing to a more inclusive and sustainable energy landscape. Policymakers and developers must continue prioritizing innovation at this intersection to ensure that the benefits are accessible to all.

11.6 Challenges and Opportunities of Sustainable and Equitable Energy Development

The pursuit of sustainable and equitable energy development is not without its challenges, but within these challenges lie opportunities for innovation, collaboration, and transformative change. As we delve deeper into this landscape, this section examines the multifaceted challenges and opportunities inherent in the journey toward energy justice and inclusive energy systems.

11.6.1 Economic Barriers and Financial Accessibility

Economic barriers and financial accessibility pose significant challenges to sustainable energy development. Building upon the broader theme of inclusivity in energy systems, this subsection builds on the theme of inclusivity in energy systems, exploring these challenges while identifying ways to mitigate them and foster economic inclusivity in the transition toward sustainable energy.

A key challenge lies in the upfront costs associated with adopting clean energy technologies. Despite the long-term economic benefits, the initial investment required for installing solar panels, energy storage systems, or other renewable energy infrastructure can be a substantial barrier, particularly for individuals and communities with limited financial resources. This financial hurdle disproportionately affects marginalized communities, perpetuating existing disparities in access to clean energy.

In addition to affordability, financial accessibility encompasses the availability of financing mechanisms that cater to diverse socioeconomic contexts. For many underserved areas, limited access to financing options inhibits the ability of individuals and communities to invest in sustainable energy solutions. Traditional financial institutions may perceive clean energy projects as high-risk, making it challenging for those without established credit histories or collateral to secure loans.

However, despite these challenges, there are opportunities for innovative financial models and policies. Inclusive financing mechanisms, such as community-based funds, microfinance initiatives, and pay-as-you-go systems, can address economic barriers by providing tailored solutions that align with the unique needs of different communities. By implementing these models, not only do clean energy technologies become more accessible, but local economic development is also fostered through community involvement in the financing and ownership of energy projects.

Moreover, policy interventions can play a pivotal role in enhancing financial accessibility. Subsidies, grants, and incentive programs targeted at low-income households reduce the financial burden of adopting sustainable practices. In doing so, policymakers can create an enabling environment for financial institutions to invest in clean energy projects and offer risk mitigation measures that attract private investment and stimulate economic growth in the sustainable energy sector.

To comprehensively address economic barriers, collaborative efforts between governments, financial institutions, and the private sector are essential [37]. Public–private partnerships can leverage the strengths of both sectors, developing innovative financing models, sharing risks, and scaling up sustainable energy initiatives. Additionally, by fostering financial literacy and education programs, communities can be empowered to navigate available financing options effectively, ensuring that economic barriers do not hinder their participation in the clean energy transition.

In summary, while economic barriers present substantial challenges, opportunities exist through inclusive financing models, targeted subsidies, and partnerships. These approaches can unlock economic opportunities, democratize access to clean energy technologies, and contribute to a more equitable and resilient energy future. Addressing these challenges requires a comprehensive and collaborative approach that considers the unique economic contexts of different communities.

11.6.2 Technological Accessibility and Inclusive Design

Technological accessibility and inclusive design are critical considerations in the pursuit of equitable and just energy transitions. As we explore the intersection of technology and energy justice, this subsection delves into the challenges and opportunities associated with ensuring that clean energy technologies are accessible to a broad spectrum of users, irrespective of socioeconomic status, physical abilities, or geographical locations.

Challenges: One of the primary obstacles to technological accessibility lies in addressing the digital divide. In many low-income or rural communities, limited access to technology, such as smartphones or computers, impedes the adoption of clean energy applications and solutions. This digital gap exacerbates existing inequalities, preventing communities from benefiting fully from innovations like smart grids and energy management apps. Therefore, bridging the digital divide requires strategic interventions, including infrastructure development, affordable technology access, and digital literacy programs tailored to diverse user groups.

Moreover, beyond physical infrastructure, considerations of technological accessibility must encompass user interface design and user experience. Without accessible interfaces, complex user journeys, or overcoming language barriers, digital environments can become exclusionary. For instance, individuals with disabilities may face challenges navigating applications that lack features catering to their specific needs. Inclusive design must incorporate features like voice controls, screen readers, and intuitive interfaces to ensure usability for diverse abilities.

Opportunities: While these challenges are significant, inclusive design presents a transformative opportunity to address technological accessibility challenges [38]. By designing clean energy technologies with diversity in mind, usability is enhanced, and a wide range of users can engage with these innovations effectively. This approach considers linguistic diversity, cultural context, and varying technical proficiency, fostering ownership and empowerment as users feel connected to their energy systems.

In addition to inclusive design, embracing open-source and collaborative development models can contribute to technological accessibility. Open-source solutions encourage community participation, allowing technologies to adapt to local contexts. By enabling collaborative development incorporates diverse perspectives, creating more responsive solutions while democratizing technology and knowledge sharing.

Furthermore, the rise of user-driven innovation and citizen science offers additional opportunities to bridge technological accessibility gaps. Engaging communities in codesign processes ensures technologies align with users' experiences and preferences. This participatory approach fosters agency, encouraging users to shape clean energy technologies based on their unique insights and needs.

In conclusion, addressing technological accessibility challenges in the clean energy sector requires a multifaceted approach. By prioritizing inclusive design, embracing open-source development, and fostering community-driven innovation unlocks opportunities for accessible, usable clean energy technologies. Stakeholders must collaborate to ensure innovations are not only advanced but also inclusive and accessible to all.

11.6.3 Policy and Regulatory Hurdles

Navigating policy and regulatory landscapes represents a pivotal challenge in advancing energy justice and ensuring the inclusive development of sustainable ECs [39]. As we delve into these complexities, this subsection explores the complex interplay of policy and regulatory hurdles, delineating both the challenges and opportunities embedded in these frameworks.

Challenges: One of the primary challenges lies in the fragmentation and inconsistency of energy policies across different jurisdictions. Given the diverse regulatory environments, varying regulations and standards at local, regional, and national levels can create uncertainties for energy developers, hindering the seamless deployment of sustainable energy solutions. This lack of uniformity particularly affects community-led projects, where the absence of standardized policies may impede their ability to secure financing, permits, and other necessary approvals.

In addition to fragmentation, another significant hurdle involves the often slow and rigid nature of regulatory processes. Traditional regulatory frameworks may struggle to keep pace with the dynamic and rapidly evolving landscape of clean energy technologies. Consequently, delays in approvals, permitting, and interconnection agreements can impede the timely implementation of projects, affecting the economic viability and overall success of sustainable energy initiatives. Streamlining regulatory processes and adopting adaptive policymaking are crucial for fostering a more agile energy transition.

Additionally, many regulatory frameworks fail to address the needs of marginalized communities. In some cases, existing policies may perpetuate inequalities, exacerbating energy poverty among low-income households. Achieving energy justice, therefore, necessitates a critical examination of existing regulations to identify and rectify potential inequities, ensuring that policies actively contribute to social and economic inclusivity.

Opportunities: Addressing policy and regulatory hurdles offers a transformative opportunity to create an enabling environment for sustainable ECs. By standardizing energy policies across jurisdictions, using best practices from successful case studies, can provide clarity for developers and investors. This harmonization enables the scaling of sustainable energy projects, fostering widespread adoption.

Furthermore, innovative policy frameworks present opportunities for incentivizing sustainable energy development. Embracing mechanisms such as feed-in tariffs, performance-based incentives, and community benefit agreements can encourage the development of projects that prioritize community engagement and equitable benefits. These strategies ensure that local communities actively participate in and benefit from the energy transition.

Policymakers can explore financial and regulatory tools that promote economic empowerment, social inclusion, and environmental sustainability concurrently.

Additionally, incorporating principles of energy justice into regulatory frameworks represents a crucial opportunity for transformative change. Policymakers can involve communities in decision-making and policy development, ensuring policies align with their needs and help dismantle historical disparities in energy access.

In conclusion, addressing policy and regulatory hurdles requires a strategic and collaborative approach. By standardizing energy policies, streamlining regulatory processes, and embracing innovative frameworks, opportunities arise to create a more supportive environment for sustainable ECs. Integrating energy justice principles into regulations ensures a more equitable and inclusive energy transition.

11.6.4 Social and Cultural Considerations

Understanding and integrating social and cultural considerations is paramount in the pursuit of sustainable ECs, as these factors profoundly influence the acceptance, adoption, and impact of clean energy technologies. As we explore this dimension, this subsection examines the intricate challenges and opportunities associated with social and cultural dimensions in the context of energy justice and equitable energy transitions.

Challenges: One of the primary challenges stems from cultural and social dynamics, which can introduce complexities that impact the acceptance and implementation of clean energy initiatives. For instance, resistance may arise due to cultural values, beliefs, or traditions that are incongruent with certain aspects of renewable energy projects. A specific example is when the construction of large-scale renewable energy infrastructure conflicts with cultural landscapes or sacred sites, leading to opposition from local communities. Effectively addressing these challenges requires understanding local cultures and engaging communities in a respectful manner.

In addition to cultural dynamics, social disparities, and unequal power dynamics within communities can influence the distribution of benefits and burdens associated with sustainable energy projects. Inclusive decision-making processes are critical to ensuring that the voices of marginalized groups are heard and considered. Without addressing these social inequalities, clean energy projects risk disproportionately impacting vulnerable populations, perpetuating existing disparities, and hindering the achievement of energy justice.

Furthermore, social acceptance of new technologies is influenced by factors such as education, awareness, and trust. In communities with limited awareness of clean energy technologies or skepticism about their benefits, initiatives may encounter difficulties in securing community support. To overcome these barriers, it is essential to build awareness and trust through effective communication and community engagement strategies, fostering positive social acceptance.

Opportunities: On the other hand, social and cultural considerations offer valuable opportunities for creating energy projects that are deeply embedded in the fabric of local communities [40]. By recognizing and respecting cultural values and social structures, community-led initiatives have the potential to build strong foundations for sustainable energy development. Engaging in meaningful dialogue with communities allows for the identification of shared values and the integration of cultural elements into project design, fostering a sense of ownership and connection.

Moreover, social and cultural considerations provide an opportunity to design inclusive education and awareness programs. These programs can demystify clean energy technologies, dispel misconceptions, and build a sense of collective responsibility for the transition to sustainable energy. Tailoring communication strategies to resonate with local cultures and languages enhances the effectiveness of educational efforts, empowering communities to make informed choices about their energy futures.

Another key opportunity lies in engaging with local leaders, cultural influencers, and community organizations. These stakeholders play pivotal roles in shaping community perceptions and facilitating dialogue between project developers and residents. Collaborative approaches that involve these influencers in the planning and decision-making processes can build bridges of understanding and trust, contributing to the success of sustainable energy initiatives.

In conclusion, social and cultural considerations are integral to the success of sustainable energy projects and the realization of energy justice. By recognizing the diversity of cultures within communities, fostering inclusive decision-making, and tailoring communication strategies to local contexts, stakeholders can overcome challenges and unlock the potential of social and cultural dimensions. By embracing these aspects, sustainable ECs can prioritize both environmental goals and the values of the people they serve.

11.6.5 Environmental Justice and Sustainability

The intersection of environmental justice and sustainability is a critical aspect of the transition toward sustainable ECs. This subsection will explore the challenges and opportunities embedded in ensuring that clean energy initiatives contribute to both environmental sustainability and equity.

Challenges: One of the main challenges lies in the fact that environmental justice concerns often arise when the benefits and burdens of clean energy projects are not distributed equitably across communities. In particular, vulnerable populations, including low-income communities and communities of color, may disproportionately bear the negative environmental impacts of energy projects, such as pollution, noise, or habitat disruption. As a result, this reinforces existing disparities in health and well-being, perpetuating environmental inequality.

In addition to these social concerns, there may be instances where the development of renewable energy infrastructure inadvertently leads to habitat disruption or biodiversity loss. For example, while large-scale solar or wind projects bring many benefits, they may also impact local ecosystems and wildlife. Balancing environmental sustainability with protecting biodiversity and habitats requires careful planning and community engagement to minimize adverse environmental impacts.

Opportunities: However, alongside these challenges come opportunities to redefine the relationship between communities and their environment. Adopting a holistic approach that integrates environmental, social, and economic considerations can help identify solutions addressing both environmental justice and sustainability.

One such opportunity is found in community-based renewable energy projects, which exemplify a pathway toward environmental justice and sustainability [41]. When communities are given the chance to actively participate in and have ownership of clean energy initiatives, there is a greater likelihood that projects will align with local environmental values and priorities. Empowering communities leads to renewable energy projects that minimize environmental harm while maximizing positive outcomes, fostering a more equitable energy transition.

Furthermore, there is also the opportunity to leverage renewable energy projects to address existing environmental injustices. By intentionally siting clean energy infrastructure in areas that have historically borne disproportionate environmental burdens, stakeholders can contribute to rectifying past injustices. This approach, often referred to as environmental justice mapping, involves identifying areas with a history of environmental injustice and prioritizing them for clean energy investments, creating a pathway toward environmental equity.

Incorporating these principles of environmental justice into policy frameworks and decision-making processes is crucial for ensuring that sustainability goals are aligned with equity considerations. Policymakers can contribute by implementing regulations that explicitly address the equitable distribution of both benefits and risks associated with clean energy projects. This could include conducting EIAs that consider the cumulative impacts on vulnerable communities and implementing mitigation measures to avoid or minimize negative consequences.

In conclusion, the intersection of environmental justice and sustainability presents both challenges and opportunities in the development of sustainable ECs. Addressing these challenges requires a comprehensive and inclusive approach that prioritizes equity in the distribution of environmental benefits and burdens. By empowering communities to actively participate in decision-making, adopting community-based renewable energy models, and integrating environmental justice principles into policy frameworks, stakeholders can contribute to a more just and sustainable energy future.

11.7 Dimensions of Sustainability in Energy Justice and Equity

The pursuit of sustainability within ECs is multifaceted, encompassing environmental, social, and economic dimensions that collectively contribute to the broader objectives of energy justice and equity. This section explores how these dimensions are interconnected, elucidating how each contributes to the creation of equitable and sustainable energy systems.

First, environmental sustainability serves as the foundation of ECs. This involves the prioritization of RESs such as solar and wind, which not only reduce reliance on fossil fuels but also minimize greenhouse gas emissions and ecological footprints. By focusing on environmental sustainability, ECs can ensure that future generations inherit a healthier planet. In this context, the section also delves into strategies for enhancing ecological resilience and sustainability through innovative technologies and practices that support biodiversity and reduce environmental degradation.

Building on the environmental focus, social sustainability is equally important. Social sustainability in ECs focuses on enhancing quality of life and ensuring that energy systems are fair, inclusive, and supportive of community needs. Through this lens, it involves actively engaging diverse community members in energy planning and decision-making, fostering social cohesion and empowerment. Moreover, this dimension addresses critical issues such as energy poverty and access to clean energy, ensuring that all community members benefit from energy advancements. The section will further explore how social goals can be integrated into energy projects, with discussions on job creation, health improvements, and educational opportunities.

Following from social sustainability, the economic dimension plays a crucial role in ensuring the long-term viability of energy initiatives. Economic sustainability pertains to the financial aspects of energy systems, emphasizing the creation of economic opportunities and the need to sustain energy projects financially over time. This section highlights how ECs can generate local jobs, stimulate economic development, and maintain financial health through models like community-owned projects and partnerships with local businesses.

Complementing the economic aspect, technological sustainability focuses on innovation. It involves developing and adopting advanced technologies that are both environmentally sustainable and economically feasible. By emphasizing the importance of technological advancements, this section explores innovations in energy storage, smart grids, and decentralized energy systems that can strengthen the resilience and efficiency of ECs. Furthermore, the section will highlight the significance of designing technologies that are accessible and adaptable to diverse community contexts, ensuring widespread usability and fostering ongoing improvement and innovation.

Lastly, governance and policy sustainability are essential for enabling the long-term success of ECs. This dimension centers on establishing strong regulatory and policy frameworks that promote the sustainable operation of energy systems. The section will outline how policies can promote equitable energy distribution, safeguard consumer rights, and ensure transparent and inclusive energy governance. Additionally, it will examine the role of both local and national policies in shaping sustainable energy systems and offer insights into aligning policies with the long-term goals of energy justice and equity.

11.8 Conclusion

The journey toward a sustainable and equitable energy future is a dynamic and collaborative endeavor that requires the concerted efforts of policymakers, industry leaders, communities, and individuals. As we delve deeper into this comprehensive exploration of sustainable energy development, energy justice, and inclusive practices, it becomes clear that the challenges and opportunities within the energy sector are multifaceted. First and foremost, environmental justice must be at the forefront of our collective efforts. In this context, rectifying historical injustices, minimizing the environmental impact of energy production, and prioritizing vulnerable communities are imperative steps toward building a sustainable foundation.

Transitioning from environmental concerns to the adoption of clean energy technologies, we recognize that embracing these innovations and fostering resilience in the face of climate change lays the groundwork for a future where the well-being of the planet and its inhabitants is prioritized. This chapter highlights the empowerment of groups through active participation in electricity choices, which in turn fosters better democratic governance and effectiveness.

Furthermore, moving from governance to the broader concept of equity, the chapter suggests improvements in strength justice and equity, underlining significant strides toward fair and inclusive electricity practices. The discussion then shifts to emphasize the role of progressive policy frameworks and technological interventions in supporting these changes, paving the way for a sustainable and equitable energy future.

At the core of this technological transition, technological inclusivity emerges as a cornerstone for success. This emphasizes the importance of designing clean energy solutions that are accessible, user-friendly, and culturally relevant. By intertwining technological advancements with societal needs, we ensure that the pursuit of a sustainable energy future becomes an ongoing process of learning, adaptation, and collaboration. Embracing environmental justice, technological inclusivity, community engagement, and inclusive policy frameworks will transform clean energy from a technological achievement to a societal transformation. Through collective action and a commitment to justice, we can create a resilient, inclusive, and sustainable energy landscape that benefits current and future generations.

Acronyms

EC energy community
EEP energy efficiency program
EIA environmental impact assessment
RES renewable energy source

References

1 Sefa-Nyarko, C. (2024). Ghana's National Energy Transition Framework: domestic aspirations and mistrust in international relations complicate 'justice and equity'. *Energy Research & Social Science* 110: 103465.

2 Obiora, S.C., Bamisile, O., Hu, Y. et al. (2024). Assessing the decarbonization of electricity generation in major emitting countries by 2030 and 2050: transition to a high share renewable energy mix. *Heliyon* 10: 28770.

3 Patel, N. (2024). Performing energy justice futures: how visions of just futures shape discourses and practices in the United Kingdom's community energy sector. *Energy Research & Social Science* 111: 103475.

4 Sareen, S., Smith, A., Gantioler, S. et al. Social implications of energy infrastructure digitalisation and decarbonisation. *Buildings and Cities* 4 (1): 612–628.

5 Llorca, M. and Rodriguez-Alvarez, A. (2024). Economic, environmental, and energy equity convergence: evidence of a multi-speed Europe? *Ecological Economics* 219: 108133.

6 Oyedepo, S.O., Babalola, O.P., Nwanya, S.C. et al. (2018). Towards a sustainable electricity supply in Nigeria: the role of decentralized renewable energy system. *European Journal of Sustainable development research.* 2 (4): 40.

7 Siciliano, G., Wallbott, L., Urban, F. et al. (2021). Low-carbon energy, sustainable development, and justice: towards a just energy transition for the society and the environment. *Sustainable Development* 29 (6): 1049–1061.

8 Jenkins, K., McCauley, D., Heffron, R. et al. (2016). Energy justice: a conceptual review. *Energy Research and Social Science* 11: 174–182.

9 Gui, E.M. and MacGill, I. (2018). Typology of future clean energy communities: an exploratory structure, opportunities, and challenges. *Energy Research and Social Science* 35: 94–107.

10 Dorahaki, S., Rashidinejad, M., MollahassaniPour, M. et al. (2023). A sharing economy model for a sustainable community energy storage considering end-user comfort. *Sustainable Cities and Society* 97: 104786.

11 Surya, B., Suriani, S., Menne, F. et al. (2021). Community empowerment and utilization of renewable energy: entrepreneurial perspective for community resilience based on sustainable management of slum settlements in Makassar City, Indonesia. *Sustainability* 13 (6): 3178.

12 Batidzirai, B., Trotter, P.A., Brophy, A. et al. (2021). Towards people-private-public partnerships: an integrated community engagement model for capturing energy access needs. *Energy Research and Social Science* 74: 101975.

13 Romero-Lankao, P., Rosner, N., Efroymson, R.A. et al. (2023). Community Engagement and Equity in Renewable Energy Projects: A Literature Review. Office of Scientific and Technical Information (OSTI).

14 Suryawan, I.W. and Lee, C.H. (2023). Community preferences in carbon reduction: unveiling the importance of adaptive capacity for solid waste management. *Ecological Indicators* 157: 111226.

15 Bickerstaff, K., Walker, G., and Bulkeley, H. (ed.) (2013). *Energy Justice in a Changing Climate: Social Equity and Low-Carbon Energy*. Bloomsbury Publishing.

16 Kojonsaari, A.-R. and Palm, J. (2021). Distributed energy systems and energy communities under negotiation. *Technology and Economics of Smart Grids and Sustainable Energy* 6 (1): 17.

17 Lennon, B., Dunphy, N.P., and Sanvicente, E. (2019). Community acceptability and the energy transition: a citizens' perspective. *Energy, Sustainability and Society* 9 (1): 1–8.

18 Swarnakar, P. and Singh, M.K. (2022). Local governance in just energy transition: towards a community-centric framework. *Sustainability* 14 (11): 6495.

19 Sedlacek, S., Tötzer, T., and Lund-Durlacher, D. (2020). Collaborative governance in energy regions–experiences from an Austrian region. *Journal of Cleaner Production* 256: 120256.

20 Boulanger, S.O.M., Massari, M., Longo, D. et al. (2021). Designing collaborative energy communities: a European overview. *Energies* 14 (24): 8226.

21 U.S. Department of Energy. (2013) Guide to Community Energy Strategic Planning. Energy.gov. https://www.energy.gov/scep/slsc/guide-community-energy-strategic-planning (accessed 14 September 2024).

22 Dall-Orsoletta, A., Romero, F., and Ferreira, P. (2022). Open and collaborative innovation for the energy transition: an exploratory study. *Technology in Society* 69: 101955.

23 Tzankova, Z. (2020). Public policy spillovers from private energy governance: new opportunities for the political acceleration of renewable energy transitions. *Energy Research and Social Science* 67: 101504.

24 Cuesta-Fernandez, I., Belda-Miquel, S., and Calabuig, T.C. (2020). Challengers in energy transitions beyond renewable energy cooperatives: community-owned electricity distribution cooperatives in Spain. *Innovation: The European Journal of Social Science Research.* 33 (2): 140–159.

25 Ciplet, D. (2021). From energy privilege to energy justice: a framework for embedded sustainable development. *Energy Research and Social Science* 75: 101996.

26 Mgaiwa, S.J. and Ishengoma, J.M. (2023). Financing higher education in Tanzania through students' loans scheme and its impact on equitable access. *Heliyon* 9 (3): e13943.

27 Byrne, J., Nyangon, P.D., Hegedus, S. et al. (2020). Navigating the Changing Landscape of Community Solar in Delaware: Policy Designs and Governance Frameworks to Support Community-Owned Sustainable Energy. Center for Energy & Environmental Policy Technical Report.

28 Berka, A.L., MacArthur, J.L., and Gonnelli, C. (2020). Explaining inclusivity in energy transitions: local and community energy in Aotearoa New Zealand. *Environmental Innovation and Societal Transitions* 34: 165–182.

29 Thirlwall, C. and Whitelaw, S. (2020). Utilising community engagement approaches to influence public mental health policy in a rural setting. *Journal of Public Mental Health* 19 (2): 169–178.

30 Boda, P.A., Fusi, F., Miranda, F. et al. (2023). Environmental justice through community-policy participatory partnerships. *Journal of Environmental Protection* 14 (8): 616–636.

31 Alola, A.A., Ozturk, I., and Bekun, F.V. (2021). Is clean energy prosperity and technological innovation rapidly mitigating sustainable energy-development deficit in selected sub-Saharan Africa? A myth or reality. *Energy Policy* 158: 112520.

32 Mishra, P. and Singh, G. (2023). Energy management systems in sustainable smart cities based on the internet of energy: a technical review. *Energies* 16 (19): 6903.

33 Refaat, S.S., Ellabban, O., Bayhan, S. et al. (2021). *Smart Grid and Enabling Technologies*. Wiley.

34 Idries, A., Krogstie, J., and Rajasekharan, J. (2022). Challenges in platforming and digitizing decentralized energy services. *Energy Informatics* 5 (1): 1–29.

35 Miesenberger, K., Edler, C., Dirks, S. et al. (2020). User centered design and user participation in inclusive R&D: introduction to the special thematic session. *Computers Helping People with Special Needs: 17th International Conference, ICCHP 2020*, Lecco, Italy (9–11 September 2020), Proceedings, Part I 17 2020, 3–9. Springer International Publishing.

36 Kato, I., Mlay, S., Nantege, Z. et al. (2020). Mobile technologies, digital financing and access to solar energy in Uganda. *International Journal of Technology Management* 5 (2): 1–23.

37 Solangi, Y.A., Longsheng, C., and Shah, S.A. (2021). Assessing and overcoming the renewable energy barriers for sustainable development in Pakistan: an integrated AHP and fuzzy TOPSIS approach. *Renewable Energy* 173: 209–222.

38 Patrick, V.M. and Hollenbeck, C.R. (2021). Designing for all: consumer response to inclusive design. *Journal of Consumer Psychology* 31 (2): 360–381.

39 D'Alpaos, C. and Andreolli, F. (2021). Renewable Energy Communities: the challenge for new policy and regulatory frameworks design. In: *New Metropolitan Perspectives: Knowledge Dynamics and Innovation-driven Policies Towards Urban and Regional Transition*, vol. 2, 500–509. Cham: Springer.

40 Caramizaru, A. and Uihlein, A. (2020). *Energy Communities: An Overview of Energy and Social Innovation*. Luxembourg: Publications Office of the European Union.

41 McGregor, D., Whitaker, S., and Sritharan, M. (2020). Indigenous environmental justice and sustainability. *Current Opinion in Environment Sustainability* 43: 35–40.

12

Electrification of Urban and Rural Areas Under Energy Justice and Equity Paradigms

Marziyeh Razeghi, Hossein Yousefi, Amir Naseri, and Younes Noorollahi

School of Energy Engineering and Sustainable Resources, College of Interdisciplinary Science and Technology, University of Tehran, Tehran, Iran

12.1 Introduction

As urban populations continue to grow and rural areas face ongoing neglect, the urgent need for equitable energy strategies becomes increasingly apparent. This research addresses a critical gap in understanding how demographic shifts, particularly in low- and middle-income countries, influence energy justice and access. While previous studies have predominantly focused on urban energy challenges, this work broadens the discussion by exploring the disparities in energy access between rural and urban populations, along with the resulting environmental impacts.

This research introduces innovative solutions for improving energy access in rural areas, addressing challenges stemming from demographic shifts and energy justice concerns. The primary innovation of this study lies in offering affordable and sustainable strategies to enhance energy access in low- and middle-income countries. For the first time, this research provides a comprehensive analysis of how a country's income levels affect energy access and distribution, using precise demographic and energy data. Furthermore, it identifies key factors that hinder energy access.

By emphasizing the importance of developing comprehensive energy strategies that address the distinct challenges of both rural and urban populations, this study makes a significant contribution toward achieving sustainable development and equitable energy access for all. The objectives of this research are to highlight the challenges of energy provision in the context of rapid urbanization and to propose actionable strategies that bridge the energy justice gap. In doing so, this work sets a new benchmark for comprehensive energy policymaking, one that is attuned to

Communities for Clean Energy Justice and Equity in Grid Modernization, First Edition.
Edited by Mohammadreza Daneshvar, Behnam Mohammadi-Ivatloo, and Amjad Anvari-Moghaddam.

the intricacies of geographical and income disparities, ultimately contributing to global efforts toward sustainable development and equitable energy access for all.

12.2 Energy Justice

Before elucidating the concept of energy justice, it is essential to first define the two distinct concepts of "justice" and "energy." Afterward, the amalgamation of these two domains, referred to as "energy justice," will be elaborated upon.

12.2.1 Justice

The discourse and analysis of justice boast a historical lineage spanning over two millennia. As a moral concept, justice predates even the eras of Plato and Aristotle, and it has persisted as a perennial subject of contemplation from ancient civilizations to the contemporary 21st century [1]. In ancient Greece, justice was synonymous with virtuous living, yet paradoxically, the institution of slavery remained prevalent, leading to varied interpretations of justice across societal strata. For instance, modern libertarians define justice as the autonomy of individuals in their personal choices, advocating for minimal governmental intervention. Conversely, social philosophers interpret justice as the pursuit of welfare equality, while Christians perceive it as a divine mandate overseeing human conduct and morality [2]. European philosophers of the 18th and 19th centuries, such as Thomas Hobbes, postulated justice as emanating from "natural law," akin to the immutable laws of physics or gravity, comprising a set of moral precepts and principles [3, 4]. Additionally, some believe that justice is essentially the laws and punishments determined by judicial systems to prevent crime, equating justice with the criminal justice system in most countries. However, others define justice as individual freedom and the ability of every human being to pursue and obtain their rights. Another interpretation of justice focuses on the fair distribution of goods and services, or in other words, "fairness." All these interpretations share a common feature: if justice is applied on a broader scale than the individual, it prevents feelings of humiliation in those whose personal desires are not fulfilled. This is because they recognize that justice applies universally, and rejection of a personal request does not signify personal humiliation, as others would be treated the same way under similar circumstances. In this sense, justice contributes to mental health and a sense of security [5]. The concept of justice, as viewed by social justice theorists, revolves around the equitable distribution of resources and opportunities. This concept is encapsulated in the idea of distributive justice. For instance, John Rawls, often regarded as the pioneer of social justice theory, defines justice as the fair allocation of "primary goods," which encompass rights, freedoms, opportunities, income, and wealth. Similarly, Michael J. Sandel, another prominent justice theorist, echoes Rawls' sentiments, emphasizing the importance of ensuring equitable access to resources and opportunities for all members of society, without infringing upon anyone's rights [6]. Consequently, both public and private decision-makers are urged to adopt an impartial stance and enact policies that promote fairness and equality.

Upon examining the perspectives of various experts in the realm of social justice, it becomes evident that justice comprises four fundamental elements: (1) access to information, (2) meaningful participation in decision-making processes, (3) impartiality of decision-makers, and (4) access to legal recourse for seeking redress [7]. According to the mentioned materials, there are three types of justice that are all interrelated, environmental justice, modern justice, and social justice or equality of species, which will be explained in the following. Modern justice is actually derived from environmental justice, because in environmental justice, both individual and social dimensions are considered, and this also exists in modern justice [8, 9], and in fact, the concept of environmental justice was first stated and then the concept of modern justice was raised. Environmental justice, according to Gordon Walker, a justice theorist, has two main features: (1) One aspect pertains to the differential consumption of certain vital environmental resources at the detriment of others, highlighting an imbalance in resource utilization. (2) Another facet concerns the unequal distribution of power to effect change and influence decision-making processes [10, 11]. Additionally, there exists another dimension of justice known as social justice or "equality between species," which underscores the importance of fostering equitable relationships between humans and the natural world. This concept explores the intersection of environmental justice and contemporary understandings of justice, highlighting humanity's interactions with nonhuman entities [12].

Types of Justice

Within another category, three types of justice emerge: distributive justice, recognition justice, and procedural justice. These are delineated below.

Distributional Justice Distributive justice encompasses the equitable allocation of both environmental benefits and burdens, acknowledging disparities in the distribution of associated responsibilities [13]. It examines the intricate interplay between the desirability of technologies and their localized impacts [14], advocating for the fair distribution of benefits and burdens across society, regardless of socioeconomic status. Given the inevitability of certain resource distributions (e.g. wind resource availability), claims of justice necessitate not only evidence of inequality but also a rationale for promoting fairer practices [15]. Consequently, empirical research may focus on evaluating the extent to which the placement of energy infrastructures contributes to distributive injustices. Moreover, distributive justice extends beyond infrastructure placement to encompass fair access to energy services [16].

Recognition Justice The concept of recognition justice extends beyond mere tolerance and acceptance; it encompasses the notion that individuals should enjoy equitable representation, freedom from physical intimidation, and full political rights [16–18]. Failure to recognize individuals can manifest in various forms of cultural and political dominance, leading to feelings of insult, humiliation, and marginalization. It encompasses not only identification but also, misrecognition — the distortion of individuals' perspectives in a manner that belittles or disparages them [16–18]. Hence, recognition justice advocates for the validation of diverse viewpoints stemming from social, cultural, ethnic, racial, and gender disparities [16, 19]. Incorporating the recognition perspective, Heffron [20] explicitly identify recognition justice as the third pillar of energy justice.

Procedural Justice Procedural justice pertains to the fairness of decision-making processes governing the aforementioned distributions. It advocates for procedures that engage all stakeholders in a nondiscriminatory manner [13], and has become synonymous with civil rights movements advocating for political inclusion across North America [16, 21]. Its application has expanded to various dimensions including socioeconomic status [22], gender [23], and religion [24]. Procedural justice depends on access and oversight from multilevel legal systems [25], while also being shaped by nonlegal influences such as cultural practices, norms, values, and behaviors [16, 26].

12.2.2 Energy

The term "energy" first appeared in English literature during the 16th century, courtesy of William Blake, though at that time, it lacked a scientific connotation and simply referred to power. However, by the 19th century, the concept of "energy" began to align more closely with its contemporary definition, as natural philosophers utilized the term to elucidate phenomena such as heat transfer and planetary motion. Over time, it evolved into the scientifically recognized definition: the capacity to perform work or resist force. Presently, "energy" has distinct meanings across various disciplines and specialties. Newtonian physicists, for instance, define energy within the framework of Newton's first law, asserting its conservation as it transitions between states. In engineering and science, "primary energy" refers to the energy inherent in natural resources like crude oil, sunlight, or wind, which can be harnessed for use. Conversely, "final energy consumption" measures the energy utilized by end consumers, such as gasoline or electricity, sourced from primary energy and delivered via specific processes. This consumed energy transforms into "useful energy" or "energy services," manifesting as heat, work, or temperature changes, enabling activities like transportation or heating. Energy services include the infrastructure supporting extraction, transportation, and agricultural industries, contributing to global welfare. Moreover, the term "energy" frequently appears in everyday language, reflecting experiences like lethargy. Perspectives on energy vary based on one's role and context; consumers prioritize its role in satisfying daily needs, politicians emphasize "energy security" and political ramifications, while engineers focus on efficiency and waste reduction. Thus, individuals' societal roles significantly shape the definition and utilization of energy, highlighting its extensive social dimensions [27].

According to the definitions of the concepts of "justice" and "energy," energy justice is understood as a combination of these concepts. In other words, energy justice can be seen as a subset of the three aforementioned justices and is defined as a global energy system in which all energy services and benefits must be equitably available to all members of society, with costs fairly distributed. Additionally, it encompasses both individual and social dimensions. On the consumption side, individuals are accountable to both themselves and society, while on the production and distribution side, an impartial decision-making system must ensure equitable access to energy for all. This system is also responsible to society and to itself. In essence, energy distribution and consumption carry both individual and social responsibilities, reflecting the principles of modern and environmental justice [5, 28]. For the realization of energy justice in society, the following key issues demand attention:

1) The unequal burden of costs, risks, and externalities imposed by the energy system disproportionately affects economically disadvantaged and marginalized communities.
2) The distribution of benefits and access to modern energy systems and services remains inequitable, exacerbating social disparities.
3) Reforming energy policies and projects is essential, as current decision-making processes often lack inclusivity and adequate representation.

By addressing these three critical issues, energy justice can be achieved, promoting social welfare and happiness. Transparent and accountable legal frameworks must be established, accessible to both consumers and producers. Moreover, adhering to energy justice principles will enhance environmental justice, ensuring the fair distribution of environmental and social risks associated with energy production and consumption. This fosters equitable access to energy systems, services, and decision-making processes, while empowering stakeholders with the necessary information and opportunities for meaningful

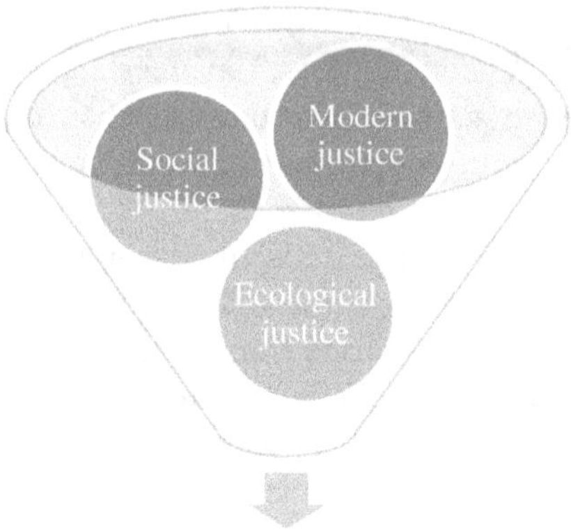

Figure 12.1 An overview of the principle of equitable energy distribution and its integral elements.

participation [5, 28]. Figure 12.1 illustrates the role and significance of energy justice within the broader framework of modern, social, and ecological justice.

As previously mentioned, energy justice encompasses multiple dimensions, including procedural, distributive, recognition, and cosmopolitan aspects. This highlights that energy justice is more than just a unified and composite concept. Theories and concepts surrounding energy justice can be categorized into eight distinct groups, with four rooted in classical theorists' perspectives and four derived from modern theorists. The classical categories include virtue, utility, human rights, and procedural justice, while the modern categories incorporate welfare, freedom, generational justice, and responsibility [5]. A summary of these concepts, along with the challenges that exist in relation to each, is provided in Table 12.1.

12.3 Energy Justice as a Decision-Making Tool

Due to the complex nature of energy systems, various factors must be carefully considered in decision-making, as energy is a multifaceted domain. If a decision-maker fails to fully recognize all its dimensions, this could lead to significant issues in the areas of energy justice, environmental concerns, and people's access to energy resources. To address this challenge, decision-making tools should serve as a roadmap for guiding effective decisions. In the energy sector, the following eight factors are essential for sound decision-making. Decision-makers must ensure these aspects are diligently observed: (1) accessibility, (2) cost-effectiveness, (3) procedural fairness, (4) effective management, (5) sustainability, (6) fairness across generations, (7) fairness within the current

Table 12.1 Analytical applications of energy justice for energy issues.

Solutions	Injustices	Utilizations related to energy	Significant philosophical impact(s)	Concept(s)	Topic
Standards for fuel economy, labeling for energy efficiency, retrofits for industries, demand-side management for utilities, pricing with ascending block rate, smart grids and advanced metering, capacity building and training, awareness and education for consumers	Shortcomings in the provision, transformation, allocation, and consumption of energy	Energy efficiency: high proportion of efficient service	Plato and Aristotle	Virtue	Energy efficiency
Tax on carbon, tax shifting and accurate price signals, and bonds for environment	Adverse societal and ecological impacts such as traffic gridlock, the mining sectors linked to energy generation, nuclear waste, air contamination, greenhouse gas release, and water usage	Well-being: reduced harm, suffering, external factors, and catastrophes associated with energy creation and consumption	Jeremy Bentham, John Stuart Mill, Henry Sidgwick	Utility	Energy externalities
Transparency initiatives for the extraction sectors, establishment of truth commissions and review boards for energy matters, enhanced evaluations of the social and environmental impacts of energy ventures, and the provision of legal assistance for disadvantaged communities	Infringement of human rights — including, in severe instances, fatalities and internal conflicts — carried out in the quest for energy resources and technologies, along with the role of energy generation in exacerbating armed disputes	Global human rights: an obligation to safeguard human rights in the generation and utilization of energy	Immanuel Kant	Human rights	Human rights and societal strife

(Continued)

Table 12.1 (Continued)

Solutions	Injustices	Utilizations related to energy	Significant philosophical impact(s)	Concept(s)	Topic
Enhanced dissemination of information, engagement, and inclusion of the wider community	Methods of energy location planning that overlook or violate principles of free, equitable, and knowledgeable consent, and/or fail to carry out comprehensive assessments of social and ecological impacts	Procedural fairness: unhindered advance knowledgeable approval for the placement of energy initiatives; equitable participation in the decision-making process of energy	Edward Coke, Thomas Jefferson, Jürgen Habermas	Procedural justice	Energy and procedural fairness
Initiatives for community-oriented pricing and support, along with collaborative ventures between the public and private sectors aimed at aiding the economically disadvantaged through micro-hydro installations, residential solar setups, enhanced cooking stoves, biogas production systems, and compact wind generators, as well as mechanical power for water pumping, crop watering, and farm produce processing	Shortfalls in the availability of electricity and modern technology, reliance on conventional solid fuels for meal preparation, and the laborious tasks of gathering fuelwood and water, as well as food processing in developing nations, predominantly carried out by women and children	Equitability and sustainability: an energy framework that ensures equitable opportunities for individuals to obtain the necessary energy, and energy infrastructures that foster economic growth and enhance livelihoods	John Rawls, Amartya Sen, Martha Nussbaum	Welfare and happiness	Energy poverty

Removal of unsuitable financial support measures, evaluation of the effects of subsidies, provisions for their expiration, and compensatory schemes for individuals reliant on such subsidies	Substantial subsidies that result in the nonconsensual redistribution of wealth to beneficiaries, effectively depleting the finances of those who have not agreed to such transactions	Libertarianism: choices regarding energy that are not excessively constrained by governmental regulations	Robert Nozick, Milton Friedman	Freedom	Energy subsidies
Enhanced energy conservation, creation of sovereign wealth funds, and implementation of large-scale renewable power and biofuel technologies	Depletion of finite energy sources and fuels	Resource fairness: a duty to reduce the use of resources and maintain sufficient supplies for the prosperity of coming generations	Ronald Dworkin, Brian Barry, Edith Brown Weiss	Posterity	Energy resources
Greenhouse Growth Entitlements, community-centric adaptation, and mitigation via stabilization segments	A formidable array of adverse effects stemming from climate change, encompassing ocean acidification, food instability, climate-induced displacement, and a rise in both the occurrence and intensity of natural and humanitarian crises	Intergenerational fairness: a duty to shield upcoming generations from the detrimental effects associated with energy	Peter Singer, Henry Shue, Paul Baer, Stephen M. Gardiner, Dale Jamieson, Simon Caney	Fairness, responsibility, and capacity	Climate change

Source: Adapted from [5].

Table 12.2 Energy justice decision-making tool.

Explanation	Principle
Individuals ought to have access to ample high-quality energy supplies.	Availability
Every individual, regardless of their economic status, should allocate no more than 10% of their earnings to energy services.	Affordability
It is essential for nations to maintain the principles of due process and human rights throughout the production and utilization of energy resources.	Due process
Every individual is entitled to obtain superior information regarding energy and environmental matters, along with equitable, lucid, and responsible approaches to making decisions about energy.	Good governance
The consumption of energy resources should be managed at a sustainable pace.	Sustainability
It is imperative that every person be granted equitable access to energy services.	Intragenerational equity
The right of future generations to lead a prosperous life, unhampered by the detrimental effects our current energy systems impose on the planet, must be upheld.	Intergenerational equity
Every country bears the duty to safeguard the natural ecosystem and mitigate the environmental hazards associated with energy.	Responsibility

Source: Adapted from [5].

generation, and (8) accountability [5]. Detailed explanations of these components are provided in Table 12.2.

Figure 12.2 shows the decision-making process in energy justice according to the effective components in energy justice decision-making.

12.4 Energization

There are three primary solutions for supplying energy to different regions: (1) Grid expansion, (2) Mini-grids, and (3) Decentralized standalone systems [31–33]. Solutions (1) and (3) are economically viable when there is a sufficient population density, which are most commonly implemented in urban areas where population concentration is higher. In contrast, rural areas are recognized for their potential to contribute to carbon reduction and mitigate the impacts of climate change due to their natural landscapes and abundant resources [31, 34, 35]. Furthermore,

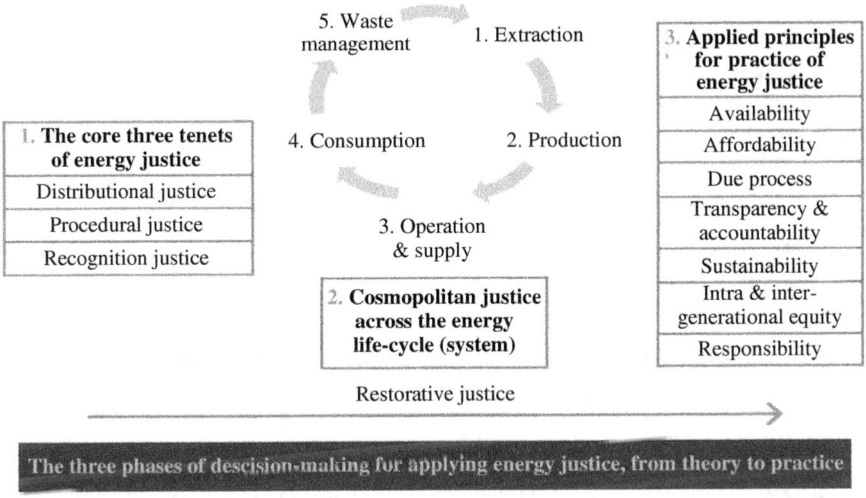

Figure 12.2 The transition of energy justice from a conceptual approach to actionable measures is depicted [29, 30].

the significance of rural areas as key sites for climate change adaptation and the reduction of its detrimental effects has been increasingly acknowledged [31, 34, 35]. For instance, Kitchen and Marsden highlight rural areas as "spaces that should contribute to adaptation and mitigation of climate change by using their renewable energy resources" [31]. Despite this recognition, the concept of a "rural energy transition" has seldom been explored in scientific literature [36]. However, only a limited number of studies have directly addressed the role of rural regions in the energy sector. Most research focuses on urban energy issues, while research on rural energy has concentrated more on topics like sustainable development within rural environments, neglecting the critical role of energy in these spaces [31]. This oversight underscores the need to clearly define the contribution of rural areas in energy provision and transmission, positioning them as a "fertile academic frontier" [37]. Despite the crucial environmental role of villages, unfortunately, the population residing in rural areas has been declining. According to statistics published by the World Bank, from 1960 to 2007, the population living in rural areas globally exceeded that of urban areas. However, after 2007, the rural population began to decrease relative to the urban population, a trend depicted in Figure 12.3.

This section delineates the issue of dwindling rural populations across various nations by categorizing countries into high-, middle-, and low-income groups. It separately assesses the populace residing in urban versus rural zones within these classifications. As depicted in Figure 12.4, it is evident that from 1960 through

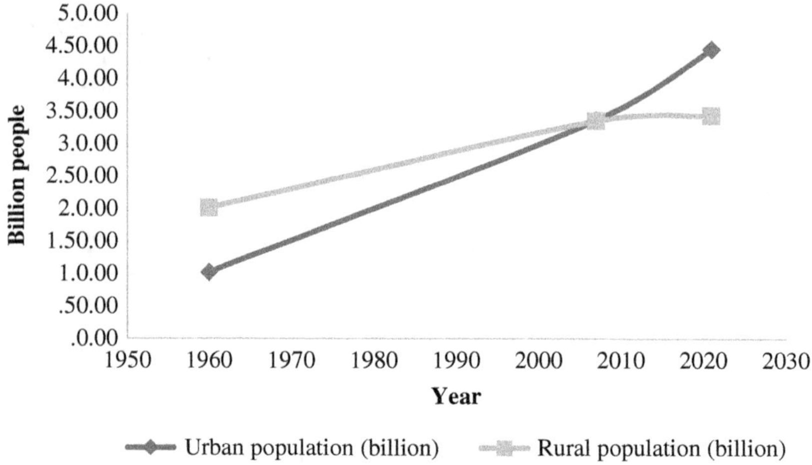

Figure 12.3 The population distribution between urban and rural regions globally. Source: Adapted from [38].

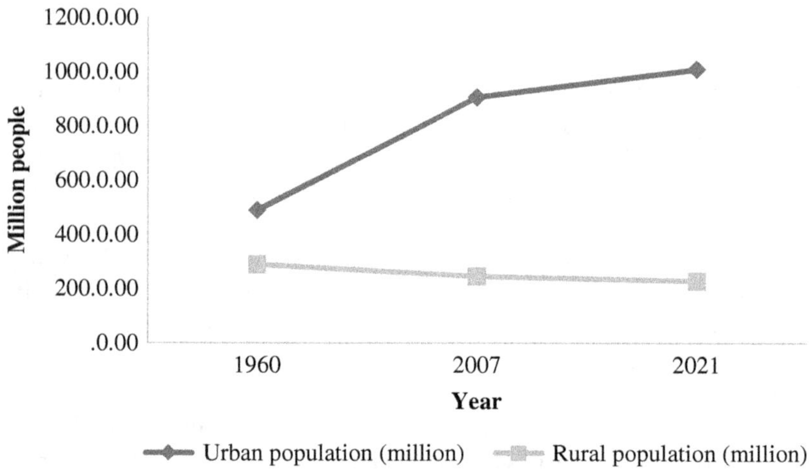

Figure 12.4 The number of residents of urban and rural areas in high-income countries. Source: Adapted from [38].

2021, the urban population in affluent nations has consistently surpassed that of rural areas, which has seen a progressive decline.

The numbers of inhabitants of urban and rural areas in middle and low-income countries are shown, respectively, in Figures 12.5 and 12.6. As can be seen in Figure 12.5, in middle-income countries, the population of urban areas surpassed the population of rural areas in 2015, which indicates the preference of the

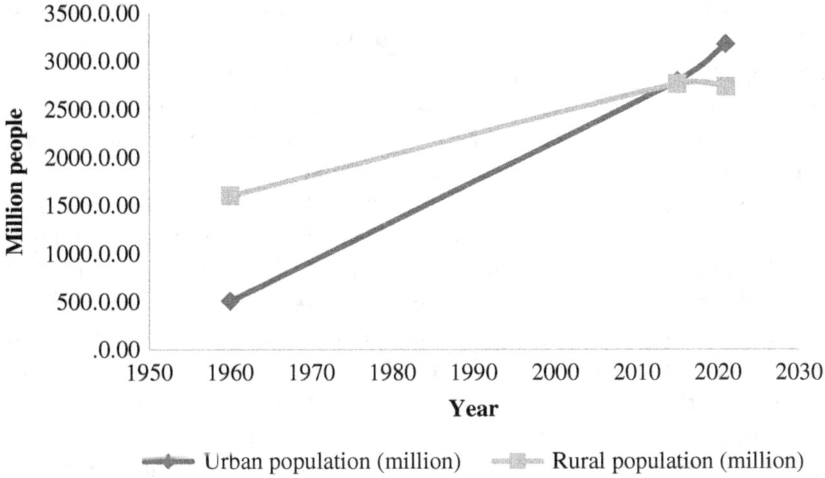

Figure 12.5 The number of residents of urban and rural areas in middle-income countries. Source: Adapted from [38].

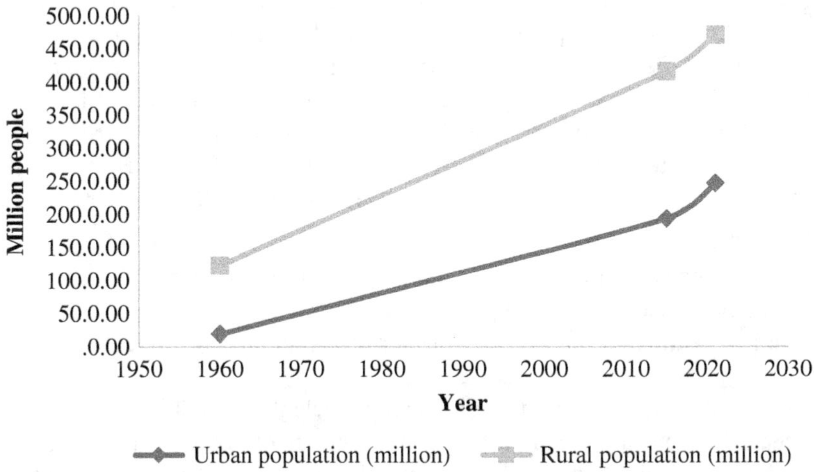

Figure 12.6 The number of residents of urban and rural areas in low-income countries. Source: Adapted from [38].

people of these countries for urbanization. However, in low-income countries, as shown in Figure 12.6, the population of rural areas still remains higher than the population of urban areas from 1960 to 2021.

This situation indicates that as the income levels of countries rise, people increasingly tend to migrate to urban areas, which subsequently leads to challenges such as energy supply issues, air pollution, and housing and healthcare concerns. Therefore, the need to prevent reverse migration from villages to cities has become more pressing than ever. One of the main reasons for this reverse migration is the pursuit of greater welfare. Since access to energy is one of the essential drivers of growth and development in any region, efforts must be intensified to improve energy access in rural areas.

Unfortunately, rural areas face significant challenges in gaining access to energy and fuel networks. One possible explanation is the lower population density in rural areas compared to urban centers. Expanding energy and fuel networks demands substantial investment, and governments — particularly in low- and middle-income countries — often refrain from such expansions due to financial constraints. As a result, around 13% of the global population, notably in Sub-Saharan Africa and South Asia, still lack access to electricity, and this extends to other energy carriers as well [39, 40]. Thus far, it has become evident that due to financial limitations, many governments have focused their attention on urban areas with higher populations, leaving rural areas isolated and disconnected from national energy grids. For governments to extend the main energy grid to rural regions, they would need to make considerable investments, which often lack economic justification. Hence, a more sustainable solution must be sought. Among the three energy supply and transmission solutions previously mentioned, one of the most effective strategies is the use of renewable energy technologies in the form of mini-grids. Micro-grids can also be utilized in tandem. Mini-grids can supply energy capacities ranging from 1 kW to 10 MW and are particularly suitable for remote, low-density areas such as villages [40–42]. However, if governments plan to implement mini-grids, they must carefully consider several important factors:

1) All energy consumption in areas where mini-grids are implemented must be electric-based. Wherever possible, all types of energy consumption, including heating, cooling, and transportation, should be directed toward electricity.
2) The use of storage equipment or energy-efficient systems is essential to mitigate the intermittent nature of renewable energy and to allow for the integration of multiple renewable energy sources. Given that renewable energy is not always available, this can create challenges regarding energy security and reliability [40, 43, 44].
3) Recognizing that renewable energy technologies, and particularly mini-grids, are currently more expensive than fossil-fuel-based alternatives — especially in Middle Eastern countries due to the availability of fossil resources and government subsidies — governments should either reduce subsidies for fossil fuels or increase subsidies for renewable energy technologies. This approach would encourage investors to enter the renewable energy market.

Therefore, to successfully implement mini-grids and renewable energy technologies, a comprehensive approach is required in both technical and policy-making arenas. As noted, an optimal energy system is necessary to minimize the intermittent nature of renewable energy sources. Based on various studies [40], it has been determined that hybrid renewable energy systems (HRESs), which incorporate batteries and diesel generators while utilizing multiple renewable energy sources simultaneously, are well-suited for mini-grid implementation and can enhance the reliability of renewable energy [40, 42, 45–48]. HRES consist of a variety of distributed generation sources, both renewable and traditional, coupled with energy storage solutions such as batteries, load management systems, and energy control systems, along with bus bars and distribution grids. In such systems, renewable energies like photovoltaic solar power, small hydroelectric turbines, wind energy, heat pumps, geothermal energy, and biomass are harnessed. Table 12.3 provides a summary of the advantages of these systems alongside their components [40]. As illustrated in Table 12.3, HRES consist of both dispatchable and non-dispatchable resources, which significantly enhance the reliability of these systems.

Figure 12.7 represents the HRES.

There are both obstacles and advantages to implementing HRES, which will be discussed below. The advantages and requirements for utilizing HRES are as follows:

- **Government Support:** For developing HRES in mini-grids, government support is essential at various levels. The government can establish policies and laws that create a favorable legal and financial environment for investment and the operation of HRES. Furthermore, government incentives can lower investment and operational costs, enhancing the return on investment and fostering greater trust and credibility within financial and commercial markets.
- **Community Organization:** Successful implementation of HRES in mini-grids requires effective community organization. Communities can take an active role in various stages of the project, including planning, design, installation, maintenance, and management of the HRES. They can also allocate responsibilities and distribute benefits among members while developing mechanisms to resolve conflicts and ensure coordination. This organization fosters increased awareness and education about HRES and promotes stronger community commitment and participation.
- **Reduction of Investment and Operational Costs:** In order to make HRES more competitive in mini-grids, it is necessary to reduce both investment and operational costs. This can be achieved by employing renewable energy technologies that have decreased in cost over time. Additionally, optimizing the design and selecting the appropriate combination of energy sources can increase efficiency and minimize energy loss. Reducing these costs will improve the return on investment and lower financial risks.

Table 12.3 The primary elements of HRESs.

Description	Component	Type
Solar photovoltaic (PV) systems are versatile and can be installed in numerous settings. They generate electricity as direct current (DC) at a low voltage level, which is capable of energizing both DC and alternating current (AC) appliances.	Solar	Non-dispatchable resources
Wind energy potential varies by location, necessitating site-specific dependency. A comprehensive assessment of these resources is required over an extended duration, typically a year or more, prior to the installation of the system. This technology produces AC electricity.	Wind	
Among various technologies, small-scale hydroelectric systems stand out as the most economical. Nonetheless, their feasibility hinges on the presence of a river that meets the necessary flow rate and volume criteria. In locales where hydro resources are abundant, these systems can independently provide a steady stream of energy without the need to integrate additional resources.	Small hydro	
Diesel generators, which supply AC power, play a crucial role in ensuring a reliable power supply, especially when the output from alternative technologies falls short of demand. Nevertheless, it is advisable to limit their usage due to the high cost of diesel fuel.	Diesel generator	Dispatchable resources
The practicality and financial viability of biomass energy remain uncertain. The inclusion of biomass can significantly influence the overall expenses of a hybrid system.	Biomass	
Batteries significantly contribute to the overall electricity costs throughout the lifespan of a project, primarily because they require periodic replacement. The charging and discharging cycles of batteries are contingent upon the state of the power generated.	Battery	Storage

Table 12.3 (Continued)

Description	Component	Type
The loads intended for integration into hybrid AC/DC systems are categorized into electrical and thermal types.	Loads	
Converters are instrumental in transforming electrical power between different forms. They facilitate the conversion from AC to DC through a process known as rectification, and from DC back to AC via inversion. Additionally, converters can adjust power from one voltage or frequency to another within the same current type.	Converters	Balance-of-plant items
Inverters are responsible for converting direct current to alternating current before delivering power to AC loads.	Inverters	
The selection between alternating current or direct current has a significant impact on the system's configuration. This choice generally hinges on the technologies chosen for incorporation into the system and the overarching strategy for energy management. Distribution networks are comprised of elements such as grid lines, transformers, and other essential infrastructures.	AC/DC Distribution networks	Bus bars and local distribution networks

Source: Adapted from [40].

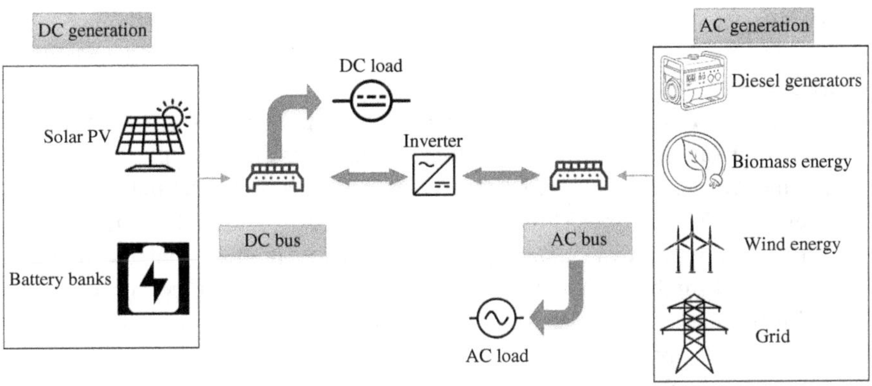

Figure 12.7 Typical schematic of HRESs.

- **Improved Reliability and Quality of Electricity Supply:** To meet the electricity needs of the community, it is crucial to enhance the reliability and quality of electricity supply. HRES systems, which can adapt to fluctuations in both energy demand and production, can play a vital role in this regard. By incorporating energy storage systems and smart controllers, stability can be improved and network disturbances minimized, resulting in increased customer satisfaction and reduced costs from power outages.
- **Reduction of Environmental Impacts from Fossil Fuel Use:** In order to protect the environment and reduce greenhouse gas emissions, the use of fossil fuels must be curtailed. This can be achieved through HRES, which relies on clean, renewable energy sources. Reducing the need for transporting fossil fuels will also diminish air and noise pollution, leading to improved community health and quality of life, while lowering costs associated with climate change.

The obstacles to implementing HRES include:

- **Lack of Supportive Policies and Appropriate Laws:** Successful implementation of HRES in mini-grids requires laws that safeguard the rights and interests of both owners and customers. For example, if the national electricity grid reaches rural areas, mini-grid owners should have the option of exchanging, integrating, or collaborating with the main grid. Without such policies, financial and commercial markets may be restricted, reducing incentives for investment.
- **Lack of Suitable Financial and Commercial Markets:** Access to financial and commercial resources is necessary for the development of HRES in mini-grids. However, in many low- and middle-income countries, these resources are scarce or nonexistent. For instance, banks and financial institutions may hesitate to lend to HRES projects due to unfamiliarity with renewable energy technologies or uncertainty regarding the return on investment. Such obstacles increase investment and operational costs and raise financial risks.
- **Limited Access to Equipment and Technical Services:** To successfully install and maintain HRES in mini-grids, access to high-quality, affordable equipment and technical services is necessary. Unfortunately, in many low- and middle-income countries, this access is limited or unavailable. For example, essential equipment and spare parts may not be readily available in local markets and might need to be imported. These barriers can compromise the reliability and quality of electricity supply and increase the demand for training and technical support.
- **Lack of Education and Awareness About Renewable Energy Technologies:** Proper use of HRES in mini-grids requires education and awareness regarding renewable energy technologies and their benefits. Yet, in many low- and middle-income countries, such education and awareness are minimal

or nonexistent. For example, owners and users of mini-grids may not fully understand how to operate and maintain HRES, or they may be unaware of the advantages of renewable energy compared to fossil fuels. These factors can lead to cultural and social resistance to change, diminishing demand for HRES.

12.5 Geographical and Contextual Differences

Geographical and contextual differences between urban and rural zones play a crucial role in shaping energy consumption. These regions differ in population size, economic development, infrastructure, cultural norms, lifestyles, and energy demands, all of which shape the structure, scale, volume, and sources of energy consumption. For instance, urban sectors in China exhibit higher energy usage than their rural counterparts, with notable disparities across provinces [49]. Similarly, the utilization of water as an energy source shows marked contrasts between urban and rural settings in Egypt [50]. Despite its importance, limited research has explored the geographical and contextual differences in energy use between urban and rural areas. Existing studies, often limited to case studies in specific regions of either developing or developed countries, have introduced various methodologies, variables, and outcomes. A thorough and critical examination of these studies is warranted. To address this gap, this section endeavors to present an exhaustive perspective on the influence of geographical shifts on energy by scrutinizing various studies focused on particular locales. Among such studies, Lian et al. [51] conducted an analysis based on the dietary intake data from inhabitants across 31 Chinese provinces from 2015 to 2021, assessing the divergence of food consumption from dietary objectives and the spatial distribution traits of urban and rural populations in China.

The findings reveal a significant divergence in the dietary patterns of Chinese citizens from the nutritional benchmarks established by the country's dietary guidelines, with pronounced disparities across urban and rural regions as well as among different provinces. Consequently, it is proposed that a novel paradigm of food security, centered on nutritional objectives, be established. The dietary habits of the populace should be directed in a scientific and judicious manner, with particular focus and bespoke measures for locales experiencing imbalanced dietary practices. The important points and achievements of this article are:

- The development of an exhaustive index to gauge the soundness of the dietary patterns among urban and rural dwellers in China.
- An analysis of the geographical variances in dietary habits between urban and rural inhabitants in China, along with an exploration of influencing factors such as economic progress, demographic spread, urban and rural advancement, and culinary traditions.

- Recommendations aimed at refining the dietary framework for both urban and rural populations in China, thereby promoting food security and sustainable progress.

In another study, Liu et al. [49] assessed the level of integrated urban–rural development across 41 cities in the Yangtze River Delta using a global principal component analysis. They explored the spatial clustering pattern and its evolution with the ESDA method. Additionally, they employed geographical identification to analyze the primary factors influencing the spatial disparities in integrated urban–rural development within the Delta, as well as their interrelationships. The study also utilized geographically weighted regression to discuss the regional variations among these factors. The findings revealed:

- A significant imbalance in the integrated urban–rural development among the 41 cities, with considerable differences between them.
- A distinct geographical clustering pattern of integrated urban–rural development in the region, with central cities acting as pivotal drivers for the development of their neighboring cities.
- The integrated urban–rural development is influenced by factors such as economic growth, basic infrastructure, governmental support, and industrial structure, which should inform policy recommendations tailored to regional realities.

In another article, Marzouk [50] studied the qualitative analysis of sustainable energy and water consumption behaviors among urban and rural customers in the Egyptian market. Marzouk [50] conducted his research using a qualitative method, which included interviewing people in urban and rural areas. Some of the important findings of this research are:

- Both urban and rural customers have a positive attitude toward sustainable consumption, but attitude is not the only factor and it should be accompanied by motivation and behavior.
- The motivation for sustainable consumption differs between urban and rural customers. Urban customers are more influenced by economic, social, and cultural reasons, while rural customers are more guided by environmental, ethical, and religious reasons.
- The behavior of sustainable consumption is also different among urban and rural customers. Urban customers use more technical and technological solutions to reduce energy and water consumption, while rural customers use more traditional and creative solutions.
- Sustainable consumption is based on two principles: the first is “choice” and the second is “action.” The principle of “choice” means that customers have to choose what kind of energy and water they consume, while the principle of “action” means that customers have to use energy and water in a sustainable way. These two principles are related and should be considered.

Table 12.4 Investigating the impact of geographical differences on energy consumption in other studies.

References	Publication year	Study location	Variables examined	Analysis method	Main findings
[51]	2023	China	Food consumption structure of urban and rural residents	Global principal analysis, cluster analysis	The dietary habits of residents in China substantially diverge from the suggested standards of the national dietary guidelines, with notable variations evident between urban and rural settings as well as across different provinces.
[49]	2023	Yangtze River Delta	Level of integrated urban–rural development	Global principal analysis, ESDA, geographical identification, geographically weighted regression	The development of 41 urban–rural areas is uneven, displaying both beneficial and detrimental patterns of spatial concentration. The key elements influencing the spatial variance in this integrated development include economic progress, societal advancement, the evolution of infrastructure, environmental enhancement, and governance growth.
[50]	2019	Egypt	Sustainable energy and water consumption behavior of urban and rural residents	Qualitative data, conceptual frameworks	The patterns of sustainable energy and water use among Egypt's urban and rural inhabitants are shaped by a variety of factors, including their values, convictions, attitudes, emotional responses, understanding, capabilities, dedication, societal pressures, motivators, and obstacles.
[52]	2021	Finland, Italy, Switzerland, Germany France, and Austria	Sustainable energy culture in the industrial sector	Case study, four-dimensional model	The sustainable energy culture in the industrial sector can be analyzed from the aspects of values, beliefs, attitudes, and behaviors. Factors such as leadership, policies, goals, structure, communication, education, motivation, feedback, tools, and working conditions affect the sustainable energy culture.

The previous section presented some articles related to the impact of geography on energy consumption, and they all showed well that changing the country and also changing the geographical position from urban and rural perspectives in one country will alter the parameters affecting energy consumption. To clarify the role of geography in energy consumption and better compare the related research with each other, Table 12.4 shows a comparative view of the articles that have been published in the field of the impact of geographical locations on energy consumption and the parameters that have been examined in them.

Of course, the income level of countries and other related factors such as gross domestic product and development level will also affect energy consumption. As mentioned in the previous sections, if divided the countries into three categories of high, medium, and low income, we can compare the access to electricity in rural and urban areas. The result of this comparison is shown in Table 12.5.

As shown in Table 12.5, the growth rate of access to electricity in rural areas of low-income countries has been the highest from 2003 to 2020. However, access to electricity in these rural areas remains significantly lower compared to high- and middle-income countries. For example, by 2020, approximately 70% of rural areas in low-income countries still lacked access to electric energy. The reasons for this situation could include the absence of supportive policies and appropriate laws, inadequate financial and commercial markets, limited access to equipment and technical services, and insufficient education and awareness about renewable energy technologies. Therefore, considering the points mentioned in this section, we can observe the impact and role of geographical location and income levels of countries on energy consumption.

12.6 Comparative Analysis of Energy Justice and Equity in Rural and Urban Areas

The concept of energy justice has emerged as a crucial approach to examining the uneven distribution of energy resources and services. It is essential to explore the distinct ways these imbalances manifest in rural versus urban areas, taking into account various socioeconomic and geographic factors [16, 27, 54].

12.6.1 Rural Energy Justice

Energy justice in rural regions often intersects with challenges related to access and cost-effectiveness. The remote nature of these areas can lead to higher expenses and reduced access to energy services. Moreover, these communities may lack the infrastructure necessary for modern energy solutions and may rely on more traditional, less efficient energy sources [16, 25, 27, 55].

Table 12.5 The level of access to electricity in urban and rural areas of high-, medium-, and low-income countries [53].

	Access to electricity, urban (% of urban population)				Access to electricity, rural (% of rural population)				Population (historical estimates)
Country or region	**2003**	**2020**	**Absolute change**	**Relative change**	**2003**	**2020**	**Absolute change**	**Relative change**	**Relative change**
High-income countries	99.99%	100.00%	+0.01 pp	0%	99.77%	99.94%	+0.17 pp	0%	11%
Low-income countries	53.26%	70.10%	+16.84 pp	32%	8.87%	30.44%	+21.57 pp	243%	60%
Middle-income countries	95.27%	98.40%	+3.13 pp	3%	69.15%	89.62%	+20.47 pp	30%	–

12.6.2 Urban Energy Justice

In urban environments, which typically have more developed energy infrastructures, distinct issues still arise. Energy fairness in urban settings often focuses on the urban heat island effect, air pollution, and the disproportionate energy costs borne by economically disadvantaged households. The dense population in cities can exacerbate these problems, making them more apparent and urgent [56–58].

12.6.3 Factors Creating Inequality of Energy Justice in Urban and Rural Areas

Several factors contribute to the energy justice gap between rural and urban areas [25, 59, 60]:

- **Infrastructure:** Cities benefit from robust energy infrastructures, unlike the often inadequate facilities in rural areas.
- **Economic Status:** Economic disparities between city dwellers and rural communities can lead to uneven distribution of energy resources.
- **Policy and Governance:** Energy regulations may unintentionally favor urban growth, thus widening the energy gap between urban and rural settings.
- **Technological Advancements:** Cities typically adopt new energy technologies more quickly, which can contribute to the growing disparity.

12.6.4 Comparative Viewpoints

From an ecological perspective, cities face challenges related to high energy consumption and emissions, while rural areas may suffer from reliance on inefficient and environmentally harmful energy sources. Socially, disparities in energy access can exacerbate the divide between urban and rural communities, affecting quality of life and economic opportunities [27, 57].

Table 12.6 presents a comparative analysis of various indicators related to energy and the environment across rural and urban areas. It highlights the differences in access to electricity, energy affordability, renewable energy adoption rates, and air quality between these two distinct regions. This overview provides insight into the challenges and advancements characterizing rural and urban landscapes in the context of sustainable development.

Therefore, this comparative study highlights the necessity for sophisticated energy strategies that cater to the distinct obstacles encountered by rural and urban populations. To close the gap in energy justice, a collaborative approach is essential, involving policymakers, interested parties, and the communities themselves [27, 57, 65].

Table 12.6 Energy justice indicators in rural and urban areas.

Indicator	Rural areas	Urban areas	References
Electricity availability	Constrained by infrastructural limits	Widely accessible	[16, 55, 59, 61]
Cost of energy	Elevated due to transportation fees	More affordable, yet higher usage cost	[55, 59, 62]
Uptake of renewable energy	Progress hindered by expenses	Advancement aided by policies	[27, 55, 57, 59]
Quality of air	Typically superior	Commonly inferior due to congestion	[55, 59, 60, 63, 64]

12.7 Conclusion

This study has shown that since 1960, there has been a significant increase in urban populations in high-income countries, a trend that extended to middle-income countries after 2015, while low-income countries continue to have a larger rural population. These demographic shifts pose challenges in energy provision, air pollution, and access to housing and healthcare. To address the urban influx and its environmental impacts, this research advocates for improved energy access in rural areas, proposing HRES, specifically mini-grids, as feasible solutions for low- and middle-income countries. The study also highlights the disparity in energy consumption and access influenced by geographical and income differences, revealing that in 2020, approximately 70% of rural areas in low-income countries lacked electricity, compared to urban areas with 44.4% more access. The decline in access is attributed to a lack of supportive policies and laws, inadequate financial and commercial markets, limited access to equipment and technical services, and insufficient education and awareness about renewable energy technologies. In conclusion, the study underscores the necessity for comprehensive energy strategies that address the unique challenges faced by both rural and urban populations. Bridging the energy justice gap requires a concerted effort from policymakers, stakeholders, and communities to implement sustainable development practices that ensure equitable energy access for all.

Abbreviations

HRES	hybrid renewable energy systems
PV	solar photovoltaic
DC	direct current
AC	alternating current
ESDA	spatial data analysis

References

1 Melé, D. (2024). Ethical theories in business ethics: a critical review. *Journal of Human Values* 30 (1): 15–25. https://doi.org/10.1177/09716858231201191.

2 Saad, T. (2023). A critical discourse study of Shakespeare's theological conceptions in Acts IV and V of Richard II: the 'divine mandate' of Richard Kingship Falls Apart. *Studies in Pragmatics and Discourse Analysis* 4 (1): 14–31. https://doi.org/10.48185/spda.v4i1.732.

3 Hobbes, T. (2010). *Leviathan-Revised Edition*. Broadview Press.

4 Kraynak, R. (2019). *History and Modernity in the Thought of Thomas Hobbes*. Cornell University Press.

5 Sovacool, B.K. and Dworkin, M.H. (2015). Energy justice: conceptual insights and practical applications. *Applied Energy* 142: 435–444. https://doi.org/10.1016/j.apenergy.2015.01.002.

6 Sandel, M.J. (2010). Justice: what's the right thing to do? *The Hedgehog Review, Book Review* 12: 85+. (in English). https://link.gale.com/apps/doc/A223901663/AONE?u=anon~27f5c4de&sid=googleScholar&xid=b21b7b64.

7 Lerner, R.M., Lerner, J.V., Murry, V.M. et al. (2021). Positive youth development in 2020: theory, research, programs, and the promotion of social justice. *Journal of Research on Adolescence* 31 (4): 1114–1134. https://doi.org/10.1111/jora.12609.

8 Schlosberg, D. (1999). *Environmental Justice and the New Pluralism: The Challenge of Difference for Environmentalism*. OUP Oxford.

9 Atteberry-Ash, B.E. (2023). Social work and social justice: a conceptual review. *Social Work* 68 (1): 38–46. https://doi.org/10.1093/sw/swac042.

10 Walker, G. (2012). *Environmental Justice: Concepts, Evidence and Politics*. Routledge.

11 Aidonojie, P.A. (2023). Environmental hazard: the legal issues concerning environmental justice in Nigeria. *Journal of Human Rights, Culture and Legal System* 3 (1): 17–32. https://doi.org/10.53955/jhcls.v3i1.60.

12 Fragkou, M.C. (2019). Environmental justice. In: *The Wiley Blackwell Encyclopedia of Urban and Regional Studies* (ed. A.M. Orum), 1–6. https://doi.org/10.1002/9781118568446.eurs0091.

13 Walker, G. (2009). Beyond distribution and proximity: exploring the multiple spatialities of environmental justice. *Antipode* 41 (4): 614–636. https://doi.org/10.1111/j.1467-8330.2009.00691.x.

14 Owens, S. and Driffill, L. (2008). How to change attitudes and behaviours in the context of energy. *Energy Policy* 36 (12): 4412–4418. https://doi.org/10.1016/j.enpol.2008.09.031.

15 Eames, M. and Hunt, M. (2013). Energy justice in sustainability transitions research. In: *Energy Justice in a Changing Climate* (ed. K. Bickerstaff,

G. Walker, and H. Bulkeley), 46–61. Permalink. http://digital.casalini.it/9781780325804.

16 Jenkins, K., McCauley, D., Heffron, R. et al. (2016). Energy justice: a conceptual review. *Energy Research & Social Science* 11: 174–182. https://doi.org/10.1016/j.erss.2015.10.004.

17 Schlosberg, D. (2003). The justice of environmental justice: reconciling equity, recognition, and participation in a political movement. *Moral and Political Reasoning in Environmental Practice* 77: 106.

18 Nazmul Haque, M. and Sharifi, A. (2024). Justice in access to urban ecosystem services: a critical review of the literature. *Ecosystem Services* 67: 101617. https://doi.org/10.1016/j.ecoser.2024.101617.

19 Rawls, J. (2001). *Justice as Fairness: A Restatement*. Harvard University Press.

20 Heffron, R. (2013). Advancing energy justice: the triumvirate of tenets. *International Energy Law Review* 3: 107–110.

21 Gibson-Wood, H. and Wakefield, S. (2013). "Participation", white privilege and environmental justice: understanding environmentalism among hispanics in Toronto. *Antipode* 45 (3): 641–662. https://doi.org/10.1111/j.1467-8330.2012.01019.x.

22 Taylor, D.E. (2000). The rise of the environmental justice paradigm: injustice framing and the social construction of environmental discourses. *American Behavioral Scientist* 43 (4): 508–580. https://doi.org/10.1177/0002764200043004003.

23 Buckingham, S. and Kulcur, R. (2009). Gendered geographies of environmental injustice. *Antipode* 41 (4): 659–683. https://doi.org/10.1111/j.1467-8330.2009.00693.x.

24 Schlosberg, D. and Carruthers, D. (2010). Indigenous struggles, environmental justice, and community capabilities. *Global Environmental Politics* 10 (4): 12–35. https://doi.org/10.1162/GLEP_a_00029.

25 Walker, G. and Day, R. (2012). Fuel poverty as injustice: integrating distribution, recognition and procedure in the struggle for affordable warmth. *Energy Policy* 49: 69–75. https://doi.org/10.1016/j.enpol.2012.01.044.

26 Hall, S.M. (2013). Energy justice and ethical consumption: comparison, synthesis and lesson drawing. *Local Environment* 18 (4): 422–437. https://doi.org/10.1080/13549839.2012.748730.

27 Sovacool, B.K. and Dworkin, M.H. (2014). *Global Energy Justice*. Cambridge University Press https://doi.org/10.1017/CBO9781139026444.

28 Antypas, A., Atkins, L., Borthwick, F. et al. (2007). Making the case for environmental and social justice in Central and Eastern Europe. http://hdl.handle.net/20.500.14018/13577 (accessed 9 January 2025).

29 Heffron, R.J. (2022). Applying energy justice into the energy transition. *Renewable and Sustainable Energy Reviews* 156: 111936. https://doi.org/10.1016/j.rser.2021.111936.

30 Goldthau, A. and Sovacool, B.K. (2012). The uniqueness of the energy security, justice, and governance problem. *Energy Policy* 41: 232–240. https://doi.org/10.1016/j.enpol.2011.10.042.

31 Naumann, M. and Rudolph, D. (2020). Conceptualizing rural energy transitions: energizing rural studies, ruralizing energy research. *Journal of Rural Studies* 73: 97–104. https://doi.org/10.1016/j.jrurstud.2019.12.011.

32 Bailey, I. and Darkal, H. (2018). (Not) talking about justice: justice self-recognition and the integration of energy and environmental-social justice into renewable energy siting. *Local Environment* 23 (3): 335–351. https://doi.org/10.1080/13549839.2017.1418848.

33 Balta-Ozkan, N., Watson, T., and Mocca, E. (2015). Spatially uneven development and low carbon transitions: insights from urban and regional planning. *Energy Policy* 85: 500–510. https://doi.org/10.1016/j.enpol.2015.05.013.

34 Woods, M. (2012). Rural geography III: rural futures and the future of rural geography. *Progress in Human Geography* 36 (1): 125–134. https://doi.org/10.1177/0309132510393135.

35 Argent, N. (2019). Rural geography II: scalar and social constructionist perspectives on climate change adaptation and rural resilience. *Progress in Human Geography* 43 (1): 183–191. https://doi.org/10.1177/0309132517743115.

36 Rehman, I.H., Kar, A., Raven, R. et al. (2010). Rural energy transitions in developing countries: a case of the Uttam Urja initiative in India. *Environmental Science & Policy* 13 (4): 303–311. https://doi.org/10.1016/j.envsci.2010.03.012.

37 Calvert, K. (2016). From 'energy geography' to 'energy geographies' perspectives on a fertile academic borderland. *Progress in Human Geography* 40 (1): 105–125. https://doi.org/10.1177/0309132514566343.

38 World Bank Group (2024). "World Bank based on data from the UN Population Division – processed by Our World in Data. "Urban population" [dataset]. World Bank based on data from the UN Population Division [original data]." https://datacatalog.worldbank.org/search/dataset/0037712/World-Development-Indicators (accessed 9 January 2025)

39 World Health Organization (2018). Tracking SDG7: The Energy Progress Report 2018.

40 Come Zebra, E.I., van der Windt, H.J., Nhumaio, G., and Faaij, A.P.C. (2021). A review of hybrid renewable energy systems in mini-grids for off-grid electrification in developing countries. *Renewable and Sustainable Energy Reviews* 144: 111036. https://doi.org/10.1016/j.rser.2021.111036.

41 Williams, N.J., Jaramillo, P., Taneja, J., and Ustun, T.S. (2015). Enabling private sector investment in microgrid-based rural electrification in developing countries: a review. *Renewable and Sustainable Energy Reviews* 52: 1268–1281. https://doi.org/10.1016/j.rser.2015.07.153.

42 Ahmad, J., Imran, M., Khalid, A. et al. (2018). Techno economic analysis of a wind-photovoltaic-biomass hybrid renewable energy system for rural electrification: a case study of Kallar Kahar. *Energy* 148: 208–234. https://doi.org/10.1016/j.energy.2018.01.133.

43 Sims, R., Mercado, P.E., and Krewitt, W. (2011). *Integration of Renewable Energy Into Present and Future Energy Systems*. Cambridge University Press. ISBN: 978-1-107-60710-1.

44 Metz, B. (2007). *Climate Change 2007: Mitigation: Contribution of Working Group III to the Fourth Assessment Report of the Intergovernmental Panel on Climate Change*. Intergovernmental Panel on Climate Change. http://www.ipcc.ch/pdf/assessment-report/ar4/wg3/ar4-wg3-spm.pdf.

45 Shahzad, M.K., Zahid, A., ur Rashid, T. et al. (2017). Techno-economic feasibility analysis of a solar-biomass off grid system for the electrification of remote rural areas in Pakistan using HOMER software. *Renewable Energy* 106: 264–273. https://doi.org/10.1016/j.renene.2017.01.033.

46 Shi, B., Wu, W., and Yan, L. (2017). Size optimization of stand-alone PV/wind/diesel hybrid power generation systems. *Journal of the Taiwan Institute of Chemical Engineers* 73: 93–101. https://doi.org/10.1016/j.jtice.2016.07.047.

47 Al Busaidi, A.S., Kazem, H.A., Al-Badi, A.H., and Khan, M.F. (2016). A review of optimum sizing of hybrid PV–Wind renewable energy systems in Oman. *Renewable and Sustainable Energy Reviews* 53: 185–193. https://doi.org/10.1016/j.rser.2015.08.039.

48 Khan, F.A., Pal, N., and Saeed, S.H. (2018). Review of solar photovoltaic and wind hybrid energy systems for sizing strategies optimization techniques and cost analysis methodologies. *Renewable and Sustainable Energy Reviews* 92: 937–947. https://doi.org/10.1016/j.rser.2018.04.107.

49 Liu, N., Chen, Y., and Su, F. (2023). Research on the spatial differences and influencing factors of integrated urban-rural development in the Yangtze River Delta. *Frontiers in Sustainable Cities* 4: 1077653. https://doi.org/10.3389/frsc.2022.1077653.

50 Marzouk, O.A. (2019). A qualitative examination of urban vs rural sustainable consumption behaviours of energy and water consumers in the emerging Egyptian market. *Journal of Humanities and Applied Social Sciences* 1 (2): 98–114. https://doi.org/10.1108/JHASS-07-2019-0016.

51 Lian, Y., Gu, L., Yang, L. et al. (2023). The reasonableness and spatial differences of the food consumption structure of urban and rural residents in China, 2015–2021. *Foods* 12 (10): 1997. https://doi.org/10.3390/foods12101997.

52 Oksman, V., Reda, F., Karjalainen, S. et al. (2021). Towards sustainable energy culture in the industrial sector: introducing an interdisciplinary method for understanding energy culture in business industries. *Energy, Sustainability and Society* 11 (1): 28. https://doi.org/10.1186/s13705-021-00303-7.

53 World Bank Group (2024). "Data compiled from multiple sources by World Bank – processed by Our World in Data. "Access to electricity, urban (% of urban population)" [dataset]. Data compiled from multiple sources by World Bank [original data]." https://datacatalog.worldbank.org/search/dataset/0037712/World-Development-Indicators (accessed 10 January 2025).

54 Hua, W., Wang, L., Fang, X., and Luo, L. (2023). Urbanization and energy equity: an urban-rural gap perspective. *Environmental Science and Pollution Research* 30 (16): 46847–46868. https://doi.org/10.1007/s11356-023-25139-w.

55 Taylor, M. (2024). The power of energy justice for rural communities. In: *The Power of Energy Justice & the Social Contract* (ed. R.J. Heffron and L. de Fontenelle), 129–135. Cham: Springer Nature Switzerland. https://doi.org/10.1007/978-3-031-46282-5_18.

56 Shabliy, E.V. and Kurochkin, D. (2022). Energy and environmental justice. In: *Energy Justice: Climate Change Mitigation and Adaptation* (ed. E.V. Shabliy, M.J. Crawford, and D. Kurochkin), 1–20. Cham: Springer International Publishing. https://doi.org/10.1007/978-3-030-93068-4_1.

57 Sidortsov, R. and McCauley, D. (2023). Energy justice. In: *Theorising Justice* (ed. J. Ohlsson and S. Przybylinski), 171–190. Bristol University Press. ISBN: 978-1-5292-3223-3 ePdf.

58 Bouzarovski, S. (2014). Energy poverty in the European Union: landscapes of vulnerability. *Wiley Interdisciplinary Reviews: Energy and Environment* 3 (3): 276–289. https://doi.org/10.1002/wene.89.

59 Lyu, A. and Li, J. (2024). Urban–rural income gap and economic sustainability: how does energy transition matters? *Economic Change and Restructuring* 57 (2): 51. https://doi.org/10.1007/s10644-024-09603-2.

60 Bouzarovski, S. and Petrova, S. (2015). A global perspective on domestic energy deprivation: overcoming the energy poverty–fuel poverty binary. *Energy Research & Social Science* 10: 31–40. https://doi.org/10.1016/j.erss.2015.06.007.

61 Cozzi, L., Gould, T., Bouckart, S. et al. (2020). *World Energy Outlook 2020*, vol. 2050. Paris, France: International Energy Agency.

62 Bouzarovski, S. and Tirado Herrero, S. (2017). The energy divide: integrating energy transitions, regional inequalities and poverty trends in the European Union. *European Urban and Regional Studies* 24 (1): 69–86. https://doi.org/10.1177/0969776415596449.

63 Pollution, W.A.A. (2016). *A Global Assessment of Exposure and Burden of Disease*, vol. 360. Geneva, Switzerland: World Health Organization. ISBN: 9789241511353.

64 World Health Organization (2019). *Global Action Plan on Physical Activity 2018–2030: More Active People for a Healthier World*. World Health Organization.

65 Mathieu, J.L. (2023). Algorithms for energy justice. In: *Women in Power: Research and Development Advances in Electric Power Systems* (ed. J.S. Tietjen, M.D. Ilic, L. Bertling Tjernberg, and N.N. Schulz), 67–83. Cham: Springer International Publishing https://doi.org/10.1007/978-3-031-29724-3_4.

13

Social Justice in Electrical Energy Outage Management Through the Controllability Index

Hamidreza Arasteh[1,2*], *Salah Bahramara*[3], *Mahyar Gholizadeh*[4], *Niki Moslemi*[1], *Zohreh Kaheh*[5], *Milad Biazar*[4], *and Miadreza Shafie-khah*[5]

[1] *Power Systems Operation and Planning Research Department, Niroo Research Institute, Tehran, Iran*
[2] *Center for Research on Microgrids (CROM), Huanjiang Laboratory, Zhuji, Shaoxing, Zhejiang, 311800, China*
[3] *Department of Electrical Engineering, Sanandaj Branch, Islamic Azad University, Sanandaj, Iran*
[4] *Tavanir Company, Deputy of Power Distribution, Tehran, Iran*
[5] *School of Technology and Innovations, University of Vaasa, Vaasa, Finland*

13.1 Introduction

13.1.1 Aims and Motivations

The effective management of network load during peak times emerges as a pivotal challenge for network operators [1]. The intricacies intensify in the face of pronounced imbalances between the supply and demand of electric energy, network faults, or the disruptive impact of natural disasters. Addressing the network load during peak times constitutes a primary challenge for network operators. This challenge becomes more aggravated when there is a substantial imbalance between the supply and demand of electric energy, or when there are network faults, or natural disasters. While network operators endeavor to mitigate these issues using available network assets and capacities, there are scenarios where curtailing the network load becomes an unavoidable recourse.

In such instances, the system operator sends load-shedding signals to large-scale customers connected to the transmission network and simultaneously communicates to distribution network operators for the specified load reduction for each distribution system [2]. Subsequently, distribution operators are tasked with implementing load-shedding measures, necessitating the disconnection of feeders under their purview.

*Corresponding author: Hamidreza Arasteh

Communities for Clean Energy Justice and Equity in Grid Modernization, First Edition.
Edited by Mohammadreza Daneshvar, Behnam Mohammadi-Ivatloo, and Amjad Anvari-Moghaddam.

A noteworthy portion of medium-voltage feeders within distribution networks, termed priority feeders, remains exempt from load shedding and demand response (DR) due to their role in supplying sensitive loads such as hospitals, government, and military centers, as well as security centers [3]. Consequently, this exemption contributes to more frequent and prolonged outages on other non-prioritized feeders.

Under these circumstances, residential consumers situated in the exempted feeders are spared from curtailments, leading to a phenomenon commonly known as "free riding." Consequently, public dissatisfaction within this sector becomes evident. Hence, there arises a necessity to introduce an economically, technically, and justly formulated plan for executing load shedding operations during periods of insufficient production in the power system. This plan aims to achieve a fair distribution of both the frequency and duration of outages among all customers.

13.1.2 Research Background

Numerous studies have delved into the realm of operation and planning within power systems, globally considering concepts associated with social justice [4, 5]. These investigations have yielded a multitude of indicators, enhancing the understanding of the intricate interplay between power systems and social equity [6].

Energy poverty embodies a form of transnational environmental injustice with three discernible facets [7]. Initially, individuals grappling with energy poverty lack access to clean and affordable energy, even within nations endowed with abundant energy resources, the utilization of which remains unclear [8]. Secondly, those affected by energy poverty experience political marginalization, impeding their participation in governmental energy policy decisions [9]. Lastly, energy poverty is intricately connected to various other social challenges, encompassing economic inequality, gender bias, child labor, and limited access to essential services like healthcare and education.

Reference [10] examines energy poverty in Latin America and the Caribbean, emphasizing the role of energy efficiency (EE) in alleviating this issue. It discusses the adverse environmental and social effects of energy subsidies in Mexico and investigates billing practices by Mexico's state electricity company, potentially infringing on consumers' rights. The authors recommend three actions for consumers to contest problematic billing. Additionally, they propose the implementation of a distributed electricity system in Mexico as a solution to address energy poverty and enhance electricity users' rights.

Akbari et al. [7] explored the potential of microgrids (MGs) in addressing energy poverty challenges. MGs encompass a collection of loads and distributed energy resources (DERs) within a defined geographical area that could be operated in grid-connected or islanded modes [11–15]. MGs could be considered as a viable and effective solution to alleviate energy poverty providing a reliable and cost-effective source of electricity for those residing in underserved areas. Particularly in remote rural and residential regions, where access to electrical energy is often restricted and relies on costly diesel units, MGs offer an alternative that mitigates the challenges associated with expensive grid expansion efforts.

Reference [16] explores the distribution of shared rooftop photovoltaics on multiapartment buildings, aiming to ensure fairness and maximize local use to reduce global CO_2 emissions and local electricity grid load. Combining technical analysis with philosophical perspectives on energy justice, it models distribution patterns in a German case study building and evaluates them through various energy justice frameworks. The findings highlight the importance of integrating diverse perspectives to inform the development of shared rooftop photovoltaics.

Inderberg et al. [17] explore policies aimed at shifting household electricity load across different countries to enhance grid efficiency and manage challenges such as rising peak loads. Their analysis, centered on recent advancements in capacity charges, contrasts the policy formulation procedures in Norway and Sweden. The study evaluates the incorporation of consumer viewpoints and its impact on energy justice. The results emphasize the necessity of equitable policy development within the energy domain.

In [18], community energy cells (CECs) are proposed as a solution to address the energy cost burden and promote energy justice for low-to-moderate income households. CECs consist of DERs and controllable loads managed by a cell aggregator. The potential of CECs to provide renewable energy to underserved communities while offering economic benefits is demonstrated through financial analysis and case studies. Despite interdisciplinary challenges, CECs are deemed crucial for climate change mitigation, resiliency, and achieving energy justice.

Reference [19] introduces the concept of energy equity within wholesale electricity markets and proposes a framework to incorporate it into locational marginal pricing models. It addresses existing concerns in current pricing models and demonstrates the proposed framework using modified power system models.

Various demand side management (DSM) programs have been proposed to address the complexities associated with peak loads. Two categories within DSM are EE and DR programs [20]. EE initiatives focus on optimizing load levels by enhancing the efficiency of appliances. However, DR programs encompass a range of strategies designed to either decrease load levels or modify load patterns, strategically mitigating peak period loads [21, 22].

However, the challenge of load shedding persists in many developing countries, primarily attributed to constraints on the production side. This issue frequently results in electricity shortages for diverse customers, particularly affecting residential loads. The absence of justice in the implementation of load-shedding programs may intensify dissatisfaction levels among certain customers.

In the study presented in [23], load shedding was implemented in accordance with principles of justice. This was accomplished by employing a solution that distributes electricity to customers based on distinct justice indicators, taking into account variables such as electrification time, customers' electricity needs, and their demand. In this study, modeling the load-shedding problem is centered on justice, employing justice indicators (social welfare), including utilitarian, egalitarian, and envy-freeness. Subsequently, each of these indicators is defined.

- **Utilitarian Index:** The aggregate of individual profits across all customers within a system.
- **Egalitarian:** The shared benefit derived from the least favorable situation.
- **Envy-Freeness:** An equitable allocation index ensuring that resources are distributed among customers in a way that prevents jealousy among them.

Energy stands as an indispensable and pivotal element for attaining sustainable development. Nations striving toward this objective are reassessing their energy systems in alignment with energy planning and diverse strategies that align with the goals of sustainable development [24]. Social justice emerges as a vital concept in sustainable development, encompassing the degree of justice and inclusiveness in the distribution of energy resources, accessibility to energy systems, and the formulation of pricing structures to guarantee availability and affordability. The overarching principle is to make energy accessible to all at an equitable cost [24].

In the research conducted for the Sweden [25], various studies have been conducted using a framework for energy justice, which focuses on different elements of justice.

- **Distributional Justice:** This form of justice centers on evaluating the level of equity in the distribution of costs and benefits arising from the advancement of the energy system at the community level. This index specifically scrutinizes which segments of society bear the greatest impact from the costs associated with energy production, transmission, and distribution, and how the benefits achieving from the development of the energy system are apportioned among the individuals within that society.
- **Procedural Justice:** This index aims to establish an equitable decision-making process that involves all stakeholders in a nondiscriminatory manner.
- **Recognition Justice:** The aim of this index is to identify those overlooked in the decision-making of the energy system and whose influence is not acknowledged in this process.

In [26], diverse viewpoints on the fair distribution of energy in the future are discussed. The study delves into the interplay between energy justice and historical contexts, political landscapes, and other social movements. Additionally, it explores the implementation of place-based, frontline-centered, and spatially suited methods for achieving energy justice. Furthermore, the research introduces six principles for shaping the future of energy justice.

Moreover, this research introduces six principles for the future of equitable energy, encompassing: (1) being place-based, (2) addressing the root causes of inequality, (3) shifting the balance of power in existing forms of energy governance, (4) establishing novel, coordinated, and collaborative energy ownership and governance systems, (5) adopting a rights-based method, and (6) rejecting incorrect solutions.

In a research endeavor employing surveys and interviews among a population residing on the outskirts of major African cities [27], a hybrid theoretical framework drawing on urban political ecology and energy justice concepts, was applied to analyze the distribution of electricity access.

The findings from this study illustrate the impact of strategic planning and design in solar energy projects within marginalized areas on the benefit of local system customers. The research also offers guidance for national and international policymakers and executives in renewable energy companies, and also tools for assessing social justice.

In [28], eight principles are introduced for the assessment of energy justice, encompassing factors such as availability, affordability, due process, good governance, sustainability, intragenerational equality, intergenerational equality, and responsibility. Reference [29] then evaluated the common concepts of DR in connection with these eight energy justice principles from [28]. Although DR presents opportunities for enhancing energy justice, the associated risks of injustice depend significantly on the design of these programs.

In [30], four smart grid pilot projects are investigated, with data collection accomplished through interviews and content analysis. The outcomes are centered on energy justice concerning the design of distributional, procedural, and recognition justice. These considerations underscore the significance of fairness in data governance, participatory design, user control and autonomy, technology inclusiveness, and design for expansion and replication.

13.1.3 Research Gap and Innovations

To the best of the authors' knowledge, while various studies have presented indicators for assessing social justice in electrical energy systems, there remains a gap in investigating and implementing practical solutions within distribution companies to enhance social justice in electric energy distribution. To address this gap, this research focuses on proposing solutions aimed at effectively plan and operate electrical energy distribution networks to achieve social justice. The primary objective is to optimize the planning and operation of distribution networks, considering the technical, economic, and equitable distribution of curtailments among all customers. This research aims to offer actionable implementation plans to distribution network operators in order to mitigate customer dissatisfaction arising from uneven curtailments.

To accomplish this objective, a review and potential revision of criteria for identifying prioritized feeders and sensitive loads within distribution companies are necessary. Subsequently, a technically efficient and cost-effective process needs to be designed, enhancing automation in distribution feeders. This would enable the feeding of sensitive loads during critical periods, allowing curtailment to other consumers in the feeder. Implementing this plan not only increases the total curtailable load, aiding distribution companies in load shedding and consumption management but also fosters a fairer distribution of the adverse effects of curtailments among all customers.

Within this research framework, a metric termed the "controllability index" is introduced. This index quantifies the ratio of feeders outside of emergency load management programs to the total number of feeders, serving as a yardstick to evaluate distribution companies.

Hence, the key innovations of this research can be outlined as follows:

- Conducting operational (optimal load shedding) and planning (optimal placement of automation switches) studies within electric energy distribution networks with the objective of enhancing justice and equity.
- Introducing a controllability index in electrical energy distribution networks to monitor justice and equity in the execution of planned curtailments during peak periods.

13.1.4 Chapter Organization

The subsequent section provides problem descriptions and outlines the proposed method for obtaining the controllability index, along with the relevant formula. The third segment delves into the outcomes derived from implementing the proposed approach and offers a future prediction of the proposed controllability index for a representative distribution company. The next part provides final concluding remarks and outlines directions for future work.

13.2 Problem Description

13.2.1 Main Concept

As mentioned, various indicators for evaluating social justice in electrical energy systems have been presented in numerous studies. However, there remains a pressing need to conduct further research and devise practical solutions to enhance social justice in electrical energy distribution systems, particularly within distribution companies.

This chapter goes beyond the operational aspect by delving into optimal load shedding and extends its purview to the planning aspect by investigating the optimal placement of automation switches. Moreover, the chapter introduces a controllability index in electrical energy distribution networks to scrutinize justice and equity in the execution of planned curtailments during peak periods.

The primary objective of this chapter is to address the planning and operation problems of distribution networks, considering both technical and economic aspects, considering the equitable curtailments among all customers connected to the electrical distribution network. This comprehensive approach aims to provide practical plans for implementation to distribution network operators. The ultimate goal is to mitigate customer dissatisfaction arising from the uneven curtailments of consumers, even if it requires economic investments.

To attain this objective, the criteria for determining prioritized feeders and sensitive loads within distribution companies should be thoroughly reviewed and revised if deemed necessary (hereinafter, the term "prioritized feeder" is employed to refer to feeders that fall outside the emergency load management program). Subsequently, there is a requirement to design a technical and cost-effective process by enhancing the level of automation in distribution feeders. This augmentation facilitates the feeding of sensitive loads through adjacent feeders using maneuver actions when necessary, allowing for the curtailment of other consumers in the same feeder.

The implementation of this strategy, on the one hand, increases the total load that can be curtailed, significantly aiding distribution companies in the processes of load shedding and consumption management. On the other hand, it contributes to a more equitable distribution of the adverse effects of curtailments among all customers.

13.2.2 Controllability Index

In line with this research, the ratio of feeders outside emergency load management programs to the total number of existing feeders is introduced as an evaluative metric for distribution companies. This metric is defined as the controllability index, as outlined below:

$$\text{Controllability index} = 1 - \left(\frac{\text{Number of feeders outside emergency load management programs}}{\text{Total number of feeders}}\right) \times 100 \tag{13.1}$$

The controllability index employed in this study is based on three significant indices linked to social justice and equity: the envy-freeness index [23], the index of distributive justice [25], and the index of being place-based [26]. These indices have been established and applied in previous studies, forming the basis for the controllability index employed in this research.

13.2.3 Information Collection from Distribution Companies

The methodology employed in this research is illustrated in Figure 13.1. Initially, a reporting template is crafted, and distribution companies are subsequently requested to complete the requisite information for project completion. The essential information sought from distribution companies includes:

- Number of feeders outside emergency load management programs
- Total number of feeders
- Network peak load [MW]
- Number of sensitive loads
- Number of feeders outside emergency load management programs related to water facilities
- Names of relevant sensitive loads
- Available distributed generation capacity at the location of each sensitive load [MW]
- Total peak value of uninterruptible loads [MW]
- Feeder peak load [MW]
- Expected potential to be released and available for load-shedding programs after implementing maneuver actions [MW]
- Reason for the difference between the maximum releasable capacity and the expected releasable capacity.

Please be aware that in the subsequent sections of this chapter, the term "releasable capacity" is utilized to denote the power of nonsensitive loads situated on prioritized feeders. Due to their location and the presence of other sensitive loads on those feeders, they do not participate in load-shedding programs. Consequently, if certain actions and plans are implemented on the relevant prioritized feeders, it is possible to separate all or a portion of their capacity from the adjacent sensitive loads, making them available for inclusion in load shedding programs. This potential, once unlocked, is referred to as "released capacity."

13.2.4 Problem Formulation

Utilizing the provided input information, the subsequent variables are calculated individually for each feeder, as well as collectively for all feeders:

- **Total Maximum Releasable Capacity [MW]:** This parameter equals the cumulative maximum releasable capacity of all prioritized feeders within the domain of a power distribution company, calculated as follows:

$$P_d^{\mathrm{Max}} = \sum_{f}^{\mathrm{F}^d} p_{f,d}^{\mathrm{m}} \tag{13.2}$$

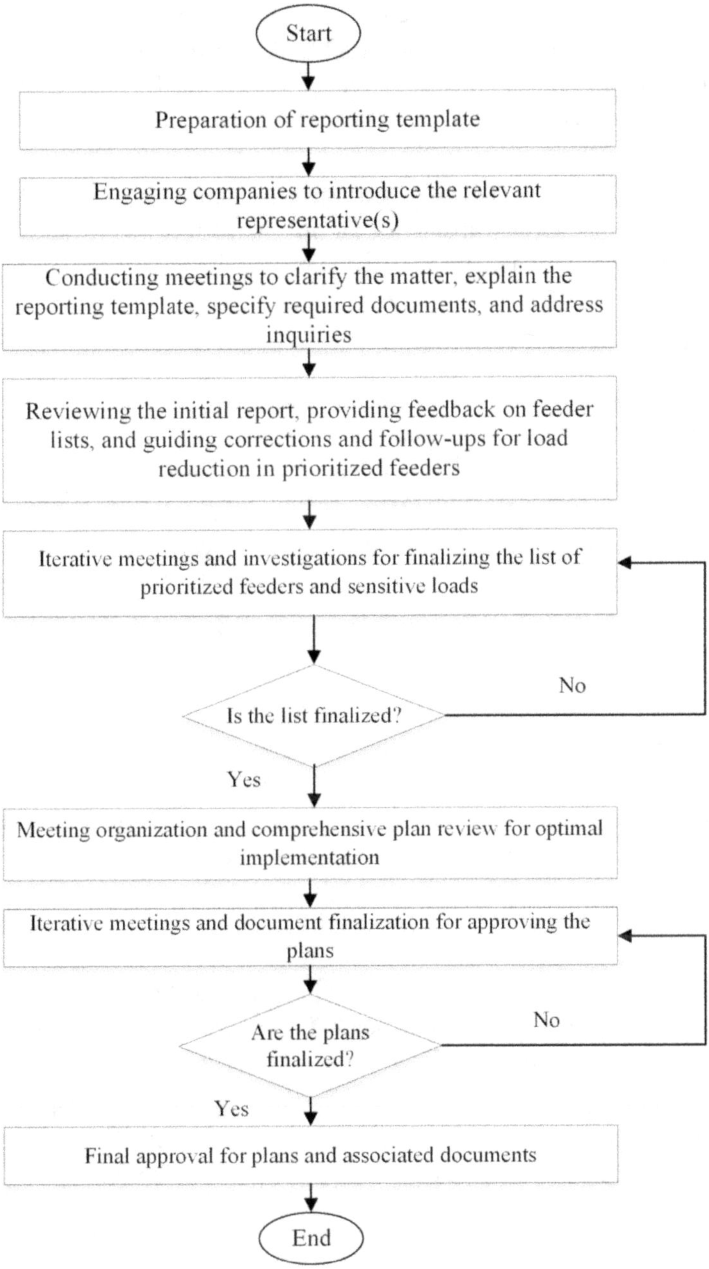

Figure 13.1 Reporting workflow from electricity distribution companies.

In Eq. (13.2), f, F^d, $p^m_{f,d}$, and P^{Max}_d represent the indicator of the number of prioritized feeders, the total number of prioritized feeders of the dth distribution company, the maximum releasable capacity on the prioritized feeder f of the power distribution company d, and the total maximum releasable capacity in the power distribution company d, respectively.

- **Total Expected Releasable Capacity [MW]:** This parameter equals the cumulative expected releasable capacity of all prioritized feeders within the domain of a power distribution company, calculated as follows:

$$PE_d = \sum_{f}^{F^d} pe_{f,d}. \tag{13.3}$$

In Eq. (13.3), $pe_{f,d}$ and PE_d denote the expected releasable capacity of prioritized feeder f within power distribution company d, and the total expected releasable capacity within the domain of power distribution company d, respectively.

- **Total Capacity of Sensitive Loads [MW]:** This parameter equals the total amount of sensitive loads in the distribution network and can be expressed as follows:

$$PS_d = \sum_{f}^{F^d} ps_{f,d} \tag{13.4}$$

In Eq. (13.4), $ps_{f,d}$ and PS_d represent the load value of the sensitive loads on the fth prioritized feeder within power distribution company d, and the sum of the sensitive loads within power distribution company d, respectively.

- **Total Load of Prioritized Feeders [MW]:** This parameter equals the total load on prioritized feeders and can be calculated using (13.5).

$$PD_d = \sum_{f}^{F^d} pd_{f,d} \tag{13.5}$$

In Eq. (13.5), $pd_{f,d}$ and PD_d represent the load of the prioritized feeder f within power distribution company d, and the total load of prioritized feeders in power distribution company d, respectively.

- **Total Cost of the Proposed Plan:** This is the estimated cost for the implementation of the proposed plans within the domain of each electricity distribution company and can be calculated as follows:

$$C_d = \sum_{i}^{I^d} c_{i,d} \tag{13.6}$$

In Eq. (13.6), I^d, $c_{i,d}$, and C_d represent the number of plans presented for power distribution company d, the estimated cost of implementing the ith proposed plan in power distribution company *d*, and the total estimated cost of the proposed plans in power distribution company *d*, respectively.

Subsequently, the companies' performance is assessed, and their proposals are scrutinized to enhance the controllability index. Ultimately, the list of their activities is finalized.

13.2.5 Budget Allocation for the Installation of Automation Switches

Strategically placing automation switches, such as Sectionalizer and reclosers, within distribution networks can enhance the controllability index and empower operators to execute diverse maneuvers for effective load-shedding management. Consequently, it might be assumed that companies with more switches would exhibit better controllability levels [31, 32]. However, the situation is more complex. A crucial consideration is that the condition of distribution networks varies significantly concerning factors such as network load levels and geographic features.

Hence, determining the equitable allocation of budget for the provision of automation switches among electric power distribution companies emerges as a multi-criteria decision-making challenge. The factors recommended for consideration in this decision-making process are depicted in Figure 13.2.

13.3 Numerical Analyses

In accordance with the flowchart depicted in Figure 13.2, the chosen representative distribution company endeavored to minimize the number of prioritized

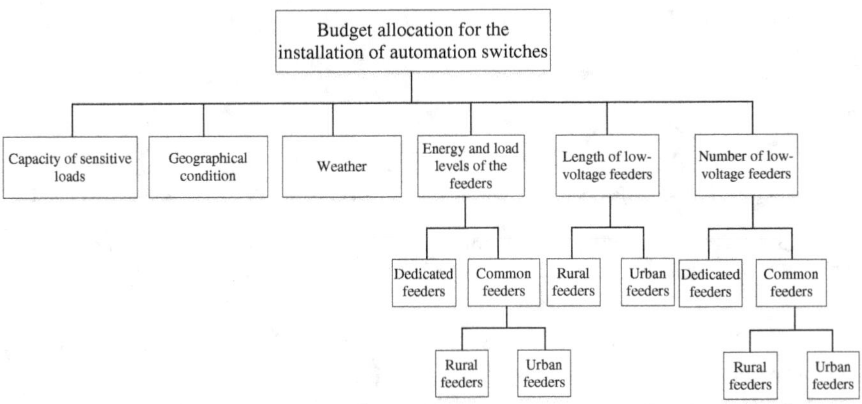

Figure 13.2 Recommended factors for multi-criteria decision-making inequitable budget allocation for automation switches.

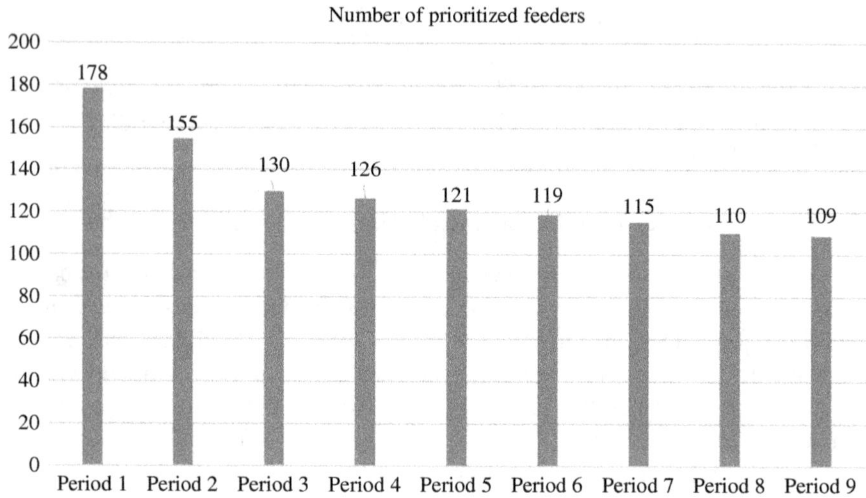

Figure 13.3 Changes in the number of prioritized feeders in the respective distribution company.

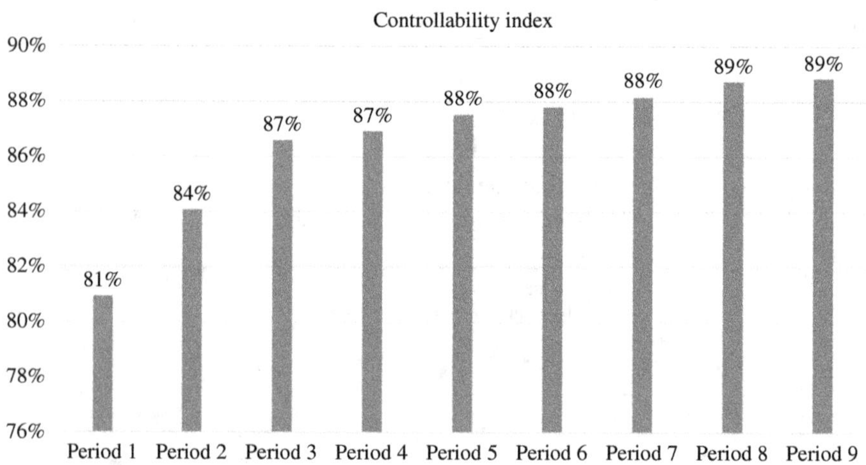

Figure 13.4 Variations in the controllability index in the respective distribution company.

feeders, as illustrated in Figure 13.3. The outcome of these efforts is reflected in the controllability index value, as depicted in Figure 13.4.

It should be noted that the initial step in the examination process involved compiling a list of all sensitive loads and then reevaluating them to avoid the inclusion of nonsensitive loads. It was observed that numerous loads conflicted with the definition of sensitive loads despite being classified as such by distribution companies. Following the reevaluation of the sensitive load list within the distribution company, a significant number of these loads were removed, resulting in the release of a substantial portion of the available capacity. Various interpretations of the sensitive load concept, along with factors such as interventions by local authorities, contributed to the absence of a standardized procedure for identifying sensitive loads. As a result of this initial investigation, a significant number of loads were able to be released by reducing the number of prioritized feeders.

As depicted in Figure 13.3, the representative distribution company implemented various measures successfully, leading to a notable reduction in the number of prioritized feeders within its distribution system. Several approaches have been adopted by distribution companies, according to the specifications of their individual systems, to increase the released capacity. These include strategies like switch placement and the development of maneuver actions. The count decreased from 178 feeders to 109 feeders. It is essential to highlight that this evaluation is based on monthly reports encompassing nine reporting time periods. Additionally, Figure 13.4 illustrates that, as a result of the implemented actions within this distribution system, the controllability index experienced an approximate increase from 81% to 89%. These outcomes signify the positive impact of the company's strategies in optimizing the distribution network's controllability.

Taking into account that the available budget is a pivotal constraint in plan implementation, an analysis was conducted on the extent of controllability index variations relative to the available budget, as depicted in Figure 13.5. This figure illustrates the estimated values of the controllability index corresponding to different available budgets (budget is considered as the ratio of total cost to the expected releasable capacity for each feeder). The sensitivity analysis focuses on examining the impact of changes in the available budget on the controllability index. Distribution companies strategically plan actions for prioritized feeders to incorporate their releasable capacities into load management programs. The analysis involves estimating the potential of releasable capacity across all distribution systems by adjusting the available budget. This estimation considers the specifications of both the distribution system and sensitive loads. Through extensive meetings and studies, distribution companies identify various feasible actions. The aim of the sensitivity analysis is to discern how the controllability index varies concerning the available budget. At every budget tier, the analysis focuses on selecting plans that yield the most cost-effective results, determined by the ratio of the required cost of each plan to its potential for releasing capacity for participation in load management programs. The objective is to select plans with minimal costs and maximum released capacity upon implementation, thereby ensuring optimal resource utilization.

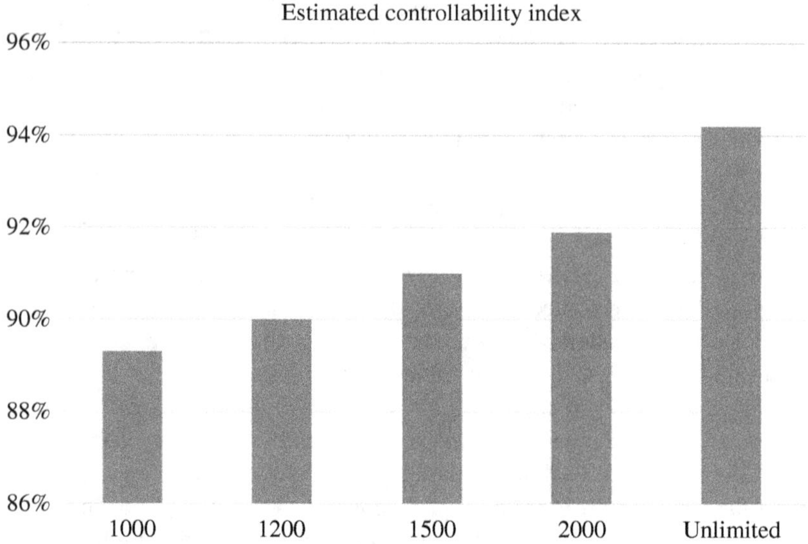

Figure 13.5 Estimated controllability index value in relation to the available budget (ratio of total cost to the expected releasable capacity for each feeder).

The analysis results, as illustrated in Figure 13.5, provide valuable insights into the relationship between the estimated controllability index value and the available budget. The findings suggest that while improvements in the controllability index can be achieved with an increased budget, there exists a limiting threshold beyond which unlimited financial allocations fail to yield further enhancements. This limitation is attributed to various factors, such as geographical and operational constraints, highlighting the complex interplay between budgetary considerations and the controllability index. These insights underscore the need for a balanced and strategic approach to budget allocation, taking into account the inherent limitations and ensuring effective decision-making for optimal operational efficiency within electric energy distribution networks.

13.4 Conclusion

The primary contribution of this chapter is considering the indices of social justice and equity in the operation and planning studies of the electrical distribution networks. The introduced controllability index, rooted in three indicators – lack of envy, distributive justice, and rule-based fairness – serves this purpose. To assess electrical distribution companies, the initial step involves defining the required information that distribution companies should provide. This allows for the calculation of the controllability index for distribution companies. Afterward, corrective measures will be implemented, and the extent of improvement in the controllability index resulting from these actions will be determined. In this chapter, an analysis of the controllability index results was carried out for a representative electrical distribution company. Following this, a sensitivity analysis was conducted on the controllability index concerning the available budget.

The results showed that, although the controllability index could be improved by increasing the available budget, specifying an unlimited budget cannot surpass a certain limit due to various factors, including geographical and operational constraints.

As the installation of automation switches in the network significantly contributes to enhance the controllability index, the allocation of an appropriate budget for the installation of these switches in various distribution companies was posed as a multi-criteria decision-making problem. Criteria such as the number and length of medium-voltage feeders, weather conditions, and the quantity of sensitive loads in the network will be crucial factors influencing this matter.

In conclusion, the outcomes of this chapter underscore the imperative need for integrating principles of justice into various decision-making processes within the electricity industry, particularly those directly impacting customers. Various indicators may be employed depending on the nature of the problem at hand; however, the final objective of these indicators remains the establishment of social justice through the equitable distribution of electricity among consumers while considering technical, economic, and environmental criteria.

Acronyms

CEC community energy cells
DER distributed energy resources
DR demand response
DSM demand side management
EE energy efficiency
MG microgrids

References

1 Vahidinasab, V., Tabarzadi, M., Arasteh, H. et al. (2020). Overview of electric energy distribution networks expansion planning. *IEEE Access* 8 (1): 34750–34769.

2 Shirley, J., Angela, J., Manivelan, H. et al. (2024). Dynamic load shedding and system restoration using wide area distribution management system. *IET Generation, Transmission & Distribution* 18: 2860–2874.

3 Hassan, Y., Merzah, A., Hassan, S.A., and Tuaimah, F. (2021). Optimal reduction of electrical loads based on priorities developed using a genetic algorithm. *IOP Conference Series: Materials Science and Engineering* 1067 (1): 012127.

4 Eikeland, P.O., Taranger, K.K., Inderberg, T.H.J., and Gulbrandsen, L.H. (2023). A wind of change in Norway: explaining shifts in municipal stances on wind power by policy feedback and energy justice. *Energy Research & Social Science* 104: 103231.

5 Sovacool, B.K., Carley, S., and Kiesling, L. (2024). Energy justice beyond the wire: exploring the multidimensional inequities of the electrical power grid in the United States. *Energy Research & Social Science* 111: 103474.

6 Diezmartínez, C.V. and Zhang, A. (2023). Powering just energy transitions: a review of the justice implications of community choice aggregation. *Energy Research & Social Science* 103: 103221.

7 Akbari, A., Vahidinasab, V., Arasteh, H., and Kazemi-Robati, E. (2021). Rural/residential microgrids: concepts, status quo, model and application. In: *Residential Microgrids and Rural Electrifications*, Chapter 7, 1e (ed. P. Sanjeevikumar, C. Sharmeela, S. Padmanaban, and J.B. Holm-Nielsen), 131–161. Elsevier.

8 Oliveira, R.S.d. (2007). *Oil and Politics in the Gulf of Guinea*. New York, NY: Columbia University Press.

9 Guruswamy, L. (2010). Energy justice and sustainable development. *Colorado Journal of International Environmental Law and Policy* 21: 231.

10 Cervantes, M.Á.M. and Solís, L.R. (2024). Energy poverty and social justice in Mexico: the rights of electricity consumers. *The Electricity Journal* 37 (2): 107372.

11 Hashemi, S.M., Arasteh, H., Shafiekhani, M. et al. (2023). Multi-objective operation of microgrids based on electrical and thermal flexibility metrics using the NNC and IGDT methods. *International Journal of Electrical Power and Energy Systems* 144: 108617.

12 Jabari, F., Zeraati, M., Sheibani, M., and Arasteh, H. (2024). Robust self-scheduling of PVs-wind-diesel power generation units in a standalone microgrid under uncertain electricity prices. *Journal of Operation and Automation in Power Engineering, University of Mohaghegh Ardabili* 12 (2): 152–162.

13 Kia, M., Shafiekhani, M., Arasteh, H. et al. (2020). Short-term operation of microgrids with thermal and electrical loads under different uncertainties using Information gap decision theory. *Energy* 208: 118418.

14 Bahramara, S., Sheikhahmadi, P., Mazza, A., and Chicco, G. (2022). Provision of the Flexible Ramping Product in a microgrid considering the trading strategies in the energy markets. *Sustainable Energy, Grids and Networks* 32: 100953.

15 Bahramara, S. (2022). The microgrids' participation problem in the wholesale markets considering different viewpoints about the ownership and the management of the battery energy storages. *Journal of Energy Storage* 55: 105432.

16 Galvin, R. (2024). Re-thinking energy justice to achieve a fair distribution of shared electricity from rooftop photovoltaics in a typical multi-apartment building in Germany: an interdisciplinary approach. *Energy Research & Social Science* 112: 103531.

17 Inderberg, T.H.J., Palm, J., and Matthiasen, E.H. (2024). Flexible electricity consumption policies in Norway and Sweden: implications for energy justice. *Energy Research & Social Science* 110: 103466.

18 Abdelbary, A.M., Manglicmot, L., Kanwhen, O., and Mohamed, A.A. (2024). Community-centric distributed energy resources for energy justice and decarbonization in dense urban regions. *Energy Reports* 11: 1742–1751.

19 Zhang, Q. and Li, F. (2024). Securing energy equity with multilayer market clearing. *IEEE Transactions on Energy Markets, Policy and Regulation* 18: 1–12, Early Access.

20 Arasteh, H., Vahidinasab, V., Sepasian, M.S., and Ghaderi, A. (2019). Integrated generation, transmission and energy efficiency planning. *34th International Power System Conference (PSC)*, Tehran, Iran, 571–578.

21 Shekari, M., Arasteh, H., Fini, A.S., and Vahidinasab, V. (2021). Demand response requirements from the cultural, social, and behavioral perspectives. *Applied Sciences* 11 (23): 1–23.

22 Arasteh, H., Vahidinasab, V., Sepasian, M.S., and Aghaei, J. (2019). Stochastic system of systems architecture for adaptive expansion of smart distribution grids. *IEEE Transactions on Industrial Informatics* 15 (1): 377–389.

23 Oluwasuji, O.I., Malik, O., Zhang, J., and Ramchurn, S.D. (2020). Solving the fair electric load shedding problem in developing countries. *Autonomous Agents and Multi-Agent Systems* 34 (1, 1): –35.

24 Vera, I. and Langlois, L. (2007). Energy indicators for sustainable development. *Energy* 32 (6): 875–882.

25 Ramasar, V., Busch, H., Brandstedt, E., and Rudus, K. (2022). When energy justice is contested: a systematic review of a decade of research on Sweden's conflicted energy landscape. *Energy Research & Social Science* 94: 102862.

26 Elmallah, S., Reames, T.G., and Spurlock, C.A. (2022). Frontlining energy justice: visioning principles for energy transitions from community-based organizations in the United States. *Energy Research & Social Science* 94: 102855.

27 Cantoni, R., Caprotti, F., and de Groot, J. (2022). Solar energy at the peri-urban frontier: an energy justice study of urban peripheries from Burkina Faso and South Africa. *Energy Research & Social Science* 94: 102884.

28 Sovacool, B.K. and Dworkin, M.H. (2015). Energy justice: conceptual insights and practical applications. *Applied Energy* 142: 435–444.

29 Calver, P. and Simcock, N. (2021). Demand response and energy justice: a critical overview of ethical risks and opportunities within digital, decentralised, and decarbonised futures. *Energy Policy* 151: 112198.

30 Milchram, C., Künneke, R., Doorn, N. et al. (2021). Designing for justice in electricity systems: a comparison of smart grid experiments in the Netherlands. *Energy Policy* 147: 111720.

31 Fayyazi, S., Azad-Farsani, E., and Haghighi, A.A. (2024). Resilience-oriented sectionalizing and tie switches sitting in distribution networks with complex topologies. *Reliability Engineering & System Safety* 243: 109919.

32 Reiz, C., De Lima, T.D., Leite, J.B. et al. (2022). A multiobjective approach for the optimal placement of protection and control devices in distribution networks with microgrids. *IEEE Access* 10: 41776–41788.

14

The Role of Artificial Intelligence in Advancing Energy Justice and Equity

Nastaran Gholizadeh, Noon Hussein, and Petr Musilek

Department of Electrical and Computer Engineering, University of Alberta, Edmonton, Alberta, Canada

Acronyms

AE-BiLSTM	Autoencoder and Bidirectional LSTM
AI	Artificial Intelligence
ANFIS	Adaptive Neuro-Fuzzy Inference System
ANN	Artificial Neural Network
ANSI	American National Standard Institute
CNN	Convolutional Neural Networks
DNN	Deep Neural Network
DRL	Deep Reinforcement Learning
DSO	Distribution System Operator
LSTM	Long Short-Term Memory
MADDPG	Multi-Agent Deep Deterministic Policy Gradients
ML	Machine Learning
POMDP	Partially Observable Markov Decision Process
RL	Reinforcement Learning
SVM	Support Vector Machine
XAI	Explainable AI

14.1 Introduction

Artificial intelligence (AI) has played a fundamental role in the digitization of energy systems and facilitation of the energy transition. As the driving force behind the second wave of digitization, AI technologies address a wide range of challenges faced by energy systems in terms of planning, operation, control, and

Communities for Clean Energy Justice and Equity in Grid Modernization, First Edition.
Edited by Mohammadreza Daneshvar, Behnam Mohammadi-Ivatloo, and Amjad Anvari-Moghaddam.

maintenance, from the integration of renewable energy to the regulation of tariffs and management of increasing consumer load demand.

Power system planning employs AI technologies in a multitude of applications, including distribution system planning, reactive power optimization, capacitor placement, and customer modeling and control. On the other hand, power system operations cover AI optimization methods for economic load dispatch, unit commitment, and customer and retail operations. The primary applications of AI in system control are power system stability, voltage, and load frequency control. These applications facilitate the transition from centralized to coordinated decentralized control, focusing on consumer participation, sustainability, and economic efficiency on the dynamic grid. Although there are many cases of power and energy applications of AI in the broad sense, most instances are based on a subfield of AI called machine learning (ML). To maintain a broad perspective, we employ the term AI, but it is important to note that ML is used more frequently.

Despite the broad opportunities offered by AI, various concerns about its implications on energy justice and equity have not yet been addressed. To demonstrate the data-driven nature of many AI applications raises privacy infringement concerns, especially when forecasting consumer demand to a high level of accuracy, as it requires different types of personal data related to daily schedules and habits. Additionally, the incorporation of AI raises concerns about transparency as certain technologies inherently operate as black boxes. In addition, autonomous AI-driven tasks reduce user participation and accountability due to increased complexity and speed, which also challenge energy regulators in monitoring and controlling the development of prices. Consequently, the advanced energy solutions offered by AI mandate reflection on energy justice, particularly on the potential implications of AI integration in energy systems as well as the implications of design and regulation.

Energy justice has been predominantly investigated in social science research on the energy transition to produce just energy-related decisions. A key issue here is that the changes introduced into the energy system can have a negative impact on certain social groups. Therefore, in addition to standard technical and economic constraints, energy justice aims to prioritize ethical and social concerns in decision-making processes. As such, injustices can be identified along with affected communities in order to design remediation processes to recognize and reduce these injustices.

To showcase the effects of integrating AI in energy justice and equity, this chapter presents a review of opportunities and challenges, followed by an examination of the prevailing three-tenet energy justice framework, which focuses on three aspects: distribution justice, procedural justice, and recognition justice. Next, AI transparency, fairness, and bias mitigation are investigated through the lens of energy justice. The case study on AI-based justice-oriented emergency load shedding is then presented using the IEEE 33-bus test system. Finally, the chapter highlights the need for an effective development of justice-oriented policies by bridging energy law and AI regulations.

The primary objective of this chapter is to examine the integration of AI in the energy sector, specifically focusing on its role in promoting justice and equity. This entails:

1) Exploring the multifaceted impact of AI on energy justice, particularly within the frameworks of distributive, procedural, and recognition justice.
2) Evaluating design principles of AI, emphasizing the necessity of transparency, fairness, and bias mitigation to ensure equitable outcomes.
3) Demonstrating the practical application of these principles through a case study on reinforcement learning (RL)-based justice-oriented emergency load shedding.
4) Promoting the development of policy frameworks that bridge energy law and AI regulations, thereby aligning AI advancements with the overarching goals of justice and equity in the energy sector.

14.2 AI and the Energy Transition: Opportunities and Challenges

In energy systems, digitization facilitates the integration of increasing shares of variable renewable energy sources, thereby improving the reliability of grids. On the other hand, end users benefit from digitization by improving energy and material efficiency while also reducing emissions.

The first wave of digitization began in the 1970s with semiconductors as foundational drivers. As a result, the 1990s and early 2000s marked the rise of connected devices and cloud computing, which allowed the storage and access to information at a fraction of local infrastructure costs [1]. As referred to by MIT's Erik Brynjolfsson and Andrew McAfee, the second wave of digitization was driven by the second machine age [2], and involved the integration of software-driven machines to automate various cognitive tasks performed originally by humans. At the heart of this second wave is AI, which aims to make decisions with minimal human intervention. With machines joining humans as intelligent participants in today's AI-driven world, their engagement in the energy transition presents notable prospects, delineated in Table 14.1. At the same time, they face a number of challenges, as highlighted in Table 14.2.

AI-aided power system planning encompasses applications such as distribution system planning and capacitor placement, in addition to customer modeling and control. Within this scope, long short-term memory (LSTM) model has been used

Table 14.1 Opportunities of integrating AI in the energy transition (ET).

Aspect of ET	Opportunity
Predictive maintenance	Artificial intelligence (AI) can analyze sensor data to predict equipment failures, reducing downtime and maintenance costs.
Renewable energy optimization	AI can optimize the generation, storage, and distribution of renewable energy based on weather forecasts and demand patterns.
Energy efficiency improvement	AI-powered systems can identify inefficiencies in energy consumption and implement optimizations to reduce waste.
Grid management	AI can optimize the distribution grid by balancing supply and demand in real-time and preventing blackouts.
Demand response	AI can analyze consumer behavior and adjust energy consumption to match supply, reducing peak demand and costs.
Smart grids	AI can enable the integration of various energy sources and devices into a unified, responsive grid infrastructure.
Energy trading	AI can optimize energy trading strategies in real-time markets, thereby maximizing profit and grid stability.
Decentralized energy systems	AI can coordinate decentralized energy systems, such as microgrids and distributed generation to improve reliability and efficiency.
Policymaking	AI can analyze vast amounts of data to inform evidence-based energy policies, driving sustainable and equitable energy transitions.
Environmental impact assessment	AI can assess the environmental impact of energy projects and recommend mitigation strategies to minimize harm.

to forecast residual demand curves with the goal to improve bidding strategy assessment in market-clearing prices. Additionally, RL is adopted for end user decision-making in deregulated retail markets to optimize pricing selection plans and consumer satisfaction [3], while deep neural networks (DNNs) are used to forecast short-term residential loads, focusing on spatiotemporal correlations of appliances load data [4].

Applications of power system operation include economic load dispatch, reactive power compensation, unit commitment, optimal power flow, and customer and retail operations. Examples of these applications include deep-learning-based economic load dispatch [5] as well as reactive power compensation through graph convolutional network-based deep RL (DRL) [6]. In addition, RL is used to solve the unit commitment problem for the next day [7], increase renewable energy generation [8], and autonomously make decisions related to energy trading and scheduling in conjunction with attention mechanisms [9]. Additionally,

Table 14.2 Challenges of integrating AI in the energy transition.

Challenge	Description
Integration complexity	Ensuring seamless AI integration into existing energy infrastructure, including legacy systems and diverse energy sources.
Data quality and availability	Dependence on high-quality real-time data, which may be limited or unavailable for certain energy sources or regions.
Scalability	Scaling AI solutions to handle complex and diverse large-scale energy systems.
Interoperability	Ensuring interoperability and compatibility between different AI systems and technologies used by various stakeholders.
Regulatory hurdles	Navigating complex and evolving regulatory frameworks that may not be designed to accommodate rapid AI advancements.
Cybersecurity	Addressing potential cybersecurity threats and vulnerabilities associated with AI applications in critical energy infrastructure.
Stakeholder engagement	Achieving collaboration and acceptance among diverse stakeholders, including government bodies, industry players, and the public.
Ethical considerations	Addressing ethical concerns related to AI decision-making, transparency, and the impact on employment.
High initial costs	Overcoming initial financial barriers associated with the adoption and implementation of AI technologies.
Skill gap	Developing and retaining a skilled workforce to implement, manage, and optimize AI applications.

ML is a powerful tool for anomaly detection in balancing electricity markets [10], optimizing demand response aggregator decisions [11], and addressing supply–demand imbalances through dynamic differential control [12]. At the same time, ML plays a pivotal role in increasing the participation of consumers in energy management, such as for electric vehicles and thermostatically controlled load aggregators [10] or for air conditioning systems through occupancy forecasting and automation [13].

Power system control is capable of establishing efficient automation management while minimizing risks of failure with AI. In this scope, stability control utilizes artificial neural networks (ANNs) for automatic restoration, fault management, and diagnosis of network faults [14]. On the other hand, dynamic line rating forecasting, aimed at improving transmission line capacity and mitigating renewable energy curtailment, is achieved through k-means clustering [15], and autoencoder and bidirectional LSTM (AE-BiLSTM) to improve grid efficiency and reliability [16]. In voltage control, DRL and multi-agent deep deterministic policy gradients (MADDPG) enable real-time volt-var control in active distribution networks [17], while load frequency control often utilizes RL [18] to actively coordinate fault tolerance load frequency control and prevent sudden frequency changes caused by unit failures.

ML plays a pivotal role in advancing power system maintenance, namely, predictive-, preventative-, and condition-based maintenance. To predict the remaining useful life of low-voltage contractors, Fu and Liu [19] employ support vector machines (SVMs), convolutional neural networks (CNNs), and LSTM. For anomaly detection, Yuan and Liu [20] use DNN-based vision analysis for timely power supply restoration, while Zemali et al. [21] employ LSTM and adaptive neuro-fuzzy inference system (ANFIS) to isolate and classify faults for wind turbines. Routine inspections, as demonstrated by Di Renzo et al. [22], are aided by automatically identifying and classifying images of healthy and defective PV cells using deep learning (YOLOv5) and CNNs. Condition-based maintenance can take advantage of partially observable Markov decision processes (POMDPs) to improve traditional traction power supply equipment maintenance by optimizing cost and reliability [23].

14.3 Evaluating the Impact of AI Through the Lens of Energy Justice

14.3.1 Defining the Role of AI in Advancing Energy Justice

While no global consensus has been reached in the literature on the definitive meaning of "energy justice," the ideas commonly used to explain what falls under this concept are widely accepted in academic literature. For example, the framework proposed by Sovacool et al. [24] centers the definition of energy justice on the societal allocation of benefits and burdens, with the aim of achieving fairness. This framework categorizes energy justice based on factors such as sustainability, intergenerational equity, due process, accountability, affordability, availability, transparency, resistance, intragenerational equity, and intersectionality. The most predominant framework is the "triumvirate of tenets," which identifies three foundational tenets of energy justice: *distributive* (or *distributional) justice*, *procedural justice*, and *recognition justice* [25]. Recently, these tenets have been expanded to five, additionally incorporating *restorative justice*, which refers to the correction of energy sector injustices and *cosmopolitan justice*, aimed at emphasizing the need for cross-border effect considerations in energy-related matters [26]. This chapter focuses on the three main "triumvirate of tenets."

In the constantly evolving landscape of energy systems, the role of AI stands as a transformative force in advancing the crucial concept of energy justice.

At its core, energy justice seeks to ensure that all individuals, regardless of socioeconomic status or geographical location, have equal access to affordable, reliable, and sustainable energy resources. In this context, AI plays a pivotal role in optimizing energy distribution, improving grid efficiency, and promoting the integration of renewable sources, in addition to other benefits. Through predictive analytics and smart technologies, AI is capable of allocating resources in a more equitable manner, identifying underserved communities, and addressing their specific energy needs. Correspondingly, AI empowers the active participation of individuals in energy decision-making processes, therefore fostering a more inclusive and participatory energy landscape.

Given the three foundational pillars of energy justice, substantial considerations for the integration of AI include transparency, fairness, and bias mitigation. Transparent AI algorithms improve clarity in decision-making processes, allowing accountability and promoting trust among stakeholders. Fairness, however, is a cornerstone for ensuring that these algorithms do not perpetuate or exacerbate existing social inequalities, instead aiming at equitable resource and benefit distribution, while bias mitigation is fundamental in preventing unintended consequences and discriminatory outcomes, particularly in energy pricing and resource allocation. Embracing these principles is not only an ethical imperative, but is also instrumental in cultivating a sustainable and just energy landscape. As such, defining the role of AI reforms from a technological imperative to an ethical and moral obligation on the path toward a just and sustainable energy future is critical.

14.3.2 AI Design and Distributive Justice: Balancing Benefits and Burdens

Distributive justice refers to considerations surrounding the equitable allocation of benefits, burdens, expenses, and obligations among stakeholders involved in an energy system [27]. Although costs and benefits may have financial implications, they can also manifest themselves in terms of the distribution of risks, externalities, and access to energy services. The concept of energy justice therefore encompasses discussions on the fair distribution of disparities, both locally (such as disparities between rural and urban areas) and globally (such as disparities between developing and developed countries). Within the concept of energy justice, common challenges of distributive justice include decisions related to infrastructure placement for energy production and addressing energy poverty [28]. In this context, it is acknowledged that some issues identified here have multidimensional implications on energy justice.

AI plays a key role in ensuring distributive justice, namely by:

1) Facilitating more informed decision-making processes in energy infrastructure development.

2) Mitigating bias in energy systems.
3) Optimizing resource allocation in energy systems.
4) Improving energy affordability for marginalized communities.
5) Promoting transparency and accountability in energy governance.

In addition to the novel contributions of AI to promote energy justice, AI design can present challenges and limitations to real-world implementation, as described by Donti and Kolter [29]. Within the context of distributive justice, AI heavily relies on the data it is trained on, hence potentially inheriting and amplifying existing biases, inequities, and market failures through its data, design, and applications.

Although the immediate societal and ethical impacts of specific algorithms or models might not be evident, their selection can have lasting effects on various stakeholders within the realm of energy justice. These technologies influence aspects such as the equitable distribution of energy and the participation of specific individuals in decision-making processes related to energy systems. In essence, those with greater access to knowledge, skills, and technology are likely to have a considerable advantage in using emerging technologies such as PV panels, electric vehicles, and batteries to their benefit, therefore potentially leaving those with limited access to these resources at a disadvantage. Similarly, AI technologies that facilitate smart grids and electricity networks, while contributing to increasingly just access for citizens, may also disproportionately favor wealthier segments of society, potentially harming more vulnerable groups.

To address the role of AI design in balancing distributive justice benefits and burdens, several reports published in the early 2000s, such as the World Energy Assessment [30, 31] and the Global Energy Assessment [32], acknowledged the need to promote equitable energy access for long-term sustainability by considering the global disparities in income, access to resources, and energy utilization among developed and developing nations. Recommendations included the adoption of renewable energy sources, locally managed renewable energy, and intelligent power grids to mitigate energy poverty and promote equity. Other studies have focused primarily on identifying and assessing injustices, such as those related to the siting of wind farms or the unequal distribution of energy resources [33, 34]. Nevertheless, energy poverty, described by Rubio-Bellido and Solis-Guzman [35] as an injustice disproportionately affecting consumers who are economically vulnerable, older adults, or people with disabilities, has been highlighted in the literature as a prominent challenge in energy distribution, since benefits tend to favor well-resourced consumers in areas with strong municipal finances.

While the traditional concept of energy justice is centered around supply and demand, this understanding has expanded to include smart grids, thereby indicating an increased convergence of information and communication technologies within the energy sector. Notably, studies referencing distributive justice examine the advantages and disadvantages of smart meters, smart homes, and demand-side response. Tensions arise between consumers and energy companies, particularly with respect to the disproportionate cost sharing for smart metering and the burden of responsibility for energy conservation. Furthermore, the growing dependence on information and communication systems brings up discrimination concerns, impacting the elderly, the disabled, and those with less technological proficiency. Despite the presence of literature evaluating decision-making and policy-design concerns [36–38], there is yet a lack of theoretical development on energy justice, particularly considering the design and integration of AI.

Moreover, concerns surrounding distributive justice are evident in debates criticizing the uneven distribution of financial gains between consumers and energy companies, as economic advantages of the energy sector are tilted in favor of energy companies, leaving consumers at a disadvantage. Perceptions of injustice are heightened by the mechanisms used to pass on costs associated with smart metering investments. In addition, reliability issues in AI technologies can disrupt productivity and innovation, further hindering how economic growth translates into a fairer distribution of wealth.

Distribution issues extend beyond consumers and energy companies to different consumer groups. Focusing on novel and complex technologies, smart grids are often seen as exclusive, as they particularly impact economically vulnerable groups. Exclusion concerns are also related to demand response, where not every consumer has an equal opportunity to shift their demand, potentially leaving certain segments financially hampered. Meanwhile, monetary aspects revolving around the role of distribution system operators (DSOs) in smart grid investments have reflected a conflict with regulatory authorities. That is, DSOs claim to be logical leaders in the development of smart grids, whereas regulatory authorities, policymakers, and energy suppliers criticize this leadership role, considering it a deviation from the core tasks of DSOs and an inefficient use of public funds.

Finally, the unfair distribution of knowledge in smart grids arguably challenges municipal autonomy [39]. Since this is linked to the value of cooperation, it emphasizes increased collaboration between stakeholders. Although public–private collaboration is generally seen in a positive light [40], concerns have been raised about the distribution of knowledge between private companies and municipalities. Regarding novel smart grid technologies, the necessary specialized knowledge also raises doubts that concentrated knowledge in private corporations could wield undue influence over municipalities. To illustrate its multifaceted role in balancing benefits and burdens, Table 14.3 reviews the impact of AI in promoting equitable energy solutions.

Table 14.3 The role of AI in balancing benefits and burdens in energy justice.

AI role	Impact on energy justice
Enhanced decision-making	AI facilitates more informed and equitable decision-making processes in energy infrastructure development, considering factors such as socioeconomic disparities and environmental justice.
Bias mitigation	AI technologies can help identify and mitigate biases in energy systems, promoting fairness in resource allocation and access to energy services.
Optimized resource allocation	AI optimizes resource allocation in energy systems, ensuring equitable distribution of benefits and burdens among diverse stakeholders.
Improved energy affordability	AI-driven solutions can improve energy affordability for marginalized communities by optimizing energy consumption, reducing costs, and providing targeted assistance programs.
Transparency and accountability	AI promotes transparency and accountability in energy governance by enabling real-time monitoring, auditing, and reporting of energy-related decisions and actions.
Community engagement	AI tools facilitate meaningful community engagement in planning and decision-making, empowering local communities and ensuring their interests are considered.
Addressing energy poverty	AI initiatives target energy poverty by identifying vulnerable populations, implementing targeted interventions, and promoting inclusive access to clean and affordable energy resources.
Mitigating environmental impact	AI-powered solutions optimize energy production and consumption to minimize environmental harm, promoting environmental justice and sustainability.
Ethical considerations	AI design incorporates ethical principles to ensure justice and equity in energy transitions, addressing concerns related to privacy, discrimination, and human rights.
Policy support	AI analytics provide valuable insights for policymakers to develop evidence-based energy policies that prioritize equity, social justice, and human well-being.

14.3.3 AI Design and Procedural Justice: Ensuring Transparency and Participation

Procedural justice encompasses equitable access and participation in decision-making processes related to the distribution aspects mentioned above [27]. Key components of procedural justice therefore include:

1) Enhancing control, transparency, stakeholder participation and democracy.
2) Providing equitable market access and supporting underprivileged consumers.

In addition to these components, this aspect of justice also involves ensuring that individuals, both citizens and consumers, can obtain information and use legal channels to address or question decision-making processes related to energy matters [41]. As a result, researchers outline the responsibility and contribution of the populace in decisions, such as the siting of nuclear waste disposal or transmission line infrastructure in light of procedural injustices.

AI design faces difficulties in enforcing physical constraints, potentially escalating operational costs, or causing blackouts. Meanwhile, AI techniques, closely linked to statistics, share long-standing limitations with statistical methods. Similarly to those that affect procedural and recognition justice, these limitations include the tendency to identify correlations rather than causal relationships [42]. Additionally, many of these methods lack interpretability, therefore posing a barrier to procedural justice.

Energy justice serves various functions, from a conceptual tool for philosophers to an analytical tool for researchers and a decision-making tool for planners. Since design choices directly impact energy justice, algorithms such as RL offer adaptability but challenge accountability, raising concerns about automated systems that limit citizen control and decision-making [43].

Positively influencing procedural justice by enhancing control, transparency, and democracy, AI can be seen in self-control and codetermination applications of voluntary smart metering rollouts. Likewise, transparency positively impacts both procedural and distributive justice by providing equitable market access and supporting underprivileged consumers with budget planning. Nevertheless, AI smart grid pilot project procedures are criticized for limiting equitable access through biased participant selection. Therefore, transparency and stakeholder representation are crucial for addressing concerns related to accountability and biased participation, and can be improved by adopting hybrid control models.

14.3.4 AI Design and Recognition Justice: Acknowledging Diverse Stakeholder Needs

Recognition justice involves determining which stakeholders have valid concerns and claims. It focuses on the equitable acknowledgment of various stakeholder groups within energy systems, considering social, ethnic, racial, and gender differences [25], as this dimension addresses both lack of recognition and misrecognition. Within the context of recognition justice, the key roles of AI integration therefore encompass:

1) Recognizing diverse stakeholder needs by incorporating various perspectives into energy system decision-making.
2) Ensuring accessibility by designing technologies that accommodate different user groups within the energy system.
3) Empowering participation by facilitating stakeholder inclusion in discussions related to algorithmic benefits and burdens.

For example, failing to recognize specific energy needs, such as those of elderly individuals or people with disabilities, illustrates the former. Thus, misrecognition occurs when concerns are dismissively labeled, as seen with the "Not-In-My-Backyard" label in opposition to wind park projects [41].

Recognition justice is a crucial tenet in selecting communities, participants, and ensuring accessibility to technologies for diverse user groups in smart grid projects. It also extends to the development of fair algorithms for renewable energy networks, thereby emphasizing the inclusion of voices in decision-making processes related to the distribution of algorithmic costs and benefits.

To showcase the effect of the triumvirate of tenets on energy justice, one can investigate the judicial implications of solar curtailment, the limitation of solar power generation or transmission for economic or grid-capacity reasons. Analyzing AI techniques for solar curtailment reveals the importance of recognizing disparities, as approaches can exclude relevant stakeholders and disregard the diverse impacts on different groups, including homeowners with PV panels, social housing estates or elderly homes. Such oversights limit the consideration of disparities in the effects of curtailment on various stakeholders, hence highlighting the need for regular interrogation of underlying assumptions in AI development for energy management.

In-depth analyses of AI techniques applied to solar curtailment demonstrate how design choices, including data selection and fairness metrics, impact energy justice across its three identified axes. That is, distributive justice is affected by the way costs and benefits are distributed among stakeholders, while procedural justice is linked to stakeholder participation in the decision-making process, where the opaqueness and complexity of AI algorithms influence accountability and intervention possibilities. Moreover, modeling choices in AI techniques can affect recognition justice by including or excluding certain groups and their unique needs [43]. As a result, balancing these considerations requires careful attention to the trade-offs involved, such as privacy concerns associated with fine-grained data.

14.4 Transparency, Fairness, and Bias Mitigation in AI for Energy Systems

14.4.1 Ensuring Interpretability in Transparent Decision-Making for Energy Justice

Despite significantly contributing to task automation and energy efficiency improvement, the inherent black-box nature of AI models poses a challenge in understanding the underlying logic governing their decision-making processes and creates a feeling of mistrust in these models. In response to this, explainable AI (XAI) methods have gained popularity in recent years. These methods aim to improve trust in AI-driven decisions by providing explanations and without compromising performance [44]. The significance of explainability is particularly pronounced in vital domains like healthcare applications. Considering the widespread role of electricity in human life and the growing dependence on it, power systems have emerged as another critical domain. Therefore, it is imperative to ensure the explainability of the AI models utilized in the energy sector. The benefits of AI interpretability in the energy sector include:

- Compliance assurance: Explainability can assist in checking whether AI solutions comply with various power system requirements and standards by providing interpretations.
- Bias detection: XAI can be used to detect and mitigate biases in AI models. Users can understand how certain features impact decisions, helping to address issues related to fairness and discrimination in energy systems.
- Fairer decision-making: With insights into the decision-making process, developers can work toward building fairer and more equitable AI models for energy systems.
- Model debugging: In cases where AI models generate unexpected or incorrect predictions, explainability helps to quickly identify and resolve issues.
- Ethical considerations: XAI contributes to ethical considerations in AI development within energy sector. Model developers can ensure that AI systems align with ethical guidelines in energy systems, and users can be informed about the ethical considerations embedded in the models.
- Risk identification: XAI can play a crucial role in recognizing and addressing the potential risks related to the implementation of AI systems in the energy sector, promoting a more responsible and secure utilization of this technology.

Various XAI methods have been developed and described in the literature. Scoop-based explainers use feature importance analysis that determines how model outputs relate to inputs and are divided into local explainers, which explain a single output, and global explainers [45] that provide an explanation for the entire model. In a different approach, a separate second model is developed to explain the actions of the original model [46, 47]. Finally, some AI models, such as decision trees, are self-explanatory, and there is no need to develop a model to interpret them [45].

14.4.2 Detecting and Addressing Bias in AI Models for Equity

AI is now widely employed in various energy management and power systems applications, with the potential to revolutionize power systems. Nevertheless, bias remains a significant challenge in the implementation of these models. Bias denotes systematic inaccuracies inherent in decision-making processes, leading to inequitable outcomes. These biases may arise during data collection, algorithm design or human interpretation phases [48]. During the data collection phase, bias can emerge if the data is obtained from biased sources, rendering it unrepresentative or incomplete. With technological advancements, human biases in various subjects become more evident. As the use of human-like technology increases, it is important to study how unintentional transfer of human biases can impact the design of AI models in the energy sector [49]. Algorithmic bias can occur during the design phase of AI models due to various factors, such as the application of biased criteria in decision-making (e.g. building electric vehicle charging stations solely based on minimizing cost). Lastly, user bias occurs when individuals introduce their own biases into the AI model by misinterpreting the outputs or interacting with the AI model in a biased manner. In the energy sector, any of these biases can lead to an unfair distribution of resources and benefits.

Research on mitigating bias in the field of AI is actively progressing. Numerous methods, including user feedback mechanisms, data set augmentation, and bias-aware algorithms, have been developed and documented in the literature. When bias originates from an inherently biased dataset, such as the bias in the distribution of electric vehicles in urban areas compared to remote rural areas or between various socioeconomic groups [50], approaches like synthetic data generation can be used to address it. If bias is the result of an imbalanced dataset [51], such as the credit default prediction dataset in the energy industry [52], methods such as over-sampling, under-sampling, or collecting additional data can be effective. Regardless of the scenario, engaging in preprocessing and identifying the presence of bias in the data before training AI models is crucial to prevent and minimize such biases.

An alternative strategy to address bias in AI involves adopting model selection approaches that emphasize fairness. This may involve using adversarial training methods [53], regularization techniques that penalize models for making discriminatory predictions [54] or leveraging ensemble methods that fuse several AI algorithms to diminish bias [55]. Post-processing methods are applied after the successful training of an AI model. These techniques involve modifying the results generated by AI models to eliminate bias [56] or developing new models after detecting bias by adjusting the features or weights [57]. Although these methods have found application in various AI domains, their utilization within the energy sector remains relatively limited, necessitating further investigation and research. Furthermore, while existing literature presents multiple strategies to reduce bias in AI, there is no singular universal solution that is effective in every context, thus, highlighting the necessity for continued research in this field.

14.4.3 Ensuring Fair Distribution of Benefits and Costs

Since AI models are increasingly integrated into energy systems for tasks, such as predictive maintenance, demand forecasting, and optimization, fairness of the generated outputs and decisions by AI needs to be examined. However, defining fairness in the context of energy systems poses unique challenges. The definitions provided in the literature can be divided into two groups: awareness-based fairness and rationality-based fairness. Awareness-based fairness can be further divided into two groups: fairness through awareness and fairness through unawareness. Fairness through unawareness involves achieving fairness in algorithmic decision-making by excluding sensitive attributes such as demographic information from the process, thus generating outcomes perceived as fair [58]. However, this approach can lead to indirect unfairness when other correlated attributes are considered to impact equitable access to energy resources and services. Fairness through awareness treats individuals with similar value assignments, including sensitive attributes, in a similar manner referred to as individual fairness. This approach relies on elaborately defined similarity and distance functions for attributes and outcomes. However, the practical implementation of this notion is challenging, particularly in energy systems where measuring the distance between individuals in specific tasks is difficult, especially when fine-grained features are scarce. The difficulty is aggravated by the likelihood that individuals often appear in groups within real-world energy data, making it difficult to ensure fair treatment between protected and unprotected groups [59].

The rationality-based fairness is divided into statistical-based fairness and causality-based fairness. Statistical-based fairness when applied to energy systems, involves ensuring that all demographic groups receive fair treatment in terms of resource allocation and access to energy services. However, disparities may arise, as seen in risk assessments where factors such as race influence outcomes. Demographic parity seeks equal predictions for different groups but weakens when the relationship between attributes and outcomes is strong. Given the ground truth, statistical fairness evaluates error rates, including equalized odds, equal opportunity, and fairness of the test. These measures ensure fair treatment but may overlook hidden relationships between attributes and outcomes. However, causal-based fairness defines unfairness as the impact of sensitive attributes on energy system outcomes. Unlike statistical methods, it does not depend solely on observational data but considers additional causal relationships, often using structural causal models. These models intervene in causal graphs, allowing fairness analysis without directly changing sensitive attributes. Causality-based fairness focuses on eliminating unfair effects while keeping fair aspects intact, employing interventions, path-specific fairness, and counterfactual fairness [59]. By incorporating these definitions, stakeholders can navigate the complexities of AI-driven energy systems, striving for fairness and justice in resource allocation and decision-making processes.

14.4.4 Balancing Safety, Privacy, and Transparency in AI Applications

In the evolving landscape of AI, achieving a balance between safety, privacy, and transparency is necessary for responsible deployment and building trust in energy systems. To ensure safe use of AI models within these systems, a robust testing and validation process is needed to identify and address potential risks. Furthermore, using diverse datasets covering a wide range of energy-related scenarios could help minimize potential biases and build fair and reliable algorithms. However, achieving fairness may involve trade-offs with model accuracy, necessitating adaptable strategies to maintain equilibrium in different energy system scenarios. Using AI models that are interpretable and provide explanations could increase trust while giving a sense of safety with respect to decisions made within the energy sector. Additionally, developing AI models for energy justice and equity adds complexity to energy systems. Ideally, systems should use AI to promote equity while remaining transparent and user-friendly. Explainable AI can also be beneficial in this context, as it aids transparency by providing understandable explanations for outputs or actions. Designing and regulating AI models must prioritize user rights and accessibility in energy schemes. Finally, human oversight and intervention are crucial, especially in critical energy applications where real-time monitoring and decision-making are vital to ensure system stability and reliability.

The widespread use of AI models, ranging from virtual assistants to recommendation systems, emphasizes the importance of safeguarding user privacy through transparent communication about data collection practices and proactive measures against potential privacy breaches by utilities. Still, a minimum amount of data needs to be collected and anonymized from energy customers to protect customer privacy. Thus, data ownership rights and responsibilities should be clearly defined between utilities, energy customers, and model developers [60].

In addition to using XAI models to increase trust in AI models within energy sector, AI developers should be open and honest about the capabilities and limitations of the AI system and clearly state the purpose of the AI application and how it will impact users. Prioritizing ethical considerations, such as fairness and accountability, is essential for user trust and the responsible evolution of AI technologies in energy sector. The establishment of clear regulatory frameworks and standards is crucial for guiding the development and deployment of AI in energy systems, necessitating collaboration between governments, industries, and organizations. A well-balanced regulatory approach ensures that AI technologies evolve responsibly and accountably within energy sector, facilitating ethical deployment and sustainable development [61].

14.5 Case Study: Justice-Oriented Emergency Load Shedding with RL

To ensure safe operation of power systems, voltage must be maintained within a defined range, typically 0.95–1.05 per unit (p.u.) based on the American National Standard Institute (ANSI) standard [62]. Large voltage variations can cause damage to both the customer assets and the system equipment. Demand response programs and reserves are used in power systems to ensure that the voltage remains in the standard range. However, during a large disturbance, these measures may prove insufficient, resulting in a more substantial voltage drop. Load shedding is the last control option during such an emergency situation to restore the system voltage and prevent frequency drop [63].

The amount and location of load shedding are two important factors that can be determined in traditional or adaptive ways. Traditional methods are formulated primarily through offline research; consequently, their solutions tend to be too conservative and may not effectively accommodate random real-time fluctuations in the power system. The vast majority of electric utilities use this method because of its simplicity and practicality [64]. In contrast, adaptive methods demonstrate superior responsiveness to variations in the condition of the system [65].

RL, a subfield of ML, is used primarily in control scenarios. In this method, an agent undergoes training to learn and execute the most optimal actions within a given system through a process of trial and error. The agent interacts with the environment by performing actions and, in return, receives rewards that serve as indicators of the efficacy of its actions [66]. Essentially, functioning similar to an objective function in an optimization problem, the reward function guides the agent toward refining its decision-making. Considering the rewards received when selecting a specific action a in a particular state s of the environment, the agent is systematically trained to optimize its future actions. Since the agent is trained to determine the appropriate action in a given state, it exhibits enhanced responsiveness in emergency situations. Moreover, the agent can be trained to serve various objectives, including but not limited to equity and justice.

Model-based RL focuses on simulating outcomes with an accurate model of the environment, leading to sample efficiency but complexity in implementation. Policy-based RL learns a direct policy from states to actions and is suitable for complex systems with high-dimensional actions, although it requires more interactions with the environment and is less sample-efficient. Model-based RL can be used for optimizing smart grids, considering renewable sources, demand response, and storage for equitable energy distribution. Policy-based RL is useful in direct control tasks such as frequency regulation or load balancing, where the immediate response to state changes is crucial for maintaining equity in energy distribution. Both approaches aim to maximize the cumulative reward, which, in the context of energy systems, could be aligned with goals like minimizing costs, reducing emissions, or ensuring reliable access to energy for all communities. The choice between model-based RL and policy-based RL depends on the specific requirements of the energy system and the desired outcomes related to justice and equity.

In this section, policy-based RL is used as an adaptive control approach for emergency load shedding while considering justice as an additional objective. Specifically, this study addresses recognition justice by categorizing loads into groups based on their importance level and priority. Treating all loads equally in load-shedding scenarios could lead to the shutdown of essential emergency services, with potentially severe consequences and unintended injustice. Furthermore, distributive justice is accounted for by evenly distributing the load shedding across various regions of the system. This approach ensures that load shedding is not disproportionately imposed on any particular region based on its socio-economic status.

14.5.1 Problem Formulation

A schematic representation of the proposed RL-based load shedding is shown in Figure 14.1, where the environment is the power system that experiences an emergency load-shedding condition, and the agent is the entity that we aim to train for optimal load-shedding actions during such emergencies. In this study, a deep Q-learning algorithm is employed to train the agent. To formulate the RL environment, it needs to be defined as a Markov decision process consisting of an action space, state space, reward function, transition probability function,

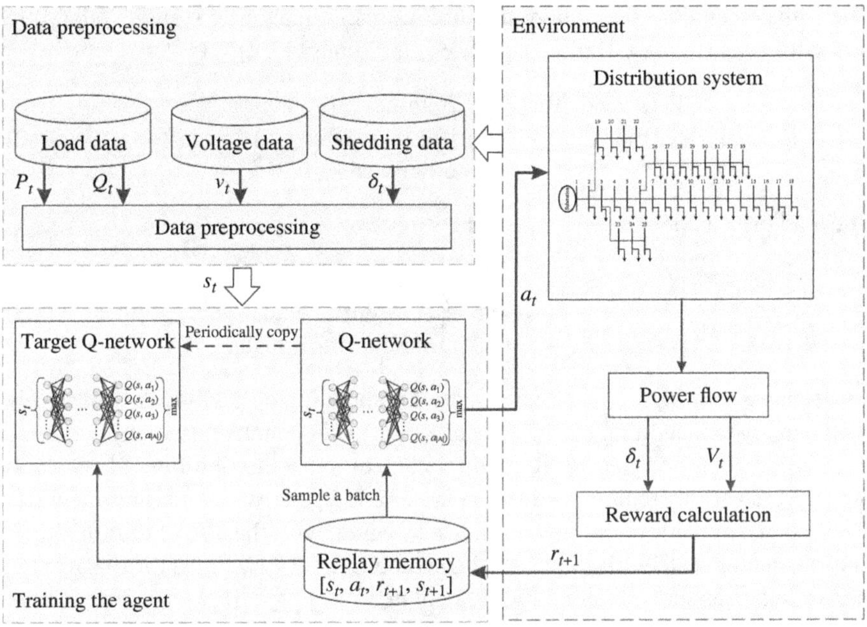

Figure 14.1 Schematic representation of the RL-based emergency load shedding.

discount factor, and a time horizon. In the context of the emergency load-shedding problem, actions are defined by specifying the bus number for load reduction (a^b) and the percentage of the load to be shed (a^r). The state of the environment consists of the active ($p_t = [p_{1t}, p_{2t}, \dots, p_{nt}]$) and reactive ($q_t = [q_{1t}, q_{2t}, \dots, q_{nt}]$) power consumption of the buses, bus voltages ($v_t = [v_{1t}, v_{2t}, \dots, v_{nt}]$), and the amount of load shedding in various regions of the network ($\delta_t = [\delta_{1t}, \delta_{2t}, \dots, \delta_{mt}]$). In summary, the state of the system is denoted as $s_t = (p_t, q_t, v_t, \delta_t)$.

The system loads are categorized into three groups: Critical loads, such as hospitals and emergency services, are indispensable for the continuous operation of the system; essential loads, including some industrial processes and transportation systems, can tolerate temporary interruptions that are less severe than critical loads; and noncritical loads, such as residential lighting, play a significant role in day-to-day operations but can withstand temporary interruptions without causing substantial harm.

The reward function consists of a penalty for voltage violations (R^v) and a penalty for load priority (R^l) as follows:

$$\mathcal{R}(s_t, a_t) = -R^v(s_{t+1}) - R^l(s_{t+1}). \tag{14.1}$$

The voltage penalty exists when the minimum and maximum system voltage are not within the standard limits (0.95–1.05 p.u.); otherwise, it is 0:

$$R^v(s_{t+1}) = \begin{cases} \kappa(0.95 - \min(v_{t,n})) & \text{if } \min(v_{t,n}) < 0.95 \\ \kappa(\max(v_{t,n}) - 1.05) & \text{if } 1.05 < \max(v_{t,n}) \\ 0 & \text{otherwise} \end{cases}. \tag{14.2}$$

The load priority penalty is defined as:

$$R^l(s_{t+1}) = \begin{cases} \beta & \text{if } a^b \in \text{critical loads} \\ \lambda & \text{if } a^b \in \text{essential loads} \\ \zeta(L_m - \delta_m - \mu)^2 & \text{if } a^b \in \text{noncritical loads} \end{cases}, \tag{14.3}$$

and consists of a highly negative penalty for shedding critical loads, a relatively high penalty for shedding essential loads, and a lower penalty for noncritical loads to ensure recognition justice. To maintain an equitable distribution of load shedding across various regions within the system, the penalty calculation integrates the variance of the remaining loads in each region, with the aim of minimizing it and improving distributive justice. The symbol μ in (14.3) stands for the average remaining loads across regions and is calculated as $\mu = \sum_{i=1}^{m}(L_m - \delta_m)/m$ where m is the number of regions, L_m is the number of loads in each region, and δ_m denotes the amount of load shedding in each region. It is worth noting that δ_m is calculated considering all shedding actions (a^r) until the voltage violation problem is addressed.

The transition probability is related to the random power injections and loads in this study. The discount factor, denoted by γ, is a constant value that balances immediate rewards with future gains, guiding the agent to make decisions that maximize long-term rewards. In the iterative learning process of a deep Q-learning algorithm, each action is assigned a q-value indicating how good that action is. These q-values are the outputs of a policy network (Q^π) that is responsible for determining the optimal action by selecting the one associated with the highest q-value. The input of this network is the state of the system. A target neural network (Q^{targ}) is a copy of the policy network that is synchronized with the policy network every few iterations to ensure a stable learning process. The policy neural network is updated by minimizing the temporal difference error:

$$\mathcal{L} = \begin{cases} Q^\pi(s_t, a_t) - (r(s_t, a_t) + \gamma \max_a Q^{targ}(s_{t+1}, a_t)) & \text{if } s_{t+1} \neq \text{terminal state} \\ Q^\pi(s_t, a_t) - r(s_t, a_t) & \text{if } s_{t+1} = \text{terminal state} \end{cases}. \tag{14.4}$$

A pseudocode of Q-learning is described in Algorithm 14.1. In this RL problem, the time horizon varies and each episode terminates upon resolving the voltage violation issue.

Algorithm 14.1: Deep Q-learning algorithm.

```
Input: Environment, RL algorithm hyperparameters
Output: Optimal policy π
for Episode ∈ {1, 2, ..., N} do
    Initialize environment
    for Step ∈ {0, 1, .., T − 1} do
        Policy selects action a_t
        Action a_t is executed in the environment
        Store the resulting transition (s_t, a_t, r_{t+1}, s_{t+1}) in the memory
        Select a batch of transitions from the memory
        Update the policy minimizing Eq. (14.4)
```

14.5.2 Experimental Setup and Data

The proposed justice-oriented emergency load-shedding strategy is implemented in the IEEE 33-bus system divided into four distinct regions. A schematic representation of the system including critical, essential, and noncritical loads is shown in Figure 14.2. The base voltage for this system is 12.66 kV with parameter values summarized in Table 14.4. Both the policy and target neural networks feature a two-layer structure, with 512 and 256 neurons in the first and second layer, respectively. For AC power flow analysis, the Newton–Raphson method is employed. Active and reactive power values at the buses in input states are normalized to p.u.

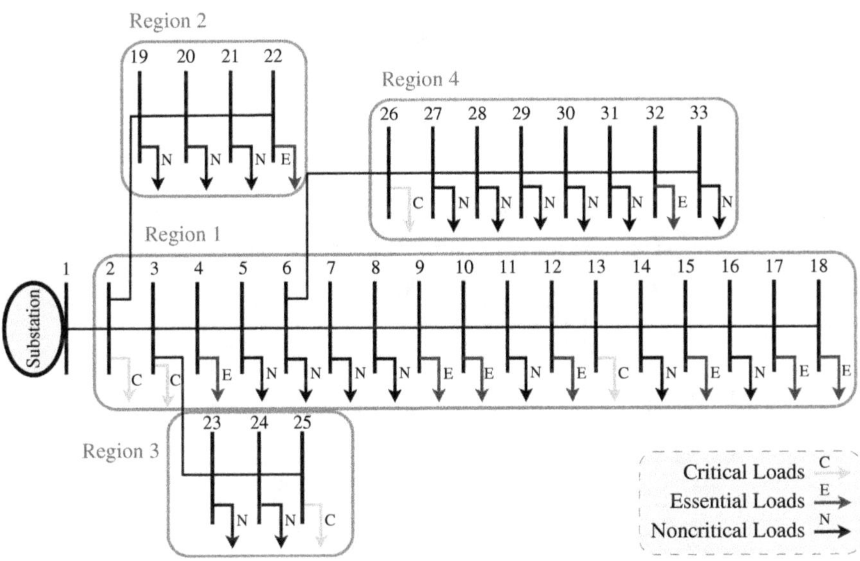

Figure 14.2 IEEE 33-bus case study.

Table 14.4 Parameter values.

Parameters	Value	Parameters	Value
κ	1000	β	1000
λ	500	ζ	0.001
α	5×10^{-6}	b	256
γ	0.99	N	30,000
Neural network structure	$\{512, 256\}$	Experience replay memory size	30,000,000

values considering a base power of 100 MW, thus ensuring a comparable range across input states. The load reduction is assumed to range from 0% to 100%, in increments of 20%, for all buses except the substation bus.

14.5.3 Results and Discussion

The obtained episode rewards and their 60-step moving average during agent training are illustrated in Figure 14.3. Notably, the rewards exhibit a steady increase over time, which is indicative of successful training. As training progresses, the agent improves its ability to navigate its environment and make

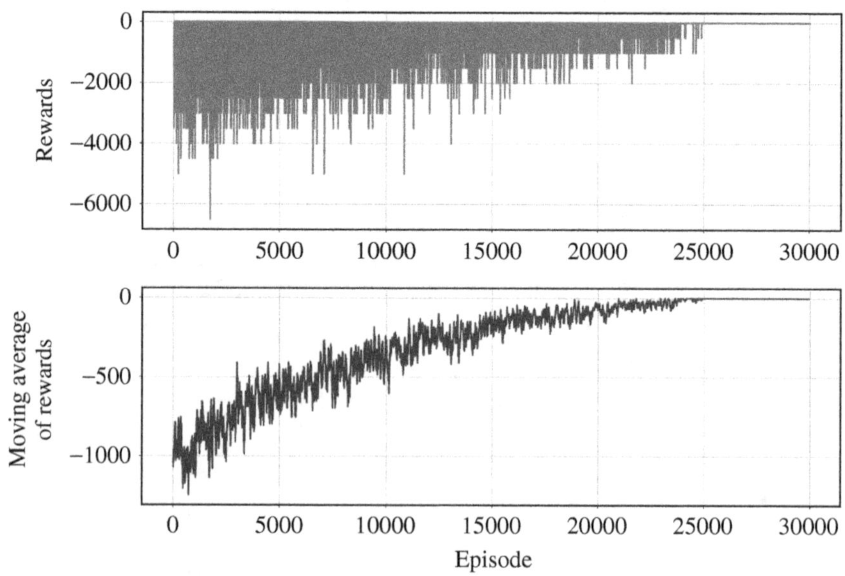

Figure 14.3 Episode rewards (top) and their 60-step moving average (bottom) during training.

informed decisions resulting in higher cumulative rewards per episode. This upward trend in rewards is important in assessing the agent's learning progress. Initially, the agent may display erratic behavior as it explores its environment and learns through trial and error. However, as training continues, the agent refines its strategies and policies, leading to more consistent and favorable outcomes. The gradual increase in the rewards reflects the iterative improvement in the agent's decision-making, indicating effective learning. Furthermore, employing a 60-step moving average offers a smoothed representation of the rewards trend, filtering out short-term fluctuations and providing a clearer insight into the overall learning trajectory.

The number of steps that the agent takes in each episode to tackle the voltage violation issue is depicted in Figure 14.4. Over the course of training episodes, the decreasing trend in this plot indicates that the agent becomes increasingly proficient in addressing the voltage-violation problem with greater speed. This progression highlights the agent's capacity to learn from its experiences and refine its decision-making process over time.

To evaluate the agent's performance post-training, it is used to make decisions for a set of 75 emergency data samples. The agent's actions, encompassing both the designated bus numbers and the extent of load shedding, are illustrated in Figure 14.5. Notably, the agent opts to shed load from buses 30, 7, 14, 8, and 16, all of which are categorized as noncritical loads. Furthermore, in an effort to minimize the standard deviation of the outages, it targets regions with a higher concentration of noncritical loads, such as regions 1 and 4. The degree of load reduction varies between 40% and 100%.

Minimum system voltages for the selected sample data both before and after load shedding are illustrated in Figure 14.6. As evident from the figure, after load shedding, all voltages have been adjusted to the acceptable range of 0.95-1.05 p.u, hence ensuring stable power system operation.

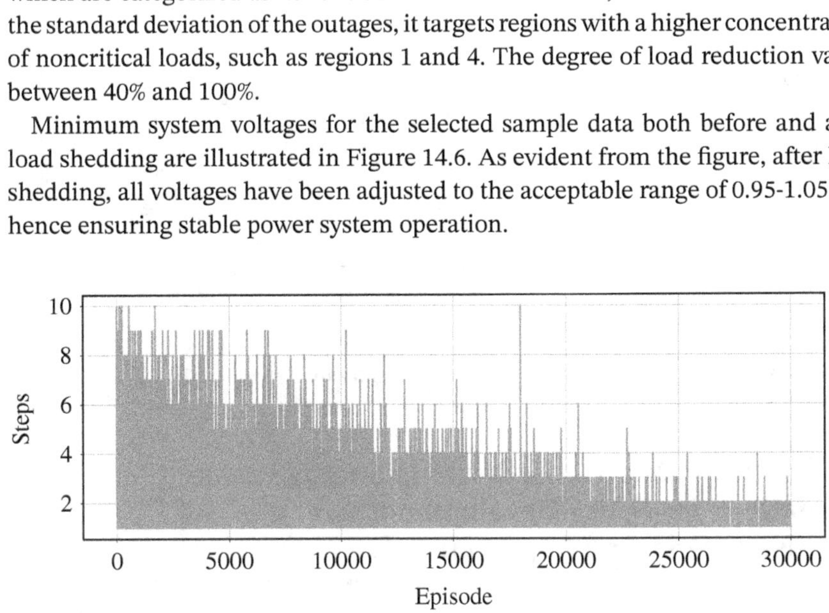

Figure 14.4 Number of steps taken in each episode during training.

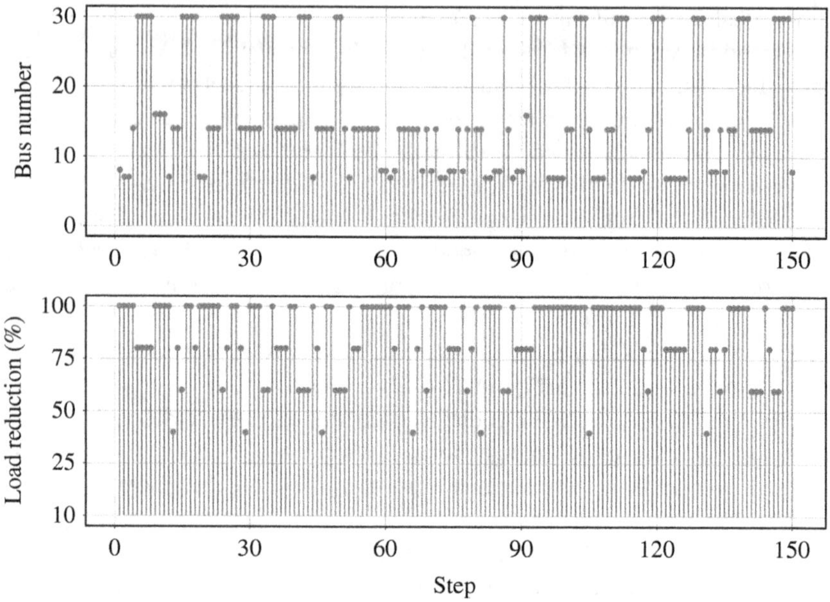

Figure 14.5 Agent actions including bus number and load shedding amount observed post-training using a sample dataset.

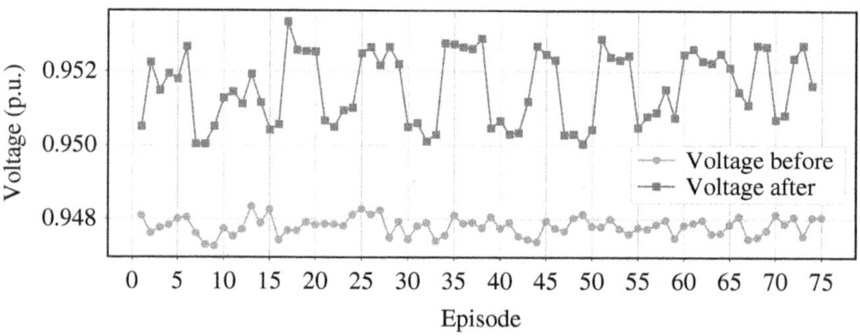

Figure 14.6 Minimum system voltage before and after load shedding by RL agent for a sample data.

14.6 Future Directions: Bridging Energy Law and AI Regulations for Policy Development and Justice-Oriented Approaches

The integration of AI into energy law frameworks offers potential benefits for regulatory processes, administrative efficiency, and data-driven decision-making.

Traditionally, energy law governs production, distribution, and consumption of energy, while AI regulations address the ethical, legal, and technical aspects of AI systems. Bringing these areas together presents opportunities to develop policies that optimize energy systems while upholding the principles of justice and fairness.

An effective energy policy that aligns with sustainability principles and integrates AI must include crucial elements such as stakeholder engagement, decision-making processes, innovation, energy efficiency, digitization, sustainability, regulation, access, privacy, accessibility, governance, resilience, smart grid technologies, equity, data management, emissions reduction, incentives, participation, security, renewable energy, and energy storage. These components, identified by organizations like the International Energy Agency and the New York Power Authority, are crucial for transitioning toward sustainable energy practices [67].

Despite challenges in classifying energy policy methodologies due to national requirements, future perspectives, and financial constraints, eight primary areas have been identified for planning energy development. These include energy efficiency, renewable energy promotion, reduction of carbon emissions, restructuring of the energy market, data management, promotion of electric vehicles, research and development, and international collaboration. Integrating AI into these domains requires a detailed research plan that is tailored to specific policy objectives and that includes critical policy components [67].

Effective policy development at the intersection of energy law and AI regulations requires collaborative governance models that engage diverse stakeholders, including government agencies, industry stakeholders, civil society organizations, and academic institutions. Multi-stakeholder dialogue and knowledge-sharing platforms can harness collective expertise to address complex challenges and cocreate inclusive regulatory frameworks. Furthermore, international cooperation and knowledge exchange are essential for harmonizing energy law and AI regulations across geopolitical boundaries. Transnational partnerships enable the exchange of best practices, capacity-building initiatives, and harmonization efforts, promoting consistency and coherence in regulatory approaches. Initiatives such as joint research projects, standardization efforts, and regulatory sandboxes facilitate cross-border collaboration and support the development of globally aligned policies [68].

Ethical considerations are at the heart of the adoption of AI in energy systems, necessitating proactive measures to support the principles of fairness, accountability, and transparency. Regulatory frameworks must integrate ethical guidelines and impact assessments to evaluate the societal implications of AI-driven interventions in energy governance. By embedding values such as human dignity, social justice, and environmental sustainability in AI regulations, policymakers can mitigate risks and maximize the societal benefits of technological advancements [69].

Finally, promoting responsible innovation involves nurturing a culture of continuous learning and adaptation within regulatory bodies and industry stakeholders. Capacity-building initiatives, training programs, and knowledge-sharing platforms can enhance the ethical literacy of decision makers and enable them to navigate complex ethical dilemmas inherent in AI-enabled energy systems. Ultimately, interdisciplinary collaborations between technologists, legal experts, and ethicists facilitate the co-creation of ethically aligned solutions and promote a shared understanding of ethical principles.

14.7 Conclusion

This chapter provides a comprehensive overview of the transformative role of AI in advancing the energy transition, along with the challenges that come with it. It starts by exploring AI applications in various domains, including predictive maintenance, energy efficiency enhancement, grid management and control, demand response, energy trading, and management of decentralized energy systems. Next, it discusses the complexities of embedding AI within energy systems, such as data quality, scalability, interoperability, and cybersecurity. Ethical-, financial-, and skill-related challenges that stakeholders face in integrating AI into energy systems are also explored. Subsequently, the chapter emphasizes the importance of upholding three key aspects of justice in the application of AI to energy systems: distributive justice, procedural justice, and recognition justice. The chapter then continues by investigating the origins of bias in AI algorithm development and offers strategies to mitigate these biases, including XAI as a valuable instrument for ensuring transparent decision-making and identifying biases. Further, the chapter reviews techniques documented in scholarly works aimed at achieving algorithmic fairness and highlights the necessity of achieving a balance between safety, privacy, fairness, model accuracy, and system complexity to navigate the ethical landscape of AI in energy systems effectively.

A case study examines the application of justice within the framework of emergency load shedding, in the IEEE 33-bus system, using RL. The objective is to achieve both recognition and distributive justice by stratifying customers into distinct priority groups and minimizing the variance of outages in different regions of the system. In scenarios where load shedding is applied uniformly to customers, there is a risk of inadvertently disrupting critical emergency services and causing unintended injustice. Similarly, if load shedding is unfairly concentrated in a single area, it may result in injustice against the socioeconomic standing of that region. An RL agent is programmed to execute two actions: selecting the bus for load shedding and determining the shedding percentage ranging from 0% to 100%. Results confirm that the developed RL algorithm targets only the lower-priority loads and areas with a denser customer base, effectively equalizing the incidence of outages across different regions. Finally, the chapter concludes by presenting possible avenues for integrating energy law with AI regulations.

References

1 Duvaut, P., Dalloz, X., Menga, D. et al. (2020). *Internet of Augmented Me, I.AM: Empowering Innovation for a New Sustainable Future.* Wiley. ISBN: 9781786305084. https://books.google.ca/books?id=HLD_DwAAQBAJ.

2 Brynjolfsson, E. and McAfee, A. (2014). *The Second Machine Age: Work, Progress, and Prosperity in a Time of Brilliant Technologies.* W. W. Norton & Company. ISBN: 9780393239355. https://books.google.ca/books?id=WiKwAgAAQBAJ.

3 Lu, T., Chen, X., McElroy, M.B. et al. (2021). A reinforcement learning-based decision system for electricity pricing plan selection by smart grid end users. *IEEE Transactions on Smart Grid* 12 (3): 2176–2187. https://doi.org/10.1109/TSG.2020.3027728.

4 Hong, Y., Zhou, Y., Li, Q. et al. (2020). A deep learning method for short-term residential load forecasting in smart grid. *IEEE Access* 8: 55785–55797. https://doi.org/10.1109/ACCESS.2020.2981817.

5 Lara, F. and Deckers, J. (2024). *Ethics of Artificial Intelligence*, The International Library of Ethics, Law and Technology. Springer Nature Switzerland. ISBN: 9783031481352. https://books.google.ca/books?id=D7jrEAAAQBAJ.

6 Wu, H., Xu, Z., Wang, M. et al. (2023). Two-stage voltage regulation in power distribution system using graph convolutional network-based deep reinforcement learning in real time. *International Journal of Electrical Power & Energy Systems* 151: 109158. https://doi.org/10.1016/j.ijepes.2023.109158. https://www.sciencedirect.com/science/article/pii/S0142061523002156.

7 Qin, J., Gao, Y., Bragin, M., and Yu, N. (2022). An optimization method-assisted ensemble deep reinforcement learning algorithm to solve unit commitment problems. *IEEE Access* 11: 100125–100136.

8 Li, Y., Wu, J., and Pan, Y. (2024). Deep reinforcement learning for online scheduling of photovoltaic systems with battery energy storage systems. *Intelligent and Converged Networks* 5 (1): 28–41. https://doi.org/10.23919/ICN.2024.0003.

9 Gao, G., Wen, Y., and Tao, D. (2022). Distributed energy trading and scheduling among microgrids via multiagent reinforcement learning. *IEEE Transactions on Neural Networks and Learning Systems* 34 (12): 10638–10652. https://doi.org/10.1109/TNNLS.2022.3170070.

10 Tao, Y., Qiu, J., Lai, S. et al. (2022). A human-machine reinforcement learning method for cooperative energy management. *IEEE Transactions on Industrial Informatics* 18 (5): 2974–2985. https://doi.org/10.1109/TII.2021.3105115.

11 Hossain, M.B., Pokhrel, S.R., and Choi, J. (2024). Orchestrating smart grid demand response operations with URLLC and MuZero learning. *IEEE Internet of Things Journal* 11 (4): 6692–6704. https://doi.org/10.1109/JIOT.2023.3314832.

12 Yin, L. and Qiu, Y. (2022). Neural network dynamic differential control for long-term price guidance mechanism of flexible energy service providers. *Energy* 255: 124558. https://doi.org/10.1016/j.energy.2022.124558.

13 Gao, Y., Shi, S., Miyata, S., and Akashi, Y. (2024). Successful application of predictive information in deep reinforcement learning control: a case study based on an office building HVAC system. *Energy* 291: 130344. https://doi.org/10.1016/j.energy.2024.130344. https://www.sciencedirect.com/science/article/pii/S0360544224001154.

14 Yousuf, H., Zainal, A.Y., Alshurideh, M., and Salloum, S.A. (2020). Artificial intelligence models in power system analysis. In: *Artificial Intelligence for Sustainable Development: Theory, Practice and Future Applications*, 231–242. https://doi.org/10.1007/978-3-030-51920-9-12.

15 Lawal, O.A. and Teh, J. (2023). Dynamic line rating forecasting algorithm for a secure power system network. *Expert Systems with Applications* 219: 119635. https://doi.org/10.1016/j.eswa.2023.119635. https://www.sciencedirect.com/science/article/pii/S0957417423001367.

16 Moradzadeh, A., Mohammadpourfard, M., Genc, I. et al. (2022). Deep learning-based cyber resilient dynamic line rating forecasting. *International Journal of Electrical Power & Energy Systems* 142: 108257. https://doi.org/10.1016/j.ijepes.2022.108257.

17 Sun, X. and Qiu, J. (2021). Two-stage volt/var control in active distribution networks with multi-agent deep reinforcement learning method. *IEEE Transactions on Smart Grid* 12 (4): 2903–2912. https://doi.org/10.1109/TSG.2021.3052998.

18 Li, J., Zhou, T., and Cui, H. (2023). Brain-inspired deep meta-reinforcement learning for active coordinated fault-tolerant load frequency control of multi-area grids. *IEEE Transactions on Automation Science and Engineering* 21 (3): 2518–2530. https://doi.org/10.1109/TASE.2023.3263005.

19 Fu, H. and Liu, Y. (2022). A deep learning-based approach for electrical equipment remaining useful life prediction. *Autonomous Intelligent Systems* 2 (1): 16. https://doi.org/10.1007/s43684-022-00034-2.

20 Yuan, Q. and Liu, X. (2023). Research on anomaly detection technology of power equipment based on AI vision analysis technology. *2023 IEEE 5th International Conference on Civil Aviation Safety and Information Technology (ICCASIT)*, 1321–1326. https://doi.org/10.1109/ICCASIT58768.2023.10351537.

21 Zemali, Z., Cherroun, L., Hadroug, N., and Hafaifa, A. (2023). Intelligent techniques employing LSTM and ANFIS models for fault detection and isolation in a wind turbine machine. *2023 2nd International Conference on Electronics, Energy and Measurement (IC2EM)*, volume 1, 1–6. https://doi.org/10.1109/IC2EM59347.2023.10419592.

22 Di Renzo, A.B., Zamarreo, C.R., Martelli, C., and Da Silva, J.C.C. (2023). Edge device for ultraviolet fluorescence inspection of photovoltaic panels. *2023 IEEE SENSORS*, 1–4. https://doi.org/10.1109/SENSORS56945.2023.10325174.

23 Lin, S., Fan, R., Feng, D. et al. (2022). Condition-based maintenance for traction power supply equipment based on partially observable Markov decision process. *IEEE Transactions on Intelligent Transportation Systems* 23 (1): 175–189. https://doi.org/10.1109/TITS.2020.3009175.

24 Sovacool, B.K., Burke, M., Baker, L. et al. (2017). New frontiers and conceptual frameworks for energy justice. *Energy Policy* 105: 677–691. https://doi.org/10.1016/j.enpol.2017.03.005.

25 McCauley, D., Heffron, R., Stephan, H., and Jenkins, K. (2013). Advancing energy justice: the triumvirate of tenets. *International Energy Law Review* 32: 107–110.

26 Heffron, R.J. (2022). Applying energy justice into the energy transition. *Renewable and Sustainable Energy Reviews* 156: 111936. https://doi.org/10.1016/j.rser.2021.111936. https://www.sciencedirect.com/science/article/pii/S1364032121012016.

27 Milchram, C., Hillerbrand, R., van de Kaa, G. et al. (2018). Energy justice and smart grid systems: evidence from the Netherlands and the United Kingdom. *Applied Energy* 229: 1244–1259. https://doi.org/10.1016/j.apenergy.2018.08.053. https://www.sciencedirect.com/science/article/pii/S0306261918312091.

28 Lippert, I. and Sareen, S. (2023). Alleviation of energy poverty through transitions to low-carbon energy infrastructure. *Energy Research & Social Science* 100: 103087. https://doi.org/10.1016/j.erss.2023.103087. https://www.sciencedirect.com/science/article/pii/S2214629623001470.

29 Donti, P.L. and Kolter, J.Z. (2021). Machine learning for sustainable energy systems. *Annual Review of Environment and Resources* 46: 719–747. https://doi.org/10.1146/annurev-environ-020220-061831.

30 Goldemberg, J. (2000). *World Energy Assessment: Energy and the Challenge of Sustainability*. United Nations Development Programme.

31 Goldemberg, J., Johansson, T.B., and Anderson, D. (2004). *World Energy Assessment: Overview: 2004 Update*. United Nations Development Programme, Bureau for Development Policy.

32 Johansson, T.B., Patwardhan, A., Nakićenović, N., and Gomez-Echeverri, L. (2012). *Global Energy Assessment: Toward a Sustainable Future*. Cambridge University Press.

33 Bennett, J.A., Ogland-Hand, J.D., Middleton, E.J. et al. (2023). The transmission ramifications of social and environmental siting considerations on wind energy deployment. *Frontiers in Energy Research* 10: 1040957. https://doi.org/10.3389/fenrg.2022.1040957. https://www.frontiersin.org/articles/10.3389/fenrg.2022.1040957.

34 Park, S., Yun, S.-J., and Cho, K. (2024). Energy justice: lessons from offshore wind farm siting conflicts in South Korea. *Energy Policy* 185: 113972. https://doi.org/10.1016/j.enpol.2023.113972. https://www.sciencedirect.com/science/article/pii/S0301421523005578.

35 Rubio-Bellido, C. and Solis-Guzman, J. (2023). *Energy Poverty Alleviation: New Approaches and Contexts*. Springer.

36 Necula, S.-C. (2023). Assessing the potential of artificial intelligence in advancing clean energy technologies in Europe: a systematic review. *Energies* 16 (22): 7633. https://doi.org/10.3390/en16227633.

37 Brandstedt, E., Busch, H., Lycke, E., and Ramasar, V. (2024). Winds of change: an engaged ethics approach to energy justice. *Energy Research & Social Science* 110: 103427. https://doi.org/10.1016/j.erss.2024.103427. https://www.sciencedirect.com/science/article/pii/S2214629624000185.

38 Halsband, A. (2022). Sustainable AI and intergenerational justice. *Sustainability* 14 (7): 3922. https://doi.org/10.3390/su14073922. https://www.mdpi.com/2071-1050/14/7/3922.

39 De Jong, M., Joss, S., and Taeihagh, A. (2024). Smart cities as spatial manifestations of 21st century capitalism. *Technological Forecasting and Social Change* 202: 123299. https://doi.org/10.1016/j.techfore.2024.123299. https://www.sciencedirect.com/science/article/pii/S0040162524000957.

40 Mirzaee, A.M. and Sardroud, J.M. (2022). *Public-Private-Partnerships (PPP) Enabled Smart City Funding and Financing*. Elsevier. ISBN: 978-0-12-819130-9. https://doi.org/10.1016/B978-0-12-819130-9.00011-5.

41 McHarg, A. (2020). Energy justice: understanding the 'ethical turn' in energy law and policy. In: *Energy Justice and Energy Law* (ed. I. del Guayo, L. Godden, D.D. Zillman, et al.). Oxford: Oxford University Press. ISBN: 978-0-19-886075-4.

42 Chou, Y.-L., Moreira, C., Bruza, P. et al. (2022). Counterfactuals and causability in explainable artificial intelligence: theory, algorithms, and applications.

Information Fusion 81: 59–83. https://doi.org/10.1016/j.inffus.2021.11.003. https://www.sciencedirect.com/science/article/pii/S1566253521002281.

43 Machlev, R., Heistrene, L., Perl, M. et al. (2022). Explainable artificial intelligence (XAI) techniques for energy and power systems: review, challenges and opportunities. *Energy and AI* 9: 100169. https://doi.org/10.1016/j.egyai.2022.100169. https://www.sciencedirect.com/science/article/pii/S2666546822000246.

44 Ahmed, I., Jeon, G., and Piccialli, F. (2022). From artificial intelligence to explainable artificial intelligence in Industry 4.0: a survey on what, how, and where. *IEEE Transactions on Industrial Informatics* 18 (8): 5031–5042. https://doi.org/10.1109/TII.2022.3146552.

45 Ali, S., Abuhmed, T., El-Sappagh, S. et al. (2023). Explainable artificial intelligence (XAI): what we know and what is left to attain trustworthy artificial intelligence. *Information Fusion* 99: 101805. https://doi.org/10.1016/j.inffus.2023.101805. https://www.sciencedirect.com/science/article/pii/S1566253523001148.

46 Kolekar, S., Gite, S., Pradhan, B., and Alamri, A. (2022). Explainable AI in scene understanding for autonomous vehicles in unstructured traffic environments on Indian roads using the inception U-Net Model with Grad-CAM visualization. *Sensors* 22 (24): 9677. https://doi.org/10.3390/s22249677. https://www.mdpi.com/1424-8220/22/24/9677.

47 Atakishiyev, S., Salameh, M., Yao, H., and Goebel, R. (2023). Explainable artificial intelligence for autonomous driving: a comprehensive overview and field guide for future research directions. *IEEE Access* 12: 101603–101625. https://api.semanticscholar.org/CorpusID:245385604.

48 Elmoudden, S. and Wrench, J.S. (2024). *The Role of Generative AI in the Communication Classroom*, Advances in Educational Technologies and Instructional Design. IGI Global. ISBN: 9798369308325. https://books.google.ca/books?id=1r_1EAAAQBAJ.

49 Varsha, P.S. (2023). How can we manage biases in artificial intelligence systems–a systematic literature review. *International Journal of Information Management Data Insights* 3 (1): 100165. https://doi.org/10.1016/j.jjimei.2023.100165. https://www.sciencedirect.com/science/article/pii/S2667096823000125.

50 Gemassmer, J., Daam, C., and Reibsch, R. (2021). Challenges in grid integration of electric vehicles in urban and rural areas. *World Electric Vehicle Journal* 12 (4): 206. https://doi.org/10.3390/wevj12040206. https://www.mdpi.com/2032-6653/12/4/206.

51 Zhang, X., Han, Y., Xu, W., and Wang, Q. (2021). HOBA: a novel feature engineering methodology for credit card fraud detection with a deep learning architecture. *Information Sciences* 557: 302–316. https://doi.org/10.1016/j.ins.2019.05.023. https://www.sciencedirect.com/science/article/pii/S002002551930427X.

52 Wang, K., Wan, J., Li, G., and Sun, H. (2022). A hybrid algorithm-level ensemble model for imbalanced credit default prediction in the energy industry. *Energies* 15 (14): https://doi.org/10.3390/en15145206. https://www.mdpi.com/1996-1073/15/14/5206.

53 Yang, J., Soltan, A.A.S., Eyre, D.W. et al. (2023). An adversarial training framework for mitigating algorithmic biases in clinical machine learning. *npj Digital Medicine* 6: 55. https://doi.org/10.1038/s41746-023-00805-y.

54 Chen, Z., Guo, W., Zhao, R. et al. (2022). Deep learning optimization of microgrid economic dispatch and wireless power transmission using blockchain. *Wireless Communications and Mobile Computing* 2022 (1): 2050031. https://doi.org/10.1155/2022/2050031. https://onlinelibrary.wiley.com/doi/abs/10.1155/2022/2050031.

55 Fletcher, R.R., Nakeshimana, A., and Olubeko, O. (2021). Addressing fairness, bias, and appropriate use of artificial intelligence and machine learning in global health. *Frontiers in Artificial Intelligence* 3: 561802. https://doi.org/10.3389/frai.2020.561802.

56 Obaid, A.J., Bhushan, B., PdI, M.S., and Rajest, S.S. (2023). *Advanced Applications of Generative AI and Natural Language Processing Models*, Advances in Computational Intelligence and Robotics. IGI Global. ISBN: 9798369305041. https://books.google.ca/books?id=79LrEAAAQBAJ.

57 Mehrabi, N., Gupta, U., Morstatter, F. et al. (2022). Attributing fair decisions with attention interventions. In: *Proceedings of the 2nd Workshop on Trustworthy Natural Language Processing (TrustNLP 2022)* (ed. A. Verma, Y. Pruksachatkun, K.-W. Chang, et al.), 12–25. Seattle, WA: Association for Computational Linguistics. https://doi.org/10.18653/v1/2022.trustnlp-1.2. https://aclanthology.org/2022.trustnlp-1.2.

58 Kallus, N., Mao, X., and Zhou, A. (2022). Assessing algorithmic fairness with unobserved protected class using data combination. *Management Science* 68 (3): 1959–1981. https://doi.org/10.1287/mnsc.2020.3850.

59 Wang, X., Zhang, Y., and Zhu, R. (2022). A brief review on algorithmic fairness. *Management System Engineering* 1: 7. https://doi.org/10.1007/s44176-022-00006-z.

60 Rodrigues, R. (2020). Legal and human rights issues of AI: gaps, challenges and vulnerabilities. *Journal of Responsible Technology* 4: 100005. https://doi.org/10.1016/j.jrt.2020.100005. https://www.sciencedirect.com/science/article/pii/S2666659620300056.

61 Alsaigh, R., Mehmood, R., and Katib, I. (2023). AI explainability and governance in smart energy systems: a review. *Frontiers in Energy Research* 11: 1071291. https://doi.org/10.3389/fenrg.2023.1071291. https://www.frontiersin.org/articles/10.3389/fenrg.2023.1071291.

62 Tasnim, S., Sarimuthu, C.R., Lan, B.L., and Tan, C.P. (2023). A game theory approach for OLTC voltage control operation in an active distribution network. *Electric Power Systems Research* 214: 108861. https://doi.org/10.1016/j.epsr.2022.108861. https://www.sciencedirect.com/science/article/pii/S0378779622009142.

63 Sundarajoo, S. and Soomro, D.M. (2023). Under voltage load shedding and penetration of renewable energy sources in distribution systems: a review. *International Journal of Modelling and Simulation* 43 (6): 1002–1020.

64 Madiba, T., Bansal, R.C., Mbungu, N.T. et al. (2022). Under-frequency load shedding of microgrid systems: a review. *International Journal of Modelling and Simulation* 42 (4): 653–679.

65 Zhu, L. and Luo, Y. (2021). Deep feedback learning based predictive control for power system undervoltage load shedding. *IEEE Transactions on Power Systems* 36 (4): 3349–3361. https://doi.org/10.1109/TPWRS.2020.3048681.

66 Ran, M. and Xie, L. (2022). Adaptive observation-based efficient reinforcement learning for uncertain systems. *IEEE Transactions on Neural Networks and Learning Systems* 33 (10): 5492–5503. https://doi.org/10.1109/TNNLS.2021.3070852.

67 Danish, M.S.S. and Senjyu, T. (2023). Shaping the future of sustainable energy through AI-enabled circular economy policies. *Circular Economy* 2 (2): 100040. https://doi.org/10.1016/j.cec.2023.100040. https://www.sciencedirect.com/science/article/pii/S2773167723000171.

68 Ala-Pietilä, P. and Smuha, N.A. (2021). *A Framework for Global Cooperation on Artificial Intelligence and its Governance*. Springer. https://ssrn.com/abstract=3696519.

69 Hagendorff, T. (2020). The ethics of AI ethics: an evaluation of guidelines. *Minds and Machines* 30 (1): 99–120. https://doi.org/10.1007/s11023-020-09517-8.

15

Artificial-Intelligence-Based Marketing Analysis for Realizing Renewable Energy Justice and Equity

Vikas Khare and Monica Bhatia

School of Technology, Management and Engineering, NMIMS (Deemed to be University), Indore, India
School of Business Management, NMIMS (Deemed to be University), Indore, India

Acronyms

AI	artificial intelligence
CS	case studies and best practices
ED	equity dimensions
EGS	enhanced geothermal system
FD	future directions
IP	interdisciplinary perspectives
JP	justice principles
ME	measurement and evaluation
OTEC	ocean thermal energy conversion
RER	renewable energy resources
SH	stakeholders
SS	solutions and strategies
SWOT	strength, weakness, opportunity, and threats
UNSDG	United Nations Sustainable Development Goal

15.1 Introduction

A renewable energy system, also known as a renewable energy infrastructure or renewable energy technology, is a system that harnesses naturally replenishing resources to generate energy. These systems are designed to produce electricity, heat, or mechanical power without depleting finite resources or emitting greenhouse gases and other pollutants associated with traditional fossil-fuel-based

Communities for Clean Energy Justice and Equity in Grid Modernization, First Edition.
Edited by Mohammadreza Daneshvar, Behnam Mohammadi-Ivatloo, and Amjad Anvari-Moghaddam.

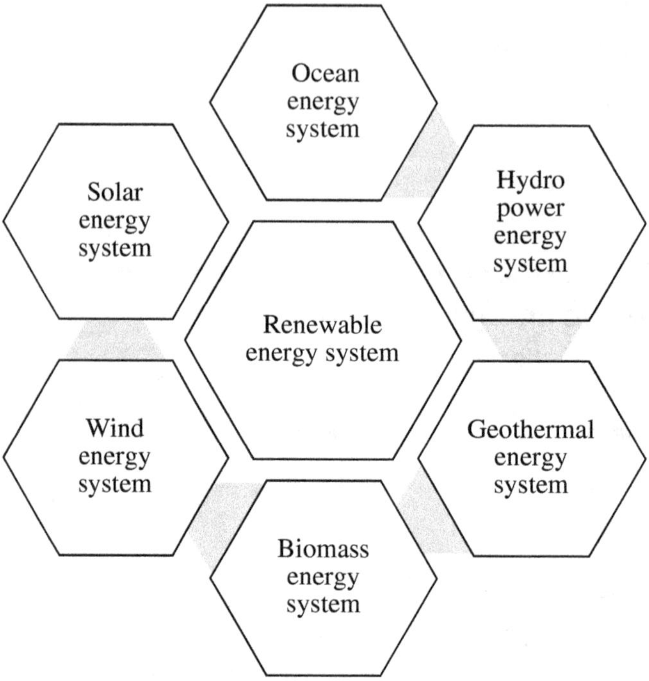

Figure 15.1 Types of renewable energy systems.

energy sources [1]. Figure 15.1 shows the types of renewable energy source. Solar power systems capture sunlight using photovoltaic (PV) panels or solar thermal collectors to convert it into electricity or heat. Solar PV systems directly convert sunlight into electricity using semiconductor materials, while solar thermal systems use sunlight to heat a fluid, typically water or oil, which then produces steam to drive a turbine and generate electricity [2]. Wind turbines harness the kinetic energy of the wind to generate electricity. As the wind blows, it causes the blades of the turbine to rotate, which spins a generator to produce electricity. Wind power systems can range from small-scale residential turbines to large-scale wind farms. Hydropower systems generate electricity by capturing the energy of flowing water, such as rivers or streams [3]. This is typically achieved by damming a river to create a reservoir and controlling the flow of water through turbines, which convert the hydraulic energy into electrical energy. Hydropower can also include smaller-scale systems like micro-hydro turbines installed in water channels. Geothermal energy utilizes heat from the Earth's interior to generate electricity or heat buildings directly [4].

Geothermal heat pumps use the relatively stable temperature of the ground to heat or cool buildings. Biomass energy systems convert organic materials, such as wood, agricultural residues, and organic waste, into heat, electricity, or biofuels. Biomass can be burned directly for heat or electricity generation, converted into biofuels like ethanol and biodiesel, or used in biogas production through anaerobic digestion. Ocean energy technologies harness the energy from waves, tides, and ocean currents to generate electricity [5].

Renewable energy justice and equity refer to ensuring fair access to and distribution of renewable energy resources and their benefits across all communities, especially those that have been historically marginalized or disadvantaged. Renewable energy justice involves ensuring that all communities have equal opportunities to access clean and sustainable energy sources like solar, wind, hydro, and geothermal power [6]. This includes both physical access to renewable energy technologies and the ability to afford them. Equity in renewable energy distribution means that the benefits of renewable energy projects, such as cost savings, employment opportunities, and environmental improvements, are fairly distributed among all members of society [7]. On the other hand, recognition justice in the context of renewable energy systems refers to ensuring that all individuals and communities affected by or contributing to renewable energy projects are acknowledged, respected, and included in decision-making processes. This includes recognizing the rights, concerns, and contributions of diverse stakeholders, including Indigenous peoples, local communities, and marginalized groups. Figure 15.2 shows the parameters of renewable energy equity and justice.

This includes addressing disparities in the location of renewable energy infrastructure and ensuring that communities that host such infrastructure also benefit from it. Renewable energy justice also entails fostering inclusive decision-making processes that involve all stakeholders, including communities that may be disproportionately affected by energy policies and projects. This means empowering communities to participate in the planning, development, and implementation of renewable energy initiatives. There's often an overlap between renewable energy justice and environmental justice, which focuses on addressing the unequal distribution of environmental benefits and burdens. This includes mitigating the disproportionate impacts of pollution and climate change on marginalized communities and ensuring that renewable energy solutions contribute to overall environmental well-being without exacerbating existing inequalities [8].

Renewable energy justice and equity aim to create a more just and sustainable energy system that benefits all members of society, particularly those who have been historically marginalized or underserved [9]. There are lots of researchers already work on this field. ***Sankaran and McIntyre-Mills*** [10] explained effect of energy justice in the renewable energy project. This article targets organizations and researchers, advocating for the adoption of systemic approaches in addressing energy justice concerns within renewable energy projects worldwide, aligning

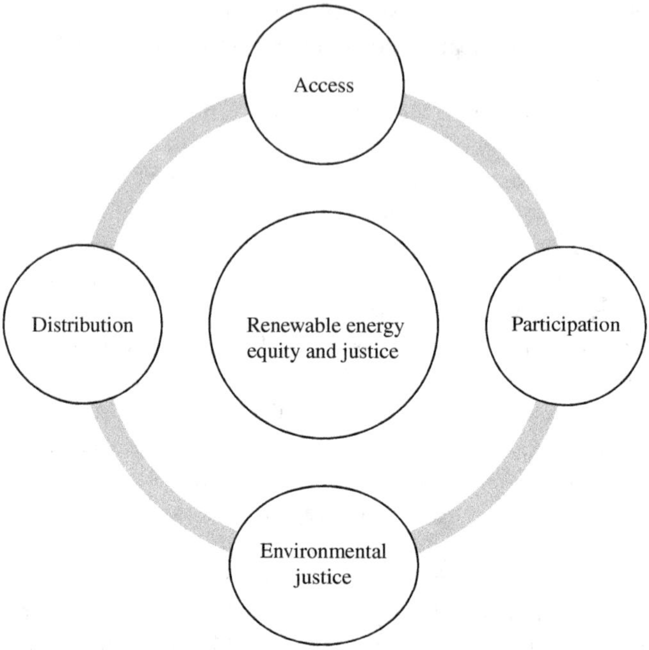

Figure 15.2 Parameters of renewable energy justice and equity.

with the United Nations Sustainable Development Goal (UNSDG). Specifically focusing on solar and wind farms, these projects play a crucial role in advancing UNSDG 7 by providing affordable clean energy. However, they also give rise to various justice issues that require attention. While recent efforts have been made to mitigate these issues, we argue that applying systemic thinking and practices could optimize the benefits and mitigate negative impacts more effectively by addressing the underlying root causes instead of solely relying on short-term fixes. **Sadat-Razavi et al.** [11] describe the energy justice perspective on the hydrogen import. This study explores the potential synergies or conflicts arising from the socioeconomic impacts of hydrogen imports. Employing a multimethod approach, it evaluates media discourses and public attitudes toward hydrogen importation. The study findings encompass an analysis of 37 newspaper articles and a comprehensive survey involving 2752 participants. Media coverage predominantly focuses on the reasons behind importing hydrogen and the terms of potential partnerships with other countries, with limited attention given to the consequences for exporting nations. The survey reveals public support for Power-to-X technologies rooted in renewable sources for importation purposes. However, there exists a significant level of uncertainty, as evidenced by the considerable number of "do not know" responses in the survey. Consequently, the study underscores the necessity for enhanced scientific communication and dialogue concerning strategies for hydrogen importation. **Fraser and Chapman** [12] describe the different drivers of social equity in renewable energy at the municipal level. On average, the introduction of renewable energy enhances social equity, with any increase in electricity prices primarily affecting wealthier residents rather than the impoverished, while towns stand to benefit from renewables regardless of the quantity implemented. To enhance these effects, municipalities can either expand their solar capacity or adjust variables such as CO_2 emissions, particulate matter emissions, tax revenue, job creation, or the presence of unpopular renewable energy facilities within their jurisdiction. However, demographic factors and preferences also play a significant role. Factors such as age, education level, and local preferences favoring employment and community development are closely linked to potential equity outcomes. Therefore, policymakers should contemplate adjusting their local energy agendas using these mechanisms if they aim to orchestrate a renewable energy transition that garners both positive outcomes and widespread support from their constituents. **O'Shaughnessy et al.** [13] describe different drivers of energy justice in the renewable energy project in the United States. This study examines the factors influencing the location selection for solar and wind projects in the United States and investigates their potential implications for energy justice. Our analysis reveals that the choice of project sites is predominantly influenced by techno-economic factors, particularly the quality of resources and the availability of open, undeveloped land. These factors tend to direct projects toward rural areas with lower population density and, to a lesser degree, regions characterized by lower income levels. We refrain from oversimplifying the assessment of the broader justice implications associated with these siting patterns and approach our findings from various angles. The research aims to investigate how artificial intelligence (AI)-based marketing analysis can be leveraged to advance renewable energy justice and equity. Following are the objectives of this chapter:

- Identify the equitable energy system.
- Identify the role of AI in renewable energy justice and equity.
- Create knowledge representation of renewable energy justice and equity.
- Create a semantic graph representation of renewable energy justice and equity.
- Apply the data-driven approach to the different parameters of renewable energy equity and justice.
- AI-based Strengths, Weaknesses, Opportunities, and Threats (SWOT) analysis of renewable energy equity and justice.

By achieving these objectives, the research aims to contribute to the development of more inclusive and equitable renewable energy systems that benefit all members of society.

15.2 How Can We Achieve Equitable Energy System?

Achieving an equitable energy system involves addressing disparities in energy access, affordability, participation, and environmental justice across various social, economic, and geographic dimensions. Figure 15.3 shows the parameters to achieve equitable. Following are several strategies to work toward this goal:

- **Universal Energy Access:** Ensure that everyone has access to reliable, affordable, and clean energy services, regardless of their socioeconomic status or geographic location. This may involve expanding infrastructure, such as grid connections, off-grid solutions (like solar home systems), and mini-grids in remote areas.

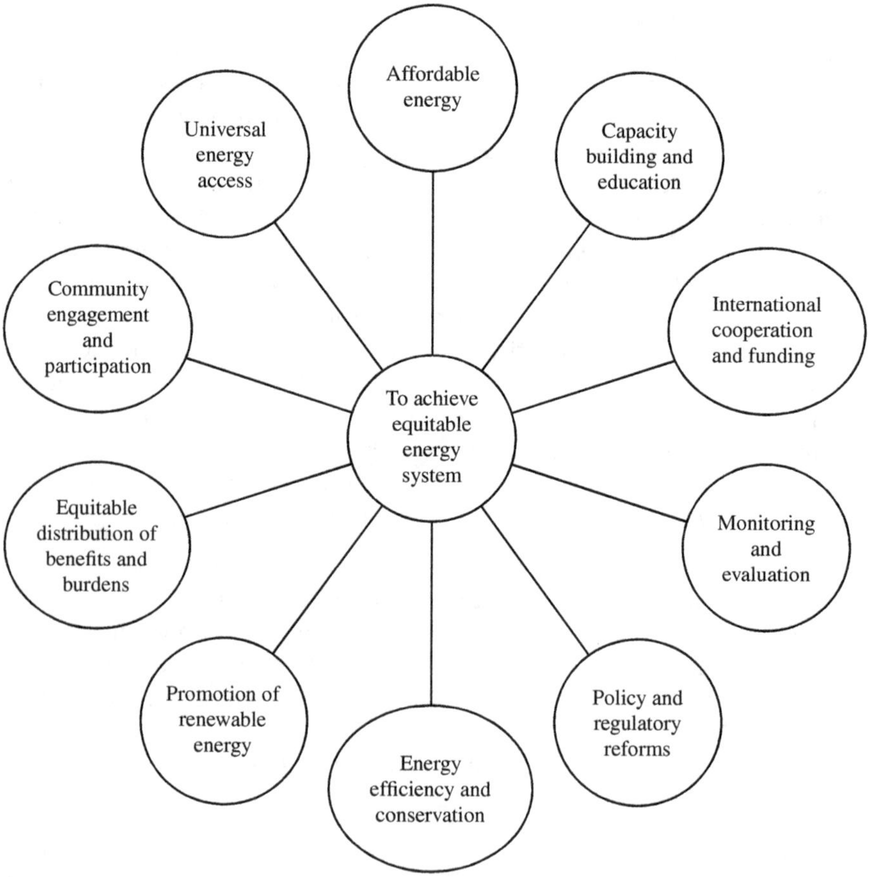

Figure 15.3 Parameters to achieve an equitable energy system.

- **Affordable Energy:** Implement policies and programs to make energy services affordable for all, particularly for low-income households. This could include subsidies, tariff structures that consider income levels, and energy efficiency programs to reduce overall energy consumption and costs [14].
- **Community Engagement and Participation:** Involve communities, especially marginalized and vulnerable groups, in decision-making processes related to energy planning, policy development, and project implementation. This ensures that their needs, preferences, and concerns are taken into account, leading to more inclusive and effective energy solutions.
- **Equitable Distribution of Benefits and Burdens:** Ensure that the benefits of energy projects (such as job creation, improved health, and economic development) are equitably distributed among all stakeholders, including local communities while minimizing negative impacts on vulnerable populations and the environment [15].
- **Promotion of Renewable Energy:** Prioritize the development and deployment of renewable energy sources, which are often more distributed, cleaner, and can be more easily accessible to local communities. This includes solar, wind, hydroelectric, biomass, and geothermal energy, among others.
- **Energy Efficiency and Conservation:** Promote energy efficiency measures and conservation practices to reduce overall energy demand, lower costs, and minimize environmental impacts. This can include energy-efficient technologies, building codes, and public awareness campaigns.
- **Policy and Regulatory Reforms:** Implement policies and regulations that promote equity and justice in the energy sector, such as renewable energy targets, feed-in tariffs, net metering policies, land-use regulations, and environmental standards.
- **Capacity Building and Education:** Invest in capacity-building programs and educational initiatives to empower communities, local organizations, and policymakers with the knowledge and skills needed to advocate for equitable energy solutions, participate in decision-making processes, and adopt sustainable energy practices.
- **International Cooperation and Funding:** Collaborate with international organizations, donors, and other countries to access funding, share best practices, and leverage expertise to support the transition to a more equitable energy system, especially in developing countries and regions with limited resources.
- **Monitoring and Evaluation:** Establish mechanisms for monitoring, evaluating, and reporting on progress toward achieving equitable energy access and justice goals. This involves collecting data, conducting impact assessments, and engaging stakeholders to ensure accountability and continuous improvement.

By implementing these strategies in a coordinated manner, policymakers, energy stakeholders, and communities can work together to create a more equitable, inclusive, and sustainable energy system for all.

15.3 Role of Artificial Intelligence in Renewable Energy Justice and Equity

AI holds significant promise for advancing renewable energy justice and equity. By leveraging AI technologies, we can optimize the deployment of renewable energy resources to benefit underserved communities, improve energy affordability, enhance community engagement in decision-making processes, and ensure equitable access to clean energy solutions. Figure 15.4 shows the attributes of renewable energy equity and justice.

AI-driven tools can analyze data to identify optimal locations for renewable energy projects, prioritize investments in disadvantaged communities, and develop targeted interventions to address energy poverty. Additionally, AI algorithms can optimize the operation of renewable energy systems and the electric grid to maximize efficiency and reliability while minimizing costs and

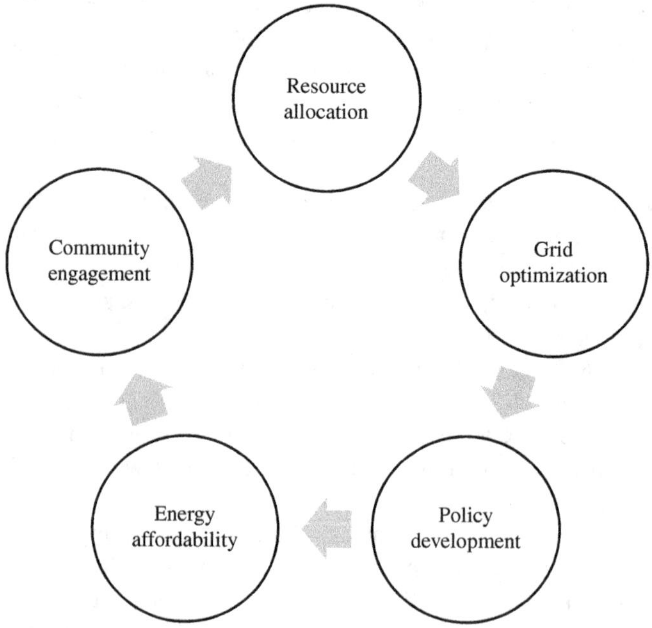

Figure 15.4 Attributes of renewable energy equity and justice.

emissions. By harnessing the power of AI, we can build a more inclusive and sustainable energy future that benefits all members of society [16].

- **Resource Allocation:** AI can analyze data to identify optimal locations for renewable energy infrastructure, taking into account factors such as sunlight exposure, wind patterns, and proximity to communities in need. By ensuring that renewable energy projects are strategically placed in areas that benefit underserved communities, AI can contribute to greater equity in access to clean energy.
- **Community Engagement:** AI-powered tools can facilitate inclusive and participatory decision-making processes by providing communities with accessible information and opportunities for input. For example, AI-driven platforms can gather feedback from community members about their energy needs and preferences, helping to shape renewable energy projects that are responsive to local priorities.
- **Energy Affordability:** AI algorithms can analyze energy usage patterns and socioeconomic data to identify households that may be struggling with energy costs. This information can inform targeted interventions, such as subsidies, energy efficiency programs, or community solar initiatives, aimed at reducing energy burdens and promoting energy justice.
- **Grid Optimization:** AI can optimize the operation of renewable energy systems and the electric grid to maximize efficiency, reliability, and affordability. For example, AI algorithms can forecast renewable energy generation, predict electricity demand, and dynamically adjust grid operations to integrate variable renewable energy sources while minimizing costs and emissions. By improving the performance of renewable energy systems and grid infrastructure, AI can enhance energy access and reliability for all communities.
- **Policy Development:** AI-driven analytics can inform evidence-based policy-making by analyzing vast amounts of data to assess the social, economic, and environmental impacts of different energy policies and interventions. This can help policymakers identify strategies that promote renewable energy justice and equity, such as targeted investments in disadvantaged communities, equitable financing mechanisms, or regulatory reforms that prioritize environmental justice considerations [17].

AI has the potential to be a powerful tool for advancing renewable energy justice and equity by improving decision-making, enhancing community engagement, optimizing energy systems, and informing policy development. However, it's important to ensure that AI applications are designed and implemented in ways that prioritize equity, fairness, and inclusivity, and that potential biases and ethical concerns are carefully addressed.

15.4 Knowledge Representation of the Renewable Energy Equity and Justice

Creating an AI-based knowledge representation of renewable energy equity and justice involves structuring information in a way that AI systems can understand and manipulate to make informed decisions and recommendations. Following are the approach of knowledge representation of renewable energy equity and justice.

15.4.1 Ontology Development

Develop an ontology that represents the concepts, relationships, and properties relevant to renewable energy equity and justice. This ontology should include entities such as communities, renewable energy resources, socioeconomic indicators, environmental impacts, policies, and interventions. Figure 15.5 shows the parameters of the ontology development of renewable energy equity and justice.

Creating an AI-based ontology for "Renewable Energy Equity and Justice" involves organizing and categorizing concepts related to the equitable distribution of renewable energy resources and the promotion of justice within the renewable energy sector [18]. Following are the elements of the elements of the ontology of the renewable energy equity and justice:

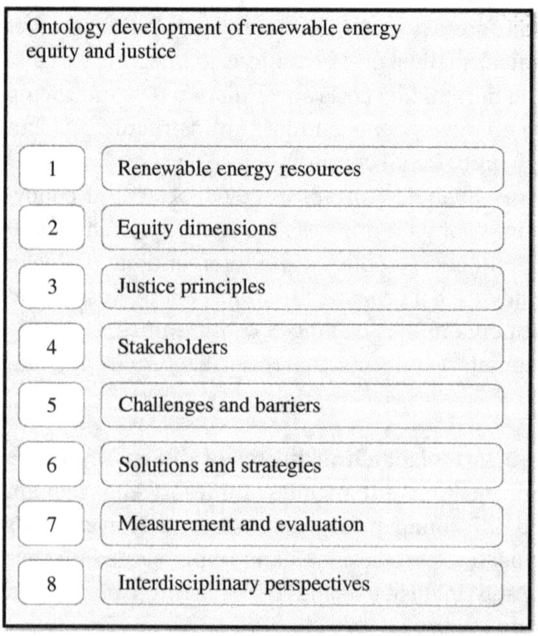

Figure 15.5 Ontology parameters of the renewable energy equity and justice.

- **Renewable Energy Resources**
  - **Solar Energy:** Solar energy harnesses the power of sunlight to generate electricity or heat water for various applications. It's a renewable and abundant energy source, with technologies like solar panels and solar thermal systems converting sunlight into usable energy. Solar energy offers environmental benefits by reducing greenhouse gas emissions and dependence on fossil fuels.
  - **Wind Energy:** Wind energy captures the kinetic energy of moving air to generate electricity through wind turbines. It's a clean and renewable energy source that does not produce greenhouse gas emissions during operation. Wind farms, often located in windy areas onshore or offshore, contribute significantly to electricity generation worldwide, providing a sustainable alternative to traditional fossil fuel-based power plants.
  - **Hydroelectric Energy:** Hydroelectric energy harnesses the gravitational force of flowing or falling water to generate electricity. It's one of the oldest and most widely used forms of renewable energy, with large-scale hydroelectric dams and smaller run-of-river projects providing power to millions of people globally. Hydroelectricity offers advantages such as reliability, flexibility, and low operating costs, but it also raises concerns about environmental impacts like habitat disruption and changes in water flow.
  - **Geothermal Energy:** It relies on natural geothermal reservoirs or enhanced geothermal systems (EGSs) to access heat trapped in rocks and fluids deep underground. Geothermal energy is a reliable and constant renewable resource, offering environmental benefits and reducing reliance on fossil fuels, particularly in regions with high geothermal potential.
  - **Biomass Energy:** Biomass can be burned directly for heat or converted into biogas, bioethanol, or biodiesel through processes like combustion, fermentation, and gasification. Biomass energy is renewable, carbon-neutral when sustainably managed, and contributes to waste reduction and energy independence. However, it also raises concerns about land use competition, emissions, and sustainability of feedstock sources.
- **Equity Dimensions**
  - **Access to Renewable Energy:** Ensuring equitable access to renewable energy involves providing all communities, regardless of socioeconomic status or geographic location, with the opportunity to benefit from clean energy resources. This includes initiatives to deploy renewable energy technologies in underserved areas, such as rural or low-income communities, and addressing barriers like upfront costs, grid connectivity, and infrastructure limitations. By improving access to renewable energy, equitable outcomes can be achieved, leading to improved health, economic opportunities, and quality of life for marginalized populations.

- **Affordability of Renewable Energy:** Equity in renewable energy also requires addressing affordability issues to ensure that all individuals and communities can afford to adopt clean energy solutions. This involves implementing policies and programs to reduce the costs of renewable energy technologies, such as subsidies, incentives, and financing options targeted at low-income households. Additionally, energy efficiency measures and demand-side management strategies can help lower energy bills and improve affordability for vulnerable populations, ultimately promoting social equity and energy justice.
- **Participation in Decision-making Processes:** Renewable energy equity and justice require inclusive and participatory decision-making processes that involve affected communities in the planning, development, and implementation of renewable energy projects. This includes meaningful engagement, consultation, and collaboration with diverse stakeholders, including community organizations, Indigenous groups, and marginalized populations. By empowering communities to participate in decision-making, their needs, concerns, and preferences can be addressed, leading to more equitable outcomes and sustainable solutions that reflect local priorities and values.
- **Distribution of Benefits and Burdens:** Achieving equity in renewable energy also entails ensuring an equitable distribution of benefits and burdens associated with clean energy development. This includes addressing disparities in the distribution of economic, environmental, and social impacts across different communities and demographic groups. Policies and initiatives should prioritize the equitable sharing of benefits, such as job creation, economic development, and environmental improvements, while mitigating potential negative consequences, such as land use conflicts, environmental degradation, and displacement [19]. By promoting fairness and justice in the distribution of benefits and burdens, renewable energy projects can contribute to inclusive and sustainable development that benefits all members of society.

• **Justice Principles**
  - **Distributive Justice:** Distributive justice in the context of renewable energy equity and justice refers to the fair allocation and distribution of benefits and burdens associated with renewable energy development. This principle emphasizes the importance of ensuring that the benefits of clean energy technologies, such as job opportunities, access to affordable energy, and environmental improvements, are distributed equitably among all communities, including marginalized and underserved populations. Similarly, distributive justice requires that the burdens and negative impacts of renewable energy projects, such as pollution, land use conflicts, and

displacement, are also distributed fairly and not disproportionately borne by vulnerable or disadvantaged groups.
- **Procedural Justice:** Procedural justice focuses on the fairness and transparency of decision-making processes related to renewable energy development. This principle emphasizes the importance of inclusive and participatory processes that involve affected communities in the planning, development, and implementation of renewable energy projects. Procedural justice ensures that all stakeholders have a voice in decision-making, access to relevant information, and opportunities to express their concerns and preferences. By promoting transparency, accountability, and democratic participation, procedural justice helps build trust, legitimacy, and social acceptance for renewable energy initiatives, leading to more equitable outcomes and sustainable solutions.
- **Intergenerational Justice:** Intergenerational justice considers the ethical responsibility to ensure that the needs and interests of future generations are taken into account in decision-making about renewable energy development. This principle recognizes that current actions and policies regarding energy use and environmental stewardship have long-term implications for future generations, including their access to natural resources, environmental quality, and overall well-being. Intergenerational justice calls for sustainable and responsible energy practices that prioritize the preservation of natural resources, the mitigation of climate change impacts, and the promotion of intergenerational equity, ensuring that future generations inherit a liveable planet with equitable access to renewable energy resources.
- **Environmental Justice:** Environmental justice focuses on the fair treatment and distribution of environmental benefits and burdens, particularly with regard to marginalized and vulnerable communities. In the context of renewable energy equity and justice, environmental justice addresses disparities in the distribution of environmental risks and impacts associated with clean energy development. This principle seeks to prevent environmental injustices, such as siting renewable energy projects in or near low-income neighborhoods or communities of color without their consent, which can exacerbate existing inequalities and health disparities. Environmental justice advocates for equitable access to clean air, water, and land, as well as meaningful participation in environmental decision-making processes, to ensure that all communities have the right to a healthy and sustainable environment.

- **Stakeholders**
  - **Government Agencies:** Government agencies play a crucial role in promoting renewable energy equity and justice through policy development,

regulation, and funding initiatives. They establish frameworks and incentives to support renewable energy deployment, ensure fair access to resources, and address environmental and social impacts. Government agencies also facilitate public participation and engage with stakeholders to incorporate diverse perspectives into decision-making processes, ultimately striving to achieve equitable and sustainable energy transitions.

- **Energy Companies:** Energy companies, including renewable energy developers and utilities, have a significant impact on equity and justice outcomes in the renewable energy sector. These entities design, finance, and operate renewable energy projects, influencing factors such as project siting, pricing, and community engagement. Energy companies have a responsibility to consider the needs and interests of local communities, mitigate negative impacts, and maximize the distribution of benefits from clean energy investments, contributing to more inclusive and equitable energy systems.
- **Communities:** Local communities are key stakeholders in renewable energy equity and justice, as they are directly affected by the development and operation of renewable energy projects. Communities provide valuable input on project siting, land use, environmental concerns, and socioeconomic impacts, advocating for their interests and rights. Community engagement processes are essential for building trust, fostering collaboration, and ensuring that renewable energy initiatives align with community priorities and values, ultimately promoting equitable outcomes and social acceptance [20].
- **NGOs and Advocacy Groups:** Nongovernmental organizations (NGOs) and advocacy groups play a critical role in advancing renewable energy equity and justice by advocating for the interests of marginalized communities, promoting policy reforms, and raising awareness about environmental and social justice issues. These organizations conduct research, provide technical assistance, and mobilize grassroots efforts to address inequities, hold decision-makers accountable, and empower communities to participate in energy decision-making processes, contributing to more equitable and inclusive energy transitions.
- **Indigenous Communities:** Indigenous communities play a crucial role in discussions surrounding renewable energy equity and justice. Historically marginalized and disproportionately impacted by environmental injustices, these communities often bear the brunt of unsustainable energy practices. In the context of renewable energy, ensuring equity and justice means recognizing Indigenous rights, sovereignty, and traditional knowledge. Engaging Indigenous communities in decision-making processes, respecting their land rights, and fostering partnerships for renewable energy projects are vital

steps toward achieving equity and justice in the transition to sustainable energy systems.

- **Researchers and Academics:** Researchers and academics are key stakeholders in advancing the discourse on renewable energy equity and justice. Their contributions include conducting interdisciplinary studies, evaluating policy frameworks, and assessing the socioeconomic impacts of renewable energy initiatives. Through rigorous research and analysis, they provide insights into the distributional effects of renewable energy projects, identify barriers to equitable access, and propose strategies for inclusive energy transitions. Collaboration between researchers, policymakers, and affected communities is essential to translate research findings into actionable policies and practices that promote fairness and justice in renewable energy development.

- **Challenges and Barriers**
  - **Socioeconomic Disparities:** One of the foremost challenges to achieving renewable energy equity and justice lies in addressing socioeconomic disparities. Marginalized communities often lack access to renewable energy technologies due to financial constraints, perpetuating energy poverty, and widening the gap between affluent and disadvantaged populations. Addressing affordability issues, providing financial assistance for renewable energy adoption, and implementing inclusive financing mechanisms are essential steps toward mitigating socioeconomic disparities in accessing clean energy resources.
  - **Technological Barriers:** Technological barriers pose significant challenges to renewable energy equity and justice, particularly in remote or underdeveloped areas. Limited infrastructure, inadequate grid connectivity, and the lack of suitable renewable energy technologies hinder the widespread adoption of clean energy solutions. Overcoming technological barriers requires innovation in decentralized energy systems, the development of appropriate storage solutions, and the adaptation of renewable technologies to diverse geographical and climatic conditions.
  - **Policy and Regulatory Challenges:** Policy and regulatory frameworks play a crucial role in shaping the equitable distribution of renewable energy benefits. However, inconsistent policies, bureaucratic hurdles, and regulatory barriers often impede the deployment of renewable energy projects in marginalized communities. Strengthening supportive policies, streamlining permitting processes, and ensuring inclusive decision-making are necessary to overcome policy and regulatory challenges and promote equitable access to renewable energy resources.
  - **Cultural Considerations:** Cultural considerations are fundamental to renewable energy equity and justice, particularly concerning Indigenous and

minority communities. Disregarding cultural values, traditional practices, and sacred lands can lead to conflicts and resistance against renewable energy projects. Building trust, fostering meaningful engagement, and incorporating Indigenous knowledge systems are essential for navigating cultural complexities and ensuring that renewable energy initiatives respect the cultural heritage and rights of affected communities.

  - **Environmental Impacts:** While renewable energy is often touted as environmentally friendly, it is not without its environmental impacts, which can disproportionately affect marginalized communities. The siting and development of renewable energy projects may encroach upon ecologically sensitive areas, disrupt ecosystems, and threaten biodiversity, exacerbating environmental injustices. Implementing robust environmental impact assessments, adopting sustainable siting practices, and prioritizing community-based conservation efforts are essential for minimizing the adverse environmental effects of renewable energy projects and promoting environmental justice for all stakeholders.
- **Solutions and Strategies**
  - **Community-owned Renewable Energy Projects:** Community-owned renewable energy projects empower local stakeholders by allowing them to actively participate in and benefit from the renewable energy transition. These projects foster ownership, generate local employment, and ensure that the economic benefits of clean energy stay within the community. By promoting community ownership models, such as cooperatives and community solar initiatives, renewable energy equity and justice can be advanced, ensuring that marginalized communities have a stake in the renewable energy economy [21].
  - **Subsidies and Incentives for Marginalized Communities:** Providing subsidies and incentives specifically targeted at marginalized communities can help overcome financial barriers to renewable energy adoption. These measures can include financial assistance for renewable energy installations, tax incentives, and discounted energy programs. By making renewable energy more affordable and accessible, subsidies and incentives contribute to narrowing the energy gap and promoting equity in access to clean energy resources.
  - **Capacity Building and Education Programs:** Capacity building and education programs are essential for empowering marginalized communities to actively participate in renewable energy initiatives. These programs provide training, technical assistance, and educational resources to enhance community skills and knowledge in renewable energy technologies, project development, and energy management. By building capacity and raising awareness, marginalized communities can effectively engage in

decision-making processes, advocate for their energy needs, and contribute to renewable energy equity and justice.

 - **Policy Reforms Promoting Equity and Justice:** Policy reforms are critical for addressing systemic inequities and promoting justice in renewable energy development. This includes implementing policies that prioritize equitable distribution of renewable energy benefits, support community participation, and address the needs of underserved populations. Policy reforms may include mandates for community engagement, renewable energy targets for marginalized areas, and incentives for inclusive project development. By aligning policy frameworks with principles of equity and justice, governments can create an enabling environment for sustainable and equitable renewable energy transitions.
 - **Collaborative Governance Models:** Collaborative governance models involve stakeholders from diverse backgrounds in decision-making processes, fostering inclusive and participatory approaches to renewable energy development. These models facilitate dialogue, consensus-building, and shared decision-making among government agencies, communities, industry stakeholders, and civil society organizations. By promoting collaboration and co-creation, collaborative governance models ensure that renewable energy initiatives reflect the diverse needs, priorities, and aspirations of all stakeholders, thereby advancing equity and justice in the renewable energy sector.

- **Measurement and Evaluation**
 - **Metrics for Assessing Equitable Access:** Developing comprehensive metrics is essential for assessing equitable access to renewable energy resources. These metrics should consider factors such as geographical distribution, socioeconomic indicators, and demographic characteristics to evaluate the accessibility of clean energy technologies and services. Metrics can include measures of energy affordability, availability of renewable energy infrastructure, and participation in renewable energy programs. By quantifying access disparities, stakeholders can identify areas for intervention and monitor progress toward achieving equitable access to renewable energy.
 - **Impact Assessments on Marginalized Communities:** Conducting impact assessments on marginalized communities is crucial for understanding the differential effects of renewable energy initiatives. These assessments evaluate the socioeconomic, environmental, and cultural impacts of renewable energy projects on vulnerable populations, including Indigenous peoples, low-income households, and minority communities. Impact assessments provide insights into potential benefits and risks, inform decision-making processes, and ensure that renewable energy development respects the rights and interests of marginalized groups.

- **Tracking Progress Toward Justice Goals:** Tracking progress toward justice goals requires establishing clear benchmarks, targets, and indicators to monitor the implementation of equity-focused initiatives. This involves measuring outcomes related to energy access, affordability, community participation, and socioeconomic empowerment. By regularly monitoring progress and collecting disaggregated data, stakeholders can assess the effectiveness of interventions, identify areas of improvement, and hold decision-makers accountable for advancing renewable energy equity and justice.
- **Evaluation of Policy Effectiveness:** Evaluating the effectiveness of policies is essential for assessing their impact on renewable energy equity and justice. This involves analyzing the outcomes of policy interventions, such as incentives, subsidies, and regulatory frameworks, in promoting equitable access to renewable energy resources. Evaluation methods may include qualitative assessments, stakeholder surveys, and cost–benefit analyzes to determine the efficiency, equity, and sustainability of policy measures. By evaluating policy effectiveness, policymakers can identify successful strategies, address implementation challenges, and refine policies to better align with equity and justice objectives.

- **Interdisciplinary Perspectives**
  - **Economics of Renewable Energy Equity:** Economic perspectives shed light on the distributional impacts of renewable energy initiatives and the economic barriers to equitable access. Analyzing factors such as cost–benefit ratios, investment flows, and market structures helps identify strategies to ensure that the benefits of renewable energy are equitably distributed across society. Economic theories guide policymakers in designing incentive mechanisms, subsidy programs, and financing models to promote inclusive renewable energy transitions.
  - **Social Justice Theories:** Social justice theories provide conceptual frameworks for understanding and addressing inequities in renewable energy development. Concepts such as distributive justice, procedural justice, and recognition justice highlight the importance of fair distribution, participatory decision-making, and respect for diversity in renewable energy initiatives. Applying social justice theories informs policies and practices that prioritize the needs and rights of marginalized communities, thereby advancing renewable energy equity and justice.
  - **Environmental Ethics:** Environmental ethics offer moral principles and values to guide decision-making in renewable energy development. Ethical considerations such as intergenerational equity, environmental stewardship, and environmental justice inform discussions on the ethical implications of renewable energy projects. Integrating environmental ethics into renewable

energy planning ensures that environmental benefits are balanced with social equity considerations, fostering sustainable and just energy transitions.

- **Engineering Solutions for Inclusivity:** Engineering perspectives contribute technical solutions to enhance inclusivity in renewable energy systems. Innovations in technology design, infrastructure planning, and energy storage address barriers to access and affordability for marginalized communities. Engineering solutions also prioritize considerations such as reliability, scalability, and compatibility with existing infrastructure to ensure that renewable energy solutions are accessible and sustainable for all.
- **Policy and Governance Frameworks:** Policy and governance frameworks provide the institutional mechanisms for translating principles of equity and justice into actionable strategies. These frameworks encompass laws, regulations, and governance structures that shape decision-making processes and resource allocation in renewable energy development. By integrating principles of equity and justice into policy frameworks, governments can create enabling environments for inclusive renewable energy transitions, foster stakeholder participation, and address systemic barriers to equity.

• **Future Directions**
  - **Emerging Technologies for Enhancing Equity:** Advancements in technology hold promise for enhancing equity in renewable energy access. Emerging technologies such as blockchain, AI, and decentralized energy systems offer innovative solutions to address barriers to access and affordability, particularly in underserved communities. By leveraging these technologies, stakeholders can develop scalable and inclusive renewable energy solutions that empower marginalized populations and promote energy justice.
  - **International Cooperation on Renewable Energy Justice:** International cooperation is essential for advancing renewable energy justice on a global scale. Collaborative efforts among nations, multilateral organizations, and civil society groups can facilitate knowledge sharing, capacity building, and resource mobilization to support renewable energy initiatives in developing countries and marginalized regions. By fostering international partnerships, stakeholders can address transboundary challenges, promote technology transfer, and promote equitable energy transitions worldwide.
  - **Addressing Intersectional Issues (e.g. gender, race, class):** Recognizing and addressing intersectional issues is critical for achieving renewable energy equity and justice. Intersectionality acknowledges that individuals experience overlapping forms of discrimination based on factors such as gender, race, class, and ethnicity. By integrating intersectional perspectives into renewable energy policies and programs, stakeholders can address the unique needs and

vulnerabilities of diverse communities, promote inclusive decision-making processes, and advance social justice in the renewable energy sector.

- **Anticipating Challenges and Opportunities:** Anticipating challenges and opportunities is essential for navigating the complexities of renewable energy equity and justice. As the renewable energy transition progresses, stakeholders must remain vigilant against potential pitfalls such as gentrification, land grabbing, and environmental injustices. At the same time, they should seize opportunities to leverage renewable energy as a tool for poverty alleviation, job creation, and community empowerment. By proactively addressing challenges and capitalizing on opportunities, stakeholders can steer the future direction of renewable energy toward greater equity, justice, and sustainability.

This ontology provides a structured framework for understanding and addressing the complex issues surrounding renewable energy equity and justice, facilitating research, policy development, and practical interventions in this critical field.

15.4.2 Semantic Graph Representation

Represent the knowledge in the form of a semantic graph, where nodes represent entities and edges represent relationships between them. Use semantic annotations to capture the meaning and context of each node and edge. For example, nodes could represent communities, renewable energy projects, and energy poverty rates (EPRs), while edges could represent connections such as "beneficiary of," "affected by," or "implemented in." Creating an AI-based semantic network for "Renewable Energy Equity and Justice" involves mapping out the interconnected concepts and relationships within this domain. Semantic network representing by the following key elements (Figure 15.6):

- **Nodes (Concepts)**

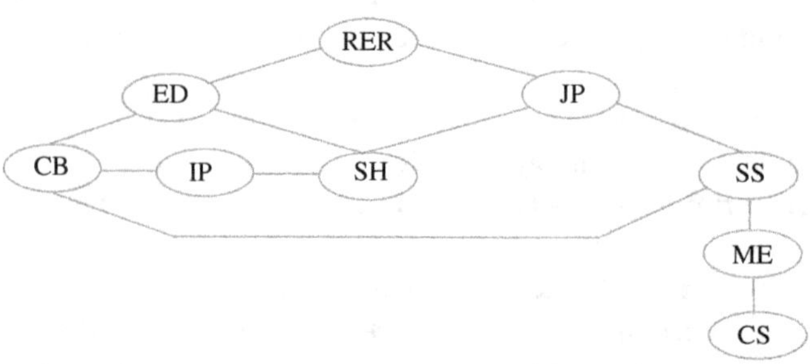

Figure 15.6 Semantic graph of the renewable energy equity and justice.

- Renewable Energy Resources (RER)
- Equity Dimensions (ED)
- Justice Principles (JP)
- Stakeholders (SH)
- Challenges and Barriers (CB)
- Solutions and Strategies (SS)
- Measurement and Evaluation (ME)
- Case Studies (CS) and Best Practices
- Interdisciplinary Perspectives (IP)
- Future Directions (FD)

- **Edges (Relationships)**
  - Renewable Energy Resources → Equity Dimensions (RER → ED)
  - Renewable Energy Resources → Justice Principles (RER → JP)
  - Equity Dimensions → Stakeholders (ED → SH)
  - Equity Dimensions → Challenges and Barriers (ED → CB)
  - Justice Principles → Stakeholders (JP → SH)
  - Justice Principles → Solutions and Strategies (JP → SS)
  - Stakeholders → Solutions and Strategies (SH → SS)
  - Challenges and Barriers → Solutions and Strategies (CB → SS)
  - Solutions and Strategies → Measurement and Evaluation (SS → ME)
  - Measurement and Evaluation → Case Studies and Best Practices (ME → CS)
  - Interdisciplinary Perspectives → Stakeholders (IP → SH)
  - Interdisciplinary Perspectives → Challenges and Barriers (IP → CB)
  - Future Directions → Solutions and Strategies (FD → SS)

- **Data Integration:** Integrate structured and unstructured data sources relevant to renewable energy equity and justice, such as demographic data, energy consumption statistics, policy documents, and environmental impact assessments. Use natural language processing (NLP) techniques to extract and classify information from textual sources and map it to the ontology.
- **Machine Learning Models:** Train machine learning models to analyze and classify data based on patterns and correlations within the knowledge representation. For example, develop models to predict EPRs based on demographic and socioeconomic factors, identify communities most in need of renewable energy interventions, or assess the environmental impact of renewable energy projects.
- **Inference Engines:** Develop inference engines that can reason over the knowledge representation to derive new insights, make recommendations, and support decision-making processes. Use techniques such as deductive reasoning, probabilistic inference, and causal modeling to infer relationships and outcomes from the available data.

- **User Interfaces:** Design user interfaces that allow stakeholders to interact with the AI-based knowledge representation, explore data visualizations, and access actionable insights. Provide tools for querying the knowledge base, visualizing relationships, and exploring alternative scenarios to support informed decision-making and policy formulation.

By developing an AI-based knowledge representation of renewable energy equity and justice, we can leverage the power of AI to analyze complex data, identify opportunities for intervention, and promote more equitable and sustainable energy systems. This approach can help policymakers, researchers, and practitioners address pressing challenges in the transition to renewable energy while ensuring that the benefits are shared equitably across communities [22].

15.5 Data Collection and Assessment

In the transition toward a sustainable energy future, ensuring equity and justice in the distribution and utilization of renewable energy resources is paramount. Data collection and assessment play a crucial role in understanding and addressing disparities in access, benefits, and impacts across various demographic groups and geographic regions [23]. Data collection involves gathering information on key variables such as energy consumption patterns, demographic characteristics, socioeconomic indicators, and environmental factors. This data enables researchers, policymakers, and stakeholders to analyze the distributional impacts of renewable energy initiatives and identify areas of inequity. Assessment methodologies help evaluate the extent to which renewable energy policies and projects contribute to or mitigate disparities in access, affordability, and environmental justice [24]. This includes assessing the distribution of benefits such as job creation, energy cost savings, and improved air quality, as well as the potential for adverse impacts on marginalized communities. Table 15.1 provides a comparative overview of various countries based on several key indicators related to renewable energy adoption, energy poverty, investment, access, employment, and environmental impact. In terms of renewable energy capacity, China leads with 600,000 MW, followed by the United States with 250,000 MW. However, when it comes to renewable energy consumption (REC), China again leads with 1,500,000 MWh, followed by the US with 750,000 MWh. EPRs varies significantly among the countries, with Nigeria having the highest rate at 30%, followed by India at 20%, while Canada and Australia have the lowest rates at 4% and 3%, respectively. The Gini coefficient, which measures income inequality, ranges from 0.29 in Germany to 0.63 in South Africa, indicating varying levels of income distribution across these nations. Public investment in renewables

Table 15.1 Data collection and data assessment.

Country	Renewable energy capacity (MW)	Renewable energy consumption (MWh)	Energy poverty rate (%) (EPR)	Gini coefficient (GC)	Public investment in renewables (%) (PIR)	Energy Access Index (EAI)	Renewable energy job (REJ)	Environmental Impact Index (EII)
USA	250,000	750,000	8	0.41	20	85	150	7.2
China	600,000	1,500,000	12	0.47	15	70	250	6.8
India	400,000	900,000	20	0.35	25	60	180	7.5
Germany	120,000	300,000	5	0.29	30	95	70	8.1
Brazil	200,000	500,000	15	0.53	18	75	120	7.0
Japan	100,000	250,000	6	0.38	22	90	80	8.0
Nigeria	50,000	150,000	30	0.49	10	55	100	6.5
Russia	300,000	700,000	10	0.42	28	80	110	7.8
Indonesia	150,000	400,000	18	0.39	17	65	140	6.9
Mexico	80,000	200,000	12	0.43	21	78	90	7.3
South Africa	70,000	180,000	25	0.63	13	50	60	6.2
France	110,000	280,000	7	0.32	27	88	75	8.3
Canada	130,000	320,000	4	0.34	23	92	65	8.5
Australia	90,000	220,000	3	0.37	24	94	55	8.7
Saudi Arabia	60,000	150,000	22	0.45	14	58	45	6.7
Argentina	100,000	250,000	14	0.46	19	72	85	7.1
Turkey	110,000	270,000	11	0.37	26	87	70	7.9
Iran	70,000	180,000	19	0.40	16	68	65	7.4
Thailand	80,000	200,000	16	0.36	20	77	80	7.6
Egypt	60,000	150,000	21	0.41	11	57	55	6.6

as a percentage of GDP is highest in Germany at 30%, followed by Russia at 28%, while the lowest investment is seen in Nigeria at 10%. The energy access index (EAI) measures the overall access to energy resources, with Canada and Australia scoring the highest at 92 and 94, respectively, while South Africa has the lowest score at 50 [25]. Renewable energy jobs (REJs) are most prevalent in Germany with 95 jobs/MW, followed by France with 88 jobs/MW, indicating the employment potential of renewable energy sectors in these countries. The Environmental Impact Index (EII) assesses the environmental sustainability of energy systems, with Canada and Australia having the highest scores at 8.5 and 8.7, respectively, indicating lower environmental impact, while South Africa has the lowest score at 6.2, suggesting a higher environmental impact [26]. So that, Table 15.1 illustrates the diverse landscape of renewable energy adoption and its associated socioeconomic and environmental factors across different countries. Table 15.2 presents a descriptive analysis of the dataset provided in Table 15.1.

15.5.1 Multi-regression Analysis

Multiple regression analysis can play a crucial role in the field of renewable energy equity and justice by helping researchers and policymakers understand the complex relationships between various factors that influence access to and distribution of renewable energy resources. Multi-regression analysis enables predictive modeling, which can help forecast future trends in renewable energy equity and justice. This can inform policymakers and stakeholders about potential challenges or opportunities in promoting equitable access to renewable energy [27]. Eq. (15.1) represents a regression model used to calculate the Gini coefficient based on two variables: REC (measured in MWh) and EPR (expressed as a percentage). Table 15.3 shows the predicted values with prediction limits of individuals of the Gini coefficient. Table 15.4 shows the analysis of variance of

Table 15.2 Descriptive analysis of the given dataset.

Parameter	Mean	SD	SE	Median	Min.	Max.	Range	Variance
EPR	13.9	7.454423	1.66686	13	3	30	27	55.56
GC	0.4135	0.07788487	0.01741558	0.405	0.29	0.63	0.34	0.00606
PIR	19.95	5.735164	1.282422	20	10	30	20	32.89
EAI	74.8	14.08844	3.150271	76	50	95	45	198.48
REJ	97.25	50.2749	11.24181	80	45	250	205	2527.56
EII	7.405	0.6999812	0.1565206	7.35	6.2	8.7	2.5	0.4899

Table 15.3 Predicted values with prediction limits of individuals of GC.

Row	Actual GC	Predicted GC	Standard error of predicted	95% lower pred. limit of individual	95% upper pred. limit of individual
1	0.41	0.3866667	0.06933811	0.2403761	0.5329574
2	0.47	0.4359185	0.08299374	0.2608171	0.61102
3	0.35	0.4686953	0.07230984	0.3161349	0.6212558
4	0.29	0.353249	0.06954249	0.2065272	0.4999708
5	0.53	0.4239535	0.06707777	0.2824318	0.5654752
6	0.38	0.3581285	0.06927527	0.2119705	0.5042866
7	0.49	0.5096663	0.07446378	0.3525614	0.6667712
8	0.42	0.3979908	0.06845552	0.2535623	0.5424193
9	0.39	0.4401571	0.06736666	0.2980259	0.5822883
10	0.43	0.3952305	0.06758517	0.2526383	0.5378228
11	0.63	0.4783829	0.07070943	0.329199	0.6275667
12	0.32	0.365512	0.06864708	0.2206793	0.5103447
13	0.34	0.3474305	0.07002994	0.1996802	0.4951807
14	0.37	0.3378561	0.07122475	0.187585	0.4881272
15	0.45	0.4581104	0.069286	0.3119297	0.6042911
16	0.46	0.4096844	0.0671434	0.2680242	0.5513446
17	0.37	0.3909769	0.06740151	0.2487722	0.5331817
18	0.4	0.439716	0.06807949	0.2960808	0.5833511
19	0.36	0.4210085	0.06744515	0.2787116	0.5633053
20	0.41	0.451666	0.06888189	0.3063379	0.596994

Table 15.4 Analysis of variance of REC and EPR.

Source	DF	R^2 lost if term(s) removed	Sum of squares	Mean square	F-ratio	Probability level
Intercept	1		3.419645	3.419645		
Model	2	0.3724	0.04291951	0.02145975	5.043	0.0191
REC (MWh)	1	0.0180	0.002073241	0.002073241	0.487	0.4946
EPR	1	0.3709	0.04275101	0.04275101	10.047	0.0056
Error	17	0.6276	0.07233549	0.004255029		
Total (adjusted)	19		0.115255	0.006066053		

Table 15.5 Regression coefficients confidence intervals of REC and EPR.

Independent variable	Regression coefficient b(i)	Standard error Sb(i)	Lower 95% conf. limit of β(i)	Upper 95% conf. limit of β(i)
Intercept	0.311637	0.03845167	0.2305111	0.392763
REC (MWh)	3.129848E-08	4.483835E-08	−6.330217E-08	1.258991E-07
EPR	0.006444484	0.002033135	0.002154944	0.01073402

REC and EPR. Table 15.5 shows the regression coefficients confidence intervals of REC and EPR (Figures 15.7–15.9).

$$\begin{aligned}\text{Gini coefficient} = {} & 0.311637020332772 + 3.12984832086939 \times e^{-8} \\ & \times \text{renewable energy consumption (MWh)} \\ & + 0.00644448381351189 \times \text{energy poverty rate (\%)}\end{aligned} \tag{15.1}$$

$$\begin{aligned}\text{Environmental impact index} = {} & 4.30009626068767 + 0.0453173374438273 \\ & \times \text{public investment in renewables} \\ & + 0.0294227654720318 \times \text{energy access index}\end{aligned} \tag{15.2}$$

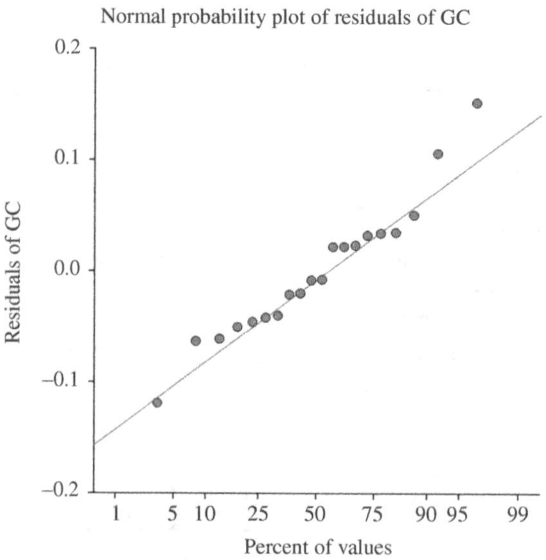

Figure 15.7 Normal probability plot of the residual of GC.

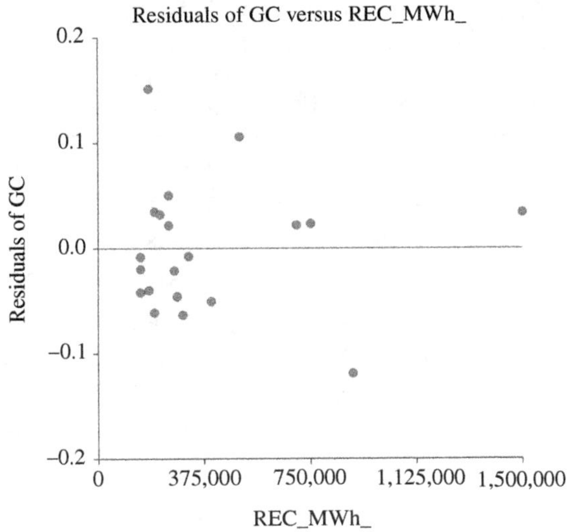

Figure 15.8 Residuals of GC versus REC.

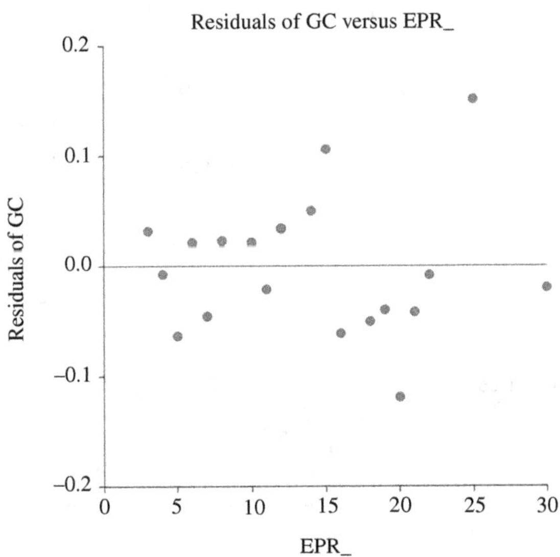

Figure 15.9 Residuals of GC versus EPR.

Table 15.6 shows the predicted values with prediction limits of individuals of the Gini coefficient. Table 15.7 shows the analysis of the variance of REC and EPR. Table 15.8 shows the regression coefficients confidence intervals of REC and EPR (Figures 15.10–15.12).

Table 15.6 Predicted values with prediction limits of individuals of EII.

Actual EII	Predicted EII	Standard error of predicted	95% lower pred. limit of individual	95% upper pred. limit of individual
7.2	7.707378	0.3174771	7.03756	8.377196
6.8	7.03945	0.3151334	6.374577	7.704323
7.5	7.198396	0.3692398	6.419368	7.977424
8.1	8.454779	0.3299581	7.758628	9.150929
7	7.322516	0.3096381	6.669236	7.975795
8	7.945127	0.3207915	7.268316	8.621938
6.5	6.371522	0.3294986	5.676341	7.066703
7.8	7.922803	0.3324474	7.2214	8.624206
6.9	6.982971	0.3108481	6.327138	7.638803
7.3	7.546736	0.3075181	6.89793	8.195542
6.2	6.36036	0.3307204	5.662601	7.058119
8.3	8.112867	0.3185629	7.440759	8.784977
8.5	8.04929	0.3220635	7.369795	8.728784
8.7	8.153452	0.3237755	7.470345	8.836559
6.7	6.641059	0.3180579	5.970016	7.312103
7.1	7.279565	0.3074263	6.630952	7.928178
7.9	8.038128	0.3155914	7.372288	8.703967
7.4	7.025922	0.3108271	6.370134	7.681709
7.6	7.471996	0.3075774	6.823064	8.120928
6.6	6.475685	0.3253447	5.789268	7.162102

Table 15.7 Analysis of variance of PIR and EAI.

Source	DF	R^2 lost if term(s) removed	Sum of squares	Mean square	F-ratio	Probability level
Intercept	1		1096.681	1096.681		
Model	2	0.8360	7.782362	3.891181	43.316	0.0000
PIR	1	0.0518	0.4824446	0.4824446	5.371	0.0332
EAI	1	0.1318	1.227213	1.227213	13.661	0.0018
Error	17	0.1640	1.527139	0.08983168		
Total (adjusted)	19		9.3095	0.4899737		

Table 15.8 Regression coefficients confidence intervals of PIR and EAI.

Independent variable	Regression coefficient b(i)	Standard error Sb(i)	Lower 95% conf. limit of β(i)	Upper 95% conf. limit of β(i)
Intercept	4.300096	0.3797535	3.498887	5.101306
PIR	0.04531734	0.01955489	0.004060121	0.08657455
EAI	0.02942277	0.007960462	0.01262766	0.04621787

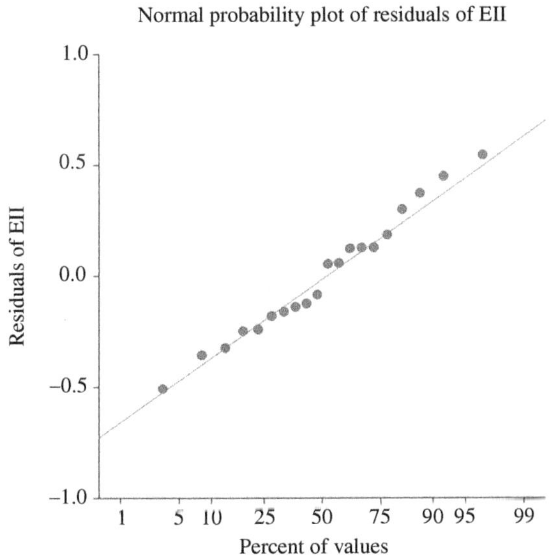

Figure 15.10 Normal probability plot of residuals of EII.

15.6 AI-based Strengths, Weaknesses, Opportunities, and Threats (SWOT) Analysis of the Renewable Energy Equity and Justice

15.6.1 Strengths

- **Data Analytics:** AI enables the analysis of vast amounts of data to identify patterns and disparities in renewable energy access and distribution, helping to address equity gaps [28].
- **Predictive Modelling:** Machine learning algorithms can predict EPRs, income inequality, and environmental impacts, guiding policymakers in developing targeted interventions.

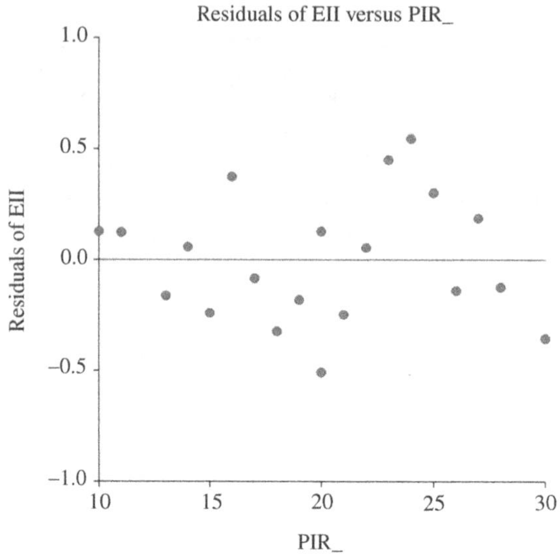

Figure 15.11 Residuals of EII versus PIR.

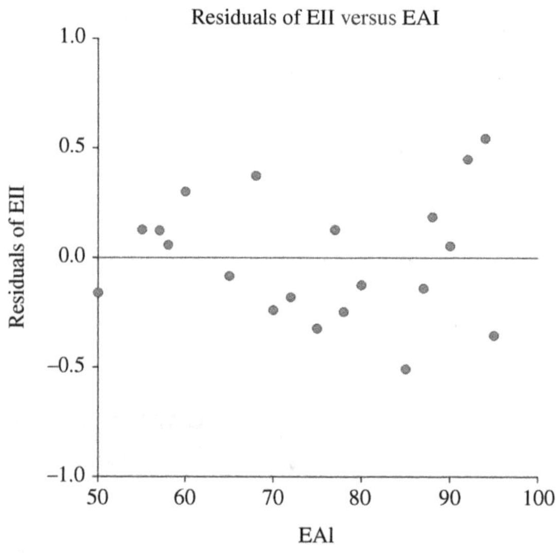

Figure 15.12 Residuals of EII versus EAI.

- **Decision Support:** AI-based systems provide decision-makers with actionable insights and recommendations to optimize renewable energy planning and investment strategies, promoting fairness and inclusivity.
- **Continuous Improvement:** AI allows for continuous learning and improvement of equity and justice initiatives through feedback loops and iterative processes, enhancing effectiveness over time.

15.6.2 Weaknesses

- **Data Bias:** AI algorithms may perpetuate biases present in training data, leading to unfair outcomes or exacerbating existing inequities if not properly addressed.
- **Complexity:** Implementing AI-based solutions requires expertise in data science, machine learning, and renewable energy policy, posing challenges for organizations with limited technical capacity.
- **Interpretability:** The black-box nature of some AI models can make it difficult to interpret their decisions and understand the rationale behind recommendations, raising transparency concerns [29].
- **Resource Intensive:** Developing and maintaining AI-based systems for renewable energy equity and justice requires significant resources, including funding, technical expertise, and data infrastructure.

15.6.3 Opportunities

- **Policy Innovation:** AI can inform the development of innovative policies and programs that prioritize equity and justice in renewable energy deployment, fostering social and environmental benefits.
- **Community Engagement:** AI-based tools can facilitate meaningful engagement with communities affected by renewable energy projects, empowering them to participate in decision-making processes.
- **Cross-sector Collaboration:** AI enables collaboration between stakeholders from government, industry, academia, and civil society to cocreate solutions for addressing equity and justice challenges in renewable energy [30].
- **Scalability:** Successful AI initiatives can be scaled up and replicated in different contexts, driving systemic change toward more equitable and just renewable energy systems.

15.6.4 Threats

- **Privacy Concerns:** AI-driven data collection and analysis raise privacy concerns related to the use of personal information and potential misuse of data, necessitating robust privacy safeguards.
- **Regulatory Hurdles:** Regulatory frameworks may lag behind technological advancements in AI, creating uncertainties and barriers to the implementation of equity-focused AI solutions in renewable energy.
- **Digital Divide:** Unequal access to technology and digital literacy may exacerbate disparities in renewable energy equity, leaving marginalized communities behind in the transition to clean energy.

- **Ethical Dilemmas:** AI introduces ethical dilemmas related to fairness, accountability, and transparency, requiring careful consideration of ethical principles and values in the design and deployment of AI systems for renewable energy equity and justice.

15.7 Social Acceptance for AI Usage in Energy Systems for Energy Justice and Equity

Realizing social acceptance for AI usage in energy systems to promote energy justice and equity involves several key strategies:

Transparency and Education: Transparently communicate the benefits, risks, and limitations of AI technologies in energy systems to the general public. Educate communities about how AI can contribute to achieving energy justice and equity, such as by optimizing energy distribution, reducing costs, and increasing access to clean energy.

Community Engagement: Involve communities in the design, development, and implementation of AI-powered energy systems. Solicit feedback, address concerns, and incorporate local knowledge and perspectives to ensure that AI technologies meet the specific needs and preferences of diverse communities.

Equitable Access and Benefits: Ensure that the benefits of AI-enabled energy systems are equitably distributed among all members of society. Prioritize underserved communities and marginalized groups in the deployment of AI technologies to address energy poverty, improve energy access, and reduce disparities in energy provision.

Ethical and Responsible Deployment: Implement AI technologies in energy systems in a manner that upholds ethical principles and respects human rights. Prioritize fairness, accountability, and transparency in algorithm design, data collection, and decision-making processes to mitigate the risk of bias and discrimination.

Regulatory Oversight and Governance: Establish robust regulatory frameworks and governance mechanisms to oversee the deployment and operation of AI-powered energy systems. Ensure compliance with data privacy regulations, ethical guidelines, and industry standards to build trust and confidence in AI technologies among stakeholders.

Demonstration Projects and Pilot Studies: Conduct demonstration projects and pilot studies to showcase the benefits and feasibility of AI-enabled energy systems in real-world settings. Highlight successful case studies and best practices to inspire confidence and promote widespread adoption of AI technologies.

Continuous Monitoring and Evaluation: Continuously monitor and evaluate the social, economic, and environmental impacts of AI usage in energy systems. Solicit feedback from stakeholders, assess performance metrics, and adjust strategies as needed to address emerging challenges and optimize outcomes.

Communication and Outreach: Engage in proactive communication and outreach efforts to raise awareness about the role of AI in promoting energy justice and equity. Utilize multiple channels, including media campaigns, public forums, and community events, to foster dialogue, build trust, and mobilize support for AI-powered energy initiatives.

15.8 Conclusion

The application of AI in marketing analysis holds significant promise for advancing renewable energy justice and equity. By harnessing the power of AI algorithms, organizations and policymakers can gain deeper insights into the socioeconomic and demographic factors influencing renewable energy access and distribution. This enhanced understanding enables the development of targeted marketing strategies and policy interventions aimed at promoting fairness and inclusivity in renewable energy deployment. AI-driven marketing analysis offers several advantages, including the ability to analyze vast amounts of data, identify patterns and disparities, and predict the impacts of different interventions on equity outcomes. By leveraging machine learning algorithms, decision-makers can tailor marketing campaigns to reach underserved communities, engage with stakeholders, and address barriers to renewable energy adoption. Furthermore, AI-based marketing analysis facilitates continuous learning and improvement by enabling organizations to adapt their strategies based on real-time feedback and evolving market dynamics. This iterative approach enhances the effectiveness and relevance of marketing efforts, ultimately driving greater adoption of renewable energy technologies among diverse populations. However, it is essential to recognize and mitigate potential challenges associated with AI-based marketing analysis, such as data privacy concerns, algorithmic biases, and the digital divide. By implementing robust privacy safeguards, ensuring algorithmic fairness, and promoting digital inclusion, stakeholders can maximize the benefits of AI while minimizing risks. In summary, AI-based marketing analysis has the potential to revolutionize efforts to realize renewable energy justice and equity. By leveraging data-driven insights and innovative approaches, organizations can accelerate the transition to a more inclusive and sustainable energy future for all. Following are the main achievements of this work:

- This framework serves as a blueprint for leveraging AI technologies to address socioeconomic disparities in renewable energy adoption.

- Through the application of machine learning algorithms and data analytics, the research achieved a deeper understanding of consumer behavior, preferences, and sociodemographic factors influencing renewable energy adoption.
- By analyzing diverse datasets, the research identified patterns and barriers related to renewable energy usage, shedding light on key challenges and opportunities in promoting renewable energy justice and equity.
- The research explored the potential impact of AI-based marketing on policy-making, aiming to influence regulatory frameworks that promote renewable energy justice.

References

1 Pearl-Martinez, R. and Stephens, J.C. (2016). Toward a gender diverse workforce in the renewable energy transition. *Sustainability: Science, Practice, and Policy* 12: 8–15.

2 Walker, C. and Baxter, J. (2017). Procedural justice in Canadian wind energy development: a comparison of community-based and technocratic siting processes. *Energy Research and Social Science* 29: 160–169.

3 Zarate-Toledo, E., Patino, R., and Fraga, J. (2019). Justice, social inclusion and indigenous opposition: a case study of wind energy development on the Isthmus of Tehuantepec. *Mexico Energy Research and Social Science* 54: 1–11.

4 Baker, S.H. (2016). Mexican energy reform, climate change, and energy justice in indigenous communities. *Natural Resources Journal* 56: 369–390.

5 Daley, D.M. and Reames, T.G. (2015). *Failed Promises: Evaluating the Federal Government's Response to Environmental Justice* Chapter 6. MIT Press.

6 Borenstein, S. and Davis, L.W. (2016). The distributional effects of US clean energy tax credits. *Tax Policy and the Economy* 30: 191–234.

7 Mullen, C. and Marsden, G. (2016). Mobility justice in low carbon energy transitions. *Energy Research and Social Science* 18: 109–117.

8 Mitra, S. and Buluswar, S. (2015). Universal access to electricity: closing the affordability gap. *Annual Review of Environment and Resources* 40: 261–283.

9 Sunter, D., Castellanos, S., and Kammen, D.M. (2019). Disparities in rooftop photovoltaics deployment in the United States by race and ethnicity. *Nature Sustainability* 2: 71–76.

10 Sankaran, S. and McIntyre-Mills, J. (2022). Energy justice in renewable energy projects: how learning about indigenous knowledge systems could inform systemic practice. *Systems Research and Behavioral Science* 39 (5).

11 Sadat-Razavi, P., Karakislak, I., Hildebrand, J. et al. (2024). German media discourses and public perceptions on hydrogen imports: an energy justice perspective. *Energy Technology*.

12 Fraser, T. and Chapman, A.J. (2020). Drivers of social equity in renewable energy at the municipal level: the case of local Japanese energy policy and preferences. *Journal of Environmental Policy & Planning* 22 (3).

13 O'Shaughnessy, E., Wiser, R., Hoen, B. et al. (2023). Drivers and energy justice implications of renewable energy project siting in the United States. *Journal of Environmental Policy & Planning* 25 (3).

14 Khare, V. (2016). Solar–wind hybrid renewable energy system: a review. *Renewable and Sustainable Energy Reviews*, Elsevier 58: 23–33.

15 Khare, V. (2013). Status of solar wind renewable energy in India. *Renewable and Sustainable Energy Reviews*, Elsevier 27: 1–10.

16 Khare, V. (2016). Optimization of hydrogen based hybrid renewable energy system using HOMER, BB-BC and GAMBIT. *International Journal of Hydrogen Energy*, Elsevier 41 (38): 16743–16751.

17 Khare, V. (2017). Optimization of the hybrid renewable energy system by HOMER, PSO and CPSO for the study area. *International Journal of Sustainable Energy*, Taylor and Francis 36 (4): 326–343.

18 Chan, G., Evans, I., Grimley, M. et al. (2017). Design choices and equity implications of community shared solar. *The Electricity Journal* 30: 37–41.

19 Balta-Ozkan, N., Davidson, R., Bicket, M., and Whitmarsh, L. (2013). Social barriers to the adoption of smart homes. *Energy Policy* 63: 363–374.

20 Wada, K., Akimoto, K., Sano, F. et al. (2012). Energy efficiency opportunities in the residential sector and their feasibility. *Energy* 48: 5–10.

21 Reames, T.G., Reiner, M.A., and Stacey, M.B. (2018). An incandescent truth: disparities in energy-efficient lighting availability and prices in an urban US county. *Applied Energy* 218: 95–103.

22 Judge, D. (2023). They offered me nothing for what they had destroyed: how the scramble for clean energy minerals is hurting African communities. In: *Views & Voices*. London: Oxfam.

23 Zhang, M., Lyu, H., Bian, H. et al. (2024). Improved chaos grasshopper optimizer and its application to HRES techno-economic evaluation. *Heliyon* 10 (2): e24315.

24 Kharrich, M., Selim, A., Kamel, S. et al. (2023). An effective design of hybrid renewable energy system using an improved Archimedes optimization algorithm: a case study of Farafra, Egypt. *Energy Conversion and Management* 283: 116907.

25 Alluraiah, N.C. and Vijayapriya, P. (2023). Optimization, design and feasibility analysis of a grid-integrated hybrid AC/DC microgrid system for rural electrification. *IEEE Access*.

26 Hasan, S., Zeyad, M., Ahmed, S.M.M. et al. (2023). Techno-economic feasibility analysis of an electric vehicle charging station for an international airport in Chattogram, Bangladesh. *Energy Conversion and Management* 293: 117501.

27 Mansouri, A., El Magri, A., Lajouad, R. et al. (2023). Optimization strategies and nonlinear control for hybrid renewable energy conversion system. *International Journal of Control, Automation and Systems* 1–8.

28 Mansouri, A., El Magri, A., Lajouad, R. et al. (2023). Nonlinear observer with reduced sensors for WECS involving vienna rectifiers theoretical design and experimental evaluation. *Electric Power Systems Research* 225: 109847.

29 Mansouri, A., El Magri, A., Lajouad, R. et al. (2023). Nonlinear control strategies with maximum power point tracking for hybrid renewable energy conversion systems. *Asian Journal of Control*.

30 Hassan, Q. (2023). Techno-economic assessment of green hydrogen production by an off-grid photovoltaic energy system. *Energies* 16 (2): 744.

16

Artificial-Intelligence-Based Equitable Energy Investment Prioritization Analytics

Ashkan Safari

Faculty of Electrical and Computer Engineering, University of Tabriz, Tabriz, Iran

Nomenclature

AI	artificial intelligence
LReg	lasso regressor
MAE	mean absolute error
ML	machine learning
MSE	mean square error
RMSE	root mean square error
n	number of data
y	target variable
α	regularization hyper parameter
β	coefficient vector
χ	feature matrix
$\overline{\omega}_{Predicted}$	predicted values
$\overline{\omega}_{Observed}$	real values

16.1 Introduction

Energy investment prioritization is an important aspect of transition to sustainable and efficient energy systems that can meet the global demand while mitigating environmental impacts. Prioritizing energy investments involves decision-making to allocate resources effectively, proposing to maximize energy accessibility, reliability, and affordability. By identifying and directing investments toward high-impact projects, such as renewable energy infrastructure and energy efficiency initiatives, countries and organizations can contribute to reducing greenhouse gas emissions, enhancing energy quality, and enhancing

Communities for Clean Energy Justice and Equity in Grid Modernization, First Edition.
Edited by Mohammadreza Daneshvar, Behnam Mohammadi-Ivatloo, and Amjad Anvari-Moghaddam.

economic development. Moreover, energy investment prioritization helps address disparities in energy access, ensuring that resources are directed to areas with the greatest need. In an era of rapid technological advancements and a transition toward cleaner energy sources, the importance of thoughtful and data-driven energy investment prioritization cannot be overstated.

By the side of recent works, the study [1] highlights the potential for cost-effective energy efficiency improvements in residential buildings in the United States, emphasizing a data and physics-based approach to accurately predict energy savings. The study, focused on university-owned student residences, employs machine learning (ML) and physics models to estimate natural gas consumption, showcasing a sequential adoption strategy leading to a 36% total energy reduction. The authors in [2] addresses resource shortage and environmental pollution by developing a scientific investment model for new energy projects. The study uses a multilinear regression model to forecast total investment and incorporates evaluation indices based on the external environment, internal development, and social perspectives. The model proves effective in predicting investment, with recommendations for accurate investment prioritization. The work [3] investigates hybrid energy systems for flexibility in commercial buildings, focusing on rooftop photovoltaics (PVs), cooling storage, and batteries. The research presents an optimization model, showcasing cooling storage as economically better regarding batteries and emphasizing the impact of PV integration.

The research done in [4] challenges the fervor for removing fossil fuel subsidies, arguing for a redirection toward carbon pricing. The study synthesizes global studies to show that subsidies contribute minimally to climate change, proposing a focus on unpriced externalities and advocating for carbon pricing as a more effective climate mitigation strategy. The study [5] investigates the relationship between clean energy, pollution reduction, and economic growth in the Gulf Cooperation Council (GCC) countries. Using autoregressive distributed lag (ARDL) methods, the study reveals the positive impacts of energy production on gross domestic product (GDP) growth and emphasizes the need for sustainable energy production and integrated urbanization policies. The work [6] investigates the nexus between natural resource scarcity, fossil fuel consumption, and greenhouse gas emissions. The study suggests policy interventions to promote sustainable development in agriculture, forestry, and fishing sectors, emphasizing technology transfer, capacity building, and market-based instruments [7]. Examines the integration of green hydrogen into various sectors and analyzes strategies and policies employed by different countries. The study identifies challenges and barriers, including technological limitations, infrastructure development, and economic feasibility, providing recommendations for overcoming these challenges and accelerating green hydrogen adoption.

Additionally, a connection is presented in [8], between green total factor energy productivity (GTFEP), economic opening-up, environmental regulations, rent of natural resources, and green innovation. The study uses ARDL techniques to analyze the long- and short-term dynamics among these factors, environmental control, and eco-friendly technological development. The work [9] highlights the significance of investing in renewable energy and green financing in achieving green economic development. Using panel data and dynamic regression techniques, the study emphasizes the role of clean energy and environmentally friendly funding in achieving carbon-neutrality goals and improving environmental quality. On the work [10], it focuses on assessing energy efficiency measures in residential and small commercial sectors in the European Union. The study develops a methodology and software tool for prioritizing measures based on primary energy consumption and initial investment cost, highlighting the inverse relationship between these two factors. The work performed in [11] introduces a multi-objective approach for optimal planning and management of energy storage systems (ESSs) in smart distribution systems. The study combines optimal ESS location, sizing, and scheduling using a genetic algorithm, aiming to identify configurations that justify ESS installation. The study [12] identifies barriers to energy efficiency in small-scale industries (SSIs) clusters. Using the analytic hierarchy process (AHP), the study prioritizes financial and economic barriers and behavioral and personal barriers as the top impediments to energy efficiency improvements. On the work [13], it examines the interconnected pillars of resources, innovation, globalization, and green growth in Brazil, Russia, India, China, South Africa (BRICS) countries. The study uses dynamic system panel data estimations to reveal the positive impact of natural resources, globalization, financial development, and research and development on green growth while recommending stricter environmental regulations to mitigate negative globalization impacts.

The research performed in [14] introduces a solar–wind biomass multi-generation system, evaluating its efficiency in terms of energy, exergy, economy, and environment. The study conducts multi-objective optimization across different scenarios, emphasizing the importance of the first scenario in achieving favorable outcomes in terms of system outputs. As well, [15] explores the integration of renewable energy sources in smart urban homes, focusing on solar panels and employing the Polar Bear Optimization Algorithm (PBOA) for efficient device scheduling. The study emphasizes the reduction of energy costs and peak-to-average ratio (PAR) for economic growth and environmental sustainability.

Relying on accurate forecasting and predictive analysis in energy investment prioritization, this chapter presents the considerable role of AI and ML in energy investment prioritization. The work highlights the substantial impact of intelligent models, including ML and AI, on energy investment. It highlights the significant changes brought about by these technologies through the use of data-driven viewpoints, improving the ability to forecast future outcomes,

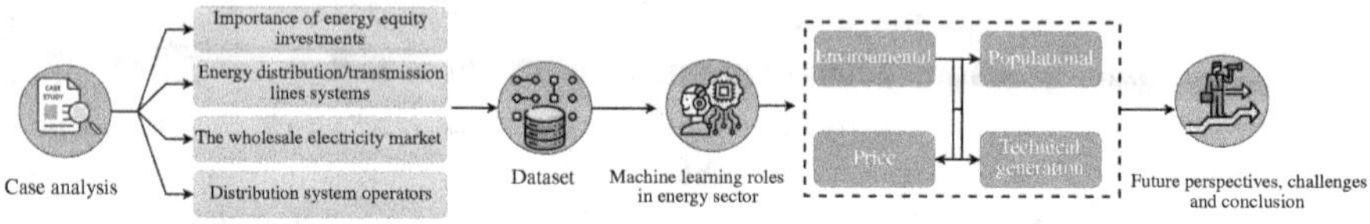

Figure 16.1 The overall process of the chapter.

and optimizing the process of making decisions in the energy industry. The chapter uses the comprehensive dataset from the NREL to demonstrate how ML techniques, specifically the lasso regression model, can be applied to predictively analyze energy equity investments in the United States. In addition, the chapter explores correlation analytics to reveal complex links within the dataset and uses clustering approaches to uncover various clusters with different properties. The ML models are used to extract key performance indicators (KPIs) that are crucial for making well-informed energy investment decisions. By combining theoretical frameworks with practical applications, this analysis perspectives into the energy investment strategies. It also provides the opportunities and obstacles involved in incorporating ML and AI to advance fair and sustainable energy investments.

As well, lasso regression is introduced as a simulation test bench using the NREL dataset. The chapter structure unfolds and is organized as follows (Figure 16.1): Section 16.2 presents a case analysis highlighting the importance of intelligent energy investment prioritization, particularly in the United States. In Section 16.3, the anticipated properties of the utilized dataset are elucidated. Section 16.4 expounds on the indispensable roles played by ML in the field of energy investment. Looking forward, Section 16.5 outlines future perspectives and challenges ahead. Finally, Section 16.6 provides a conclusive summary, wrapping up the findings from the case analysis and discussions on the use of intelligent models in energy investment prioritization.

16.2 Case Analysis: Importance of Energy Investment Prioritization Equity in USA

As demand for electrical power continues to grow globally, utility companies have to make strategic investments in energy infrastructure to ensure a reliable and sustainable supply of electricity. However, resources for expanding generation, transmission, and distribution are limited. This necessitates prioritizing investments where they can produce the greatest value and impact. Several factors should be considered when prioritizing electrical energy investments. Projects that improve grid efficiency and reduce transmission losses should rank highly. Upgrading aging infrastructure and equipment can help reduce blackouts and outages while lowering operational costs. Investments in renewable energy and energy storage can boost the grid's flexibility and resilience against disruptions. This supports the transition to cleaner sources of power generation.

Meeting rising peak demand is another key priority area. Building new generation capacity ensures there is adequate power available during periods of heavy usage. Investments may focus on locations experiencing the most severe congestion or constraints on the grid. Improving interconnectivity between regions allows utilities to share reserves and balance supply and demand more cost-effectively across a wider area.

Social and economic development goals also factor into prioritization. Expanding access to electricity remains important in underserved communities. Investments can stimulate local job creation and productivity growth.

16.2.1 Energy Distribution

The electricity grid in the United States consists of three main interconnected systems. The Eastern Interconnection spans most of the eastern states. The Western Interconnection covers the West Coast and Rocky Mountain region. Texas makes up its own isolated grid to maintain reliability in its deregulated market. High-voltage transmission lines carry power over long distances from centralized generation plants to local communities. There, lower-voltage distribution lines deliver electricity to homes and businesses, as depicted in Figure 16.2, conceptualized from [16].

According to Figure 16.1, the United States sources its electricity from a variety of resources, primarily relying on fossil fuels for over 60% of its total generation.

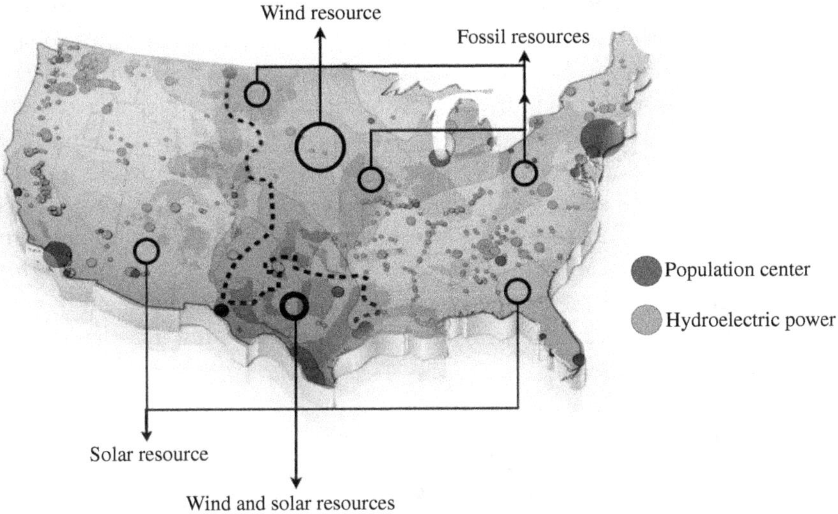

Figure 16.2 The electricity production resources in USA.

Coal, historically the dominant source, now contributes around 23% due to plant closures, while natural gas has risen to 38%, supporting the integration of renewables. Nuclear power supplies about 20% of the country's electricity from over 100 reactors. Large hydroelectric dams provide around 7% of the nation's needs, mainly in regions like the Pacific Northwest and Appalachians. Wind power, particularly strong in the Midwest and coastal areas, now accounts for over 8%, while solar has grown significantly but remains small at 2.3%, mostly in sunny desert regions. The focus is shifting toward more renewable sources, alongside efforts to reduce emissions from fossil fuels. The natural gas network relies on underground pipelines to transport fuel from production to consumers, with major pipelines connecting shale regions like the Marcellus and Permian Basins to population centers. Regional distribution lines deliver gas to various users for heating, cooking, and electricity. The infrastructure is maintained by utilities, with independent system operators ensuring reliability by balancing supply and demand.

16.2.2 Energy Transmission

The United States boasts one of the world's largest and most intricate electric transmission systems, comprised of three major interconnected grids spanning the contiguous states: the Eastern Interconnection, Western Interconnection, and Electric Reliability Council of Texas (ERCOT) Interconnection in Texas. This vast network includes over one million miles of high-voltage transmission lines, towers, and poles that transport electricity from centralized generation sources to local distribution systems. These transmission lines operate at voltages between 345 and 765 kV to minimize transmission losses over long distances, as illustrated in Figure 16.3, which conceptualizes the United States' electricity transmission [17].

The infrastructure of the national grid is made up of several thousand substations that step voltage up or down using transformers. Independent system operators, such as Pennsylvania-New Jersey-Maryland Interconnection (PJM) and Midcontinent Independent System Operator (MISO), coordinate transmission across multiple state jurisdictions. This increases reliability through load balancing across broad regions. While generally reliable, aging infrastructure poses challenges as extreme weather impacts increase. Proactive grid modernization through smart grids, storage integration, and DC lines will help transport greater amounts of clean energy from remote renewable resources regions like the Midwest or Southwest.

16.2.3 Electricity Market

The US electricity market operates through several regional wholesale markets, each managed by independent system operators and regional transmission

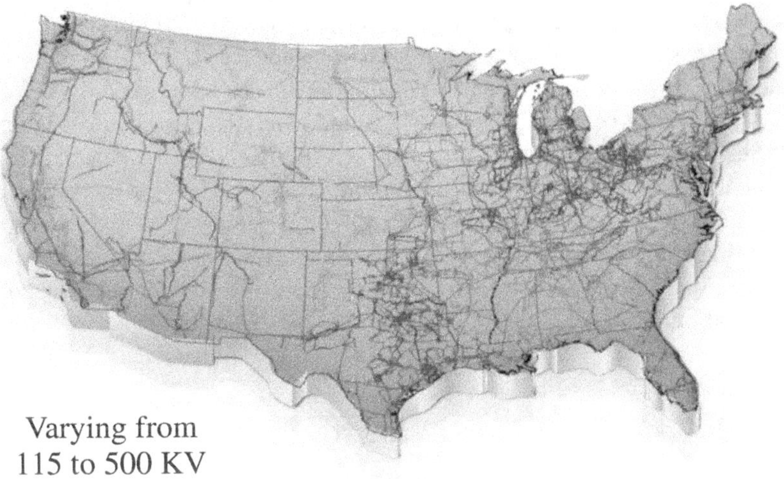

Figure 16.3 The USA electricity transmission lines.

organizations. These markets, including PJM, Independent System Operator New England (ISO-NE), New York Independent System Operator (NYISO), MISO, Southwest Power Pool (SPP), and California Independent System Operator (CAISO), integrate generation and transmission across multiple states within their areas. In these organized markets, electricity generation resources compete

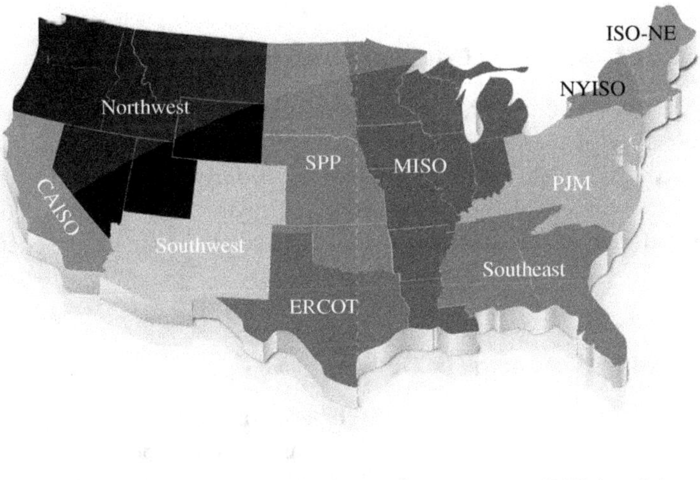

Figure 16.4 The electricity distribution system operators (DSOs) and the electricity market of the United States.

to sell power through day-ahead and real-time auctions, using locational marginal pricing to reflect supply/demand at different points on the grid. Currently, coal and natural gas dominate dispatch due to their low fuel costs, however, renewable portfolio standards are driving increased adoption of clean energy across regional grids. Additionally, some capacity markets and a mix of regulated/deregulated structures for retail sales exist, overseen at the state level to ensure nation-wide reliability and customer choice in competitive regions, as in Figure 16.4 conceptualized from [18–20].

Accordingly, the United States relies on competitive wholesale markets organized regionally to efficiently allocate generation and ensure reliable power supplies nationally.

16.3 Utilized Dataset: NREL Energy Investment Prioritization

The equitable energy investment prioritization (EEIP) dataset, from Department of Energy (DOE)/National Renewable Energy Laboratory (NREL)/ ALLIANCE, [21], which is used in this chapter, is a compilation of metrics related to energy justice, environmental justice, and renewable energy deployment potential. This dataset includes a wide range of variables, including sociodemographic characteristics, environmental hazard exposure, and technical generation potential for various renewable energy technologies. In this section, an overview of key variables and perspectives derived from the dataset are manifested as:

16.3.1 Population and Energy Generation Potential

The dataset includes county-level population metrics, providing a foundational understanding of the demographic landscape. The technical generation potential for six renewable energy technologies (utility PV, residential PV, land-based wind, commercial PV, geothermal, and hydro) is assessed through quintile scores, offering insights into the spatial distribution of renewable energy resources.

16.3.2 Levelized Cost of Energy (LCOE)

For each renewable energy technology, the dataset incorporates the levelized cost of energy (LCOE) quintile scores, as well as the corresponding median capital costs. This information is crucial for evaluating the economic feasibility of renewable energy projects at the county level.

16.3.3 Electric Vehicle Growth and Projections

Moreover, the dataset has metrics related to the projected growth of electric vehicles (EVs) categorized into battery electric vehicles (BEV), hybrid electric gasoline vehicles (HEV), plug-in hybrid electric vehicles (PHEV), and internal combustion engine/conventional gasoline vehicles (ICEV). The growth quintile scores provide insights into the evolving landscape of the transportation sector.

16.3.4 Cost Savings

The other section of the dataset introduces national quintile scores for cost savings estimate solutions for medium office building load profiles in each county of the United States. This information is valuable for assessing the potential economic benefits of renewable energy integration.

16.3.5 Environmental Justice Indicators

Various environmental justice indicators, such as exposure to particulate matter (pm25), ozone, lead paint, diesel particulate, and other hazards, are presented as quintile scores. These indicators shed light on potential environmental disparities across different regions.

16.3.6 Socioeconomic Factors

Socioeconomic factors, including unemployment rates, mining employment percentages, and rural–urban continuum codes, offer insights into the economic landscape of each county. Additionally, indicators of population loss and persistent poverty provide a comprehensive view of regional challenges.

Consequently, more details of the utilized dataset are available in Tables 16.1 and 16.2, retrieved from [21].

The EEIP provides a source of information for comprehending the integration of energy and environment. The diverse set of variables allows for deep analyses, assessments, and strategic planning aimed at developing equitable and sustainable energy development.

16.4 Machine Learning and Predictive Analysis in EEIP

As the energy sector addresses challenges around decarbonization, aging infrastructure, and rising demand, strategic investment prioritization has become increasingly taken into account. Utility companies and grid operators must evaluate thousands of potential projects across generation, transmission, distribution,

Table 16.1 The affecting factors in energy investment prioritization of the United States.

Factors	Description
county_fips	• Five digits (implied leading 0 if four digits) • First two digits: State code • Last four digits: County code
county	• Name of county
state	• Two-letter state abbreviation
county_pop	• County population Summed block group population across each county
_techpot_quint	• National quintile score for technical generation potential • Calculated using specified technology only, using the *_technical_generation_potential_MWh variable • Available for: utility_pv, residential_pv, landbasedwind, commercial_pv
_technical_generation_potential_MWh	• Technical generation potential for given technology measured in MWh • Taken directly from SLOPE (unaltered) • Available for: utility_pv, residential_pv, landbasedwind, commercial_pv
_lcoe_cap_quint	• National quintile score for LCOE capital costs • Calculated using specified technology only • Available for: commercial_pv, geothermal, hydro, landbasedwind, utility_pv, residential_pv
_lcoe_capital_price_per_MW_median	• Capital costs for given technology • Taken directly from SLOPE (unaltered) • Available for: commercial_pv, geothermal, hydro, landbasedwind, utility_pv, residential_pv
_lcoe_quint	• National quintile score for levelized cost of energy (LCOE) • Calculated using specified technology only • Available for: commercial_pv, geothermal, hydro, landbasedwind, utility_pv, residential_pv
_lcoe_price_per_MWh_median	• LCOE for a given technology • Taken directly from SLOPE (unaltered) • Available for: commercial_pv, geothermal, hydro, landbasedwind, utility_pv, residential_pv
_growth_quint	• Electric vehicle projected growth metric • National quintile score calculated from the *_prop_change variable. Calculated by the following process: 1) Calculate the proportion of vehicle type for each county (vehicle type count divided by total vehicle count for 2018 and for 2050 in a reference case) 2) Subtract the proportion of 2018 vehicle reference counts from the proportion of 2050 vehicle reference counts 3) Create national quintiles from those scores

Table 16.1 (Continued)

Factors	Description
	• Available for: bev (battery electric vehicle), hev (hybrid electric gasoline vehicle), phev (plug-in hybrid electric vehicle), icev (internal combustion engine/conventional gasoline vehicle)
**_prop_change*	• Proportion of 2050 vehicle reference counts minus the proportion of 2018 vehicle reference counts • Available for: bev (battery electric vehicle), hev (hybrid electric gasoline vehicle), phev (plug-in hybrid electric vehicle), icev (internal combustion engine/conventional gasoline vehicle)
**_2018_reference_vehicle_counts*	• 2018 vehicle reference counts • Taken directly from SLOPE (unaltered) • Available for: bev (battery electric vehicle), hev (hybrid electric gasoline vehicle), phev (plug-in hybrid electric vehicle), icev (internal combustion engine/conventional gasoline vehicle)
**_2050_reference_vehicle_counts*	• 2050 vehicle reference counts • Taken directly from SLOPE (unaltered) • Available for: bev (battery electric vehicle), hev (hybrid electric gasoline vehicle), phev (plug-in hybrid electric vehicle), icev (internal combustion engine/conventional gasoline vehicle)
reopt_savings_quint	• National quintile score for cost saving estimates assuming a medium office building load profile from solar + storage for each county (based on cost_savings_per_county variable)
cost_savings_per_county	• Cost savings estimates from solar + storage • Reaggregated from utility subdivisions to the county by intersecting the United States. county shapefile, calculating the percent overlap between reference sites and counties and scaling the cost savings for each site within the county by percent overlap
energyburden_1_prop	• Proportion of census tracts within each county with less than 4% energy burden
energyburden_2_prop	• Proportion of census tracts within each county with greater than or equal to 4% and less than 7% energy burden
energyburden_3_prop	• Proportion of census tracts within each county with greater than or equal to 7% and less than or equal to 10% energy burden
energyburden_4_prop	• Proportion of census tracts within each county with greater than 10% energy burden

(Continued)

Table 16.1 (Continued)

Factors	Description
energy_burden_indicator	• Weighted average of energyburden_3_prop and energyburden_4_prop • (energyburden_3_prop*1.1 + energyburden_4_prop*1.2)/2
unemployment_quint	• National quintile score for the percentage of individuals unemployed
unemprate2020	• Unemployment rate • Taken directly from Rural Atlas (unaltered)
mining_quint	• National quintile score for the percentage of individuals employed in mining, quarrying, or oil and gas
pctempmining	• Percent of the workforce employed in mining, quarrying, or oil and gas extraction • Taken directly from Rural Atlas (unaltered)
ruralurbancontinuumcode 2013	• Rural–urban continuum code • Taken directly from Rural Atlas (unaltered) • Metro counties: • 1: metro population of 1 million or more • 2: metro population of 250,000 to 1 million • 3: metro population of fewer than 250,000 • Nonmetro counties: • 4: urban population of 20,000 or more adjacent to metro area • 5: urban population of 20,000 or more, not adjacent to the metro area • 6: urban population of 2500 to 19,999, adjacent to metro area • 7: urban population of 2500 to 19,999, not adjacent to metro area • 8: Completely rural or less than 2500 urban population, adjacent to metro area • 9: Completely rural or less than 2500 urban population, not adjacent to metro area
type_2015_farming_no	• Farming-dependent county code • Taken directly from Rural Atlas (unaltered) • 0: No; Not a farming-dependent county • 1: Yes; Farming-dependent county
population_loss_2015_update	• Population loss county code • Taken directly from Rural Atlas (unaltered) • 0: No; Not a population-loss county • 1: Yes; Population-loss county
perpov_1980_0711	• Persistent poverty county code • Taken directly from Rural Atlas (unaltered) • 0: No; Not a persistent-poverty county • 1: Yes; Persistent-poverty county

Table 16.1 (Continued)

Factors	Description
**_indicator*	For minority_indicator: • Weighted sum of 4th quintile score minus 0.20 and 5th quintile score minus 0.20 • *(minority_4_prop−0.20) + (minority_5_prop−0.20)*1.7* For lowincome_indicator: • Weighted sum of 4th quintile score minus 0.20 and 5th quintile score minus 0.20 • *(lowincome_4_prop−0.20) + (lowincome_5_prop−0.20)*1.5* For lessthanhs_indicator: • Weighted sum of 4th quintile score minus 0.20 and 5th quintile score minus 0.20 • *(lessthanhs_4_prop−0.20) + (lessthanhs_5_prop−0.20)*2.1* For all environmental hazard indicators (pm25, ozone, leadpaint, dieselparticulate, cancerriskairtoxics, respiratoryhazardairtoxics, trafficproxvol, wastewaterdischarge, nplprox, rmpprox, tsdfprox): • 5th quintile score minus 0.20 • (*_5_prop−0.20)
*_pop_1_prop	• Proportion of census block groups within each county that fell within the 1st national quintile for the specified demographic variable • Available demographics: percent minority (minority), percent low income (lowincome), percent less than high school education (lessthanhs)
*_pop_2_prop	• Proportion of census block groups within each county that fell within the 2nd national quintile for the specified demographic variable • Available demographics: percent minority (minority), percent low income (lowincome), percent less than high school education (lessthanhs)
*_pop_3_prop	• Proportion of census block groups within each county that fell within the 3rd national quintile for the specified demographic variable • Available demographics: percent minority (minority), percent low income (lowincome), percent less than high school education
*_pop_4_prop	• Proportion of census block groups within each county that fell within the 4th national quintile for the specified demographic variable • Available demographics: percent minority (minority), percent low income (lowincome), percent less than high school education (lessthanhs)

(Continued)

Table 16.1 (Continued)

Factors	Description
*_pop_5_prop	• Proportion of census block groups within each county that fell within the 5th national quintile for the specified demographic indicator • Available demographics: percent minority (minority), percent low income (lowincome), percent less than high school education (lessthanhs)
*_1_prop	• Proportion of census block groups within each county that fell within the 1st national quintile for the specified environmental hazard indicator • Available environmental hazard indicators: pm25, ozone, leadpaint, dieselparticulate, cancerriskairtoxics, respiratoryhazardairtoxics, trafficproxvol, wastewaterdischarge, nplprox (proximity to national priority list sites), rmpprox (proximity to sites with risk management plans), tsdfprox (proximity to treatment, storage, and disposal facilities)
*_2_prop	• Proportion of census block groups within each county that fell within the 2nd national quintile for the specified environmental hazard indicator • Available environmental hazard indicators: pm25, ozone, leadpaint, dieselparticulate, cancerriskairtoxics, respiratoryhazardairtoxics, trafficproxvol, wastewaterdischarge, nplprox, rmpprox, tsdfprox
*_3_prop	• Proportion of census block groups within each county that fell within the 3rd national quintile for the specified environmental hazard indicator • Available environmental hazard indicators: pm25, ozone, leadpaint, dieselparticulate, cancerriskairtoxics, respiratoryhazardairtoxics, trafficproxvol, wastewaterdischarge, nplprox, rmpprox, tsdfprox
*_4_prop	• Proportion of census block groups within each county that fell within the 4th national quintile for the specified environmental hazard indicator • Available environmental hazard indicators: pm25, ozone, leadpaint, dieselparticulate, cancerriskairtoxics, respiratoryhazardairtoxics, trafficproxvol, wastewaterdischarge, nplprox, rmpprox, tsdfprox
*_5_prop	• Proportion of census block groups within each county that fell within the 5th national quintile for the specified environmental hazard indicator • Available environmental hazard indicators: pm25, ozone, leadpaint, dieselparticulate, cancerriskairtoxics, respiratoryhazardairtoxics, trafficproxvol, wastewaterdischarge, nplprox, rmpprox, tsdfprox

Table 16.2 The statistical details of the factors.

Labels	Mean	Variance	Data Count
county_pop	103,199	1.1E+11	3,108
utilitypv_techpot_quint	2.999035	1.999677	3,108
utility_pv_technical_generation_potential_mwh	66,540,261	8.6E+15	3,108
residentialpv_techpot_quint	2.999034	2.000321	3,107
residential_pv_technical_generation_potential_mwh	188,158.2	2.88E+11	3,107
landbasedwind_techpot_quint	2.999035	1.999677	3,108
land_based_wind_technical_generation_potential_mwh	9,832,100	2.18E+14	3,108
commercialpv_techpot_quint	2.999034	2.000321	3,107
commercial_pv_technical_generation_potential_mwh	282,689.3	9.54E+11	3,107
commercialpv_lcoe_quint	2.999034	2.000321	3,107
commercial_pv_lcoe_price_per_mwh_median	69.43428	40.46404	3,107
geothermal_lcoe_cap_quint	2.993421	1.993378	304
geothermal_lcoe_capital_price_per_mw_median	14,517,528	2.46E+13	304
geothermal_lcoe_quint	2.993421	1.993378	304
geothermal_lcoe_price_per_mwh_median	241.8553	5593.769	304
hydro_lcoe_cap_quint	3	2	3,060
hydro_lcoe_capital_price_per_mw_median	6,934,180	3.94E+11	3,060
hydro_lcoe_quint	3	2	3,060
hydro_lcoe_price_per_mwh_median	24.12712	4.997238	3,060
landbasedwind_lcoe_cap_quint	2.999033	2.000321	3,102
land_based_wind_lcoe_capital_price_per_mw_median	1,530,092	0	3,102
landbasedwind_lcoe_quint	2.999033	2.000321	3,102
land_based_wind_lcoe_price_per_mwh_median	47.94584	142.5875	3,102
utilitypv_lcoe_cap_quint	2.999033	2.000321	3,102
pv_lcoe_capital_price_per_mw_median	1,085,440	0	3,102
utilitypv_lcoe_quint	2.999033	2.000321	3,102
pv_lcoe_price_per_mwh_median	51.96728	31.98354	3,102
residentialpv_lcoe_quint	2.999034	2.000321	3,107
residential_pv_lcoe_price_per_mwh_median	109.6692	106.5463	3,107
bev_growth_quint	2.999356	2.000644	3,106
hev_growth_quint	2.999356	2.000644	3,106
phev_growth_quint	2.999356	2.000644	3,106

(Continued)

Table 16.2 (Continued)

Labels	Mean	Variance	Data Count
icev_growth_quint	2.999356	2.000644	3,106
bev_prop_change	0.043933	0.002213	3,106
hev_prop_change	0.041337	0.000744	3,106
phev_prop_change	0.011149	0.0001	3,106
icev_prop_change	−0.09642	0.004389	3,106
bev_2018_reference_vehicle_counts	661.4459	18,558,890	3,106
bev_2050_reference_vehicle_counts	9677.57	3.18E+09	3,106
hev_gasoline_2018_reference_vehicle_counts	2067.742	91,297,919	3,106
hev_gasoline_2050_reference_vehicle_counts	6590.638	5.76E+08	3,106
phev_2018_reference_vehicle_counts	311.8567	6,292,633	3,106
phev_2050_reference_vehicle_counts	2068.758	1.96E+08	3,106
icev_gasoline_2018_reference_vehicle_counts	74,208.72	4.66E+10	3,106
icev_gasoline_2050_reference_vehicle_counts	68,615.65	3.04E+10	3,106
reopt_savings_quint	1.407658	3.177766	3,108
cost_savings_per_county	7340.571	1.69E+09	3,108
energyburden_indicator	0.042894	0.011082	3,107
energyburden_1_prop	0.343176	0.123105	3,107
energyburden_2_prop	0.579142	0.117353	3,107
energyburden_3_prop	0.074312	0.03391	3,107
energyburden_4_prop	0.003369	0.001415	3,107
unemployment_quint	2.999035	1.999677	3,108
unemprate2020	6.706081	4.952264	3,108
mining_quint	2.999035	1.999677	3,108
pctempmining	1.49915	11.35863	3,108
ruralurbancontinuumcode2013	4.986808	7.305167	3,108
type_2015_farming_no	0.142857	0.122449	3,108
type_2015_mining_no	0.070142	0.065222	3,108
population_loss_2015_update	0.168919	0.140385	3,108
perpov_1980_0711	0.112934	0.10018	3,108
minority_indicator	0.10723	0.051895	3,108
minority_pop_1_prop	0.36862	0.098234	3,108
minority_pop_2_prop	0.2422	0.030434	3,108

Table 16.2 (Continued)

Labels	Mean	Variance	Data Count
minority_pop_3_prop	0.169624	0.026263	3,108
minority_pop_4_prop	0.14044	0.033081	3,108
minority_pop_5_prop	0.079117	0.027645	3,108
lowincome_indicator	0.211512	0.052307	3,108
lowincome_pop_1_prop	0.088013	0.01565	3,108
lowincome_pop_2_prop	0.175073	0.022355	3,108
lowincome_pop_3_prop	0.266454	0.030354	3,108
lowincome_pop_4_prop	0.275764	0.029017	3,108
lowincome_pop_5_prop	0.194697	0.032973	3,108
lessthanhs_indicator	0.274697	0.120821	3,108
lessthanhs_pop_1_prop	0.110867	0.020173	3,108
lessthanhs_pop_2_prop	0.188394	0.027908	3,108
lessthanhs_pop_3_prop	0.242844	0.02585	3,108
lessthanhs_pop_4_prop	0.259805	0.029191	3,108
lessthanhs_pop_5_prop	0.198089	0.045645	3,108
pm25_indicator	0.045458	0.029643	3,108
pm25_1_prop	0.342497	0.214128	3,108
pm25_2_prop	0.224056	0.147282	3,108
pm25_3_prop	0.206893	0.132423	3,108
pm25_4_prop	0.164226	0.112882	3,108
pm25_5_prop	0.062328	0.049305	3,108
ozone_indicator	0.109462	0.072478	3,108
ozone_1_prop	0.216798	0.162505	3,108
ozone_2_prop	0.21906	0.152429	3,108
ozone_3_prop	0.230168	0.153758	3,108
ozone_4_prop	0.193281	0.138078	3,108
ozone_5_prop	0.140693	0.114643	3,108
leadpaint_indicator	0.024406	0.005445	3,108
leadpaint_1_prop	0.10729	0.025927	3,108
leadpaint_2_prop	0.23749	0.038954	3,108
leadpaint_3_prop	0.304272	0.038	3,108
leadpaint_4_prop	0.259915	0.059458	3,108

(Continued)

Table 16.2 (Continued)

Labels	Mean	Variance	Data Count
leadpaint_5_prop	0.091032	0.018275	3,108
diesel_indicator	0.011175	0.005594	3,108
dieselparticulate_1_prop	0.599776	0.164835	3,108
dieselparticulate_2_prop	0.237854	0.081793	3,108
dieselparticulate_3_prop	0.097049	0.030727	3,108
dieselparticulate_4_prop	0.043618	0.015115	3,108
dieselparticulate_5_prop	0.021703	0.011516	3,108
cancer_indicator	0.076961	0.048103	3,108
cancerriskairtoxics_1_prop	0.416566	0.218766	3,108
cancerriskairtoxics_2_prop	0.199729	0.114948	3,108
cancerriskairtoxics_3_prop	0.158082	0.092631	3,108
cancerriskairtoxics_4_prop	0.118257	0.071907	3,108
cancerriskairtoxics_5_prop	0.107366	0.079091	3,108
respiratory_indicator	0.079477	0.049089	3,108
respiratoryhazardairtoxics_1_prop	0.414572	0.217883	3,108
respiratoryhazardairtoxics_2_prop	0.206453	0.116724	3,108
respiratoryhazardairtoxics_3_prop	0.149914	0.083954	3,108
respiratoryhazardairtoxics_4_prop	0.11832	0.069494	3,108
respiratoryhazardairtoxics_5_prop	0.110741	0.080924	3,108
traffic_indicator	0.006133	0.00187	2,970
trafficproxvol_1_prop	0.501492	0.07822	2,970
trafficproxvol_2_prop	0.238822	0.02369	2,970
trafficproxvol_3_prop	0.148354	0.015788	2,970
trafficproxvol_4_prop	0.081417	0.010971	2,970
trafficproxvol_5_prop	0.029914	0.0068	2,970
wastewater_indicator	0.053544	0.020125	2,889
wastewaterdischarge_1_prop	0.311566	0.097785	2,889
wastewaterdischarge_2_prop	0.227971	0.047839	2,889
wastewaterdischarge_3_prop	0.182228	0.036382	2,889
wastewaterdischarge_4_prop	0.161279	0.03672	2,889
wastewaterdischarge_5_prop	0.116955	0.041893	2,889
npl_indicator	0.031557	0.010501	3,108

Table 16.2 (Continued)

Labels	Mean	Variance	Data Count
nplprox_1_prop	0.493587	0.202227	3,108
nplprox_2_prop	0.268434	0.108113	3,108
nplprox_3_prop	0.104581	0.030365	3,108
nplprox_4_prop	0.066656	0.019226	3,108
nplprox_5_prop	0.066742	0.025951	3,108
rmp_indicator	0.059472	0.017613	3,108
rmpprox_1_prop	0.422286	0.1349	3,108
rmpprox_2_prop	0.151783	0.022021	3,108
rmpprox_3_prop	0.15005	0.021506	3,108
rmpprox_4_prop	0.137105	0.023077	3,108
rmpprox_5_prop	0.138776	0.039517	3,108
tsdf_indicator	0.009153	0.004021	3,108
tsdfprox_1_prop	0.590105	0.131835	3,108
tsdfprox_2_prop	0.190236	0.038802	3,108
tsdfprox_3_prop	0.122021	0.022396	3,108
tsdfprox_4_prop	0.072332	0.016627	3,108
tsdfprox_5_prop	0.025305	0.009548	3,108

and customer segments. Traditional evaluation methods struggle with the complexity of interacting factors and uncertainty around new technologies. This is where ML and predictive analytics can provide more results and perspectives. By using historical operational and market data, these techniques can develop advanced models to forecast the performance and cost implications of different investment portfolios over time. ML allows for automated feature engineering to discover unseen correlations. Predictive algorithms, such as random forests, long short-term memories (LSTMs), bidirectional LSTM (BiLSTM), neuro-fuzzy systems, regressors, and neural networks can outperform intuition in weighing hard-to-quantify criteria.

Key applications include predicting maintenance requirements and reliability impacts, estimating renewable energy and demand flexibility potential, optimizing asset utilization, and assessing technology adoption rates. Scenario analysis and risk quantification tools based on ML give decision-makers have perspectives into investment consequences. By the side of the mentioned models, lasso regressor (LReg) is selected to perform analysis on the NREL dataset and execute the evaluations.

16.4.1 Lasso Regressor

LReg is a linear model that performs both variable selection and regularization by imposing a constraint on the absolute sum of the coefficients, which has the effect of setting some coefficients exactly to zero and producing sparse solutions. It is useful for problems with more features than observations, as it addresses the challenge of overfitting by adding a penalty that performs automatic variable selection. The degree of penalty is controlled by a hyperparameter, and cross-validation is employed to select the optimal value. A concept of this model is illustrated in Figure 16.5.

Consequently, LReg utilized in this chapter, is formulated as:

$$\text{Min}\boldsymbol{Y}:$$

$$\frac{1}{2n}||y - \chi\beta||^2 + \alpha||\beta||_1 \tag{16.1}$$

where, n, y, and χ are the number of samples, target variable, as well as feature matrix, respectively. Coefficient vector, least squares cost function, and regularization hyperparameter are symbolized by $||y - \chi\beta||^2$, and α. Also, to assess the performance evaluation of the model, KPIs of MAE, MSE, and RMSE are defined [22–24]:

MAE (Mean Absolute Error): One noteworthy additional assessment method for evaluating the effectiveness of regression equations is MAE. The average absolute difference between the expected and actual values is measured. MAE, in contrast to RMSE, is more lenient when it comes to large errors because it solely looks at absolute deviations rather than squaring them up to increase their impact. Furthermore, the target variable and MAE have the same units of expression:

$$MAE = \frac{1}{n}\sum|(\varpi_{Observed} - \varpi_{Predicted})| \tag{16.2}$$

RMSE (Root Mean Square Error): In regression analysis, RMSE is a frequently used metric to evaluate the degree of variation between predicted and observed values. The square root of the average of the squared discrepancies

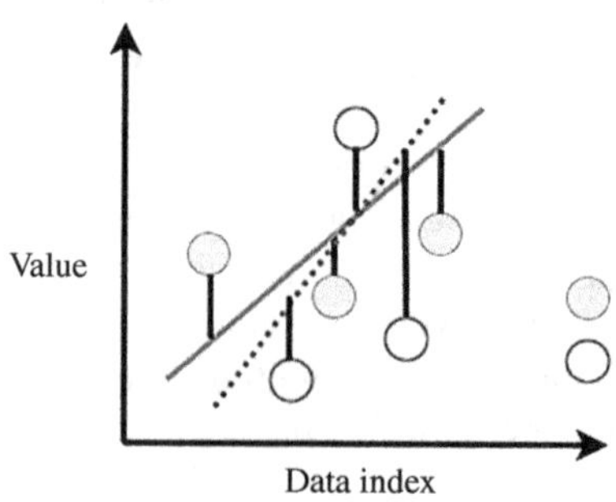

Figure 16.5 The concept of LReg.

between the expected and actual numbers is used to calculate it. A regression model's fit to the data is assessed using the RMSE measure; lower values denote better performance. Moreover, RMSE is simpler to interpret, as it keeps the same units as the target variable.

$$MSE = \frac{1}{n}\sum(\varpi_{Observed} - \varpi_{Predicted})^2 \tag{16.3}$$

$$RMSE = \sqrt{\frac{1}{n}\sum(\varpi_{Observed} - \varpi_{Predicted})^2}, \tag{16.4}$$

where n is the number of data, and $\varpi_{Observed}$, with $\varpi_{Predicted}$ are the observed and predicted values of the grid community, respectively.

For the first phase of analysis, the data clustered and classified based on their type, into five classes of *environment, generation, population, price, and quintile.* As the next steps, each class of labels and variables analyzed by the correlation matrix, as in Figures 16.6 and 16.7, a correlation matrix is a table

Figure 16.6 Zoomable versions of the correlation matrix.

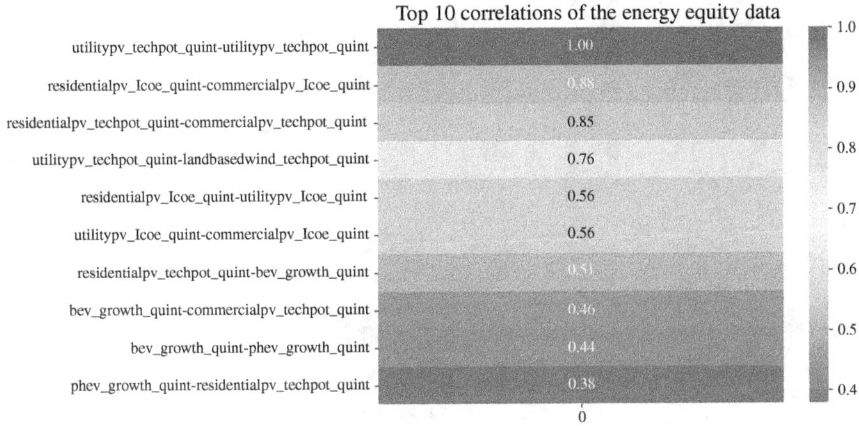

Figure 16.7 The top ten highly correlated factors of the NREL dataset.

that compactly displays the correlation between all variable pairs in a dataset. Arranged with variables along both the rows and columns, the matrix shows the strength and direction of linear relationships between predictors through values ranging [−1 to 1] along the off-diagonals. The diagonal entries are always 1, representing the entire self-correlation. Correlation matrices are tools that reveal how predictors are associated without needing multiple separate tests. This allows users to understand the influence and importance of predictors at a glance, as well as detect issues such as multicollinearity. As a concise visualization of correlations across all variables, the matrix is useful for data profiling, feature selection, and building initial perspectives prior to model development. Since the correlation matrix of figures is so large, the zoomable version of them is available in Figure 16.7. As shown in Figure 16.8, the price quintile-based factors have the highest correlation and effect with each other.

As depicted in Figure 16.9 and Table 16.3, the correlation matrix shows the relationships between various energy-related variables. Notably, L.1 manifests a strong positive correlation with L.7 (0.659) in addition to positive associations with *hydro_lcoe_capital_price_per_mw_median* (0.027) as well to L.3 (0.007). Additionally, there is a negative correlation between L.1 and L.9 (−0.151). The hydro-related variables (L.2 and L.3) present a strong positive correlation (0.899). The L.5 is positively correlated with itself (1.000). Furthermore, L.7 shows a positive correlation with L.8 (0.643), while L.8 has negative correlations with several variables, including L.2 (−0.019) and L.3 (−0.029).

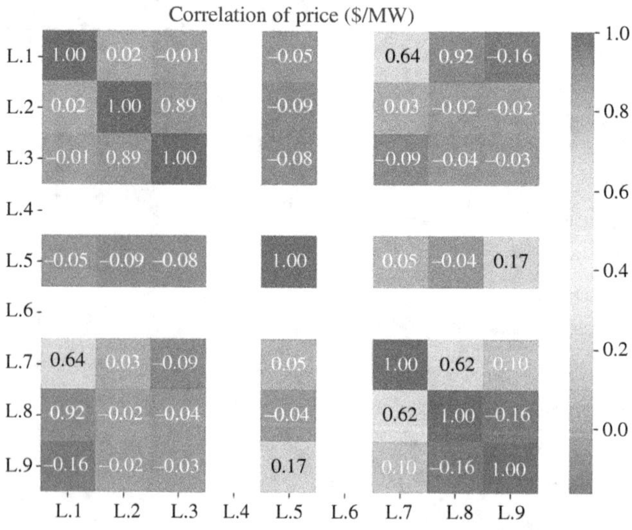

Figure 16.8 The correlation of price economic factors.

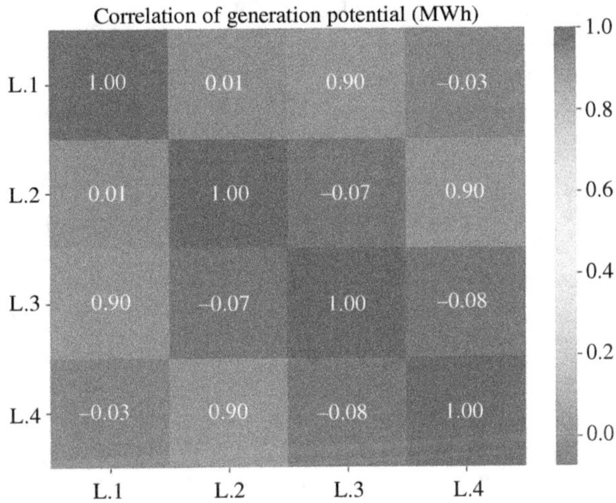

Figure 16.9 The correlation of technical generation potentials of the United States.

Table 16.3 The labels of Figure 16.8.

Labels (L)	Equivalent
L.1	commercial_pv_lcoe_price_per_mwh_median
L.2	hydro_lcoe_capital_price_per_mw_median
L.3	hydro_lcoe_price_per_mwh_median
L.4	land_based_wind_lcoe_capital_price_per_mw_median
L.5	land_based_wind_lcoe_price_per_mwh_median
L.6	pv_lcoe_capital_price_per_mw_median
L.7	pv_lcoe_price_per_mwh_median
L.8	residential_pv_lcoe_price_per_mwh_median
L.9	cost_savings_per_county

The relationships between various technological generation potential variables in the energy sector are illustrated by the data in Figure 16.9 and Table 16.4. Utility PV potential and land-based wind potential, for instance, have a substantial positive correlation (0.905), while utility PV potential and commercial PV potential have a negative correlation (−0.034). The PV potential of residential properties exhibits a robust positive connection (0.903) with that of commercial properties, while the utility PV potential (0.003) shows a weaker positive association. The utility PV potential and land-based wind potential have a substantial positive correlation (0.905), however the residential and commercial PV potentials have negative correlations (−0.080 and −0.079, respectively).

Table 16.4 The labels of Figure 16.9.

Labels (L)	Equivalent
L.1	utility_pv_technical_generation_potential_mwh
L.2	residential_pv_technical_generation_potential_mwh
L.3	land_based_wind_technical_generation_potential_mwh
L.4	commercial_pv_technical_generation_potential_mwh

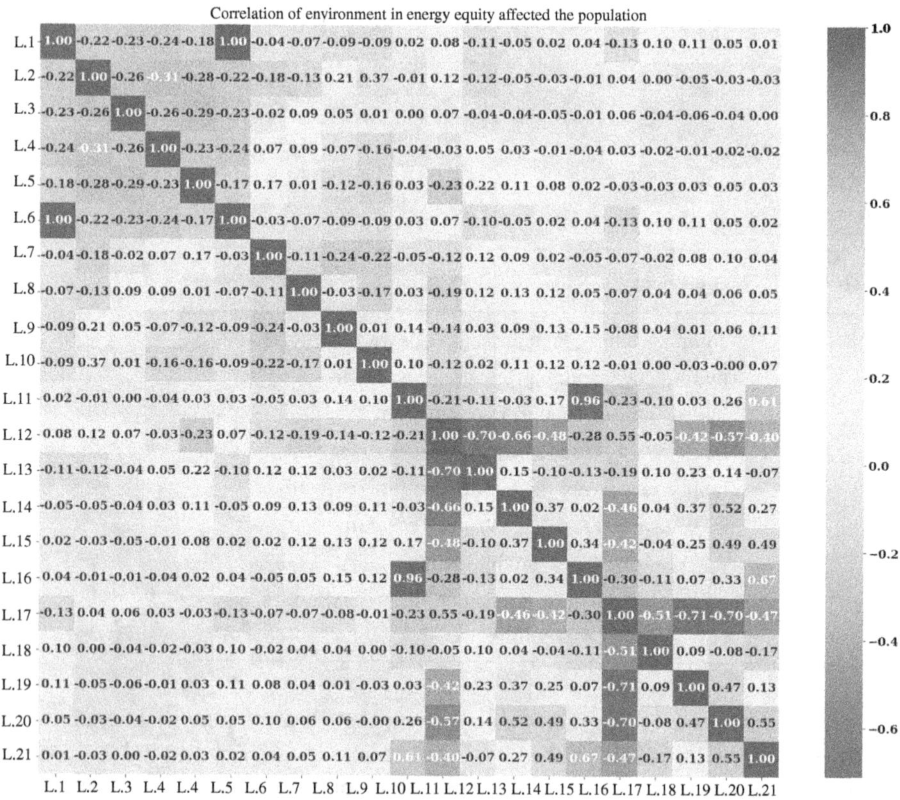

Figure 16.10 Environmental factors correlation.

Based on Figure 16.10 and Table 16.5, in general, metrics within the same category (e.g. different *ozone level indicators*) show moderate to strong positive correlations, as would be expected. Meanwhile, correlations between different categories (e.g. *ozone and diesel particulate levels*) tend to be weaker in magnitude. Some outliers include a very strong positive correlation between diesel indicator

Table 16.5 The labels of Figure 16.10.

Labels (L)	Equivalent
L.1	ozone_indicator
L.2	ozone_1_prop
L.3	ozone_2_prop
L.4	ozone_3_prop
L.5	ozone_4_prop
L.6	ozone_5_prop
L.7	cancerriskairtoxics_2_prop
L.8	cancerriskairtoxics_3_prop
L.9	cancerriskairtoxics_4_prop
l.10	cancerriskairtoxics_5_prop
L.11	diesel_indicator
L.12	dieselparticulate_1_prop
L.13	dieselparticulate_2_prop
L.14	dieselparticulate_3_prop
L.15	dieselparticulate_4_prop
L.16	dieselparticulate_5_prop
L.17	trafficproxvol_1_prop
L.18	trafficproxvol_2_prop
L.19	trafficproxvol_3_prop
L.20	trafficproxvol_4_prop
L.21	trafficproxvol_5_prop

and diesel particulate 5 prop, and a very strong negative correlation between *trafficproxvol* categories.

Regarding (Figure 16.11), the matrix shows the relationships between various socioeconomic and health indicators across US states. Some of the strongest correlations include between different measures of low-income or less-than-high school rates within the same county. This aligns with expectations as these would be linked measures. There are also moderate positive correlations between minority population proportions and some low-income or education metrics, likely reflecting related social determinants. Meanwhile, correlations between economic/demographic indicators and health metrics like cancer risk or diesel particulate levels are generally weaker in magnitude. This suggests those risk factors have a more multi-factorial relationship to county characteristics. Some outliers with weaker, however, significant correlations include between county population and certain diesel emission levels.

As of the second phase, the classified data predictively analyzed by LReg that the prediction results are manifested in Figure 16.12.

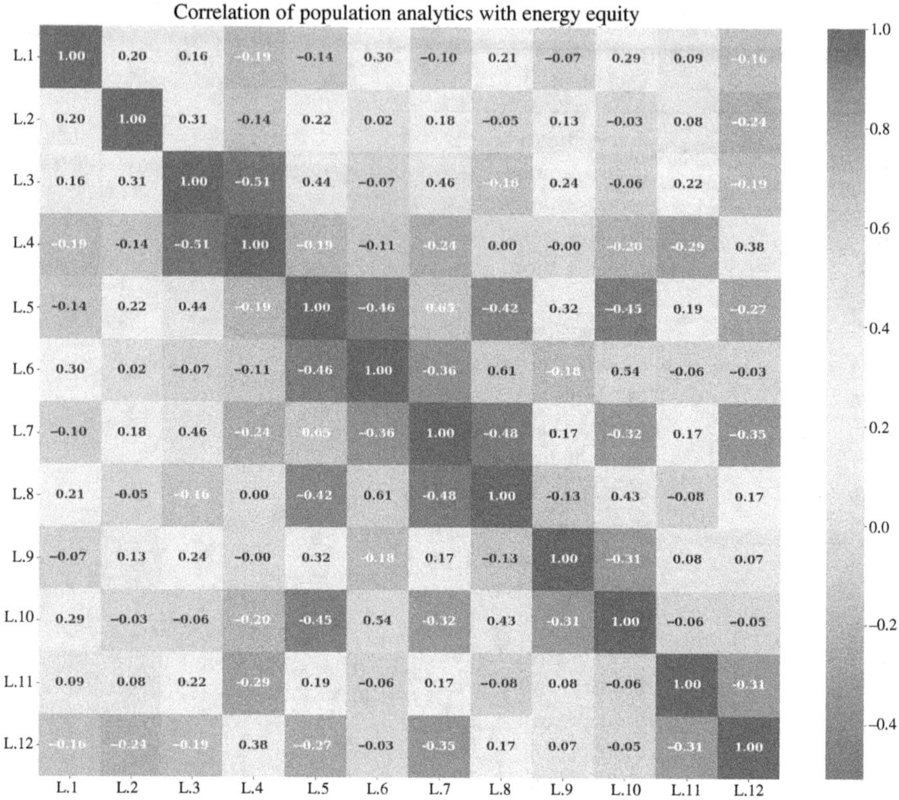

Figure 16.11 The correlation between the populational factors. The labels are provided in Table 16.6.

Table 16.6 The labels of Figure 16.11.

Labels (L)	Equivalent
L.1	county_pop
L.2	unemprate2020
L.3	minority_indicator
L.4	minority_pop_1_prop
L.5	lowincome_indicator
L.6	lowincome_pop_1_prop
L.7	lessthanhs_indicator
L.8	lessthanhs_pop_1_prop
L.9	energyburden_indicator
L.10	energyburden_1_prop
L.11	cancer_indicator
L.12	cancerriskairtoxics_1_prop

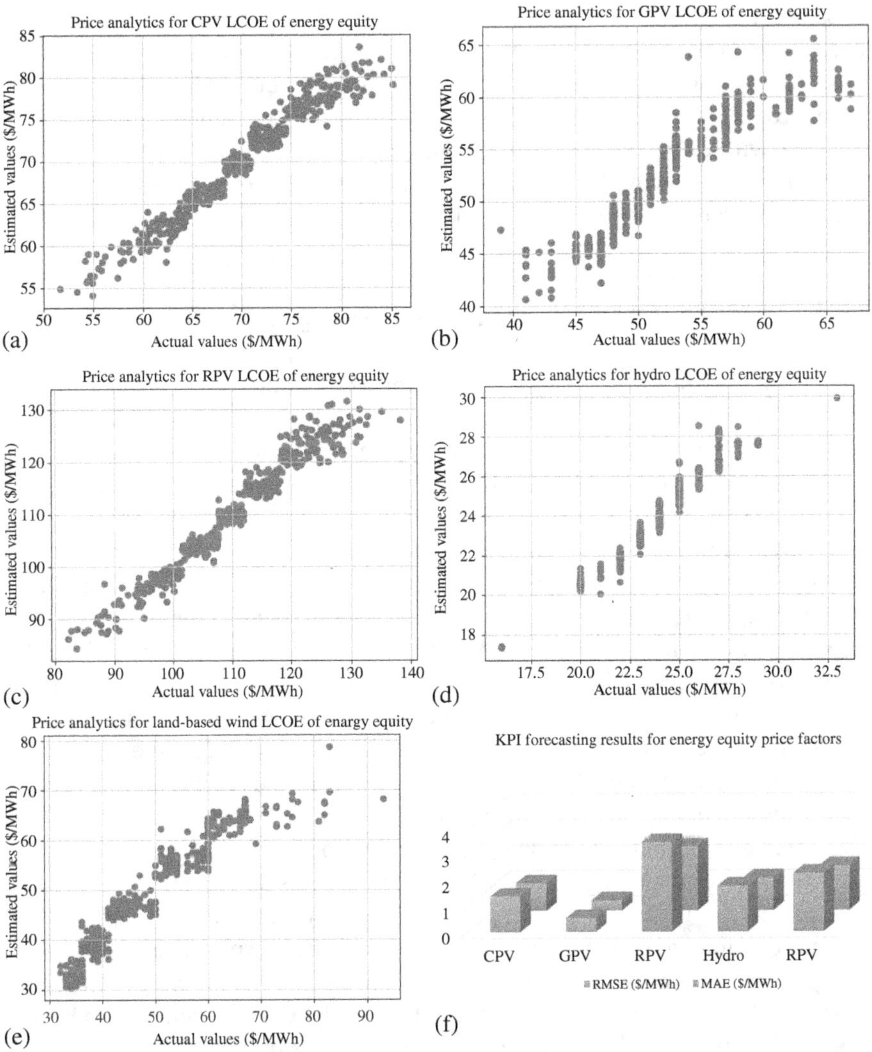

Figure 16.12 Predictive price analysis (a) commercial PV, (b) general PV, (c) residential PV, (d) hydropower, (e) land-based wind power generation, and (f) The KPI-derived results.

Regarding Figure 16.12, the KPIs show that the model has the lowest errors in predicting hydro prices with an RMSE of 0.54 and MAE of 0.41, indicating relatively accurate hydro price forecasts. In contrast, the land wind price prediction model has the highest errors with an RMSE of 3.49 and MAE of 2.5, suggesting this model is the least accurate among the options. The concentrated photovoltaics (CPV) and guaranteed purchase volume (GPV) models exhibit moderate errors for their respective predictions, with CPV having an RMSE of

1.38 and MAE of 1.07, while GPV has better results of RMSE 1.78 and MAE 1.27. The revenue per vehicle (RPV) model sits between GPV and land wind models in terms of predictive errors, with an RMSE of 2.27 and MAE of 1.73. Overall, the hydro price forecasting model performs best, while the land wind is the least accurate, with CPV, GPV, and RPV models displaying moderate predictive errors. Lower RMSE and MAE values generally imply better model performance with lesser predictive variability from actual values.

16.5 Future Perspectives and Challenges Ahead

The complex strategies effect on the prioritization of energy investment in AI models. AI-based analytics using data from large-scale datasets, such as demographic profiles, energy consumption patterns, environmental influences, and economic inputs, can present regions that have not already been taken into considerable account or targeted for the development of energy infrastructure in the past. Additionally, optimization algorithms can manifest insights for the stakeholders deliver economic benefits and address social equity challenges such as alleviating energy poverty or facilitating employment creation in underserved regions. Therefore, they help make the most vulnerable populations the least impacted negatively. Moreover, AI is important in predicting the energy requirements of the future and identifies upcoming patterns as an early sign for deploying resources into areas gaining maturity or facing issues related to energy affordability. The AI models also help in promoting inclusive dialogue on energy investments as per the wider societal objectives, by ensuring that they are accountable using data-driven decision-making processes.

Further, prioritizing equitable energy investments is necessary to ensure investments in the next generations. This ensures that inexpensive energy services continue to be delivered to every community, in spite of social status or geographical location, and maintains social fairness. Energy poverty can then be addressed by providing resources that allow for the development of renewable energy infrastructure and energy efficiency programs in more vulnerable areas, ensuring they benefit society as a whole. Prioritizing the investments for renewable technology also ties into environmental factors. By decreasing reliance on fossil fuels and curbing pollution, better living standards can be addressed. From an economic standpoint, equitable energy investment helps in incentivizing the economic empowerment of the underserved and accelerates job creation, and employment in these areas, as a growth and expediting the local economy.

While the ML future applications for energy investment prioritization are increasing, its integration will take time. ML can be optimistic in the energy sector by having data-driven views [25], managing to optimize resources and provide insights to make decisions more efficiently. These applications can help with finding the most lucrative investment opportunities, forecasting energy needs, and fine-tuning resource production and delivery. On the other hand, there are notable challenges ahead, including data size and quality, as well as the interpretability of complex models. Overcoming these challenges in the application of ML to the energy sector is considerably essential for ensuring model robustness and reliability, as well as collaboration between domain experts and data scientists. It is also taken into account that the adaption to the dynamics of the energy is important, too, as a part of industrialization [26].

16.6 Conclusion

This chapter has addressed the application of data-driven ML and AI models in EEIP with a data case study on NREL equitable energy investment periodization dataset for the United States. The ML model utilized in this chapter is lasso regression providing the perspectives into two main benefits of useful predictors inside the dataset and how regularization techniques can assist in handling feature selection. Consequently, the predictive analytics, correlation analysis, as well as cluster results, have come together to form a whole understanding of the dynamics of this dataset and uncovering patterns, and correlations. In order to evaluate the performance of the model, several KPIs are further used to develop metrics that present measurable outcomes for assessing the performance and return on investment (ROI) of intelligent models on energy investment. However, integrating ML-based models becomes applicable in optimizing and improving resource allocation, predictivity, and eventually, the framing of equitable and sustainable energy investments. However, issues related to data quality, interpretability, and social acceptability continue to be the challenges toward deploying these technologies responsibly. As such, the chapter uses the NREL dataset to develop a model of how AI and ML affect energy investment. It also uses correlation analytics to present intricate relationships within the data. Moreover, clustering techniques are employed to discover different clusters where members of each cluster have distinct characteristics and derive KPIs that are important in the decision-making process for investment. The chapter provides a thorough analysis of the strategies for investment in energy and unites general knowledge with applied experience.

References

1 Al Tarhuni, B., Naji, A., Brodrick, P.G. et al. (2019). Large scale residential energy efficiency prioritization enabled by machine learning. *Energy Efficiency* 2055–2078.

2 Tian, G., Chen, X., Chen, C. et al. (2024). Multi-scenario investment forecast of new energy projects based on multiple linear regression and comprehensive evaluation model of differentiated project priorities. *Heliyon* 10 (1): e23771.

3 Chen, Q., Kuang, Z., Liu, X., and Zhang, T. (2024). Optimal sizing and techno-economic analysis of the hybrid PV-battery-cooling storage system for commercial buildings in China. *Applied Energy* 355: 122231.

4 van den Bergh, J., van Beers, C., and King, L.C. (2024). Prioritize carbon pricing over fossil-fuel subsidy reform. *iScience* 27 (1): 108584.

5 Hazmi, A., Kort, H.M., Khallouli, W., and Raissi, N. (2024). A dynamic interrelationships among clean energy, environmental pollution, and economic growth in GCC economies: a panel ARDL approach. *International Journal of Energy Research* 2024: 5571175.

6 Wang, J. and Azam, W. (2024). Natural resource scarcity, fossil fuel energy consumption, and total greenhouse gas emissions in top emitting countries. *Geoscience Frontiers* 15 (2): 101757.

7 Hassan, Q., Algburi, S., Sameen, A.Z. et al. (2024). Green hydrogen: a pathway to a sustainable energy future. *International Journal of Hydrogen Energy* 50: 310–333.

8 Zheng, X. (2024). Driving green transformation: innovations and green innovations in natural resource markets. *Resources Policy* 89: 104540.

9 Qing, L., Abbas, J., Najam, H., and Ma, X. Investment in renewable energy and green financing and their role in achieving carbon-neutrality and economic sustainability: insights from Asian region. *Renewable Energy* 221: 119830.

10 Karmellos, M., Kiprakis, A., and Mavrotas, G. (2015). A multi-objective approach for optimal prioritization of energy efficiency measures in buildings: model, software and case studies. *Applied Energy* 139: 131–150.

11 Celli, G., Pilo, F., Pisano, G., and Soma, G.G. (2018). Distribution energy storage investment prioritization with a real coded multi-objective genetic algorithm. *Electric Power Systems Research* 163: 154–163.

12 Nagesha, N. and Balachandra, P. (2006). Barriers to energy efficiency in small industry clusters: multi-criteria-based prioritization using the analytic hierarchy process. *Energy* 31 (12): 1969–1983.

13 Huang, J. (2024). Resources, innovation, globalization, and green growth: the BRICS financial development strategy. *Geoscience Frontiers* 15 (2): 101741.

14 Korpeh, M., Asadbagi, P., Afshari, R. et al. (2024). A comprehensive analysis and multi-objective optimization of a sustainable production system based on renewable energies. *Applied Thermal Engineering* 17: 122483.

15 Zhang, M., Zhang, D., and Xie, T. (2024). Balancing urban energy considering economic growth and environmental sustainability through integration of renewable energy. *Sustainable Cities and Society* 5: 105178.

16 Solar Industry (2023). Multi-model analysis explores benefits and costs of uniting U.S. coasts' power systems. Solar Industry Magazine. https://solarindustrymag.com/multi-model-analysis-explores-benefits-and-costs-of-uniting-u-s-coasts-power-systems (accessed 22 September 2024).

17 Pennsylvania State University (2022). The electric power grid, GEOG 469: energy industry applications of GIS. https://www.e-education.psu.edu/geog469/node/222 (accessed 22 September 2024).

18 US Environmental Protection Agency (2023). U.S. electricity grid and Markets. EPA. https://www.epa.gov/green-power-markets/us-electricity-grid-markets (accessed 22 September 2024).

19 Clean Coalition (2023). Distribution system operator (DSO). Clean Coalition. https://clean-coalition.org/distribution-system-operator-dso/ (accessed 22 September 2024).

20 Greer, M. (2012). U.S. electric markets, structure, and regulations. In: *Electricity Marginal Cost Pricing* (ed. M. Greer). Butterworth-Heinemann https://doi.org/10.1016/B978-0-12-385134-5.00003-X.

21 Day, Megan, and Liz Ross (2021). "Equitable Energy Investment Prioritization Data Set." NREL Data Catalog. Golden, CO: National Renewable Energy Laboratory. https://orcid.org/0000-0002-4359-4879 (accessed 23 January 2024).

22 Safari, A., KheirandishGharehbagh, H., Nazari-Heris, M., and Oshnoei, A. (2023). DeepResTrade: a peer-to-peer LSTM-decision tree-based price prediction and blockchain-enhanced trading system for renewable energy Decentralized Markets. *Frontiers in Energy Research* 11: 1275686.

23 Safari, A., Kharrati, H., and Rahimi, A. (2023). Multi-term electrical load forecasting of smart cities using a new hybrid highly accurate neural network-based Predictive Model. *Smart Grids and Sustainable Energy* 9 (1): 8.

24 Safari, A., Kheirandish Gharehbagh, H., and Nazari, H.M. (2023). DeepVELOX: INVELOX wind turbine intelligent power forecasting using hybrid GWO–GBR algorithm. *Energies* 16 (19): 6889.

25 Safari, A. (2024). Hybrid emerging model predictive data-driven forecasting of three-phase electrical signals of photovoltaic systems using GBR-LSTM. *e-Prime-Advances in Electrical Engineering, Electronics and Energy* 8: 100549.

26 Safari, A. and Ghavifekr, A.A. (2022). Use case of artificial intelligence and neural networks in energy consumption markets and industrial demand response. In: *Industrial Demand Response: Methods, Best Practices, Case Studies, and Applications.* (ed. H.H. Alhelou, A. Moreno-Muñoz, and P. Siano), 379–399. The Institution of Engineering and Technology.

17

Intelligent Remedial Action Scheme Against False Data Injection Cyberattacks Targeting Energy Justice and Equity in Modern Power Systems

Ehsan Naderi[1] *and Arash Asrari*[2]

[1] *Department of Electrical Engineering, College of Engineering and Computer Science, Arkansas State University, Jonesboro, AR, USA*
[2] *Department of Electrical and Computer Engineering, Purdue University Northwest, Hammond, IN, USA*

Abbreviations

AI	artificial intelligence
BDD	bad data detection
CNR	customized network reconfiguration
DG	distributed generation
DNN	deep neural network
FDIA	false data injection attack
FL	federated learning
IEEE	Institute of Electrical and Electronics Engineers
MVA	mega volt-amps
MVAR	mega volt-amps reactive
PMU	phasor measurement unit
PV	photovoltaic
RAS	remedial action scheme
SCADA	supervisory control and data acquisition
UE	unserved energy
ULTC	under load tap changer
VVO	volt-var optimization

Communities for Clean Energy Justice and Equity in Grid Modernization, First Edition.
Edited by Mohammadreza Daneshvar, Behnam Mohammadi-Ivatloo, and Amjad Anvari-Moghaddam.

17.1 Introduction

17.1.1 Background and Motivation

Modern-day power systems are distinguished by their digital technology, which enables two-way communications between utilities and their clients (or aggregators on behalf of the clients) as well as sensor devices along the transmission lines and distribution branches. Hence, there is a huge amount of information that needs to be managed by markets, service providers, and grid operators paving the way for a highly distributed and hierarchical communication network to process the information. Nevertheless, a significant cybersecurity threat is posed to smart grids as a result of the rapid development of modern power systems integrated with different forms of communication networks. This is considered as a motivation for researchers to (1) scrutinize different types of cyberattacks targeting modern power systems and (2) introduce feasible remedial action schemes (RASs) to expeditiously react to cyberattacks in real time, mitigate the negative impacts, and recover the targeted power system to the normal operation, resulting in enhancement of social welfare (i.e. energy justice) for end users. In 2015, the Ukrainian power grid was targeted by a coordinated cyberattack resulting in shortage of power for a quarter of a million end-use customers for up to six hours [1]. Another real-world example is the cyberattack on the US power grid in 2018 by which attackers collected critical information about the operation of power plants and the entire generation sector [2]. Among different types of cyberattacks targeting modern power systems, false data injection attacks (FDIAs) dominate since they are easier to be launched to result in a variety of operational issues, including but not limited to shortage of power [3], voltage violation [4], market power in deregulated electricity markets [5], etc. These examples clearly indicate the link between energy insecurity and socioeconomic. According to the article published on May 3, 2022 in National Conference of State Legislatures, "The affordability of and access to *reliable energy* is at the heart of energy justice." [6], which is the motivation of this chapter. In other words, energy justice and equity have emerged as effective approaches for pushing decision-making processes toward reliability, sustainability, and resilience in the energy sector. Hence, it is essential to develop RASs to be implemented by power systems operators to (1) neutralize the efforts of cybercriminals, attacking energy justice and equity and (2) recover the targeted power grid to normal operation in a timely manner, enhancing social welfare of end users. In other words, the main objectives of this chapter are twofold, which are elaborated as follows: scrutinizing the negative impacts of FDIAs targeting energy justice and equity in the form of affecting the social welfare of end users and implementing artificial intelligence (AI) to develop a remediation framework against such FDIAs.

17.1.2 Background

17.1.2.1 Related Works on Energy Justice and Equity

In [7], social equity access to mobile charging stations for electric vehicles (EVs) was investigated on a case study in Los Angeles, where a self-scheduling model for smart parking lots was proposed to integrate scheduling of temporary mobile charging stations taking into account social equity access and optimization of energy generation and storage schedule for smart parking lots. In [8], a mathematical model for renewable energy communities was proposed for scrutinizing the impacts of participation of community members in collective photovoltaic (PV) solar panels and collective energy storage systems on the fairness and equity of such renewable energy communities. In [9], a quantitative framework was developed to support decision-making associated with equitable energy interventions. The proposed framework in [9] was modeled as a multi-objective optimization problem optimizing equity in the form of minimizing the sum of the intervention costs and energy insecurity concurrently. Moreover, comprehensive reviews from various standpoints were presented in [10–12], highlighting the significance of considering energy justice and equity in the digital era.

17.1.2.2 Related Works on Cyberattacks Targeting Energy Equity

Due to the significance of research in the field of energy justice and equity, different frameworks (e.g. cyberattacks models targeting energy systems, detection algorithms identifying the presence of false data in the advanced metering infrastructure (AMI), and RASs against such cyberattacks) are covered in this section. For example, in [13], an RAS, distinguishing cyberattacks from physical disturbances in modern power systems, was proposed in the form of a decision tree-based anomaly detection algorithm taking corrective actions to enhance the social welfare of end-users. In [14], an RAS was introduced according to a resilience-oriented optimal power flow problem to optimize the system's security against cyberattacks targeting smart power grids. In [15], an evolutionary-based optimization algorithm was developed to enhance energy access via identifying a set of countermeasures against potentially dangerous threats in a cyber–physical power system. In [16], a hybrid attack model was introduced for integrity attack and availability attack concurrently, where the impacts of such an attack were studied in case of false negatives attack and false alarm attack. In [17], a comprehensive study was performed on cyberattack detection and diagnosis for power electronics converters associated with PV plants to reduce the power shortage via classifying abnormal phenomena into normal conditions, open-circuit faults, short-circuit faults, and cyberattacks. In [18], prevention, and detection frameworks were introduced against coordinated stealthy FDIAs targeting integrated power and gas systems to result in disconnecting gas-fired power plants via attacking the measurements. In [19], a single mixed-integer linear programming model was proposed for high-stealth FDAs leading to overloading a set of power lines and disrupting the normal operation of the power grid. In [20], a mathematical basis for attacker–defender was presented, where the defender was not aware of the attack parameters, to predict the target of cyberattacks via machine-learning-based algorithms in a decentralized power system.

17.1.3 Research Gap and Contributions of This Chapter

Despite the extensive research efforts in the field of cyberattacks targeting power networks ([13–20]), the following research questions are yet to be addressed in the existing literature:

1) How to investigate the impacts of a coordinated FDIA, targeting a modern power system, to negatively affect the social welfare of end-use customers via maximizing the unserved energy (UE) and voltage violation (in the forms of both overvoltage and undervoltage) throughout the system?
2) If such an event is ever experienced (i.e. successful voltage violation and power shortage due to maximizing attackers' objective functions), how to take advantage of AI (e.g. deep learning concept generating remedial actions) for a better reaction to the FDIA to effectively minimize the UE and voltage violation and provide continuous and fair access to energy for end users?

In order to address these research questions, we have developed a framework with the following contributions:

- A coordinated FDIA to identify the vulnerable buses of a power system to voltage violation and power shortage via a multi-objective attack framework targeting smart meters in the power system, which are the input of volt-var optimization (VVO) in the control center of the power grid,
- A framework implementing federated learning (FL) concept to train deep neural networks (DNNs) (i.e. local models and global model) in order to develop a remediation model against different cyberattacks with different consequences.

Accordingly, among the six principles of energy justice and equity introduced in [21], only the following items are considered in the process of developing the proposed framework in this chapter, which are associated with the social welfare of end-use customers via providing fair access to energy while the power grid is targeted by a successful cyberattack causing operational issues (e.g. voltage violation and power shortage):

- creating participatory systems for energy governance, and
- rejecting false solutions.

17.2 Proposed Framework

The proposed framework in this chapter contains two phases. In the first phase (i.e. the attack phase), an attacker will coordinate an FDIA via simultaneously maximizing the UE and voltage violation in the daily operation of a power system by injecting false data into the AMI. The targets of such an organized cyberattack will be renewable-based generation units (i.e. PV plants), under-load tap changing (ULTC) transformers, and load centers within the network. It is noted that the ultimate goal of the FDIA is to affect energy justice and equity in the targeted system by cutting fair access to energy. Figure 17.1 illustrates the introduced first level of the proposed framework.

According to Figure 17.1, one can perceive that attackers in a coordinated FDIA target the phasor measurement units (PMUs) of generation buses, power lines, and load centers to inject false data vectors, which are obtained after solving a multi-objective optimization problem (refer to Section 17.3.1 for detailed information about this cyberattack), to push the power system toward intended operational issues. The information collected from the PMUs, including the malicious information, will be sent to the control center through supervisory control and data acquisition (SCADA) systems to be utilized by the power system operator to run a VVO problem, managing the voltage and current profiles throughout the power grid. The outcome of such a coordinated FDIA will be a deteriorated voltage profile (in the form of overvoltage, undervoltage, or both) and a huge amount of UE in the system, which will show up in the form of power

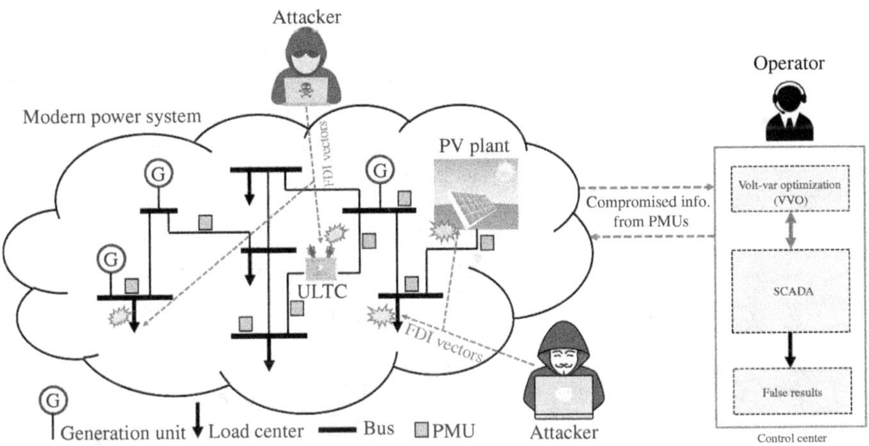

Figure 17.1 The modeled FDIAs to target a modern power system with the objectives of increasing voltage violation and unserved energy, affecting energy justice and equity as well as social welfare of end users.

shortage and/or power outage, affecting the energy justice and equity since the fair access of end users to energy will be noticeably limited. This is due to the fact that the control system of the power grid has processed malicious information, injected by attackers, since the attackers have successfully bypassed the bad data detection (BDD) mechanism embedded in state estimation.

The second phase of the proposed framework is related to real-time operation, where successful implementation of the developed attack model has resulted in the intended power shortage and deteriorated voltage profile from the attacker's perspective. In other words, attackers have been successful in maximizing the objective functions of the attack phase (i.e. UE and voltage violation). Therefore, in the second phase, which is the reaction phase in this chapter, an AI-based remedial action will be proposed to (1) ameliorate the negative impacts of the coordinated FDIAs and (2) enhance the social welfare of the end users via reducing the overall UE and voltage deviation indexes of the system. The proposed RAS against indicated FDIAs is depicted in Figure 17.2. From this figure, it can be inferred that the proposed framework takes advantage of the concept of FL to effectively react to the successful FDIAs developed in the previous phase. Hence, the entire process can be accomplished via communicating a set of trusted local agents and the main operator of the power grid. To obtain a better perspective, as illustrated in Figure 17.2, the first agent is responsible for reacting to FDIAs resulting in voltage violation, whereas the second agent is only responsible for providing reaction mechanisms against FDIAs resulting in power shortages. Therefore, each agent develops the local model independently and trains the model to reduce both training error and test error locally. It is noted that each agent maintains the information associated with the training phase confidential to increase data privacy. Hence, agents send the results of the training phase, without details, to the system operator who is responsible for the global model to react to different types of FDIAs with different objectives (e.g. voltage violation, power shortage, etc.). Hence, the system operator implements the information

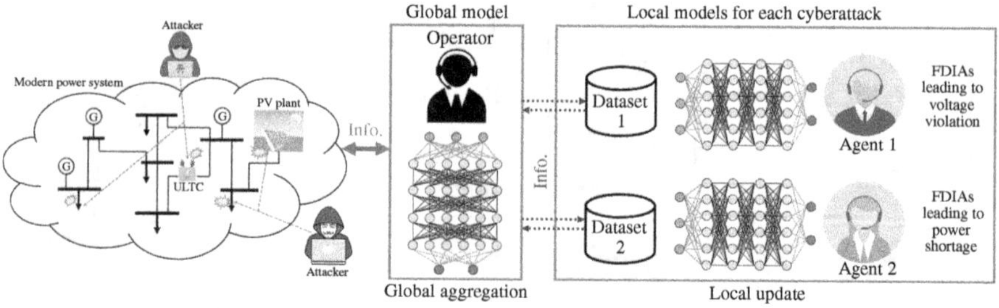

Figure 17.2 The proposed RAS against FDIAs targeting voltage profile and reliability of a modern power grid.

provided by the local agents as the input vectors of the global model. In real time, when the power system is targeted by a successful cyberattack, the status of the power grid at the time of the attack will be saved as the first input vector for the global model. The second and third input vectors of the global model will be the information provided by the first and second agents (see the three blue circles underneath the operator icon). Then, the global model, which has been trained via the concept of deep learning, will recognize proper remedial action against each cyberattack or each operational issue. The proposed remediation mechanisms need to be implemented by the power system operator in real time to recover the targeted system and mitigate the negative impacts via enhancing energy justice and equity throughout the power system (i.e. providing fair access to energy for end users).

It can be also inferred from Figure 17.2 that the main motivation for using the FL concept is to divide the burden of training the global model among different agents so that the generalized version of the proposed framework can respond to a variety of cyberattacks at a time. This is due to the fact that each agent only focuses on one type of attack; hence, collecting data for the database and maintaining the data are easier tasks for agents. To extend the idea of this chapter, each agent could learn to accomplish a task through repeated trial/error within a dynamic environment, which empowers the agent to make right decisions maximizing a reward function for the same task without or with less human intervention (i.e. the concept of reinforcement learning).

17.3 Problem Formulation

17.3.1 Formulation of Cyberattack Model

Objective functions of the coordinated FDIA are presented in Eqs. (17.1) and (17.2), where ω indicates the rate of data manipulation or weight factor showing the attackers' effort at time interval t; $\lambda_{l,t}$ denotes the binary variable for the FDIA associated with lth line at tth time interval (1 means the line including transformer is targeted); $Tap_{x,t}^{Trans}$ indicates the tap position of xth ULTC at tth time interval; T, L, and X are, respectively, the sets of time intervals, lines having transformer, and ULTC transformers; $\pm$ indicates the nature of the FDIA, meaning that + (−) refers to the situation based on which the tap positions of the ULTC transformers move upward (downward), leading to overvoltage (undervoltage) issues; TDL_d is the time duration of the dth load point; FOR_m is the forced outage rate of mth generation unit; and $P_{d,t,}$ and $P_{g,t}$ are, respectively, the average load connected to dth demand center at tth time interval and the summation of output powers associated with available generation units at dth load center at tth time interval. Interested readers are directed to [22, 23] for more information about the mentioned objective functions.

$$\max \left\{ \omega \sum_{t \in T} \left(\sum_{l \in L} \lambda_{l,t} \pm \sum_{x \in X} Tap_{x,t}^{Trans} \right) \right\} \tag{17.1}$$

$$\max \left\{ \sum_{d \in D} TDL_d \sum_{g \in G} \left(\prod_{m \in M} FOR_{m,t} \right) \times \left(\prod_{n \in N} (1 - FOR_{n,t}) \right) \times |P_{d,t} - P_{g,t}| \right\} \tag{17.2}$$

The intention of the coordinated FDIA is to maximize objective functions Eqs. (17.1) and (17.2) concurrently. However, the numerical ranges of the objective functions are not the same. To address this issue, a membership function is utilized to convert the numerical values of each objective function into [0–1]. Figure 17.3 displays the trapezoidal membership function, where $OF_k^{\min}$ and $OF_k^{\max}$ are, respectively, the minimum value and maximum value (i.e. 10 times of the minimum value) of kth objective function (i.e. $k = 2$ in this chapter); and Ψ_k denotes the fuzzy set for kth objective function [24].

To keep the cyberattack undetectable, attackers consider Eqs. (17.3)–(17.6) to ensure the (to be injected into the PMUs) false data vectors bypass the BDD, which functions based on the Euclidean norm of measurement residual vectors [25, 26].

$$z = h(x) + e = Hx + e \tag{17.3}$$

$$||z - H\tilde{x}|| \leq \varepsilon \tag{17.4}$$

$$||Z_{False} - H\tilde{x}_{malicious}|| = ||z + \Delta z + H\tilde{x}_{malicious}|| \tag{17.5}$$

$$||z_{Valid} \pm \Delta D - Hx_{Original} + H\Delta D|| \geq 0 \tag{17.6}$$

where $h(x)$ denotes the relationship between state variables (i.e. x) and measurements (i.e. z); e is the measurement noise; $\tilde{x}$ is the estimated state variables; ε is

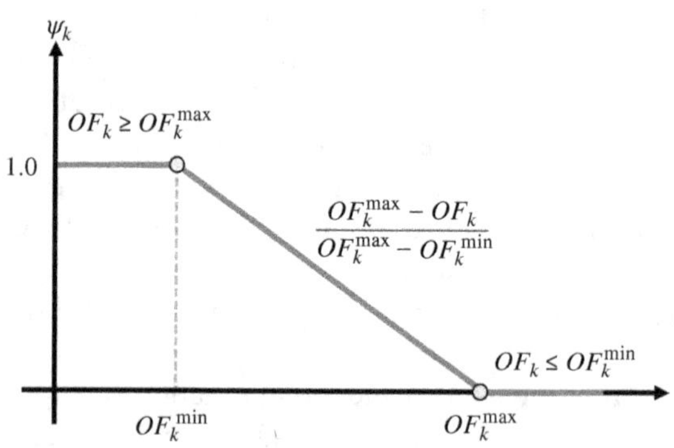

Figure 17.3 Trapezoidal membership function to normalize objective functions Eqs. (17.1) and (17.2).

the threshold of the BDD algorithm, which is oriented toward the Euclidean norm of measurement residual vectors embedded in state estimation; Z_{False} and $\tilde{x}_{malicious}$ are, respectively, the false measurement and estimated state variables that are compromised via FDIA, and ΔD is the amount of false data to be injected into the system.

It is noted that the optimal false data vector will be injected into the measurements through launching FDIAs to falsify the input of the VVO problem, performed by the power system operator in the upper level to regulate the tap positions of the ULTC transformers and power flow of branches. The main objective function of the VVO problem along with its main constraints are provided in Eqs. (17.7)–(17.9), where $V_{b,t}$ and $V_{\acute{b},t}$ are, respectively, voltage magnitudes of buses b and $\acute{b}$ at tth time interval; σ_x is the alteration associated with the tap size for xth transformer; and Λ indicates the maximum number of actions for updating the tap position [27].

$$\min \left\{ \sum_{b,\acute{b}\in B, b\neq\acute{b}} G_{b\acute{b}} \times \left(V_b^2 + V_{\acute{b}}^2 - 2 \times V_b \times V_{\acute{b}} \times \cos\theta_{b\acute{b}} \right) \right\} \tag{17.7}$$

$$V_{b,t} = V_{b,t}^{Nominal} + \sigma_x \times Tap_{x,t}^{Trans}, \forall b \in B, \forall t \in T, \forall x \in X \tag{17.8}$$

$$\Lambda \geq \left| Tap_{x,t}^{Trans} - Tap_{x,t-1}^{Trans} \right| \tag{17.9}$$

To ensure the coordinated FDIA does not violate the ratings of the power system (e.g. power flow, voltage magnitude), objective functions Eqs. (17.1) and (17.2) need to be optimized subject to satisfying a set of technical constraints, which are provided in Eqs. (17.10)–(17.17) [28].

$$P_{b,t} = \sum_{b,\acute{b}\in B} V_{b,t} \times V_{\acute{b},t} \times G_{b\acute{b}} \times \cos(\theta_{b\acute{b}} - \delta_{b,t} + \delta_{\acute{b},t}), \forall t \in T \tag{17.10}$$

$$Q_{b,t} = \sum_{b,\acute{b}\in B} V_{b,t} \times V_{\acute{b},t} \times G_{b\acute{b}} \times \sin(\theta_{b\acute{b}} - \delta_{b,t} + \delta_{\acute{b},t}), \forall t \in T \tag{17.11}$$

$$P_{g,t}^{\min} \leq P_{g,t} \leq P_{g,t}^{\max}, \quad \forall g \in G, \forall t \in T \tag{17.12}$$

$$Q_{g,t}^{\min} \leq Q_{g,t} \leq Q_{g,t}^{\max}, \quad \forall g \in G, \forall t \in T \tag{17.13}$$

$$V_b^{\min} \leq V_{b,t} \leq V_b^{\max}, \quad \forall b \in B, \forall t \in T \tag{17.14}$$

$$\left| I_{x,t}^{Trans} \right| \leq I_x^{\max}, \quad \forall x \in X, \forall t \in T \tag{17.15}$$

$$|P_{b\acute{b}}| \leq P_{b\acute{b}}^{\max}, b, \acute{b} \in B, b \neq \acute{b} \tag{17.16}$$

$$\sum_{s\in S} P_s^{Solar} \leq \rho \times \sum_{b\in B} P_b^{Demand} \tag{17.17}$$

where $P_{b,t}$ and $Q_{b,t}$ are, respectively, the active and reactive powers injected into bus b at tth time interval; $\delta_{b,t}$ and $\delta_{\acute{b},t}$ are voltage phase angles associated with buses b and $\acute{b}$ at tth time interval, respectively; $G_{b\acute{b}}$ and $\theta_{b\acute{b}}$ are, respectively, the magnitude and angle of the line admittance connecting buses b and $\acute{b}$; $P_{g,t}$, $P_{g,t}^{\min}$, and $P_{g,t}^{\max}$ are, respectively, the active power generated from gth generation unit at tth time interval as well as its minimum and maximum boundaries; $Q_{g,t}$, $Q_{g,t}^{\min}$, and $Q_{g,t}^{\max}$ are, respectively, the reactive power generated from gth generation unit at tth time interval as well as its minimum and maximum boundaries; $V_{b,t}$, $V_{b,t}^{\min}$, and $V_{b,t}^{\max}$ are, respectively, the voltage magnitude of bth bus at tth time interval as well as its minimum and maximum limits; $I_{x,t}^{Trans}$ and $I_{x}^{\max}$ are, respectively, the magnitude of current associated with xth transformer at tth time interval and its maximum limit; $P_{b\acute{b}}$ and $P_{b\acute{b}}^{\max}$, are respectively, the power flow through line connecting buses b and $\acute{b}$ at tth time interval as well as its maximum limit; P_{s}^{Solar} is the overall power generated by sth solar plant; ρ is a coefficient indicating the level of power generated by the solar plant (which can be altered from zero in supplying the demand) to one (i.e. fully serving the load); and P_{b}^{Demand} is the overall demand associated with bth bus.

17.3.2 Formulation of the Remedial Action Scheme

The proposed RAS focuses on a customized approach to network reconfiguration (CNR) problems, each with its own set of objective functions, which will be detailed in the following sections. It is noted that the CNR problems are single-objective optimization problems. In other words, the main objective function of each CNR problem (i.e. the proposed RAS in this study) focuses only on one objective function at a time. However, the proposed intelligent framework decides the objective function of the RAS based on the type of FDIA. For example, the proper RAS to mitigate the negative impacts of an FDIA targeting a power system and resulting in power shortage will be different from the situation in which the power grid is targeted by an FDIA resulting in voltage violation. In some cases, the system operator might need to perform a combination of remedial actions to fully handle the cyberattack and alleviate the negative impacts. In such situations, the remediation algorithms along with the corresponding priority list will be generated by the intelligent framework (refer to Figure 17.2).

The objective functions of the proposed RAS are introduced in Eqs. (17.18)–(17.20), where Eq. (17.18) denotes the active power loss of the power grid, Eq. (17.19) presents the accumulative voltage deviation index, and Eq. (17.20) shows the energy not supplied (or the unserved energy). It is noted that the minimum number of required switching processes to reach the optimal value of each objective function (i.e. Eqs. (17.18)–(17.20)) will be also taken into account.

Hence, the minimization problem associated with the number of switching processes, which is provided in Eq. (17.21), needs to be added to the objective functions [29].

$$\min \left\{ \sum_{t \in T} \sum_{l \in L} R_l \times |I_{l,t}|^2 + \sum_{t \in T} \sum_{x \in X} R_x \times \left| I_{x,t}^{Trans} \right|^2 \right\} \tag{17.18}$$

$$\min \left\{ \sum_{t \in T} \sum_{b \in B} \left| V_{b,t} - V_{b,t}^{Nominal} \right|^2 \right\} \tag{17.19}$$

$$\min \left\{ \sum_{d \in \acute{D}} TDL_d \sum_{g \in G} \left(\prod_{m \in M} FOR_{m,t} \right) \times \left(\prod_{n \in N} (1 - FOR_{n,t}) \right) \times |\acute{P}_{d,t} - P_{g,t}| \right\} \tag{17.20}$$

$$\min \left\{ \sum_{t \in T} \sum_{w \in W} \left| S_{w,t} - S_{w,t}^{Base} \right| \right\} \tag{17.21}$$

where R_l is the resistance of lth line; R_x is the equivalent resistance of xth transformer, which can be also a ULTC transformer; $V_{b,t}^{Nominal}$ is the nominal voltage magnitude of bus b at tth time interval; and $S_{w,t}$ and $S_{w,t}^{Base}$ are, respectively, the new and old status of wth tie-switch at tth time interval.

17.3.3 Formulation of the Unified and Intelligent Remedial Action Scheme via Federated Learning Concept

According to Figure 17.2, different blocks of DNNs receive three different and main input vectors from the corresponding local datasets, which are established by the local agents. These four inputs include (1) information about the power system when the system is operated normally (i.e. the clean history of the measurements), (2) information reflecting the status of the power grid at the time of the attack after the cyberattack is considered successful (i.e. the last valid and available set of information [e.g. voltage magnitude of system buses, power flow-related data, etc.] recorded by smart meters prior to the cyberattack), and (3) the results of agent's efforts to simulate at least two attack scenarios in addition to the corresponding remedial actions, proposed by the local DNNs, to mitigate the attack scenarios. It is noted that the information about the prior outputs of the DNNs, which were performed as responses to previous cyberattacks (i.e. the history of DNNs' outputs) will be considered, if available. As long as these input vectors are fed into the DNNs, the developed framework is expected to generate proper remedial actions to mitigate the impacts of FDIAs regardless of nature of the load, price of electricity, and level of renewable integration. To reflect the uncertainties associated with PV plants, a conservative factor is considered indicating that the total expected PV power may be increased/decreased by 15%; hence, $\rho = 0.15$, which is determined based on trial-and-error, and is considered in Eq. (17.17). The output of each local DNN block will be a remediation to be implemented by the system operator to ameliorate the negative impacts of the FDIA. The entire learning algorithm used to train the DNNs, including both the local models and the global model, is designed around the FL concept, which is a distributed machine learning approach. Therefore, the agents train themselves locally without sharing their direct information to the server located in the control center of the power system. This significantly reduces the vulnerability of modern power systems against cyberattacks, leading to data privacy enhancement, because lower amounts of information will be transmitted each time. In other words, the proposed framework promotes a decentralized training algorithm to engage all agents in the process of collaboratively learning a shared prediction model. However, this chapter introduces a centralized version since the agents are to send their data to the central server or the system operator (refer to Figure 17.2), which is responsible for aggregating the data and training the global model. In the next step of this research, we will propose a new framework by eliminating the central server from Figure 17.2. Therefore, agents can communicate and share data with each other without the central server to save computational resources and provide customizable peer-to-peer communication between agents. This strategy will be useful when there are many agents in the framework covering a big range of cyberattacks.

To obtain a better perspective, Figure 17.4 displays the detailed process of learning an FL model. From this figure, one can perceive that the whole process of learning can be divided into three sections: initiating the training task (i.e. the global model), updating the generated model locally, and global aggregation. Interested readers are directed to [30] for more information about the concept of FL and its application in different engineering fields.

According to Figures 17.2 and 17.4, the following steps need to be considered in the proposed framework:

(1) System operator initiates the global model, which is $W_{Glob,t}$.
(2) Each agent (e.g. agent responsible for FDIAs leading to power shortage), namely, agent a, starts the process of updating the model parameters (i.e. $W_{a,t}$) in order to find the weight factors resulting in the minimization of the error function (i.e. $\min\{J_a(W_{a,t})\}$).
(3) The locally updated models, developed via all agents, are shared with the system operator.
(4) System operator aggregates the models received from the agents and updates the global model to produce $W_{Glob,t+1}$.

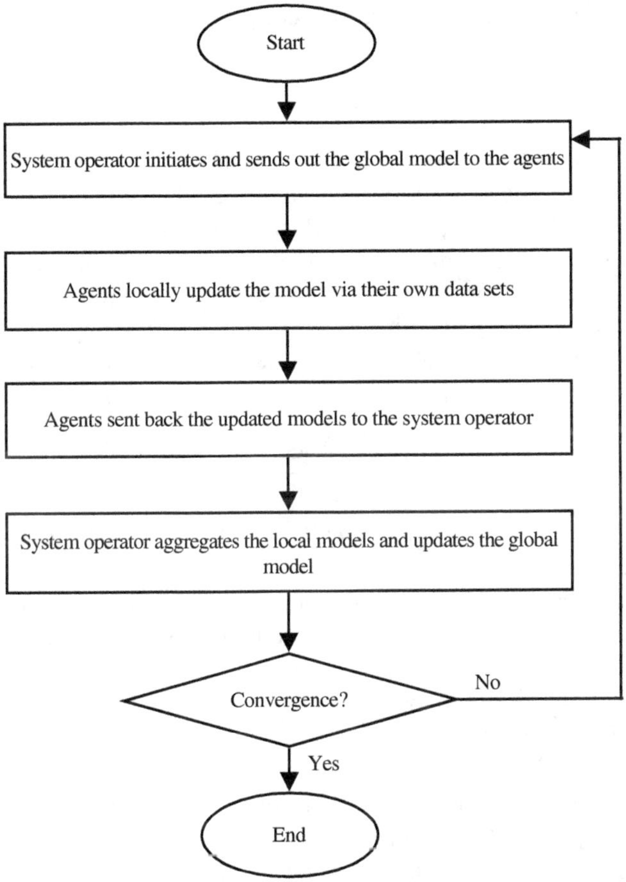

Figure 17.4 Flowchart of the proposed FL framework.

(5) The updated global model once again will be shared with the agents, and this process will continue unless the stopping criterion is met (i.e. the level of global errors, including both training error and test error, satisfies the threshold).

The main objective function of the training phase is presented in Eq. (17.22), where A is the set of agents; ξ_a indicates the relative impact coefficient of each agent to satisfy Eq. (17.23); and $J_a(W_{a,t})$ denotes the expected prediction error for a sample input associated with ath agent.

$$\min \left\{ \sum_{t \in T} \left\{ \sum_{a \in A} \xi_a \times J_a(W_{a,t}) \right\} \right\} \tag{17.22}$$

$$\sum_{a \in A} \xi_a = 1 \tag{17.23}$$

If the ath agent owns e_a examples, the relative impact coefficient can be written in Eq. (17.24).

$$\xi_a = \frac{e_a}{\sum_a e_a} \tag{17.24}$$

Equation (17.25) presents the general approach to formulate data-parallel training associated with the local models in which a typical model with parameters p optimizes the error (i.e. $\in$) using the step-size α on cluster size U and mini-batch size h with stochastic gradient descent mechanism to obtain the best fit between predicted and actual output. An important aspect of such a formulation is to mitigate the noise of common deep learning frameworks, which stems from gradient variance. Hence, a trade-off is obtained between the size of the batch and the noise [31].

$$p_{t+1} = p_t - \alpha\frac{1}{U} \times \frac{1}{h}\sum_{u\in U} \nabla \in (p_{t,u}) + \frac{E\left\{\frac{1}{K}\sum_{a\in A} \|g_{t,a}\|^2\right\}}{E\left\{\|\tilde{g}_t\|^2\right\}} \tag{17.25}$$

where g and $\tilde{g}$, respectively, indicate the gradient of the agent and the aggregated gradient of all agents.

A fitness estimation assesses how accurately the probability model represents the target information by evaluating the performance of the DNNs, which can be mathematically written in Eq. (17.26), called perplexity evaluation metric.

$$\Theta = EXP\left(\sum_{a\in A} \times \frac{1}{R} \times \sum_{r\in R} - \log \hat{P}_{a,r+1}\right) \tag{17.26}$$

where R indicates the number of attack scenarios performed by ath agent; and $\hat{P}$ denotes the forecasted probability distribution.

To effectively coordinate different agents (i.e. Agents #1 and #2, as depicted in Figure 17.2) against different FDIAs (i.e. attack leading to voltage violation and attack leading to power shortage), Eqs. (17.27) and (17.28) are applied, where η and π are coordination parameters for restricting undervoltage (UV) and unserved energy (UE) in the process of coordination of the cyberattacks and agents; $UV_{Agent\,\#1}$ is the threshold of undervoltage to assign the UV attack to Agent #1; and $UE_{Agent\,\#2}$ denotes the boundary of UE to assign UE attack to Agent #2. As an illustration, if $\eta = 0.4$, the maximum level of undervoltage should be 60% to assign the UV attack to Agent #1, otherwise, this cyberattack will be assigned to Agent #2. It is noted that η and π are determined according to the difference between the magnitude of normalized values of the objective functions Eqs. (17.1) and (17.2) via trial-and-error procedure.

$$UV \leq UV_{Agent\#1} \times (1 - \eta) \tag{17.27}$$

$$UE \leq UE_{Agent\#2} \times (1 - \pi) \tag{17.28}$$

17.4 Case Study, Simulation Results, and Analysis

17.4.1 Initialization

To evaluate the effectiveness of the developed anti-attack framework, the optimization problems, introduced in Section 17.3, were modeled/solved via the optimization toolbox in MATLAB R2020b and the available functions and system descriptions of the MATPOWER package [32] on an Intel(R) Core(TM) CPU i7-13 700 @ 2.10 GHz and 32 GB of RAM.

17.4.2 Case Study

The IEEE 30-bus test system, adopted from [28], has been modified to contain two PV plants with the capacity of 0.8 MW and 2 MW located on buses #18 and #25, respectively. Moreover, it has 41 transmission lines, 6 thermal generation units at buses #1, #2, #5, #8, #11, and #13, 2 compensator components at buses #10, and #24, and 4 transformers installed on the branches (4–12), (6–9), (6–10), and (27–28). The network demand is 283.4 MW, and bus #1 is the slack bus. It is noted that three transformers within the system have been upgraded to include tap changing features. It is also noted that the range of integer tap positions for all three ULTCs is limited to a range of [−20, 20] with the step of 0.005. The maximum of allowable actions (i.e. alteration of the tap positions) is limited to four (refer to Λ in Eq. (17.9)). The voltage magnitude of buses is limited to a range of [0.95–1.05] p.u. Figure 17.5 illustrates the case study of this chapter, including the PV plants and ULTC components. Daily active and reactive load profiles are adopted from [33], providing hourly load profiles for researchers in the field. It should be noted that all switches are closed during normal operation in this test system. Tables 17.1–17.3 present the data associated with generation units, transmission lines, and load centers, as shown in Figure 17.5. The rest of the system's data can be found in [24, 28].

The initial value of the learning rate and the batch size associated with local DNNs are, respectively, equal to 0.00001 and 3000. For the global model, the initial value of the learning rate and the batch size are set to 0.001 and 200. These numbers are selected based on trial-and-error; however, in the future steps of this research we will develop optimization problems to obtain the optimal values to balance the performance of the trainer. A neural network's learning rate affects how fast it adapts to the problem, and a batch size determines how many samples it processes before the entire network is updated in each iteration. Moreover, 80% and 20% of the samples are, respectively, utilized in the training and testing phases. The number of hidden layers and the probability of layers are selected as 100 and 0.15.

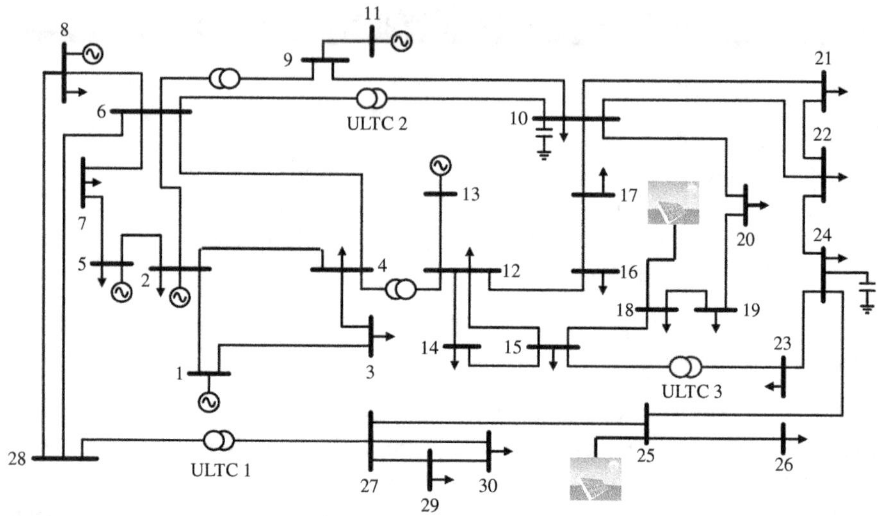

Figure 17.5 Modified IEEE 30-bus test system.

Table 17.1 Data associated with generation units in IEEE 30-bus test system.

Unit	Generator	a	b	c	d	e	P_{min} (MW)	P_{max} (MW)
1	G_1	0	2.00	0.00375	18.0	0.037	50	250
2	G_2	0	1.75	0.01750	16.0	0.038	20	80
3	G_5	0	1.00	0.06250	14.0	0.040	15	50
4	G_8	0	3.25	0.00830	12.0	0.045	10	35
5	G_{11}	0	3.00	0.02500	13.0	0.042	10	30
6	G_{13}	0	3.00	0.02500	13.5	0.041	12	40

17.4.3 Investigated Scenarios of Cyberattacks

This section provides two scenarios of FDIAs, targeting the power system illustrated in Figure 17.5 to cause voltage violations and power shortages, as well as one scenario of RAS to highlight the effectiveness of the proposed remediation mechanism to minimize the inconvenience of end-use customers and enhance the energy equity for all users. The scenarios are described as follows:

- **Scenario I: FDIAs Resulting in Voltage Violation**
  Targeting the PMUs of the branches having ULTC transformers to result in voltage violation in the form of undervoltage and overvoltage,

Table 17.2 Data associated with transmission lines in IEEE 30-bus test system.

Line no.	From-to	*R* (p.u.)	*X* (p.u.)	*C* (p.u.)	S_{max} (MVA)
1	1–2	0.0192	0.0575	0.0132	130
2	1–3	0.0452	0.1852	0.0102	130
3	2–4	0.0570	0.1737	0.0092	65
4	3–4	0.0132	0.0379	0.0021	130
5	2–5	0.0472	0.1983	0.0104	130
6	2–6	0.0581	0.1763	0.0094	65
7	4–6	0.0119	0.0414	0.0022	90
8	5–7	0.0460	0.1160	0.0051	70
9	6–7	0.0267	0.0820	0.0043	130
10	6–8	0.0120	0.0420	0.0022	32
11	6–9	0.0000	0.0208	0.0000	65
12	6–10	0.0000	0.5560	0.0000	32
13	9–11	0.0000	0.2080	0.0000	65
14	9–10	0.0000	0.1100	0.0000	65
15	4–12	0.0000	0.2560	0.0000	65
16	12–13	0.0000	0.1400	0.0000	65
17	12–14	0.1231	0.2559	0.0000	32
18	12–15	0.0662	0.1304	0.0000	32
19	12–16	0.0945	0.1987	0.0000	32
20	14–15	0.2210	0.1997	0.0000	16
21	16–17	0.0824	0.1932	0.0000	16
22	15–18	0.1070	0.2185	0.0000	16
23	18–19	0.0639	0.1292	0.0000	16
24	19–20	0.0340	0.0680	0.0000	32
25	10–20	0.0936	0.2090	0.0000	32
26	10–17	0.0324	0.0845	0.0000	32
27	10–21	0.0348	0.0749	0.0000	32
28	10–22	0.0727	0.1499	0.0000	32
29	21–22	0.0116	0.0236	0.0000	32
30	15–23	0.1000	0.2020	0.0000	16
31	22–24	0.1150	0.1790	0.0000	16

Table 17.2 (Continued)

Line no.	From-to	R (p.u.)	X (p.u.)	C (p.u.)	S_{max} (MVA)
32	23–24	0.1320	0.2700	0.0000	16
33	24–25	0.1885	0.3292	0.0000	16
34	25–26	0.2544	0.3800	0.0000	16
35	25–27	0.1093	0.2087	0.0000	16
36	28–27	0.0000	0.3960	0.0000	65
37	27–29	0.2198	0.4153	0.0000	16
38	27–30	0.3202	0.6027	0.0000	16
39	29–30	0.2399	0.4533	0.0000	16
40	8–28	0.0636	0.2000	0.0107	32
41	6–28	0.0169	0.0599	0.0032	32

Table 17.3 Data associated with load center in IEEE 30-bus test system.

Bus no.	P_D (MW)	Q_D (MVAR)	Bus no.	P_D (MW)	Q_D (MVAR)
1	0.00	0.00	16	3.50	1.80
2	21.7	12.7	17	9.00	5.80
3	2.40	1.20	18	3.20	0.90
4	7.60	1.60	19	9.50	3.40
5	94.2	19.0	20	2.20	0.70
6	0.00	0.00	21	17.5	11.2
7	22.8	10.9	22	0.00	0.00
8	30.0	30.0	23	3.20	1.60
9	0.00	0.00	24	8.70	6.70
10	5.80	2.00	25	0.00	0.00
11	0.00	0.00	26	3.50	2.30
12	11.2	7.50	27	0.00	0.00
13	0.00	0.00	28	0.00	0.00
14	6.20	1.60	29	2.40	0.90
15	8.02	2.50	30	10.6	1.90

- **Scenario II: FDIAs Leading to Power Shortage**
  Targeting load centers via injecting false load data into the PMUs to increase the UE and result in power shortage or probable power outage,
- **Scenario III: RASs Against FDIAs Resulting in Operational Issues**
  Reacting to FDIAs, introduced in Scenarios I and II, via applying the proper RASs proposed by the proposed framework (i.e. DNNs trained via FD concept).

17.4.4 Numerical Results and Analysis

17.4.4.1 Negative Impacts of the Launched Cyberattack on Energy Justice and Equity

According to Figure 17.1, attackers in a coordinated FDIA target the voltage profile of the power system via concurrently maximizing objective functions Eqs. (17.1) and (17.2) to obtain the optimal amount of false data to be injected into the PMUs to result in severe undervoltage in the system. To keep the cyberattack undetectable, attackers solve the maximization problem subject to satisfying Eqs. (17.3)–(17.6) as well as Eqs. (17.10)–(17.17) in order to, respectively, bypass the BDD algorithm embedded in state estimation and meet the rating of assets (e.g. bus voltage magnitude, line current magnitude, etc.). It is noted that satisfying the mentioned constraints makes the FDIA stealthy and significantly reduces the chance of being caught by the system operator. It is also noted that the importance degrees of the multi-objective optimization problem have been set to 0.9 and 0.1, which are, respectively, associated with Eqs. (17.1) and (17.2) to meet $\sum_k \Psi_k = 1$, as introduced in Figure 17.3. The normal and affected voltage profile of the modified IEEE 30-bus test system is illustrated in Figure 17.6. To obtain a better perspective about the impact of the FDIA, the reported voltage magnitude

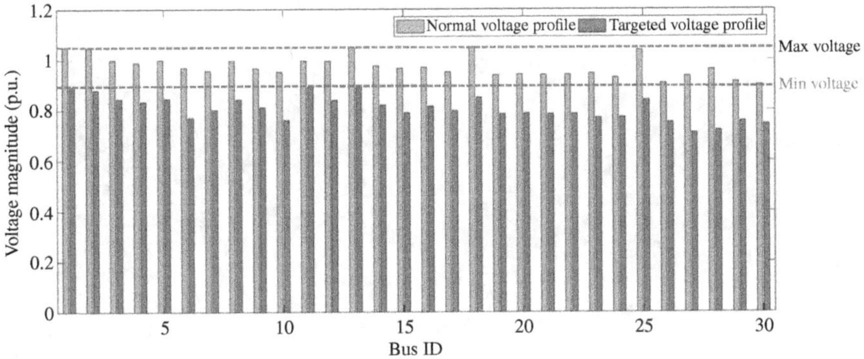

Figure 17.6 The voltage profile of the IEEE 30-bus test system before and after the coordinated FDIA, resulting in extreme undervoltage, in Scenario I.

associated with each bus in Figure 17.6 is the average of voltage magnitudes for 24 hours. From Figure 17.6, one can infer that the voltage magnitude of all 30 buses is in the acceptable range of [0.95–1.05] p.u. in normal operation (see the left bars depicted in Figure 17.6); however, the voltage magnitude of almost all 30 buses fell below the minimum threshold (i.e. 0.95 p.u.), which results in a higher amount of current absorbed by different assets, paving the way for tripping overcurrent relays and shotting down the power grid. As a result, the cyberattack significantly limits end-use customers' access to power, negatively affecting energy justice via reducing the reliability and sustainability of energy in the IEEE 30-bus test system. Another important conclusion that can be drawn from Figure 17.6 is that the affected voltage magnitude of buses #1, #11, and #13 remains on the maximum threshold after the FDIA since these three buses do not have any load centers (see Figure 17.5), the PMUs that could be compromised by the attackers. In other words, buses #1, #11, and #13 are not vulnerable to voltage violation in this case study.

To obtain a better perspective about the reason for such an undervoltage in the system, Figure 17.7 depicts the tap position of the ULTC transformers. It is noted that the negative sign has been used by attackers for the second term of Eq. (17.1) to push the system toward undervoltage. According to Figure 17.7, it can be concluded that the tap positions associated with all three ULTC transformers are significantly decreased after the FDIA (see the dashed boxes displayed in Figure 17.7, having more blue blocks) compared to tap positions of the same ULTCs for the normal operation.

The objective of the introduced FDIA can be changed from undervoltage to overvoltage via altering the negative sign of Eq. (17.1) to the positive sign. To obtain a better view about the negative impacts of the sign alteration, Figure 17.8 displays the voltage profile of the power system after being targeted by an FDIA

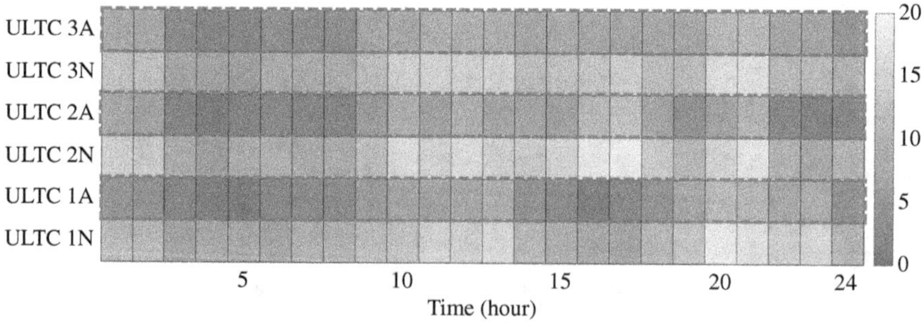

Figure 17.7 Tap position of ULTC transformers before and after the FDIA leading to undervoltage (A and N are, respectively, refer to normal operation and attack situation) in Scenario I.

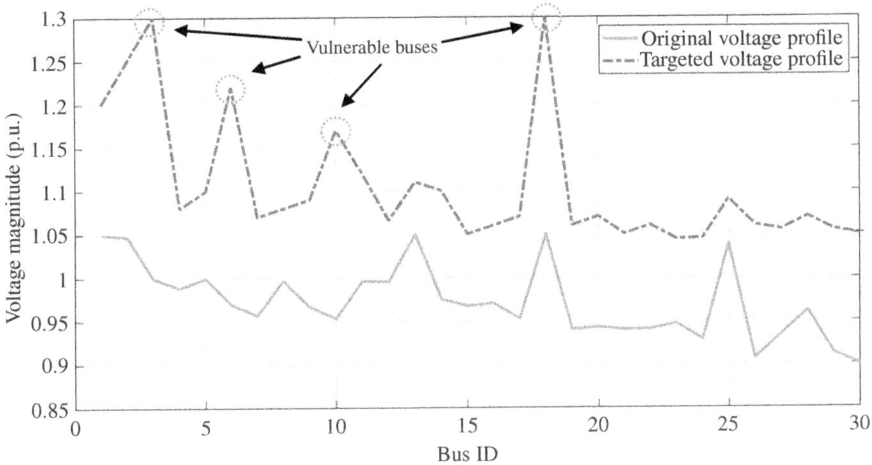

Figure 17.8 The voltage profile of the IEEE 30-bus test system before and after the coordinated FDIA, resulting in extreme undervoltage, in Scenario I.

leading to overvoltage. Another piece of critical information that can be extracted from Figure 17.8 is the identification of the vulnerable buses to overvoltage, which can be used by attackers for their future FDIAs leading to voltage violation. Comparing Figures 17.6 and 17.8 shows that these cyberattacks can drive the power system toward varying degrees of voltage violations, underscoring the importance of the proposed RAS in countering such FDIAs.

To change the objective of the coordinated FDIA from voltage violation to power shortage, attackers need to update the weight factors in such a way that the importance degree for Eq. (17.2) is increased from 0.1 to 0.9. Hence, the importance of Eq. (17.1) needs to decrease from 0.9 to 0.1 since the summation of these two numbers should be equal to one. Figure 17.9 shows the Pareto-optimal frontier associated with the optimization problem in the attack phase (i.e. maximizing objective functions Eqs. (17.1) and (17.2) simultaneously). As can be seen in Figure 17.9, the trade-off between objective functions is depicted by a red pentagram. The cyan triangle, displayed in Figure 17.9, shows the trade-off between the same objective functions in Scenario I, where the importance of objective function Eq. (17.1) was higher from the attackers' perspective.

17.4.4.2 Performance of the Remedial Actions Against Such FDIAs

To show the effectiveness of the proposed RAS, the system operator simulates different scenarios of cyberattacks/remediations in collaboration with the local agents (see Figure 17.2). The obtained results of the preparation phase are provided in Table 17.4.

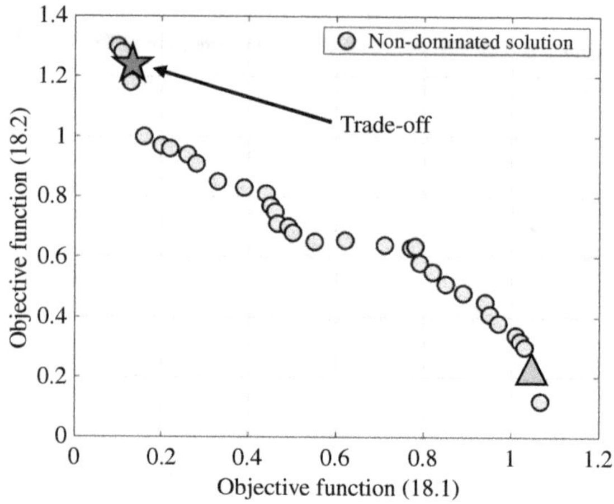

Figure 17.9 2D Pareto-optimal solutions associated with Scenario II (The unit of both horizontal and vertical axes is p.u. while the base power in the system is 100 MVA).

Table 17.4 The optimal obtained results associated with the pre-evaluated attack-RAS scenarios performed by the local agents.

FDIA scenario	Appropriate RAS	Effect of the RAS on the system
Scenario I leading to a mild undervoltage in the system up to 15% of the rated voltage	CNR with Eq. (17.18) as the objective function, taking into account minimizing Eq. (17.21) to mitigate the overall cost of switching	Mitigating the voltage deviation index of the system by 87%
Scenario I leading to severe overvoltage up to 35% of the rated voltage	CNR with Eq. (17.19) as the main objective function	Improving the voltage deviation index by 66%
Scenario II leading to 5% unserved energy in the system (base power: total demand)	CNR with Eq. (17.20) as the main objective function, considering minimizing Eq. (17.18)	Reducing the energy not supplied by 71%
Scenario II leading to a 20% power shortage in the system	CNR with Eq. (17.20) as the main objective function	Reducing the energy not supplied by 54%

To address the successful FDIAs introduced in Section 17.4.4.1 and enhance energy justice and equity within the targeted power system, the system operator employs the proposed framework shown in Figure 17.2. Table 17.5 presents the obtained results. According to this table, one can infer that the consequences of the attack scenarios are different from the pre-evaluated scenarios in Table 17.1;

Table 17.5 The RASs obtained from the proposed framework against FDIAs.

Attack scenario	Attack consequence	Recommended RAS and priority	Outcome of each RAS on the system
Scenario I	25% Undervoltage	1. CNR with Eq. (17.19) as the main objective function	Decrease the numerical value of Eq. (17.19) from 0.6876 p.u. after the FDIA to 0.0752 p.u. after the RAS
		2. CNR with Eq. (17.18) as the main objective	The voltage deviation index decreased from 0.6876 to 0.3222 p.u.
		3. CNR with two objective functions of Eq. (17.18) and (17.19)	The voltage deviation index decreased from 0.6876 to 0.5319 p.u.
Scenario II	25% Power Shortage	CNR with Eq. (17.20) as the main objective function	Unserved energy in the system decreased from 1.1855 to 0.6979 p.u.

however, the proposed framework was able to propose proper remedial actions to mitigate the FDIAs.

From Tables 17.4 and 17.5, it can be concluded that the proposed framework is able to respond to the FDIAs in real time regardless of the type and scale of the cyberattack. This will significantly increase the reliability of the IEEE 30-bus tests system since (1) the voltage profile of the system has enhanced by at least 53% and (2) the total energy not supplied (or unserved energy) throughout the system has decreased by 48%. This will help enhance energy justice and equity in the IEEE 30-bus test system by reducing interruptions in power delivery services for end use customers. In addition, the performance of the proposed framework can significantly be enhanced against a variety of FDIAs with different consequences by increasing the number of local agents and the pre-investigated attack-RAS scenarios.

17.4.5 Discussion

Figures 17.6 and 17.7, which show the results of FDIAs affecting the IEEE 30-bus test system and causing voltage violations, reveal that tampering with the load data resulted in malicious outcomes from the VVO problem. Thus, the tap position of ULTC transformers (see Figure 17.5) intentionally moved downward causing different levels of undervoltage via increasing the voltage deviation index of the system by 67.6%. This level of voltage is unacceptable from the operation standpoint since the voltage magnitude of almost all the system buses is below the minimum boundary. This operational issue results in increased current in the transmission lines, which may trigger protective devices and cause a significant power shortage in the system. This will negatively affect the reliability of the system through the concept of energy equity. On the other hand, in Tables 17.4 and 17.5, which display the performance of the developed remedial action against these FDIAs, one can infer that the developed FL-based remedial action in this chapter was successful to mitigate the operational issues in real time. For example, 25% undervoltage and power shortage decreased to 4% and 11% after implementing the remedial actions suggested by the proposed framework (refer to Figure 17.2). This is where the importance of the framework developed in this chapter becomes evident. By improving system reliability, it enhances social welfare for end-use customers, offering benefits such as fair energy access and reduced power bills.

The optimal performance of the developed RAS in this chapter for larger scale systems can be significantly improved via taking the decentralized approach into account. This will allow eliminating the central server and all the corresponding communications with the central server. Therefore, the current remediation framework, similar to any other initiative frameworks in the field, has some limitations, the most important one of which is elaborated as follows:

- It is computationally expensive while taking more than two cyberattacks and their corresponding computational loops into account at a time. To address this challenge, in our next work, we will propose a decentralized version of the developed framework in this chapter to provide a peer-to-peer communication between agents via parallel computation, leading to a generalized framework that can be applied to large-scale systems with a massive number of agents.

17.5 Acknowledgment

This research was supported in part by the National Science Foundation under Grant No. 2348420.

17.6 Conclusions

In this chapter, various attack frameworks were modeled to explore a range of operational issues in a power grid, such as voltage violations and power shortages, which impact energy justice and equity by compromising the social welfare of end-use customers. Moreover, an intelligent RAS was proposed against such attack models to mitigate their consequences and enhance the social well-being of end users. The developed RAS was oriented toward two blocks of DNNs trained in parallel based on the concept of FL to increase data privacy. The effectiveness of the introduced framework was investigated on IEEE 30 bus test system modified to encompass under load tap changing transformers. The main achievements of this chapter are itemized as follows:

- The most vulnerable buses to voltage violation are the ones having load centers. Hence, the PMUs associated with such vulnerable buses need to be equipped with a moving target strategy (i.e. the scope of our future work) to significantly reduce the chance of being targeted by FDIAs via dividing the information to several segments and transmitting one segment at a time.
- The proposed RAS effectively suggested suitable remediations for various FDIA scenarios, regardless of whether these attacks had been previously investigated by local agents and the system operator. However, when the attackers' main objective was to increase unserved energy, which directly affects the social welfare of end users, the remedial actions proposed by the intelligent framework were unable to completely mitigate the negative impacts of the cyberattack. This performance issue can be addressed by increasing the number of pre-investigated attack scenarios and/or increasing the number of objective functions considered in the network reconfiguration problem. This will be the scope of our future papers.

References

1 Whitehead, D.E., Owens, K., Gammel, D., and Smith, J. (2017). Ukraine cyber induced power outage: analysis and practical mitigation strategies. In: *Annual Conference for Protective Relay Engineers (CPRE)*. College Station, TX: IEEE. 10.1109/CPRE.2017.8090056.

2 Dvorkin, Y. (2019). Executive order shines a light on cyberattack threat to the power grid. *IEEE Spectrum*. Available Executive Order Shines a Light on Cyberattack Threat to the Power Grid - IEEE Spectrum.

3 Qu, Z., Wang, P., Li, J. et al. (2023). A globally cooperative recovery strategy for cyber-physical power system based on node importance. *IEEE Access* 12: 179–191. 10.1109/ACCESS.2023.3346939.

4 Naderi, E. and Asrari, A. (2024). Stealthy false data injection cyberattack targeting under load tap changing transformers in smart power grid causing abnormal voltage profile. In: *Third International Conference on Power, Control and Computing Technologies (ICPC2T)*, 145–150. Raipur, India: IEEE. https://doi.org/10.1109/ICPC2T60072.2024.10474991.

5 Naderi, E. and Asrari, A. (2024). A remedial action scheme to mitigate market power caused by cyberattacks targeting a smart distribution system. *IEEE Transactions on Industrial Informatics* 20 (3): 3197–3208. https://doi.org/10.1109/TII.2023.3304049.

6 Energy Justice and the Energy Transition (2022). [Online]. Available: https://www.ncsl.org/energy/energy-justice-and-the-energy-transition#:~:text=Building%20off%20the%20tenets%20of,nationality%2C%20income%20or%20geographic%20location (accessed 20 March 2024).

7 Nazari-Heris, M., Loni, A., Asadi, S., and Mohammadi-Ivatllo, B. (2022). Toward social equity access and mobile charging stations for electric vehicles: a case study in Los Angeles. *Applied Energy* 311: 118704. https://doi.org/10.1016/j.apenergy.2022.118704.

8 Guedes, W., Oliveira, C., Soares, T.A. et al. (2024). Collective asset sharing mechanisms for PV and BESS in renewable energy communities. *IEEE Transactions on Smart Grid* 15 (1): 607–616. https://doi.org/10.1109/TSG.2023.3288533.

9 Heleno, M., Sigrin, B., Popovich, N. et al. (2022). Optimizing equity in energy policy interventions: a quantitative decision-support framework for energy justice. *Applied Energy* 325: 119771. https://doi.org/10.1016/j.apenergy.2022.119771.

10 Jenkins, K., McCaŭley, D., Heffron, R. et al. (2016). Energy justice: a conceptual review. *Energy Research and Social Science* 11: 174–182. https://doi.org/10.1016/j.erss.2015.10.004.

11 Zhang, Q. and Li, F. (2024). Securing energy equity with multilayer market clearing. *IEEE Transactions on Energy Markets, Policy and Regulation* 2 (3): 301–312. https://doi.org/10.1109/TEMPR.2024.3377210.

12 Heffron, R.J. and McCauly, D. (2017). The concept of energy justice across the disciplines. *Energy Policy* 105: 658–667. https://doi.org/10.1016/j.enpol.2017.03.018.

13 Singh, V.K. and Govindarasu, M. (2018). Decision tree-based anomaly detection for remedial action scheme in smart grid using PMU data. In: *IEEE Power and Energy Society General Meeting (PESGM)*. Portland, OR: IEEE. https://doi.org/10.1109/PESGM.2018.8586159.

14 Huang, H., Kazerooni, M., Hossain-McKenzie, S. et al. (2019). Fast generation redispatch techniques for automated remedial action schemes. In: *International Conference on Intelligent System Application to Power Systems (ISAP)*. New Delhi, India: IEEE. https://doi.org/10.1109/ISAP48318.2019.9065961.

15 Nespoli, P., Mármol, F.G., and Vidal, V.J. (2021). A bio-inspired reaction against cyberattacks: AIS-powered optimal countermeasures selection. *IEEE Access* 9: 60971–60996. https://doi.org/10.1109/ACCESS.2021.3074021.

16 Tu, H., Xia, Y., Tse, C.K., and Chen, X. (2020). A hybrid cyber attack model for cyber-physical power systems. *IEEE Access* 8: 114876–114883. https://doi.org/10.1109/ACCESS.2020.3003323.

17 Guo, L., Zhang, J., Ye, J. et al. (2022). Data-driven cyber-attack detection for PV farms via time-frequency domain features. *IEEE Transactions on Smart Grid* 13 (2): 1582–1597. https://doi.org/10.1109/TSG.2021.3136559.

18 Zadsar, M., Abazari, A., Ameli, A. et al. (2023). Prevention and detection of coordinated false data injection attacks on integrated power and gas systems. *IEEE Transactions on Power Systems* 38 (5): 4252–4268. https://doi.org/10.1109/TPWRS.2022.3216118.

19 Du, M., Wang, L., and Zhou, Y. (2023). High-stealth false data attacks on overloading multiple lines in power systems. *IEEE Transactions on Smart Grid* 14 (2): 1321–1324. https://doi.org/10.1109/TSG.2022.3209524.

20 Appiah-Kubi, J. and Liu, C.-C. (2023). Cyberattack correlation and mitigation for distribution systems via machine learning. *IEEE Open Access Journal of Power and Energy* 10: 128–140. https://doi.org/10.1109/OAJPE.2023.3236429.

21 Elmallah, S., Reames, T.G., and Spurlock, C.A. (2022). Frontlining energy justice: visioning principles for energy transitions from community-based organizations in the United States. *Energy Research and Social Science* 94: 102855. https://doi.org/10.1016/j.erss.2022.102855.

22 Choeum, D. and Choi, D.-H. (2019). OLTC-induced false data injection attack on volt/var optimization in distribution systems. *IEEE Access* 7: 34508–34520. https://doi.org/10.1109/ACCESS.2019.2904959.

23 Narimani, H., Azizivahed, A., Naderi, E. et al. (2019). A practical approach for reliability-oriented multi-objective unit commitment problem. *Applied Soft Computing* 85: 105786. https://doi.org/10.1016/j.asoc.2019.105786.

24 Naderi, E., Mirzaei, L., Trimble, J.P., and Cantrell, D.A. (2023). Multi-objective optimal power flow incorporating flexible alternating current transmission systems: application of a wavelet-oriented evolutionary algorithm. *Electric Power Components and Systems* 52 (5): 766–795. https://doi.org/10.1080/15325008.2023.2234378.

25 Hug, G. and Giampapa, J.A. (2012). Vulnerability assessment of AC state estimation with respect to false data injection cyber-attacks. *IEEE Transactions on Smart Grid* 3 (3): 1362–1370. https://doi.org/10.1109/TSG.2012.2195338.

26 Bi, S. and Zhang, Y.J. (2014). Using covert topological information for defense against malicious attacks on DC state estimation. *IEEE Journal on Selected Areas in Communications* 32 (7): 1471–1485. https://doi.org/10.1109/JSAC.2014.2332051.

27 Naderi, E., Asrari, A., and Fajri, P. (2024). False data injection cyberattacks targeting electric vehicles in smart power distribution systems. In: *IEEE*

Transportation Electrification Conference and Expo (ITEC). Rosemont, IL: IEEE.

28 Naderi, E., Mirzaei, L., Pourakbari-Kasmaei, M. et al. (2023). Optimization of active power dispatch considering unified power flow controller: application of evolutionary algorithms in a fuzzy framework. *Evolutionary Intelligence* 17: 1357–1387. https://doi.org/10.1007/s12065-023-00826-2.

29 Aziz, T., Lin, Z., Waseem, M., and Liu, S. (2020). Review on optimization methodologies in transmission network reconfiguration of power systems for grid resilience. *International Transactions on Electrical Energy Systems* 31: https://doi.org/10.1002/2050-7038.12704.

30 Imteaj, A., Thakker, U., Wang, S. et al. (2022). A survey on federated learning for resource-constrained IoT devices. *IEEE Internet of Things Journal* 9 (1): 1–24. https://doi.org/10.1109/JIOT.2021.3095077.

31 Tyagi, S. (2023). Scavenger: a cloud service for optimizing cost and performance of DL training. In: *IEEE/ACM 23rd International Symposium on Cluster, Cloud and Internet Computing Workshops (CCGrid)*. IEEE https://doi.org/10.1109/CCGridW59191.2023.00081.

32 Zimmerman, R.D., Murillo-Sanchez, C.E., and Thomas, R.J. (2011). MATPOWER: steady-state operations, planning and analysis tools for power systems research and education. *IEEE Transactions on Power Systems* 26 (1): 12–19. https://doi.org/10.1109/TPWRS.2010.2051168.

33 OpenEI (2024). Commercial and residential load profiles. [Online]. Available: https://data.openei.org/ (accessed 20 March 2024).

18

Energy Policy Strategies for Energy Justice and Equity

Iman Rahmati[1] *and Asghar Akbari Foroud*[2]

[1] *Electricity Market Monitoring & Studies, Iran Grid Management Company, Tehran, Iran*
[2] *Electrical and Computer Engineering Faculty, Semnan University, Semnan, Iran*

Acronyms

COP	conference of the parties
FiT	feed-in-tariff
GDP	gross domestic product
GHG	greenhouse gas
GLP	Green Liberal Party (Switzerland)
GPS	Green Party of Switzerland
GPUS	Green Party of the United States
IPCC	Intergovernmental Panel on Climate Change
IRENA	International Renewable Energy Agency
LNG	liquefied natural gas
NGO	Nongovernmental Organization
PEB	pro-environmental behavior
REC	Renewable Energy Certificates
UNEP	United Nations Environmental Programme
UNFCCC	United Nations Framework Convention on Climate Change

18.1 Introduction

As a matter of fact, the current economic platform in the energy industry and the natural trading mechanisms are not capable to support energy justice and equity, and these highly respected goals won't be achievable without proper policymaking. Hence, the role of authorities in setting the stage for aligning the forces present

Communities for Clean Energy Justice and Equity in Grid Modernization, First Edition.
Edited by Mohammadreza Daneshvar, Behnam Mohammadi-Ivatloo, and Amjad Anvari-Moghaddam.

in this field, political and financial, is undeniable. In this chapter, we focus on this influential parameter and discuss different practices in the world.

18.2 Why Energy Justice Relies on Policy?

18.2.1 Energy as a Critical Factor in Social Welfare and Development

Based on what was explained in the previous chapters, it is clear that nowadays energy is a critical commodity in human society that nobody can afford to be without. Social welfare, individual's quality of life, and national prosperity, to a great deal, depend on the suitable and secure provision of this commodity. The need for energy is now more widespread than ever and includes a vast range of areas, including transportation, lighting, communications, medical assistance and healthcare systems, water supply facilities, air conditioning, and even hobbies (cinema, computer games, gyms, etc.). Any disruption in providing energy is widely visible to the public and projected in the media and, more importantly, is attributed to the government's incompetence. Because of such sensitivity, the energy sector has always been an area with a high level of government's intervention.

According to the microeconomic fundamentals, social welfare is at the core of social and economic development. The evidence regarding the dependency of social welfare development on energy provision is abundant and pervasive. Although with the growing general awareness regarding the environmental impact of carbon footprint, modern societies would not accept any sort of development but also seek to explore approaches to realize development in a sustainable way that interacts with nature and the environment in a friendly manner. And energy, as a key element for development, is at the center of attention.

In fact, in most cases, the major challenges regarding energy injustice stem from the policies that have been in place in the energy sector for a long time, and policy reform is inevitable to amend them. However, notwithstanding all the sustainability-oriented reforms in the energy sector, current legal frameworks and market structures in the energy sector are generally based on large-scale fossil-based energy technology and have proved their inability to spur the transition toward sustainability and energy justice [1]. Hence, the transition toward energy justice would not be achievable without a proper transition within the energy policies.

18.2.2 Role of Policy in Energy Consumption Habit

Sustainability and energy justice would not be achievable without taking proper actions both on the supply and consumption side and solely focusing on the supply side to reduce carbon emissions and stop the course of global warming will be futile. In fact, in most cases, it is the consumption pattern that defines and mandates operational activities on the supply side. From this point of view and according to the Intergovernmental Panel on Climate Change (IPCC), the role of consumption pattern has a significant role in the formation of the current climate change problem [2]. Consequently, in order to attain energy justice goals, active and wise participation of the consumption side in this transition is of the highest importance.

Based on [3], total energy consumption has registered around 139% growth during 2017–2021. This rapid increase in energy consumption has been the primary driver of fossil fuel usage and carbon emissions that contribute to global warming. It is essential to raise public awareness about consumers' role in participating in energy justice initiatives to decrease energy usage through conservation and efficiency enhancements [4, 5].

When it comes to forming consumer energy consumption habits, exploiting the influential capacity of two major factors is essential: public media and policymaking. Energy policies can form and reform consumption patterns, both in the industrial and residential sectors. For instance, residential consumers make a considerable contribution to carbon emissions [6], and due to population growth, this carbon footprint is increasing. Thus, it's important to raise the consumer's sensitivity to lowering energy consumption and to choose energy-efficient home appliances that can considerably contribute to less energy usage and thus less greenhouse gas (GHG) emissions [7]. Proper policies in the residential sector, such as proper increasing block tariff, green obligation, etc., can lead to pro-environmental behavior (PEB) of these consumers.

Consumer's responsibility in the realization of energy justice is not restricted to the residential sector. Industrial consumers contribute largely to annual energy consumption and resultantly in annual carbon emissions. Policies inciting R&D in industries to explore technological innovation and improve energy efficiency can promote sustainable development in this sector. This is of higher importance in developing economies where higher annual growth in industrial energy consumption is prevalent. Following business-as-usual scenarios in development and without suitable policies to support and incite taking environmentally friendly approaches, high annual economic growth may directly translate to a high increase in emission volume, which entrenches the sustainability goals of energy justice [8].

18.3 Nexus Between Policy and Energy Justice

18.3.1 Policymakers Promise to Bring Justice

Improving public awareness regarding energy justice goals and the environmental impacts of human activities can be considered as the greatest achievement in realizing energy justice and equity. This has been a driving force for the national and international authorities to follow public demands regarding taking serious actions in the course of realizing energy justice. During past decades, fighting energy poverty, carbon emission abatement, and renewable energy integration have been at the center of many political disputes throughout the world. In some modern economies, the pressure of social demands regarding environmental concerns has become so intense that some parties have made the pursuit of social demands in the environmental field the focal point of their election policies. They have concluded that due to the public awareness and demands regarding energy justice improvement, making such promises has a positive effect on their winning chance in elections. These days, green parties actively engage in politics and try to get into the parliaments, cabinets, senates, governmental bodies, etc. As a matter of fact, the formation of green political parties itself is a manifestation that green activists came to this conclusion that using political power was a way to win the battle against environmental degradation and global warming.

The history of registering environmental political parties goes back to 1972 when the first political ecological party, called Values, was formed in New Zealand and a year later, in 1973, a small group of environmental activists formed a political party named People, in Coventry [9]. Currently, green parties are politically active in many countries and they have succeeded in finding their way to the parliaments in Germany, Ireland, Sweden, Switzerland, Italy, Austria, Canada, Bulgaria, etc.

For example, green parties in Switzerland and Austria have had members in the national parliament since 1979 and 1986, respectively. The Green parties have 10% of seats in the parliament and there are members within several local governmental authorities [10]. The 2019 Swiss federal election was a great success for the Green parties when the Green Party of Switzerland (GPS) and Green Liberal Party (GLP) gained 13.2% and 7.8% of the votes, respectively [11]. In Belgium, Green parties have been represented in the national parliament since 1978, and in 2014, they had about 8.6% of the votes [12]. German Greens can be considered as the most successful green party through time. In late 2013, the number of Green party members at the European, national and land level reached 324 members [13]. They gained their best result in the 2021 federal election when they got 15% of the votes [14]. Green parties are very active in the United States too. Green Party of the United States (GPUS) was established in 2001 and since then has been politically active at the state and national level. GPUS claims to have "four pillars": peace, ecology, social justice, and democracy. Green parties in the United States are becoming politically influential and in November 2023, one of the GPUS leaders declared her intention for candidacy in the 2024 presidential election [15].

The extensive engagement of environmentalists in political endeavors underscores two fundamental principles: first, there exists a prevailing consensus among environmental advocates that the realization of their objectives is contingent upon wielding political influence. Second, the heightened societal awareness regarding the imperative of environmental conservation has enhanced the electoral prospects of individuals identified as proponents of environmental issues. Consequently, the dynamic between politics and environmental protection can be conceptualized as a mutually reinforcing relationship: political authority can be leveraged to advance environmental causes while being associated with environmental activism can cultivate a favorable public perception.

On the other hand, environmental policies and realizing energy justice targets likely have considerable financial implications. The need for a precise and thorough package of financial and monetary instruments, including budgeting, spending priorities, financial penalties, tax premiums, reporting systems, etc., is essential [16]. There is a strong linkage to the policymaking activities regarding all these aspects and normally necessitates multidimensional interactions between multiple national authorities to succeed.

18.3.2 Political Challenges of Energy Justice

Contrary to what is explained in the previous section, political targets may be in opposition to energy justice and equity goals. Although almost all political parties verbally support energy justice, there could be situations where the political preferences of the government or the party in power hinder boosting energy justice.

In fact, in many cases, the disparity in the availability of modern energy for marginalized communities can be attributed to political orientation and the balance between political powers. This is a general case in which political power is backed by the more resourceful communities in society. However, the economic, social, and medical burden of the disparity in the energy system is nonuniformly imposed on communities and the less well-off and marginalized part of society has to bear the consequences more severely [17].

In this context, another challenge may arise from the political influence of large energy companies. When promoting energy justice contradicts with the profitability of these companies, their lobbying power may shift the regulatory framework of the energy sector to align with the economic profit of these companies instead of following energy justice guidelines and protecting vulnerable population.

The complexity of the regulatory framework of the energy industry is another barrier to choosing and implementing proper policies. Complex tariff structures, direct and indirect subsidies, cross-subsidy, fine and taxation basics, market designs, technical and environmental standards, and social aid policies all should be organized together to form an efficient and justice-oriented policy framework regarding the energy industry. Some researchers believe that this complexity lies within the definition of energy justice that covers a wide range of humanitarian goals, and such complexity reduces the ability of governance and monitoring [18].

Furthermore, justice and equity rely on a democratic decision-making process. Energy justice promotion involves ensuring that communities have a say in energy-related decisions that affect them. This can be challenging in political systems where power is concentrated in the hands of a few decision-makers or where marginalized communities lack representation.

Also, it is important to comprehend that governments generally have to deal with multiple issues and challenges at a time, and energy justice is not the sole matter at hand. Based on their situations, different states and governments in the world may prioritize their challenges differently. Moreover, as stated earlier, energy justice itself is a multidimensional concept that encompasses several aspects. Hence, from the viewpoint of each state or country, or even each political party, some aspects may be prioritized higher than others.

All the abovementioned challenges are powerful barriers that may impede social and environmental activists from getting as much political support as they need to accelerate the transition toward a just energy system. However, as some researchers have pointed out, there is a time duality between the cost and benefit of justice-oriented actions within the energy sector, and this duality may be considered as one of the most hindering challenges in this regard. Transitioning to a just energy system requires highly orchestrated immediate actions that impose an almost instant economic burden. However, most positive outcomes of energy justice activities are remote and take time to appear to the public. The problem is that the duration between justice-oriented activities, and the appearance of their positive outcomes is generally longer than the life of governments. Governments generally follow leads that their positive outcomes can be achievable within their time in power, and they are less willing or even reluctant to bury seeds that will give fruit when their time in power is passed. Furthermore, this short-term prioritized mindset is not limited to governments [19]. Even public people, as potential voters, respond to immediate costs more furiously than they do to future benefits. Hence, it is understandable that the tendency to prioritize immediate costs over future benefits is common among governmental authorities and policymakers.

In short, major challenges regarding promoting energy justice from the viewpoint of policymaking arrangements can be summarized as follows:

- Nonuniform political power distribution among different communities can leave marginal communities politically voiceless.
- Political influence of large energy companies may hinder the transition toward justice and equity
- Inherent complexity of regulations within the energy sector makes it difficult to harmonize the efforts.
- Political objectives and challenges may downgrade the priority of justice-oriented goals within the energy sectors.

- Political entities may prioritize the short-term costs over the long-term benefits of justice-oriented policies.
- Political structures characterized by centralization and nondemocratic decision-making mechanisms pose limitations in achieving comprehensive energy justice and equity.

In order to surmount the aforementioned obstacles, it is imperative to consistently advocate for energy justice measures as integral components of universal human values and to bolster public support for policies and initiatives that align with principles of justice.

18.3.3 Energy Justice at the International Level

Particularly, in order to promote energy security, energy affordability, and energy sustainability, broadening global cooperation and harmonizing the efforts are vital. Mega projects on gas pipelines, oil and liquefied natural gas (LNG) transportation, and cross-border electricity transmission have played a significant role in addressing energy security issues and energy poverty abatement. Moreover, the energy sector is a capital-intensive and technological industry where international investment and partnership are essentially necessary for progress and it cannot be achievable without a sound and friendly atmosphere.

On the other hand, regional tensions not only diminish the cooperation capacity necessary for implementing such mega projects but also disrupt the energy markets at the regional and global levels and expose a large population of the earth's inhabitants to the risk of high energy prices and lower energy affordability.

18.4 Policies Toward Sustainability

18.4.1 Sustainability as the Most Political Aspect of Energy Justice

It is crucial to recognize that while all facets of energy justice and equity require a solid policy framework to be achievable, energy sustainability is one that particularly depends on policy backing to thrive and endure. There are several reasons for this significant reliance on policy support for sustainability:

- Most other aspects of energy justice benefit from natural social demands. For example, even long before the concept of energy justice became popular, citizens and governments were sensitive to energy prices and energy affordability. In most countries, there is a long history of expanding energy systems to provide people with modern energy forms and fight energy poverty. Hence, it is just to claim that with or without, before and after the emergence of energy justice and equity, nations have been following some aspects of energy justice.

- Some aspects of energy justice are, either fully or partially, supported by natural market mechanisms. Energy efficiency is one example, where people's economic incentives are not distorted by subsidies, they have a natural response to their energy bills and seek higher energy efficiency.
- Energy is one of the critical inputs for economic and social development, and because of that, countries have always been sensitive to ensuring their energy security and diversifying their energy sources. The modern history of the energy sector is abundant with examples showing different countries' efforts to gain an acceptable level of energy security and how this factor has formed international relations during war and peace periods. Interested readers can find numerous examples in *The Prize: The Epic Quest for Oil, Money & Power* written by Daniel Yergin [20]. Hence, it is obvious that emphasizing energy security has a history much longer than energy justice.
- Energy sustainability and addressing climate change concerns within the energy sector is a challenging task that has gained a lot of attention during the past two decades. The difficulty of this task is resulted from the fact that, in many cases, the required activities are in contradiction with the current and normal practice of financial and market mechanisms. This is evidently visible in the case of harnessing energy from renewable sources, where almost none of the major achievements could be realizable without the central intervention of the governments. In other words, the current economic structure and financial streams don't support investments in this field and the need for central intervention through proper policymaking is vital. This is mostly due to the point presented in the previous section regarding the conflict between immediate costs and future benefits of sustainability-oriented energy projects. The overall platform of the global financial structure and market mechanisms are designed to support the profitability of individuals and companies currently active in the industry, not those who suffer negative consequences in the process, nor generations to come.

Based on the explanation above, it is clear that achieving the sustainability goals of energy requires central and political support more than other energy justice issues. In situations where the role of central authorities and institutions is crucial, concerns about administrative incompetence or bureaucratic corruption become more pronounced. There is abundant and pervasive evidence supporting the fact that free information circulation and public transparency are crucial factors in minimizing the potential of bureaucratic corruption and elevating the accountability of the authorities. Providing accurate information, public transparency, and accountability are essential elements for efficient and nondiscriminatory governance, which supports clear decision-making processes and transparent financial accounting [17, 21].

18.4.2 Sustainability as Justice Toward Next Generations

Energy sustainability is a pivotal term of the energy justice concept, which is getting increasing attention worldwide. In simple words, energy sustainability refers to the concept of meeting current energy needs without compromising the ability of future generations to meet their own energy needs. It involves using energy sources and technologies that are environmentally friendly, socially acceptable, and economically viable in the long term. This includes promoting renewable energy sources, increasing energy efficiency, and lowering GHG emissions to mitigate climate change. Energy sustainability aims to create a balance between the availability of energy resources and the impact of their use on the environment and society. In principle, energy sustainability can be promoted via several factors, including:

- **Energy Efficiency Improvement:** Using less energy to achieve a certain amount of output and reducing waste contributes to less environmental impact and lower GHG emissions.
- **Utilizing Renewable Energy Sources:** Transitioning to renewable energy sources such as solar, wind, hydro, and geothermal power is essential for achieving energy sustainability. These sources are abundant, clean, and have lower environmental impacts compared to fossil fuels. Increasing the share of renewable energy in the energy mix helps reduce GHG gas emissions and dependence on finite resources.
- **Energy Conservation:** Demand-side management and energy conservation measures, such as energy audits, behavioral changes, and energy-efficient practices, help reduce energy consumption and minimize environmental impacts.

Furthermore, some other energy enhancement programs, such as smart grid technologies, microgrid and distributed generation, and energy storage, can have a positive effect on the sustainability of the energy sector.

In accordance with what was explained in the previous section, energy sustainability can be considered, inter alia, as serving justice to the next generations. This honorable description came to existence when people started to realize their commitment to nature as a whole, and the next generations as part of nature and keep the earth habitable for them. Furthermore, as the European Environment Agency (EEA) states sustainability-oriented actions can be considered "just" only when they contribute to a better quality of life for current and future generations, which can include nonhuman species too [22].

Since the next generations are not present to defend their right to live, it is the current policymakers' role to support and sustainability-oriented actions as a mean to do justice in favor of generations to come. Without this political support, it is very likely that the inherent selfishness of humans and the profiteering of economic enterprises, along with regional and international competition, will seriously endanger the quality of life of future generations and even the future livability of our planet.

18.4.3 Sustainability in National and International Policy and Challenges

In general, sustainability refers to meeting the needs of the present without compromising the ability of future generations to meet their own needs. It involves balancing environmental, social, and economic factors to ensure that resources are used in a way that preserves them for future use. And the most challenging aspect of sustainability is to align development of technology and life quality without unrecoverable damage to the environment. Sustainability is often closely related to environmental concern because until recently, this factor has been less considered in the humankind history. Over the past three decades, there have been considerable movements for supporting environmental causes and addressing climate change and global warming as well as fighting air pollution. These movements have been widely reflected in the media, pervaded in the globe and welcomed by individuals, academicians, environmental communities, and political parties. Because of that, there is a global recognition of the urgent need for action in this regard. We did a simple search on sustainability-related keywords in academic articles to understand how pervasive this issue has become over the past decades. Although this review is not comprehensive and accurate, it can serve its purpose and generally shows the widespread growth of this concept during the last few decades.[1]

As can be observed in Figure 18.1, sustainability has gained a considerable amount of attention in the last three decades. This wide range and inclusive acceptance that stems from inclusive participation from different groups of people worldwide is doubtlessly the biggest achievement in this arena that acts as an impetus for aggregating the efforts and potentials at the national and international levels.

Governments and organizations are increasingly prioritizing sustainability in their decision-making processes to mitigate climate change, promote social equity, and ensure long-term economic stability. Also, currently, there are numerous international organizations that dedicate themselves to environmental and sustainability improvement. This move toward sustainability is apparent in the implementation of national policies and programs focused on decreasing carbon emissions, safeguarding natural resources, advocating for renewable energy, and

1 We used the Google Academic service for this simple survey. The targeted keywords include: "sustainable, climate change, global warming, carbon emission, carbon footprint, energy justice, renewable."

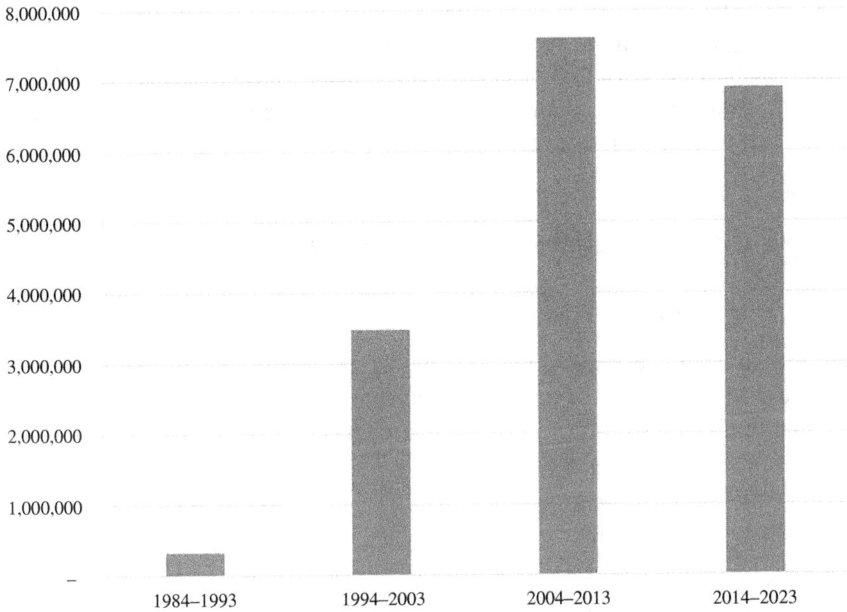

Figure 18.1 Number of academic articles with sustainability-related keywords in the past decades.

encouraging sustainable development. By incorporating sustainability into policy frameworks, nations and global organizations are striving to build a more resilient and equitable future for everyone.

Considering growing awareness regarding environmental issues and global warming, an intriguing question may come to mind that despite all this support from different groups and communities, including academicians, environmentalists, green political parties, journalists, environmental and Nongovernmental Organizations (NGOs), why are we still uncertain about solving problems related to environmental threats in the near future?

The answer to this sophisticated question lies within the challenges and contradictions discussed in Section 18.3.2. As a matter of fact, while sustainability and environmental concerns have gained significant attention worldwide, overcoming global warming remains uncertain due to various factors. These include the complexity of the issue, the need for global cooperation and action, conflicting interests among different stakeholders, technological limitations, and the inertia of existing systems. On the other hand, some international frameworks lack the necessary orientation to promote energy sustainability. For instance, international trade rules don't distinguish between fossil fuels and renewable energy, this could create challenges for the energy transition. Policies aimed at promoting clean energy or restricting fossil fuel imports might face legal hurdles under existing rules [23]. Addressing global warming requires coordinated efforts at local, national, and international levels, as well as long-term commitment and innovative solutions. Despite the challenges, continued awareness, advocacy, and collective action can pave the way toward a more sustainable and climate-resilient future.

18.5 Survey on Policy-related Attempts

18.5.1 Energy Justice in Policy

According to what is explained in the previous sections of this chapter, it is now clear that different aspects of energy justice have different timelines of existence. Some of them have a long history, and some others are newborns.

Currently, energy justice is highly projected in the energy policies both at the national and international levels. In most cases, policies supporting energy justice include the following areas:

- Energy efficiency and conservation
- Renewable energy support
- Energy access
- Environmental issues
- Standards
- Energy diversification
- Energy market regulations

It is noteworthy to acknowledge that due to the critical role of energy infrastructure in facilitating an acceptable quality of life and serving as a fundamental component for development, the history of energy policy and regulation predates that of energy justice. Nevertheless, as explained earlier in this chapter, success in achieving sustainability and climate change goals is much more dependent on policy support than of other aspects of energy justice, and the inclusion of sustainability-oriented issues and environmental regulations in energy policies has never been so prominent. This policymaking approach has led to continuous improvement of the regulations quality and this improvement is clearly projected in Regulatory Indicators for Sustainable Energy (RISE) indicators published by the World Bank. These indicators monitor the trends of policies and regulations from the viewpoint of energy sustainability [24]. As the 2022 report summarizes, most of the indicators show a positive trend compared to their values in 2010, and we look at some of them:

- Global RISE score for electricity access is 53/100 in 2021 which compared to its value in 2010 (16/100) shows more than 232% improvement.

- Global progress on sustainable energy regulation in 2021 reached 60/100 from 52/100 in 2017.
- Global RISE score for clean cooking in 2021 reached 32/100 from 6/100 in 2010.
- Global RISE score for renewable energy in 2021 reached 55/100 from 19/100 in 2010.
- Global RISE score for energy efficiency in 2021 reached 46/100 from 16/100 in 2010.

As per the findings of this report, the RISE scores concerning utility creditworthiness, which signify the quality of policies for aiding companies in transitioning toward sustainability, represent an anomaly indicative of a decline. Also, the report concludes that although RISE scores concerning energy efficiency show moderate improvement between 2019 and 2021, considering the existing energy crisis, there is a need to reinforce these policies in order to support consumer's affordability and restrict possible price spikes.

Also, it's worth saying that due to the intrinsic complexity of energy justice, as well as countries' different situations and priorities, it is not rational to expect all countries to follow one universal approach. Hence, it is understandable that countries take different passages in policymaking to reach energy justice goals.

As justified in Section 18.4.1, among energy justice pillars, energy sustainability needs policy support the most. The good news is that, in spite of all the differences, a universal consensus exists for supporting renewable energy expansion. Almost all countries in the world have applied one or several forms of renewable support mechanisms to varying degrees. The most common support mechanisms are reviewed as follows [25, 26]:

- **Feed-in Tariffs:** Governments guarantee a fixed payment for renewable energy generated and fed into the grid.
- **Premium Tariff:** Governments guarantee a minimum price, and compensate when the energy price at the market goes below the that.
- **Renewable Energy Certificates (RECs):** Tradable certificates that represent the environmental attributes of renewable energy generation.
- **Tax Benefit:** Tax credits or deductions for investments in renewable energy projects.
- **Low-interest Loan:** Direct financial assistance for covering the fixed cost of renewable energy projects.
- **Grant and Subsidies:** Direct financial aid provided by governments to support renewable energy development.
- **Net Metering:** Allows renewable energy system owners to sell excess electricity back to the grid.
- **Green Obligation:** Mandates procuring a certain share of electricity demand from renewable sources.

- **Green Bonds:** Financial instruments used to raise funds for renewable energy projects.

As a final point in this context, it should be emphasized that when it comes to policymaking regarding energy justice, the procedures should be inclusive and make capacity for active participation from all the established individuals and entities active in the energy sector. Moreover, justice will not be served unless inclusive participation is broadened to entities that are not represented in policy and decision-making that include marginalized groups such as Indigenous peoples, ethnic minorities, refugees and migrants, undocumented individuals, children, and future generations, and even other species in nature [22, 27].

18.5.2 Energy Justice in Low-income Societies

The economic and financial status of the countries is a determining factor in formulating their policies in various sectors. In addition, the financial strength of each country has a great effect on the perspective of politicians and plays a decisive role in prioritizing various goals. That's why in reviewing policies related to energy justice among different countries, we found it more sophisticated to have a look at low-income and high-income countries separately.

Low-income countries generally face a variety of problems, which can include widespread poverty, inadequate infrastructure, limited access to technology and financial resources, high unemployment rates, etc. Therefore, even without energy justice concerns, these countries struggle with quite a handful of challenges. Although, some of these challenges stem from and perhaps, at the same time, contribute to energy justice concerns. For instance, a deficiency in financial resources can result in inadequate investment in energy infrastructure, leading to energy scarcity and elevated energy prices, thereby serving as the underlying causes of energy poverty and unaffordability. Consequently, the scarcity of energy impedes industrial progress, perpetuating diminished national income levels and heightened rates of unemployment.

Although suffering from a financial deficit brings about major restrictions to the energy sector as a capital-intensive industry, it can serve as a way out aligned with energy justice targets. It is interesting that some of the low-income and lower-middle-income countries have managed to harmonize potentials around energy justice targets and incorporate them for alleviating challenges in their energy system. For example, regarding rapid improvement in providing access to modern energy forms, four lower-middle-income countries (Nigeria, Angola, Honduras, and Kenya) are among the five top countries in 2021. Also, three lower-middle-income countries (Côte d'Ivoire, Lebanon, and Vietnam), and one low-income country (Mozambique) were among the five countries having

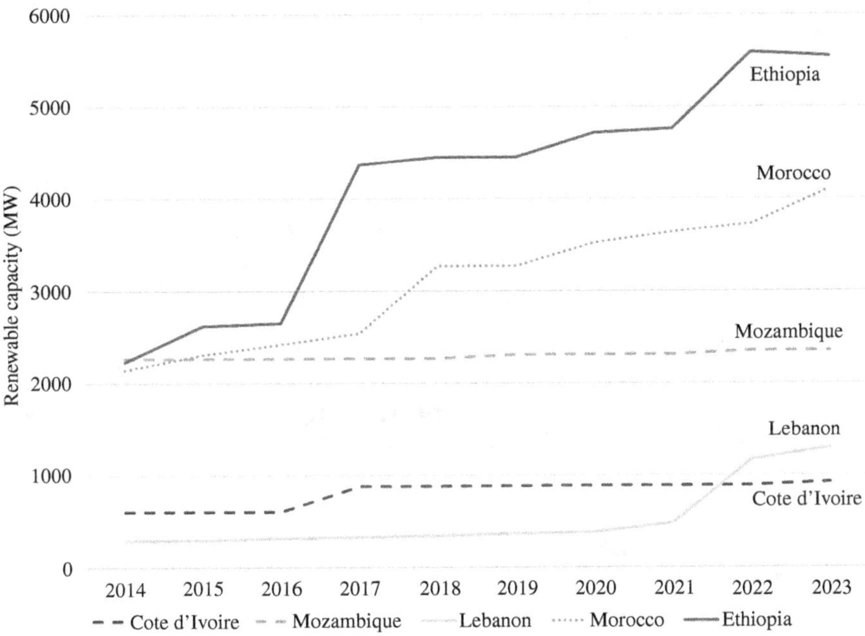

Figure 18.2 Total renewable energy capacity.

the fastest development in renewable energy sources in 2021. Similarly, regarding rapid improvement in energy efficiency, three lower-middle-income countries (Côte d'Ivoire, Senegal, and Zambia) ranked among the five countries with the best performance [24, 28]. Based on data available in [29], total renewable capacity in five low-income and lower-middle-income countries is shown in Figure 18.2. As can be seen in the figure, despite the financial challenges, these countries could include a noticeable amount of renewable energy capacity in their energy system.

India is another interesting case. In the year 2000, India's GDP per Capita stood at a mere 24% of the global average; however, by 2019, over the course of two decades, it had ascended to 43% of the global average. Notwithstanding this significant increase, India's GDP per capita remains notably below the global average [30]. As a lower-middle-income country, India could make considerable investments in solar and wind energy during the last decade and now stands at the fourth global rank in renewable installed capacity [29]. The mixture of renewable capacity in India based on the data presented in [29] is depicted in Figure 18.3. As the International Renewable Energy Agency (IRENA) explains, India successfully pioneered a novel model for low-carbon inclusive development [30].

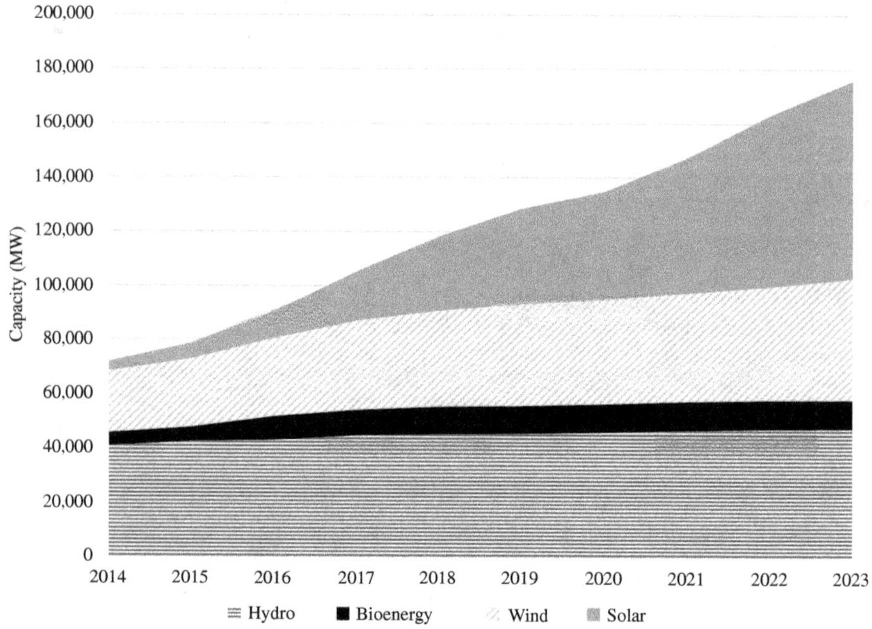

Figure 18.3 Indian renewable capacity portfolio.

18.5.3 Energy Justice in Developed Countries

At first glance, it may seem that developed countries and high-income countries have the privilege to follow the path to a carbon-free economy easily. Certainly, benefitting from a resilient economy with access to financial sources to launch large energy projects facilitates promoting energy justice and energy equity. However, due to the intrinsic complexity of energy justice and contradicting targets, having a sound financial status does not necessarily guarantee a smooth passage through the path of realizing energy justice. As a rationale for this claim, it can be pointed out that the energy justice concept originated majorly from a specific part of the developed Western world where academicians and institutions from high-income countries developed this concept to address some of the concerns of the energy systems [31].

It is not an exaggeration to claim that the most prominent advantage of developed countries is not their financial resources but their systematic structure and mature bureaucracy. As stated in the previous sections, orchestrated efforts and inclusive participation from different groups of individuals and entities are vital for the successful and lasting promotion of energy justice. Such harmony and social participation have roots in sociopolitical democracy, where people trust the central policies and take their responsibility seriously and positively respond to the call for active participation in serving energy justice. Regardless of financial status, one thing is certain: following the path toward energy justice imposes an immediate economic burden that can lead to higher consumer's costs in the short term. Prioritizing long-term benefits over short-term costs is a fundamental social capability that paves the way for big decisions.

Despite the aforementioned financial and social advantages, due to geographical distribution, many developed and high-income countries, especially European ones, struggle with low domestic natural resources and energy dependency which endangers their energy security that raises energy justice concerns.

In European Union (EU) countries, the energy dependency is tighter in the case of natural gas because unlike coal and oil, its storage and transportation are associated with many technical problems, and EU countries, to a great extent, rely on gas imports from Russia [32]. This dependency to gas import from Russia has brought energy insecurity concerns and, on several occasions, became the biggest challenge for these countries. In 2006 and 2009, Russia's gas export to Europe through Ukraine was cut off or largely reduced, resulting in supply inadequacy and high energy prices [33]. Followed by these two similar crises, "European Energy Security Strategy" was adopted in 2014 as a Communication from the Commission to the European Parliament and the Council with the aim of providing more robust energy security. Due to the significant reliance of several EU Member States on a sole gas supplier, namely Russia, the Commission aimed to collaborate with Member States, regulators, Transmission Systems Operators, and industry operators to enhance the Union's prompt readiness in anticipation of potential disruptions [34].

Later, the more recent and almost hardest case ever, is the corollary of Russian invasion of Ukraine on February 24, 2022. Before the invasion and in 2021, Russia was by far the largest energy supplier of the European countries who provided 36% of the EU's total natural gas imports, 46% of coal imports, and 25% of the EU's total petroleum oil imports [32]. As an aftermath of this war, in late 2021, natural gas price rose over 339$^{€}/_{MWh}$ from 50$^{€}/_{MWh}$ in the middle of that year which showed around 578% increase in less than a year. Such a large impact shows the EU's energy vulnerability. Triggered by this vulnerability and to contain Russian's ambitions for using its energy leverage to exert control over the EU's policy and economy, European Commission constantly took actions to diversify the energy sources within its member states as a mean to defend their autonomy. In this context, targeting less fossil-based energy industry through European Green Deal and the Paris Agreement, as well as expanding LNG import facilities, and integrating internal energy markets, accompanied by incorporating policies for moderating energy demand have been instrumental mechanisms to improve EU energy security [33, 34].

Although energy security is a major concern to the EU countries, they also share some other energy justice concerns with non-EU developed countries. The increase in GHG emissions has become a global problem for humankind and other species on the planet. The phenomenon of global warming and climate change, stemming from the emission of these gasses, has led to the melting of icebergs and the subsequent elevation of sea levels, posing a threat to the sustainability of human habitation along certain coastal regions worldwide. More frequent natural disasters such as hurricanes and wildfires, droughts, food and water scarcity, and health issues are other impacts of climate change that negatively affect people's quality of life in developed and high-income countries too [35]. For example, the financial impact of these extreme events on the US economy amounts to nearly \$150 billion annually, representing a conservative estimation that does not encompass some factors such as loss of human life, healthcare expenditures, or harm to ecosystem services [36]. Also, the analysis done by Canadian Climate Institute shows that in 2025 the Canadian economy is projected to be \$25 billion dollars less than its potential if the global warming had stabilized in 2015. Despite maintaining a rapid growth rate, the cumulative impact of this shortfall becomes increasingly significant over time [37]. Another study estimates the total economic cost to the UK economy something around 1.1% of national gross domestic product (GDP) and will reach 3.3% by 2050 and 7.4% by 2100. The study reports agriculture as the most affected sector by climate change [38].

Above examples imply that energy justice concerns are present concerns in the developed countries. Although the pattern of concerns in developed countries varies from the low-income ones. On the one hand, in low-income developing countries, challenging aspects of energy justice often stem from issues such as limited access to modern energy sources, inadequate infrastructure for just energy distribution, and disparities in energy access among different socioeconomic groups. Also, challenges related to environmental degradation and its health impacts resulting from low-energy efficiency and reliance on traditional and polluting energy sources are common among these countries.

On the other hand, in developed countries, challenging aspects of energy justice may include high energy prices for low-income households, energy efficiency, and sustainability concerns. Also, moving toward a low-carbon economy while ensuring a just transition for workers in traditional energy sectors can be considered as a socioeconomic challenge. Furthermore, as discussed in this section. Some of the developed countries may also struggle with issues of energy security and resilience.

18.5.4 International Attempts Regarding Energy Justice

As stated earlier, some of the concerns related to energy justice can be truly addressed without orchestrated efforts and combining worldwide potentials. The sustainability aspect of energy justice is the most global concern that affects all life on the planet. The inclusive nature of this concern necessitates international cooperation and in order to achieve the needed harmony, several international platforms have been developed to address sustainability and climate change. It is due to this fact that while other aspects of energy justice are also incorporated in international policy, they don't get so much emphasis as sustainability does.

Presently, a variety of international entities, encompassing both governmental and nongovernmental organizations, are actively engaged in diverse areas pertaining to the mitigation of climate change. Furthermore, certain international bodies have specifically delineated their mandate to combat climate change and enhance the sustainability of life on the Earth for both human and nonhuman entities. Additionally, select longstanding international organizations, originally established for different purposes, have incorporated sustainability concerns into their objectives. In the following passages, we are going to review some of the most meaningful attempts by international entities to put energy-justice-friendly policies in place.

The UN, as the most global international organization, has a special role to align capacities to fight climate change. In this context, the UN tries to be effective and influential through various mechanisms and initiatives as well as partnerships with other international bodies.

United Nations Framework Convention on Climate Change (UNFCCC) is a convention developed by the UN established in 1992 with the responsibility of supporting the global response to the threat of climate change. The convention with 198 members is almost universally adopted and has achieved two major successes during its time: the Kyoto Protocol in 1997 and the Paris Agreement in 2015.

Conference of the Parties (COP) is the high-level decision-making body of UNFCCC. COP is attended by all nations that have ratified the Convention. During the COP meetings, countries assess the progress made in implementing the Convention and any additional legal agreements adopted by the COP. Decisions are made to enhance the effective implementation of the Convention, which may include establishing new institutional and administrative structures. The first COP meeting was held in Germany in 1995. At COP15 of the UNFCCC in Copenhagen in 2009, developed countries committed themselves to mobilizing 100 billion dollars each year by 2020 dedicated to financing climate change projects in developing countries. [39].

Kyoto Protocol: Although it was approved in 1997 but due to a complex ratification process it didn't become effective until 2005. In short, it commits industrialized countries and economies to limit and reduce their GHG emissions. According to its principle, it is only binding for developed countries and recognizes them as largely responsible for the existing global warming problem since they have been the major source of GHG emissions. With this philosophy, the Kyoto Protocol defined binding reduction targets for GHG emissions of 37 industrialized countries and eventually resulted into consolidation of a market for trading carbon credits. Later, in 2012, the Doha amendment to the Kyoto Protocol was adopted [40, 41].

The Paris Agreement, also known as the Paris Accords or the Paris Climate Accords, is an international treaty on climate change that was adopted in 2015. This agreement covers multidimensional policies regarding climate change mitigation. Later, in COP26, the Glasgow Climate Pact was adopted in which the parties in the Paris agreement are required to expedite the energy transition process via gradually eliminating subsidies on fossil fuels and also reducing the usage of polluting coal power plants [23].

United Nations Environment Programme (UNEP) is a UN program that provides evidence-based data regarding policy decisions and facilitates cooperation for low-carbon transitions. It also provides finance mechanisms for climate actions and aids countries in adaptation climate policies and mitigation efforts [42].

The European Commission is another international organization that is tasked to harmonize the policies and actions of European countries. Since Europe has contributed greatly to the development of the concept of energy justice, the European Commission has developed guidelines and initiatives to realize both regional and global objectives of energy justice. As a more recent attempt of the Commission, the European Green Deal was approved by the European Commission in 2020 to set policy initiatives in order to make the EU climate-neutral in 2050. The European Commission has announced a series of measures aimed at aligning the EU's climate, energy, transport, and taxation policies with the goal of reducing net GHG emissions by at least 55% by 2030, as compared to levels in 1990 [43].

G20 is an intergovernmental forum composed of 19 countries as well as the EU and the African Union. This entity was founded in 1999 with the aim of economic collaboration. However, currently, it is taking a proactive role in supporting sustainable development and climate change mitigation. G20 has taken several major actions to respond to climate change. Some of these actions include setting emission reduction targets, promoting renewable energy sources, encouraging sustainable development practices, and supporting international agreements such as the Paris Agreement. Additionally, G20 has also focused on promoting green finance and investing in climate-resilient infrastructure [44].

18.6 Summary

Within this chapter, the discourse has centered on the critical necessity of a paradigm shift in policymaking to attain a satisfactory level of energy justice. Undoubtedly, the current state of energy justice within human societies directly reflects the global policies that have been implemented, notably influenced by the approaches and mindsets of policymakers. As expounded in this chapter, the interplay between energy justice initiatives and political frameworks can yield reciprocal impacts. Furthermore, the examination delves into how political dynamics within societies can either align with or diverge from energy justice objectives. Emphasis is placed on the potential conflicts between the political agendas and priorities of influential groups and the pursuit of energy justice and equity, highlighting the pivotal role of enhancing public awareness regarding the positive impacts of justice-oriented endeavors on quality of life in transitioning toward an equitable energy system. Moreover, recognizing that certain facets of energy justice necessitate coordinated global efforts, the imperative for international bodies to facilitate the development of universal policies is underscored.

References

1 Vitéz, B. and Lavrijssen, S. (2020). The energy transition: democracy, justice and good regulation of the heat market. *Energies* 13 (5): 1088.

2 Taufique, K.M.R. and Vaithianathan, S. (2018). A fresh look at understanding green consumer behavior among young urban Indian consumers through the lens of theory of planned behavior. *Journal of Cleaner Production* 183: 46–55.

3 International Energy Agency (2023). World energy balances 2023 highlights. https://www.iea.org/data-and-statistics/data-product/world-energy-statistics-and-balances (accessed 08 February 2024).

4 Pelau, C. and Acatrinei, C. (2019). The paradox of energy consumption decrease in the transition period towards a digital society. *Energies* 12 (8): 1428.

5 Valenzuela, E., Campbell, H., Montero, G. et al. (2021). Evaluation of home energy efficiency improvements in a hot desert climate in Northwestern Mexico: the energy saving vs. money saving conflict. *Energies* 14 (23): 7909.

6 Consumer, G. (2009). The impact of social factors and consumer behavior on carbon dioxide emissions in the United Kingdom. *Journal of Industrial Ecology* 14: 50–72.

7 Hossain, I., Fekete-Farkas, M., and Nekmahmud, M. (2022). Purchase behavior of energy-efficient appliances contribute to sustainable energy consumption in developing country: moral norms extension of the theory of planned behavior. *Energies* 15 (13): 4600.

8 Rahman, H.U., Zaman, U., and Górecki, J. (2021). The role of energy consumption, economic growth and globalization in environmental degradation: empirical evidence from the brics region. *Sustainability* 13 (4): 1924.

9 Rüdig, W. (1991). Green party politics around the world. *Environment: Science and Policy for Sustainable Development* 33 (8): 6–31.

10 Dolezal, M. (2016). The greens in Austria and Switzerland: two successful opposition parties. In: *Green Parties in Europe* (ed. E. Van Haute), 15–41. Routledge.

11 Ladner, A. (2020). Switzerland's successful green parties in the federal election of 20 October 2019: close to entering government? *Environmental Politics* 29 (3): 552–557.

12 Wavreille, M.C. and Pilet, J.B. (2016). The greens in Belgium's federal landscape: divergent fates. In: *Green Parties in Europe* (ed. E. Van Haute), 42–58. Routledge.

13 Bukow, S. (2016). The green party in Germany. In: *Green Parties in Europe* (ed. E. Van Haute), 112–139. Routledge.

14 Conradt, D. P. (2023). Green party of Germany. Encyclopedia Britannica. https://www.britannica.com/topic/Green-Party-of-Germany (accessed 11 April 2024).

15 Matthias, M. (2024). Green party of the United States. Encyclopedia Britannica. https://www.britannica.com/topic/Green-Party-of-the-United-States (accessed 13 April 2024).

16 Shah, A., Maken, A., Meng, T. et al. (2021). *Budgeting for Climate Change: A Guidance Note for Governments to Integrate Climate Change into Budgeting*. Bangkok: United Nations Development Programme. The Governance of Climate Change Finance Team.

17 Sovacool, B.K. and Dworkin, M.H. (2015). Energy justice: conceptual insights and practical applications. *Applied Energy* 142: 435–444.

18 Sari, R., Voyvoda, E., Lacey-Barnacle, M. et al. (2017). *Energy Justice: A Social Science and Humanities Cross-cutting Theme Report*. Cambridge: Shape Energy https://shapeenergy.eu/wp-content/uploads/2017/07/SHAPE-ENERGY_ThemeReports_ENERGY-JUSTICE.pdf.

19 Janicka, M., Sajnóg, A., and Sosnowski, T. (2021). Short-termism—the causes and consequences for the sustainable development of the financial markets. In: *Innovations and Traditions for Sustainable Development* (ed. E.V. Krasnov, W.L. Filho, and D.V. Gaeva), 485–501. Springer International Publishing.

20 Yergin, D. (2011). *The Prize: The Epic Quest for Oil, Money and Power*. Simon and Schuster.

21 Jenkins, K., McCauley, D., Heffron, R. et al. (2016). Energy justice: a conceptual review. *Energy Research and Social Science* 11: 174–182.

22 European Environment Agency (2024). Delivering justice in sustainability transitions. https://www.eea.europa.eu/publications/delivering-justice-in-sustainability-transitions#:~:text=Sustainability%20transitions%20can%20be%20considered (accessed 16 March 2024).

23 Asmelash, H. (2023). The role of International trade law in the energy transition. *The Journal of World Investment and Trade* 24 (6): 847–878.

24 The world Bank (2022). Regulatory indications for sustainable energy.(RISE). https://rise.esmap.org/ (accessed 10 April 2024).

25 European Comission (2018). Legal sources on renewable energy, renewable energy policy database and support. http://www.res-legal.eu/ (accessed 22 October 2021).

26 Abdmouleh, Z., Alammari, R.A., and Gastli, A. (2015). Review of policies encouraging renewable energy integration & best practices. *Renewable and Sustainable Energy Reviews* 45: 249–262.

27 Wijsman, K. and Berbés-Blázquez, M. (2022). What do we mean by justice in sustainability pathways? Commitments, dilemmas, and translations from theory to practice in nature-based solutions. *Environmental Science and Policy* 136: 377–386.

28 World Bank (2022). DataBank popular indicators. https://databank.worldbank.org/indicator/SP.POP.TOTL/1ff4a498/Popular-Indicators (accessed 28 January 2024).

29 IRENA (2024). Renewable capacity statistics 2024, International Renewable Energy Agency (IRENA). www.irena.org/Publications (accessed 10 April 2024).

30 IEA (2021). India energy outlook 2021. International Energy Agency (IEA). https://iea.blob.core.windows.net/assets/1de6d91e-e23f-4e02-b1fb-51fdd6283b22/India_Energy_Outlook_2021.pdf (accessed 29 March 2024).

31 Lacey-Barnacle, M., Robison, R., and Foulds, C. (2020). Energy justice in the developing world: a review of theoretical frameworks, key research themes and policy implications. *Energy for Sustainable Development* 55: 122–138.

32 Boehm, L. and Wilson, A. (2023). EU energy security and the war in Ukraine: from sprint to marathon. European Parliamentary Research Service (EPRS). https://www.europarl.europa.eu/thinktank/en/document/EPRS_BRI(2023)739362 (accessed 19 March 2024).

33 Ostrowski, W. (2022). The twenty years' crisis of European energy security: central and Eastern Europe and the US. *Geopolitics* 27 (3): 875–897.

34 Prisecaru, P. (2022). The war in Ukraine and the overhaul of EU energy security. *Global Economic Observer* 10 (1): 16–25.

35 United Nations (2024). Causes and effects of climate change. Climate Action. https://www.un.org/en/climatechange/science/causes-effects-climate-change (accessed 20 February 2024).

36 Jay, A.K., Crimmins, A.R., Avery, C.W. et al. (2023). Ch. 1. Overview: Understanding risks, impacts, and responses. In: *Fifth National Climate Assessment* (ed. A.R. Crimmins, C.W. Avery, D.R. Easterling, et al.). Washington, DC: U.S. Global Change Research Program https://doi.org/10.7930/NCA5.2023.

37 Beugin, D. and Sawyer, D. (2022). The GDP costs of climate change for Canada. Canadian Climate Institute. https://climateinstitute.ca/the-gdp-costs-of-climate-change-for-canada/#:~:text=Climate%20damages%20mean%20lower%20GDP&text=Our%20analysis%20suggests%20that%20by (accessed 26 March 2024).

38 Rising, J., Dietz, S., Dumas, M. et al. (2022). *What Will Climate Change Cost the UK? Risks, Impacts and Mitigation for the Net-Zero Transition*. London: Grantham Research Institute on Climate Change and the Environment.

39 The United Nations Climate Change (2024). Conference of the Parties (COP). https://unfccc.int/process/bodies/supreme-bodies/conference-of-the-parties-cop (accessed 19 March 2024).

40 The United Nations Climate Change (2024). What is the Kyoto Protocol? https://unfccc.int/kyoto_protocol (accessed 19 March 2024).

41 Bodnar, Z., Lanzarini, C.A., and de Mesquita, T.B. (2024). Transnational climate justice and the carbon credit market. *Beijing Law Review* 15: 828.

42 The United Nations (2024). Climate action. https://www.unep.org/topics/climate-action (accessed 19 March 2024).

43 The European Commission (2024). The European green deal. https://commission.europa.eu/strategy-and-policy/priorities-2019-2024/european-green-deal_en (accessed 18 March 2024).

44 G20 (2024). Task force for the global mobilization against climate change. https://g20.gov.br/en/tracks/sherpa-track/climate-change (accessed 02 February 2024).

19

Challenges of Energy Justice and Equity in Different Regions

Iman Rahmati[1] *and Asghar Akbari Foroud*[2]

[1] *Electricity Market Monitoring & Studies, Iran Grid Management Company, Tehran, Iran*
[2] *Electrical and Computer Engineering Faculty, Semnan University, Semnan, Iran*

Acronyms

CAD	Canadian dollar
CO_2	carbon dioxide
CO_{2e}	carbon dioxide equivalent
GDP	gross domestic product
GHG	greenhouse gas
IEA	International Energy Agency
IRENA	International Renewable Energy Agency
MENA	Middle East and North Africa
US	The United States of America
USD	United States dollar

19.1 Introduction

Energy justice is a worldwide concern that exists everywhere and all the countries in the world of one sort or another need to address it actively. Even the impacts of those aspects of energy justice that may seem local, such as energy access and energy poverty, far exceed the territory of a single country. For instance, lack of access to modern energy will force people to use solid fuels for heating and household cooking, and this leads to very high greenhouse gas (GHG)[1] and polluting emissions, which are thread to global warming.

1 Greenhouse gas

Communities for Clean Energy Justice and Equity in Grid Modernization, First Edition.
Edited by Mohammadreza Daneshvar, Behnam Mohammadi-Ivatloo, and Amjad Anvari-Moghaddam.

On the other hand, challenges regarding promoting energy justice are different from one place to another. Here in this chapter, we will have a look at the current state of energy justice in various parts of the world and discuss the hardships to promote it.

19.2 Energy Justice and Equity in Europe

19.2.1 Key Features of Europe

Europe is known for its diverse cultural heritage, advanced economies, high standards of living, and well-established democratic institutions. Overall, Europe is seen as a prosperous and influential region with a strong commitment to democratic values, human rights, and social welfare. From the viewpoint of energy industry, western and eastern parts of Europe have significant differences.

The western region of the continent includes countries such as France, Germany, the United Kingdom, Italy, Spain, Belgium, the Netherlands, and others. It is characterized by a high level of industrialization, technological innovation, and a strong emphasis on education, healthcare, and environmental sustainability.

Europe hosts a varied group of energy producers and consumers with extensive connections to global markets. The variation is also evident in the energy systems, policies, and resources of each nation. In fact, there is significant disparity among European countries in terms of their energy mix, encompassing both consumption and production structures [1].

Fossil fuels continue to be the primary source in the energy systems, with varying degrees of diversity across the region. Natural gas is significant in North-western European nations and Italy, owing to historical domestic production that facilitated the development of extensive pipeline networks. Coal remains a crucial component in the energy frameworks of Central and Eastern European countries such as Germany, the Czech Republic, and Poland. Conversely, countries like France, Norway, Sweden, and Switzerland have adopted a less carbon-intensive energy mix, largely supported by nuclear and hydroelectric power sources [1, 2].

Following Russia's invasion of Ukraine, the region's energy sector has undergone significant transformations. The conflict sparked an energy crisis, leading to record high prices and prompting European countries to prioritize energy security on their political agendas. This shift resulted in a substantial reduction in fuel imports from Russia, previously a key source. Additionally, European nations have heightened their clean energy ambitions, aiming to diversify their energy sources and make progress toward climate goals [2].

Thanks to extensive supportive initiatives, renewable energy deployment has reached unprecedented levels. However, further efforts are necessary to address persistent energy challenges, including strengthening clean energy supply chains, upgrading outdated infrastructure, and advancing energy system integration across the region.

19.2.2 Challenges of Energy Justice in Europe

Notwithstanding the economic prosperity and technological advancement, Europe, and especially the western part, has been facing one major challenge regarding having firm access to energy carriers. Historically, industrial evolution commenced in Western Europe by utilizing coal to harness steam power. But only a small portion of coal reserves are located in Europe. According to [3], in 2021, only 3.3% of worldwide coal production came from Europe. Later in the mid-20th century and particularly after World War II, oil products became the dominant source of energy. Yet, the major parts of the oil have been produced out of Europe, and European countries have been ranked among major global oil buyers. European countries on the western side of the continent had to rely on imports to meet their increasing energy demand.

The significant reliance on energy imports is historically viewed as a security issue. Within Europe, the extensive imports of Russian gas and crude oil have consistently been regarded as a potential risk to the region's energy security [1]. In 2021, oil products accounted for more than 30% of total energy consumption in Europe, and more than 72% of total crude oil supply came from import channels [4]. A similar pattern exists for natural gas: while it includes around 27% of Europe's energy supply, more than 57% of it is imported [5].

This dependence on energy imports leaves Europe vulnerable to supply disruptions and geopolitical tensions that could threaten its energy security. The energy security issue is a threat to some other justice-related targets, too. For example, due to high reliance on energy import, any major disruption in energy import high fluctuation in energy prices are likely to happen which entrenches access to energy and energy affordability. Moreover, as some researchers estimate, annually more than 41,000 people die of cold only in the United Kingdom, and in the whole Europe more than two hundred thousand lives are lost due to cold winters. Such a high number of death attributable to cold can be expressed as a major energy equity and energy poverty issue [6, 7].

In order to alleviate the energy vulnerability, the European Union's energy security strategy aims to diversify its energy sources, suppliers, and transit routes to reduce this vulnerability. This includes promoting renewable energy sources, improving energy efficiency, investing in infrastructure such as interconnectors and storage facilities, and enhancing cooperation with neighboring countries and international partners [8].

The ongoing conflict between Russia and Ukraine, as well as tensions with other energy suppliers in the Middle East, have underscored the importance of strengthening Europe's energy security. By reducing its reliance on a few dominant suppliers and increasing its energy independence, Europe can better protect itself against potential disruptions and ensure a stable and sustainable energy supply for its citizens and economy.

19.3 Energy Justice and Equity in Africa

19.3.1 Key Features of Africa

Africa is a continent rich in energy resources, with vast potential for both traditional and renewable energy sources. In 10 out of 54 countries in Africa, oil and gas account for more than 60% of total annual export value. Thanks to natural fossil fuel endowment, these countries have higher gross domestic product (GDP)[2] per capita compared to other African countries [9]. Nigeria, Algeria, Angola, and Libya rank among the top 20 largest oil producers globally, with Algeria, Egypt, and Nigeria also being recognized as part of the top 20 largest producers of natural gas in the world [10]. Presently, a multitude of significant energy corporations (including Shell, BP, Total, ExxonMobil, Chevron, ENI, among others), alongside numerous independent, junior entities, and even certain public (primarily Asian) firms, engage in oil and gas exploration and production activities within Sub-Saharan Africa, frequently through collaborative partnerships [11]. In addition to oil, a number of African nations, such as Botswana, Mozambique, South Africa, and Zimbabwe, are significant producers of coal, with South Africa being the largest among them [12].

The continent receives abundant sunlight, has strong wind resources, and is home to several major rivers and geothermal hotspots. Africa is endowed with significant potential for solar power generation on a global scale. The continent experiences an average annual solar irradiation of 2119 kWh/m^2, with the majority of countries in North, West, and Southern Africa receiving an average exceeding 2100 kWh/m^2 annually. According to estimates by the International Renewable Energy Agency (IRENA)[3], Africa's solar technical potential is calculated at 7900 GW, highlighting extensive opportunities for solar power generation. In the case of wind power, according to IRENA's analysis, the technical capacity for wind power generation is substantial, standing at a notable 461 GW when considering a 1% land utilization factor. Among the countries assessed, Algeria, Ethiopia, Namibia, and Mauritania are identified as possessing the most significant potential in this regard [9].

2 Gross Domestic Product
3 International Renewable Energy Agency

However, apart from hydropower, contemporary renewable energy sources such as solar, wind, geothermal, and modern bioenergy make a relatively small contribution to Africa's energy portfolio. Africa's share of the global installed capacity for electricity generation based on renewables is less than 3% [9]. By harnessing these renewable energy sources, African countries can diversify their energy mix, reduce GHG emissions, and improve energy security.

Although around 25% of the world's population lives in Africa, only 6% of global energy demand is associated with African countries, and only 3% of electricity is consumed in Africa. In fact, more than half of the population without access to electricity live in Sub-Saharan Africa [13, 14].

19.3.2 Challenges of Energy Justice in Africa

Despite the abundant natural energy sources, energy poverty and lack of access to modern energy forms have been a major challenge regarding energy justice throughout Africa. Lots of people in Africa do not have access to electricity, and many more rely on traditional biomass for cooking and heating. This energy poverty hinders economic development, limits social progress, and exacerbates environmental and health problems. To be specific, only less than 27% of the population in the whole continent has access to modern energy sources. More than 690 million people in Sub-Saharan Africa suffer from a deficit in access to clean cooking fuel and largely rely on unhealthy solid fuels for cooking. As for electricity, by the end of 2019, 570 million people had no access to electricity of which 37% lived in three countries: the Democratic Republic of the Congo, Ethiopia, and Nigeria [9, 14].

Apart from energy poverty, a high level of economic injustice exists within African societies. Even in the oil and gas exporting African countries where there is higher GDP per capita, higher national revenue has not led to better life quality for all and a persistent income inequality is observed.

As explained in Section 19.3.1, some of the African countries highly rely on the revenue from exporting oil and gas. Since this export revenue accounts for a large portion of annual revenue, these countries are highly vulnerable to disruptions in the fossil fuel markets and price fluctuations. The resultant fluctuations in government revenues pose a significant threat to the long-term socioeconomic stability of these nations. This instability in the revenue hinders the African governments' financial ability to constantly invest in energy sector expansion, which is desperately needed to provide people with an acceptable level of access to modern energies and diminish energy poverty.

Another challenge in Africa's energy sector is the reliance on fossil fuels, particularly coal and oil, for electricity generation. This dependence not only contributes to carbon emissions and air pollution but also makes the electricity sector of African countries vulnerable to price fluctuations and supply disruptions in the global energy market.

Moreover, Gender inequality in energy access is a significant challenge, particularly in Sub-Saharan Africa, where the burden of energy poverty disproportionately falls on women. Studies consistently demonstrate that women experience the impacts of energy poverty more acutely than men, and this disparity is amplified in rural areas [15].

In recent years, there has been a growing interest and investment in renewable energy projects in Africa, supported by initiatives such as the African Renewable Energy Initiative (AREI)[4] and the Sustainable Energy for All (SEforALL) initiative. These efforts aim to accelerate the deployment of renewable energy technologies, increase energy access, and promote sustainable development across the continent. However, it's worth mentioning that a considerable number of policymakers in African nations hold the belief that oil and gas resources will generate a revenue stream capable of revitalizing their economy and breaking the cycle of poverty prevalent in their countries. They are of the opinion that oil and gas will not only create employment opportunities but also facilitate access to food, education, healthcare, agricultural assistance, and housing [11]. Combined with a natural fossil-based endowment, this mindset that can be consistent with the short-term economic benefits may give an advantage to a fossil-based development approach and hinder renewable development projects.

19.4 Energy Justice and Equity in East Asia

19.4.1 Key Features of East Asia

The energy sector in East Asia is characterized by a mix of traditional and renewable energy sources, as well as a growing focus on energy efficiency and sustainability. Countries in East Asia have rapidly growing economies, leading to increased demand for energy. Based on the data published in [16] and as can be seen in Figure 19.1, except for Japan, the other countries have experienced almost constant rapid increase in their total annual energy consumption. The odd case of Japan is extra interesting, where despite 12.5% increase in GDP from 2000 to 2021[5] (12.5% increase), there has been more than 20% decline in energy use, showing considerable improvement in the Energy Intensity Index [17]. Oil, coal, and natural gas are the prominent energy sources in Japan and account for 38.7%, 27.4%, and 21.2% of the total energy supply, respectively [18]. Over the past two decades, Japan has extensively invested on renewable energy sources

4 For more information please visit: https://www.arei.info/eng
5 4000 in 2000 to 4500 billion dollars in 2021

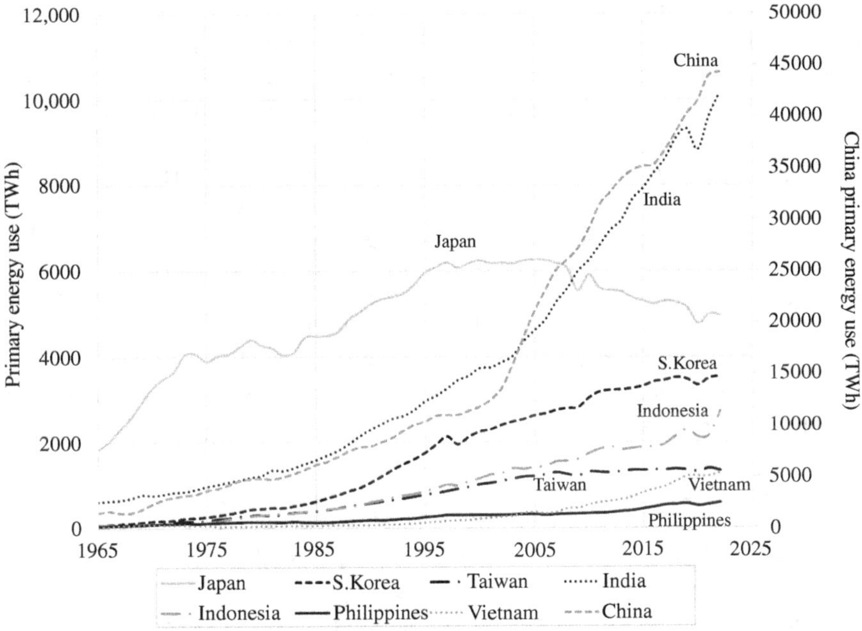

Figure 19.1 Trend of primary energy use in East Asian countries.

and around 22.6% of annual electric energy comes from these energy sources. The major part of renewable electricity in Japan is generated via solar panels and hydro power plans and these two energy sources account for 47% and 46% of total renewable electricity respectively. Share of wind power in renewable electricity is below 5% [19].

On the other hand, China is the largest energy consumer in the region, relying heavily on coal for power generation. However, the country has also been investing heavily in renewable energy sources, such as wind and solar, in an effort to reduce its carbon footprint.

Similar to Japan, South Korea is a highly industrialized country with an advanced energy infrastructure. While Japan has been increasing its reliance on renewable energy following the Fukushima nuclear disaster in 2011, South Korea has been investing in nuclear and renewable energy to diversify its energy mix.

The energy sector in India is diverse, and as can be seen in Figure 19.1, rapidly growing. It encompasses a mix of traditional sources such as coal, oil, and natural gas, as well as renewable sources such as wind, solar, and hydropower. India is one of the largest consumers of energy in the world, and the demand for energy continues to increase as the country's population and economy grow. The Indian government has made significant investments in renewable energy in recent years and now stands at the fourth global rank in renewable installed capacity [20]. This has led to the rapid expansion of solar and wind power capacity in India.

Other East Asian countries, including Taiwan, Philippines, Indonesia, and Vietnam, have limited domestic energy resources and rely heavily on imported fossil fuels. These countries have been working to increase its use of renewable energy and improve energy efficiency to reduce its dependence on imports.

19.4.2 Challenges of Energy Justice in East Asia

Almost all countries in East Asia have vast disparities in access to energy resources, with rural and marginalized communities often lacking access to reliable and affordable energy sources. Energy security is a common concern among the East Asian countries, which they intend to address by rapid growth on investment in renewable capacities.

Moreover, the rapid industrialization and urbanization in East Asia have led to severe environmental degradation, including air and water pollution, deforestation, and habitat destruction. This disproportionately affects marginalized communities that often bear the brunt of these environmental impacts.

Energy poverty is another challenge regarding energy equity. Many people in East Asia struggle to afford basic energy services, such as electricity and clean cooking fuels. This leads to health problems, economic hardship, and social inequality. Despite rapid economic growth in China and India, still a large population lives in energy poverty, particularly in rural areas. These two countries also face environmental degradation and air pollution issues due to heavily fossil-based energy systems and immense emissions [21, 22].

Climate change impacts is another major problem in this region. East Asia is particularly vulnerable to the impacts of climate change, including extreme weather events, sea-level rise, and food insecurity. Marginalized communities are often the most affected by these impacts, further exacerbating existing inequalities in access to energy and resources [13].

19.5 Energy Justice and Equity in Central Asia

19.5.1 Key Features of Central Asia

The energy sector in Central Asia is characterized by a mix of traditional and renewable energy sources, as well as a heavy reliance on fossil fuels such as oil and natural gas. The region is rich in natural resources, particularly in countries like Kazakhstan, Turkmenistan, and Uzbekistan that are major producers of oil and gas. Based on data published by the International Energy Agency (IEA)[6]

6 International Energy Agency.

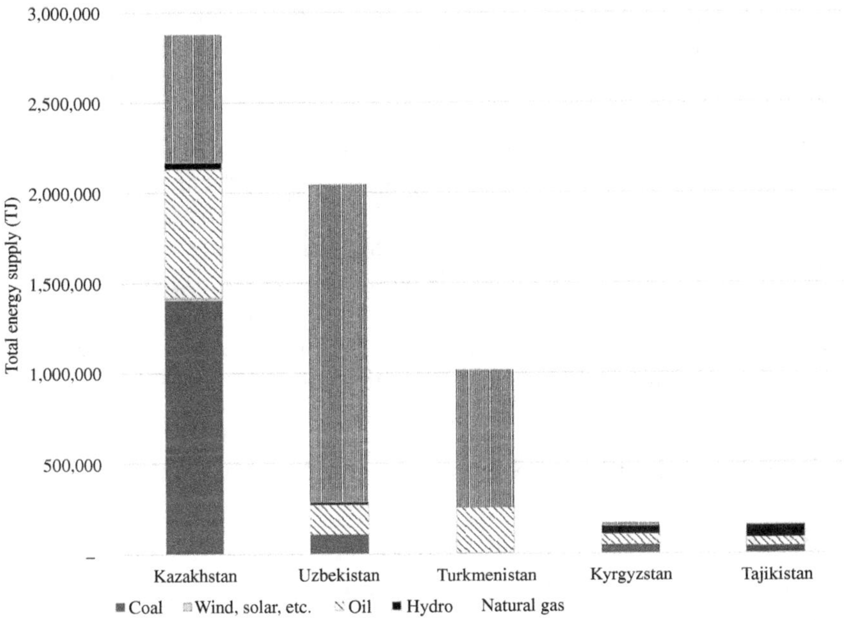

Figure 19.2 Total energy supply in 2021 of Central Asian countries.

[23–27], Figure 19.2 depicts the energy mix in Central Asian countries, and as is seen in this figure, the energy portfolios of the three largest are highly dominated by fossil-based energy forms.

Central Asia also has significant potential for renewable energy development, particularly in countries like Kyrgyzstan and Tajikistan, which have abundant hydropower resources. However, investment in renewable energy projects has been limited, and as stated earlier, the region continues to rely heavily on fossil fuels for energy production.

19.5.2 Challenges of Energy Justice in Central Asia

Heavily reliance on traditional fossil fuels is a common challenge in the territory, which underscores the need for promoting sustainability-driven initiatives. Moreover, in these countries, energy access and energy affordability indices experienced degradation in the past decade, and a considerable portion of the rural populations in these countries lack firm and affordable access to modern energy forms [28].

Apart from the abovementioned common problems, Kyrgyzstan relies on imported energy resources for more than 40% of its total energy supply [23]. The imported energy is mainly in the form of natural gas and electricity, which makes the country vulnerable to price fluctuations and supply disruptions. This dependence on imported energy sources can also limit the country's energy security and sovereignty. Similarly but less severely, Tajikistan with more than 30% reliance on energy imports suffers from energy security issues and vulnerability to disruptions in energy markets [24].

Kazakhstan has set ambitious goals to decrease GHG emissions and boost the proportion of renewable energy in electricity generation. Under an initiative introduced in 2013 known as the Green Economy Concept, Kazakhstan has set a goal to produce 50% of its electricity from "alternative or renewable" sources by 2050, which may encompass nuclear energy. Additionally, the plan includes targets for a 15% decrease in carbon dioxide (CO_2) emissions from electricity generation by 2030 and a 40% decrease by 2050. Yet, the nation's existing reliance on abundant, low-cost coal presents a challenge in meeting these goals [29].

19.6 Energy Justice and Equity in the Middle East

19.6.1 Key Features of the Middle East

The Middle East is known for its abundance of oil and natural gas reserves, making the region a major player in the global energy market. The Middle East is home to some of the world's largest oil-producing countries, such as Saudi Arabia, Iraq, and the United Arab Emirates, as well as significant natural gas reserves in countries like Qatar and Iran. Based on the data published by IEA [30], Figure 19.3 illustrates the energy supply portfolio of the Middle East, which clearly is super-dominated by oil and gas. Oil and gas production are the primary drivers of the economy in many Middle Eastern countries, providing a significant portion of government revenue and supporting various industries.

The Middle East produces almost 95% of its electricity from natural gas and oil, making it the highest share globally. Countries in the region also possess some of the world's most abundant solar resources, and at the same time, this region is at high risk for the devastating impacts of climate change [31]. That's why the region is also investing in renewable energy sources, such as solar and wind power, to diversify its energy mix and reduce dependence on fossil fuels.

19.6.2 Challenges of Energy Justice in the Middle East

Although countries may vary in their characteristics, they share common challenges, such as a youthful population with high unemployment rates, limited research capabilities, growing electricity demand, and insufficient investments in energy [32].

Furthermore, high subsidized energy tariffs are very common among the countries in the Middle East. In Egypt, Yemen, and Morocco, the subsidies on fuel are

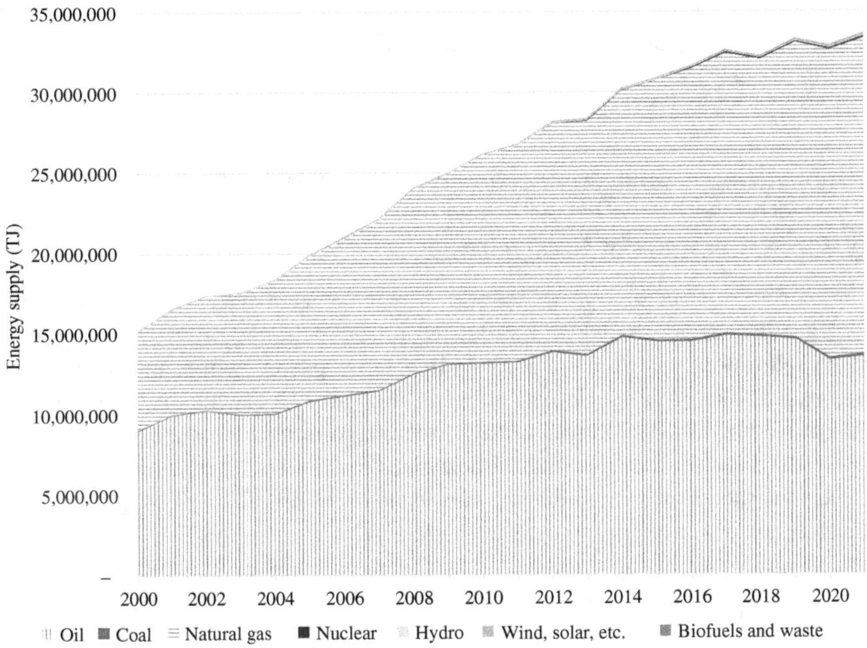

Figure 19.3 Total energy supply by category in the Middle East.

more than three times greater than the government's expenditure on health [17], and the Iranian government spends more than 36% of GDP on energy subsidies, which happens to be the highest share in the whole world [33]. High subsidy on fossil-based energy negatively impacts consumers' behavior and diminishes the willingness to invest in efficiency. Despite claiming to protect low-income families, energy subsidies are often more beneficial to wealthier individuals who own larger homes, multiple vehicles, and have higher energy-intensive lifestyles, while leaving poorer communities without access to affordable energy.

As mentioned earlier, the countries within the Middle East are highly dependent on the oil and gas exporting revenue, which makes them vulnerable to political instability in the region, geopolitical tensions, and anything else that leads to fluctuations in global oil prices. For instance, when in 2016, the average global oil price fell to 43 United States dollar (USD) per barrel, not a single country in the Middle East and North Africa (MENA) region was able to achieve fiscal breakeven [31].

Additionally, water scarcity and environmental concerns pose challenges to sustainable energy development in the region. Furthermore, as the shift to clean energy diminishes the global need for fossil fuels, producer economies in the region will have to discover fresh revenue streams. Consequently, several producers are formulating strategies to establish low-carbon energy sectors, utilizing their energy proficiency to broaden their economies and energy portfolios simultaneously [30].

19.7 Energy Justice and Equity in South America

19.7.1 Key Features of South America

The energy sector in South America shows a high level of diversity and it's rapidly evolving. The region is rich in natural resources, including oil, natural gas, and renewable energy sources such as hydroelectric power, wind, and solar. The energy portfolio of this continent is shown in Figure 19.4, which is based on the data published in [34]. Many countries in South America have significant oil and gas reserves, with countries like Venezuela, Brazil, and Argentina being major players in the petroleum industry. However, there are also growing efforts to develop renewable energy sources in the region, with countries like Chile, Brazil, and Uruguay leading the way in the adoption of clean energy technologies already covering 27%, 45%, and 55% of their final energy use from renewable resources respectively [35, 36].

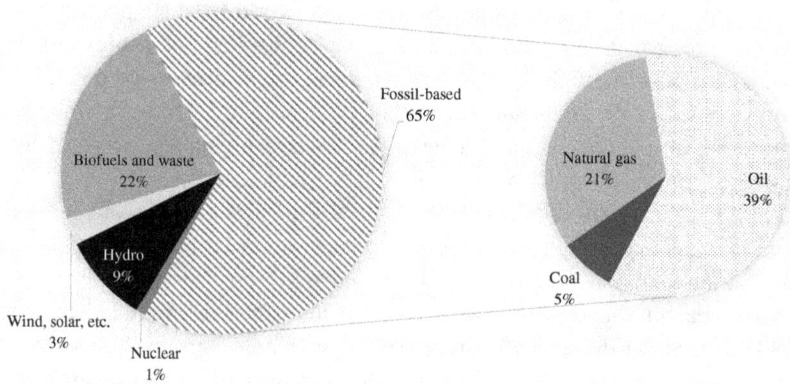

Figure 19.4 Energy mix in South America in 2021.

Although as is seen in Figure 19.4 share of hydro in the total energy portfolio is less than 10%, some of the countries in the region, such as Brazil with 80% generation from hydropower, benefit from Amazon as a huge source for hydroelectricity. While being a large world-class oil and gas supplier, Brazil is the second-largest producer of biofuels [37].

Overall, the energy sector in South America is characterized by a mix of traditional fossil-fuel-based energy production and a growing emphasis on renewable energy sources. The region faces challenges such as political instability, regulatory hurdles, and environmental concerns, but there is also great potential for growth and innovation in the sector.

19.7.2 Challenges of Energy Justice in South America

Regarding universal access to electricity In Latin America, advancements have been achieved in, thanks to government initiatives like Brazil's "Luz para Todos" program. Over the past two decades, this program has played a key role in reducing the number of people in Brazil without electricity access by 95%. However, approximately 3% of the population in Latin America and the Caribbean (17 million people) still lack access to essential electricity services, a situation that has shown no signs of improvement for nearly a decade. Additionally, concerns persist regarding the quality and reliability of service even for those who do have access [38].

Unequal access to energy carriers is one of the prominent challenges regarding energy justice. As Figure 19.5 depicts that many communities in South America, especially in rural and indigenous areas, do not have access to reliable and affordable energy sources. In some local communities and indigenous people who live

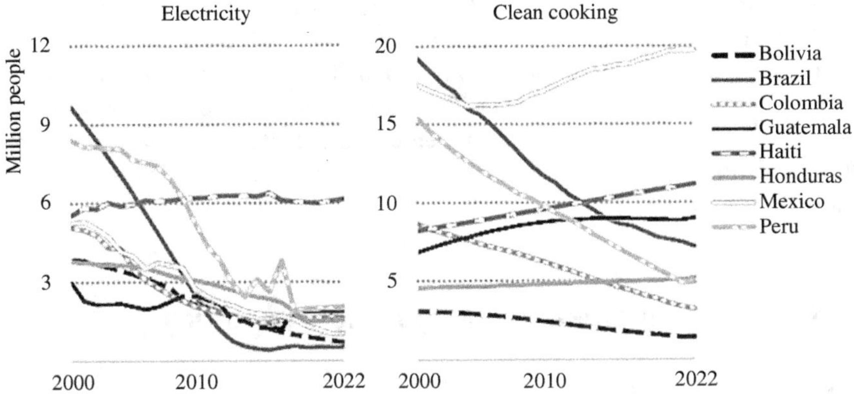

Figure 19.5 Population without access to electricity and cleaning cooking in Latin America. Source: [40]/INTERNATIONAL ENERGY AGENCY/CC by 4.0. (This figure is originally produced by IEA under License: CC BY 4.0.)

next to large renewable facilities such as solar, wind, and hydroelectric do not have access to energy or have to bear high prices. Such a condition evidently violates the distributional element of energy justice [39].

Besides electricity, the region's advancements in achieving access to clean cooking have been significantly impeded by the global energy crisis, which has resulted in higher natural gas prices. Approximately 11% of the Latin American and Caribbean population lack access to clean cooking facilities. Notably, the number of individuals without access to clean cooking has been on the rise in Mexico, Haiti, Honduras, and Guatemala [38]. The lack of clean cooking fuel notably contributes to a decline in indoor air quality, which affects the most vulnerable segments of the population, particularly women and children.

This lack of access to energy can exacerbate existing inequalities and limit economic opportunities for these communities. High energy costs disproportionately impact low-income households in South America, leading to energy poverty and limiting access to basic services such as electricity and heating. Addressing energy affordability issues is crucial for achieving energy justice in the region.

Furthermore, although Latin American countries rely on renewables for over 25% of their primary energy, which is double the global average, this shift toward renewable energy comes with a steep cost, as there have been numerous allegations of human rights abuses linked to these projects. In fact, the development of energy projects in South America can result in land displacement, loss of livelihoods, and violations of indigenous rights, raising concerns about social justice and equity. As Business & Human Rights Resource Centre reports, from January 2010 to September 2020, 501 instances of abuse, involving more than 2300 allegations of human rights violations, have been documented, committed by 156 renewable energy companies, as well as private and state banks, operating in 17 countries throughout Latin America [39]. Latin America has the highest number of abuses in renewable energy development globally, with 61% of all recorded allegations, and this number is increasing. Mexico and Central America bear the brunt of abuse, accounting for 68% of all documented cases. The countries with the highest reported instances of alleged abuse are Honduras at 28%, Mexico at 27%, and Colombia at 13% [40].

19.8 Energy Justice and Equity in North America

19.8.1 Key Features of North America

The energy sector in North America is characterized by a diverse mix of energy sources, including fossil fuels, renewable energy, and nuclear power. The region is a major producer and consumer of energy, with the United States, Canada, and Mexico being key players in the sector. Technical advancement to exploit shale gas and shale oil in 2008 in the United States has been a substantial breakthrough in the petroleum industry and transformed the United States from a net importer to the largest oil and gas producer in the world, putting Saudi Arabia in second place [41]. This has launched a series of major shifts in the oil and gas industry and its impact on the whole world was not less than to the United States itself [42]. As is seen in Figure 19.6, which is based on the date published by [17, 43], total energy demand in the United States in 2022 was almost as much as in 2000, while a sharp increase in GDP (more than 52%), which is due to success in energy improvement programs and technology innovations.

North America has abundant reserves of oil, natural gas, and coal, which have historically been the dominant sources of energy in the region. However, there is a growing emphasis on transitioning toward cleaner and more sustainable energy sources, such as wind, solar, hydroelectric, and geothermal power [44].

The energy sector in North America is highly interconnected, with energy infrastructure, such as pipelines, transmission lines, and power plants, crossing national borders to facilitate the movement of energy resources. This interconnectedness

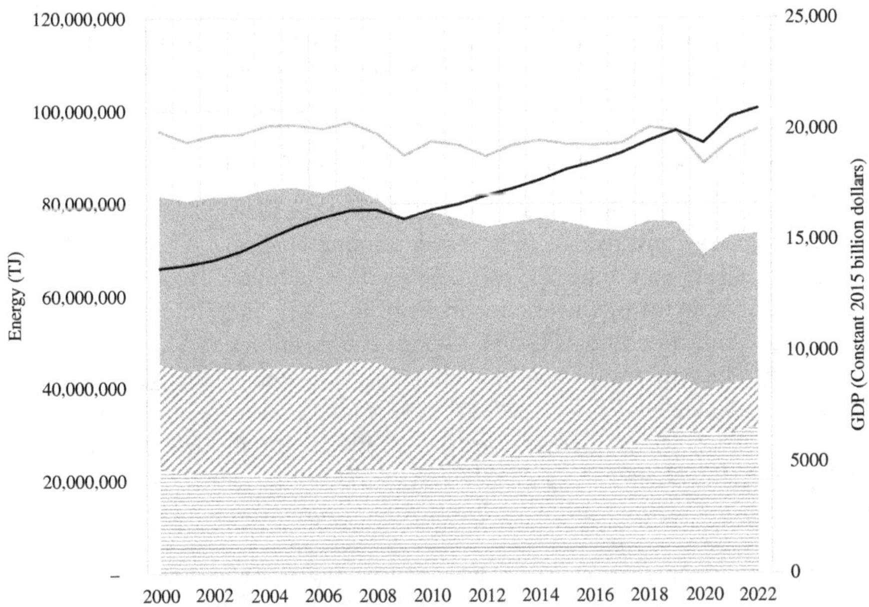

Figure 19.6 Fossil energy supply by the source, energy demand, and GDP of the United States.

allows for the efficient exchange of energy between countries and regions, contributing to energy security and stability in the region.

Regulatory frameworks and policies governing the energy sector vary across North America, with each country having its own set of rules and regulations. There is also a growing focus on energy efficiency, environmental sustainability, and reducing GHG emissions in the region, driving investments in clean energy technologies and infrastructure. However, after Donald Trump took the presidential office in the United States, the perspective of the United States regarding Climate Change changed unexpectedly, since President Trump saw it as a Chinese hoax and exited from the Paris Agreement[7] [42]. Consequently, sustainability-driven initiatives and funds became less effective in the United States. Overall, the energy sector in North America is evolving to meet the challenges of a changing climate, shifting energy landscape, and growing energy demand, with a growing emphasis on diversifying energy sources, increasing energy efficiency, and promoting sustainability.

Canada, another major country in the region, has established a challenging goal to reduce GHG emissions by 40–45% from 2005 levels by 2030 and achieve net zero emissions by 2050, while more than 75% of the total energy supply has been from fossil fuels, namely, natural gas and oil [45]. The Canadian government is allocating up to $750 million Canadian dollar (CAD)[8] to establish a new emissions reduction fund aimed at assisting workers and lowering emissions within Canada's oil and gas industry, with a specific emphasis on methane [46].

19.8.2 Challenges of Energy Justice in North America

Although numerous energy assistance programs in the United States are in place to combat energy poverty, the bulk of the funding is allocated toward offering immediate bill support rather than addressing the underlying causes of the issue. According to a report by the US Department of Health and Human Services in 2016, low-income households had an average energy burden of 18.4% during the fiscal year 2014. Any energy bills exceeding 6–11% of a household's annual gross income are considered unaffordable [47]. Yet the total amount of funding allocated for energy assistance programs is insufficient to support all low-income households in the United States [48]. Residential energy consumption survey (RECS)[9] study conducted by the US Energy Information Administration (EIA) [10]showed

7 The Paris Agreement is a legally binding international treaty on climate change. It was adopted by 196 Parties at the UN Climate Change Conference (COP21) in Paris, France, on December 12, 2015.

8 Canadian dollar

9 Residential Energy Consumption Survey.

10 U.SUS. Energy Information Administration.

that 2.1% of low-income families had to seek for medical attention due to lack of access to insufficient cooling/heating, which is a sign of social and economic disparity in energy access [47].

Moreover, while public agencies and utilities strive to move toward a sustainable energy future, a significant number of programs that advocate for energy efficiency, rooftop solar, electric vehicles, and home energy storage remain largely out of reach for low-income households, and this translates to a pervasive and persistent disparity in access to modern energy forms [49]. In other words, while there is a growing shift toward renewable energy sources in North America, not all communities have equal access to clean energy technologies. This can limit the benefits of clean energy, such as reduced emissions and lower energy costs, for marginalized communities.

To address the latter issue, in the 2023 Budget, the US Department of Agriculture has earmarked 300 million dollars to aid rural electricity providers in shifting toward cleaner energy sources. Additionally, 6.5 billion dollars has been allocated for rural electric loans. These financial resources will be utilized to fund projects facilitating the transition to clean energy, encompassing initiatives such as the expansion of renewable energy and energy storage, updates to transmission infrastructure, and the generation of employment opportunities [50].

In terms of contribution to fighting climate change and emission abatement, the United States has undergone various transformations that bolstered the nation's energy security and gradually reduced its dependency on fossil fuels to fulfill energy requirements. Nonetheless, as is described in [41], present patterns in GHG emissions indicate that climate change mitigation objectives may not be achieved. Should the present trajectories persist, the Climate Action Tracker predicts that the United States' CO_{2e} emission in 2030 will be approximately 800–900 $MTCO_{2e}$ higher than its defined target, 4148 $MTCO_{2e}$ [51].

Canada, as one of the world's major oil and gas producers, possesses significant proven reserves, with the sector playing a crucial role in both contributing to and facilitating economic growth. Canadian government is dealing with a two-face dilemma: it's a democratic nation that relies on exporting hydrocarbons for growth and meeting domestic energy needs, and simultaneously secure ambitious GHG emission reductions. While there are public and international pressures on emission abatement, "Canada seems incapable of kicking its oil and gas addiction" as Climate Action Tracker puts it [41, 52].

19.9 Summary

Throughout this chapter, we analyzed the current situation of the energy sector in different parts of the world. In every region, the energy sector plays a noticeable role in economic prosperity, social equity, political stability, and inhabitant's quality of life. Energy justice includes several frontiers such as energy security, affordability, equality, sustainability, etc., and in every part of our world the battle against injustice and inequity is ongoing. Due to different geographical characteristics and natural endowment, each region may face a different combination of challenges, but reviewing the chapter's different sections proves that despite the differences, the similarity of challenges is evident throughout the world. This chapter emphasizes that contrary to popular belief, and despite all the technological advancement, even in the 21st century and even in modern and advanced economies, vulnerable groups and marginal communities suffer from energy poverty and unaffordable energy prices. The chapter's findings suggest that despite the mobilization of public opinion toward combating climate change, challenges may arise in terms of enacting legislation and executing sustainability-focused initiatives. The intricate nature of the energy justice concept implies that pursuing a singular target without consideration for potential negative impacts on other aspects is a risk. This scenario is exemplified in Latin America, where instances of indigenous rights violations in favor of large companies operating in the renewable energy sector have been observed. Furthermore, the text underscores that possessing natural resources does not automatically ensure energy equity or access to affordable and clean energy. Presently, numerous countries with substantial oil and gas reserves encounter significant obstacles in addressing energy poverty, reducing disparities, and ensuring universal access to modern energy sources for their populations.

References

1 Hafner, M. and Raimondi, P.P. (2022). Energy and the economy in Europe. In: *The Palgrave Handbook of International Energy Economics* (ed. M. Hafner and G. Luciani). Springer Nature.

2 IEA (2022). Energy system of Europe. International Energy Agency. https://www.iea.org/regions/europe (accessed 8 May 2024).

3 IEA (2022). Coal supply in Europe. International Energy Agency. https://www.iea.org/regions/europe/coal (accessed 4 May 2024).

4 IEA (2022). Oil supply in Europe. International Energy Agency. https://www.iea.org/regions/europe/oil (accessed 4 May 2024).

5 IEA (2022). Natural gas supply in Europe. International Energy Agency. https://www.iea.org/regions/europe/natural-gas (accessed 4 May 2024).

6 Masselot, P., Mistry, M., Vanoli, J. et al. (2023). Excess mortality attributed to heat and cold: a health impact assessment study in 854 cities in Europe. *The Lancet Planetary Health* 7 (4): e271–e281.

7 Macintyre, H.L., Heaviside, C., Cai, X. et al. (2021). The winter urban heat Island: impacts on cold-related mortality in a highly urbanized European region for present and future climate. *Environment International* 154: 106530.
8 Mišík, M. (2022). The EU needs to improve its external energy security. *Energy Policy* 165: 112930.
9 IRENA, A. (2022) Renewable energy market analysis: Africa and its regions. International Renewable Energy Agency and African Development Bank, Abu Dhabi and Abidjan.
10 International Energy Agency, (2023). World energy balances 2023 highlights. https://www.iea.org/data-and-statistics/data-product/world-energy-statistics-and-balances.
11 Copinschi, P. (2022). Energy and the economy in Sub-Saharan Africa. In: *The Palgrave Handbook of International Energy Economics* (ed. M. Hafner and G. Luciani). Springer Nature.
12 IEA, (2019). Africa energy outlook 2019, International Energy Agency, Paris, www.iea.org/reports/africa-energyoutlook.
13 IEA, I., UNSD, World Bank, WHO, Tracking SDG 7: The Energy Progress Report. 2023, World Bank: Washington DC. License: Creative Commons Attribution—NonCommercial 3.0 IGO (CC BY-NC 3.0 IGO).
14 International Energy Agency, I.E.A. and W. Bank, (2014) Sustainable Energy for All 2013–2014: Global Tracking Framework Report. The World Bank.
15 Mang-Benza, C., Jodoin, L., Doubogan, Y.O. et al. (2023). Making energy justice work for women in rural Sub-Saharan Africa: a qualitative diagnostic from Benin, Senegal, and Togo. *Energy Policy* 173: 113345.
16 Ritchie, H., P. Rosado, and M. Roser, (2024). Energy production and consumption, OurWorldInData.org, https://ourworldindata.org/energy-production-consumption.
17 World Bank (2022). Data bank popular indicators.
18 IEA (2023). Energy supply in Japan. International Energy Agency. https://www.iea.org/countries/japan/energy-mix (accessed 4 May 2024).
19 IEA (2023). Share of renewables in energy consumption in Japan. International Energy Agency. https://www.iea.org/countries/japan/renewables (accessed 4 May 2024).
20 International Renewable Energy Agency (IRENA) (2024). Renewable capacity statistics 2024. www.irena.org/Publications.
21 Khanna, R.A., Li, Y., Mhaisalkar, S. et al. (2019). Comprehensive energy poverty index: measuring energy poverty and identifying micro-level solutions in South and Southeast Asia. *Energy Policy* 132: 379–391.
22 Igawa, M. and Managi, S. (2022). Energy poverty and income inequality: an economic analysis of 37 countries. *Applied Energy* 306: 118076.

23 IEA (2022). Energy supply in Kyrgyzstan. International Energy Agency. https://www.iea.org/countries/Kyrgyzstan/energy-mix (accessed 4 May 2024).

24 IEA (2022). Energy supply in Tajikistan. International Energy Agency. https://www.iea.org/countries/Tajikistan/energy-mix (accessed 4 May 2024).

25 IEA (2022). Energy supply in Uzbekistan. International Energy Agency. https://www.iea.org/countries/Uzbekistan/energy-mix (accessed 4 May 2024).

26 IEA (2022). Energy supply in Kazakhstan. International Energy Agency. https://www.iea.org/countries/Kazakhstan/energy-mix (accessed 4 May 2024).

27 IEA (2022). Energy supply in Turkmenistan. International Energy Agency. https://www.iea.org/countries/Turkmenistan/energy-mix (accessed 4 May 2024).

28 Che, X., Zhu, B., and Wang, P. (2021). Assessing global energy poverty: an integrated approach. *Energy Policy* 149: 112099.

29 IEA (2022). Kazakhstan 2022 energy sector review. International Energy Agency, https://doi.org/10.1787/73d1d69f-en.

30 IEA (2022). Energy system of Middle East. International Energy Agency. https://www.iea.org/regions/middle-east (accessed 5 May 2024).

31 Nathaniel, S., Anyanwu, O., and Shah, M. (2020). Renewable energy, urbanization, and ecological footprint in the Middle East and North Africa region. *Environmental Science and Pollution Research* 27 (13): 14601–14613.

32 Saidi, H., G.E. Montasser, and N. Ajmi,(2018). Renewable energy, quality of institutions and economic growth in MENA countries, Ecole Supérieure de Commerce de Tunis, https://mpra.ub.uni-muenchen.de/84055/.

33 IEA (2022). Value of fossil-fuel subsidies by fuel in the top 25 countries, 2022, IEA, Paris https://www.iea.org/data-and-statistics/charts/value-of-fossil-fuel-subsidies-by-fuel-in-the-top-25-countries-2022, IEA. Licence: CC BY 4.0.

34 IEA (2022). Energy supply in central and South America. International Energy Agency. https://www.iea.org/regions/central-south-america/energy-mix (accessed 4 May 2024).

35 IEA (2022). Share of renewables in energy consumption in Chile. International Energy Agency. https://www.iea.org/countries/chile/renewables (accessed 4 May 2024).

36 IEA (2022). Share of renewables in energy consumption in Uruguay. International Energy Agency. https://www.iea.org/countries/uruguay/renewables (accessed 4 May 2024).

37 IEA (2023). Energy supply in Brazil. International Energy Agency. https://www.iea.org/countries/brazil (accessed 4 May 2024).

38 IEA (2023). Latin America energy outlook 2023. International Energy Agency: https://iea.blob.core.windows.net/assets/1055131a-8dc4-488b-9e9e-7eb4f72bf7ad/LatinAmericaEnergyOutlook.pdf, Licence: CC BY 4.0.

39 Vázquez, K.H. and C. Hodgkins, (2021). Renewable energy (in)justice in Latin America. Business and Human Rights Resource Centre, https://media.business-humanrights.org/media/documents/RE_LATAM_final_English.pdf.

40 Business and Human Rights Resources Centre (2020). Renewable energy and human rights benchmark. https://www.business-humanrights.org/en/from-us/briefings/renewable-energy-human-rights-benchmark/.

41 Chow, E.C. and Rincon-Romero, J.-A. (2022). Energy and the economy in North America. In: *The Palgrave Handbook of International Energy Economics* (ed. M. Hafner and G. Luciani). Springer Nature.

42 Yergin, D. (2020). *The New Map: Energy, Climate, and the Clash of Nations.* Penguin Uk.

43 IEA (2023). Energy system of United States. International Energy Agency. https://www.iea.org/countries/united-states (accessed 4 May 2024).

44 IEA (2022). Energy system of North America. International Energy Agency. https://www.iea.org/regions/north-america (accessed 4 May 2024).

45 IEA (2023). Energy system of Canada. International Energy Agency. https://www.iea.org/countries/canada (accessed 4 May 2024).

46 IEA (2022). Emissions reduction fund. International Energy Agency. https://www.iea.org/policies/11476-emissions-reduction-fund.

47 Xu, X. and Chen, C.-f. (2019). Energy efficiency and energy justice for US low-income households: an analysis of multifaceted challenges and potential. *Energy Policy* 128: 763–774.

48 Semega, J., Fontenot, K., Kollar, M. et al., (2017). Income and Poverty in the United States: 2016. Current population reports, (P60–259).

49 Brown, M.A., Soni, A., Lapsa, M.V. et al. (2019). Low-income energy affordability in an era of US energy abundance. *Progress in Energy* 1 (1): 012002.

50 IEA, (2022). Rural clean energy initiative and support to rural energy transition. International Energy Agency: https://www.iea.org/policies/16609-rural-clean-energy-initiative-support-to-rural-energy-transition.

51 Climate Action Tracker (2023). Country summary (USA). https://climateactiontracker.org/countries/usa/ (accessed 8 May 2024).

52 Climate Action Tracker (2023). Country summary (Canada). https://climateactiontracker.org/countries/canada/ (accessed 8 May 2024).

Index

Communities for Clean Energy Justice and Equity in Grid Modernization, First Edition.
Edited by Mohammadreza Daneshvar, Behnam Mohammadi-Ivatloo, and Amjad Anvari-Moghaddam.

f

g

h

i

j

k

l

m

n

o

p

IEEE Press Series on Power and Energy Systems

Series Editor: Ganesh Kumar Venayagamoorthy, Clemson University, Clemson, South Carolina, USA.

The mission of the IEEE Press Series on Power and Energy Systems is to publish leading-edge books that cover a broad spectrum of current and forward-looking technologies in the fast-moving area of power and energy systems including smart grid, renewable energy systems, electric vehicles and related areas. Our target audi- ence includes power and energy systems professionals from academia, industry and government who are interested in enhancing their knowledge and perspectives in their areas of interest.

1. *Electric Power Systems: Design and Analysis, Revised Printing*
   Mohamed E. El-Hawary

2. *Power System Stability*
   Edward W. Kimbark

3. *Analysis of Faulted Power Systems*
   Paul M. Anderson

4. *Inspection of Large Synchronous Machines: Checklists, Failure Identification, and Troubleshooting*
   Isidor Kerszenbaum

5. *Electric Power Applications of Fuzzy Systems*
   Mohamed E. El-Hawary

6. *Power System Protection*
   Paul M. Anderson

7. *Subsynchronous Resonance in Power Systems*
   Paul M. Anderson, B.L. Agrawal, and J.E. Van Ness

8. *Understanding Power Quality Problems: Voltage Sags and Interruptions*
   Math H. Bollen

9. *Analysis of Electric Machinery*
   Paul C. Krause, Oleg Wasynczuk, and S.D. Sudhoff

10. *Power System Control and Stability, Revised Printing*
    Paul M. Anderson and A.A. Fouad

11. *Principles of Electric Machines with Power Electronic Applications, Second Edition*
    Mohamed E. El-Hawary
12. *Pulse Width Modulation for Power Converters: Principles and Practice*
    D. Grahame Holmes and Thomas Lipo
13. *Analysis of Electric Machinery and Drive Systems, Second Edition*
    Paul C. Krause, Oleg Wasynczuk, and S.D. Sudhoff
14. *Risk Assessment for Power Systems: Models, Methods, and Applications*
    Wenyuan Li
15. *Optimization Principles: Practical Applications to the Operations of Markets of the Electric Power Industry*
    Narayan S. Rau
16. *Electric Economics: Regulation and Deregulation*
    Geoffrey Rothwell and Tomas Gomez
17. *Electric Power Systems: Analysis and Control*
    Fabio Saccomanno
18. *Electrical Insulation for Rotating Machines: Design, Evaluation, Aging, Testing, and Repair*
    Greg C. Stone, Edward A. Boulter, Ian Culbert, and Hussein Dhirani
19. *Signal Processing of Power Quality Disturbances*
    Math H.J. Bollen and Irene Y.H. Gu
20. *Instantaneous Power Theory and Applications to Power Conditioning*
    Hirofumi Akagi, Edson H. Watanabe, and Mauricio Aredes
21. *Maintaining Mission Critical Systems in a 24/7 Environment*
    Peter M. Curtis
22. *Elements of Tidal-Electric Engineering*
    Robert H. Clark
23. *Handbook of Large Turbo-Generator Operation and Maintenance, Second Edition*
    Geoff Klempner and Isidor Kerszenbaum
24. *Introduction to Electrical Power Systems*
    Mohamed E. El-Hawary

25. *Modeling and Control of Fuel Cells: Distributed Generation Applications*
    M. Hashem Nehrir and Caisheng Wang

26. *Power Distribution System Reliability: Practical Methods and Applications*
    Ali A. Chowdhury and Don O. Koval

27. *Introduction to FACTS Controllers: Theory, Modeling, and Applications*
    Kalyan K. Sen and Mey Ling Sen

28. *Economic Market Design and Planning for Electric Power Systems*
    James Momoh and Lamine Mili

29. *Operation and Control of Electric Energy Processing Systems*
    James Momoh and Lamine Mili

30. *Restructured Electric Power Systems: Analysis of Electricity Markets with Equilibrium Models*
    Xiao-Ping Zhang

31. *An Introduction to Wavelet Modulated Inverters*
    S.A. Saleh and M.A. Rahman

32. *Control of Electric Machine Drive Systems*
    Seung-Ki Sul

33. *Probabilistic Transmission System Planning*
    Wenyuan Li

34. *Electricity Power Generation: The Changing Dimensions*
    Digambar M. Tagare

35. *Electric Distribution Systems*
    Abdelhay A. Sallam and Om P. Malik

36. *Practical Lighting Design with LEDs*
    Ron Lenk and Carol Lenk

37. *High Voltage and Electrical Insulation Engineering*
    Ravindra Arora and Wolfgang Mosch

38. *Maintaining Mission Critical Systems in a 24/7 Environment, Second Edition*
    Peter Curtis

39. *Power Conversion and Control of Wind Energy Systems*
    Bin Wu, Yongqiang Lang, Navid Zargari, and Samir Kouro

40. *Integration of Distributed Generation in the Power System*
    Math H. Bollen and Fainan Hassan

41. *Doubly Fed Induction Machine: Modeling and Control for Wind Energy Generation Applications*
    Gonzalo Abad, Jesús López, Miguel Rodrigues, Luis Marroyo, and Grzegorz Iwanski

42. *High Voltage Protection for Telecommunications*
    Steven W. Blume

43. *Smart Grid: Fundamentals of Design and Analysis*
    James Momoh

44. *Electromechanical Motion Devices, Second Edition*
    Paul Krause Oleg, Wasynczuk, and Steven Pekarek

45. *Electrical Energy Conversion and Transport: An Interactive Computer-Based Approach, Second Edition*
    George G. Karady and Keith E. Holbert

46. *ARC Flash Hazard and Analysis and Mitigation*
    J.C. Das

47. *Handbook of Electrical Power System Dynamics: Modeling, Stability, and Control*
    Mircea Eremia and Mohammad Shahidehpour

48. *Analysis of Electric Machinery and Drive Systems, Third Edition*
    Paul C. Krause, Oleg Wasynczuk, S.D. Sudhoff, and Steven D. Pekarek

49. *Extruded Cables for High-Voltage Direct-Current Transmission: Advances in Research and Development*
    Giovanni Mazzanti and Massimo Marzinotto

50. *Power Magnetic Devices: A Multi-Objective Design Approach*
    S.D. Sudhoff

51. *Risk Assessment of Power Systems: Models, Methods, and Applications, Second Edition*
    Wenyuan Li

52. *Practical Power System Operation*
    Ebrahim Vaahedi

53. *The Selection Process of Biomass Materials for the Production of Bio-Fuels and Co-Firing*
    Najib Altawell

54. *Electrical Insulation for Rotating Machines: Design, Evaluation, Aging, Testing, and Repair, Second Edition*
Greg C. Stone, Ian Culbert, Edward A. Boulter, and Hussein Dhirani

55. *Principles of Electrical Safety*
Peter E. Sutherland

56. *Advanced Power Electronics Converters: PWM Converters Processing AC Voltages*
Euzeli Cipriano dos Santos Jr. and Edison Roberto Cabral da Silva

57. *Optimization of Power System Operation, Second Edition*
Jizhong Zhu

58. *Power System Harmonics and Passive Filter Designs*
J.C. Das

59. *Digital Control of High-Frequency Switched-Mode Power Converters*
Luca Corradini, Dragan Maksimoviæ, Paolo Mattavelli, and Regan Zane

60. *Industrial Power Distribution, Second Edition*
Ralph E. Fehr, III

61. *HVDC Grids: For Offshore and Supergrid of the Future*
Dirk Van Hertem, Oriol Gomis-Bellmunt, and Jun Liang

62. *Advanced Solutions in Power Systems: HVDC, FACTS, and Artificial Intelligence*
Mircea Eremia, Chen-Ching Liu, and Abdel-Aty Edris

63. *Operation and Maintenance of Large Turbo-Generators*
Geoff Klempner and Isidor Kerszenbaum

64. *Electrical Energy Conversion and Transport: An Interactive Computer-Based Approach*
George G. Karady and Keith E. Holbert

65. *Modeling and High-Performance Control of Electric Machines*
John Chiasson

66. *Rating of Electric Power Cables in Unfavorable Thermal Environment*
George J. Anders

67. *Electric Power System Basics for the Nonelectrical Professional*
Steven W. Blume

68. *Modern Heuristic Optimization Techniques: Theory and Applications to Power Systems*
Kwang Y. Lee and Mohamed A. El-Sharkawi

69. *Real-Time Stability Assessment in Modern Power System Control Centers*
    Savu C. Savulescu

70. *Optimization of Power System Operation*
    Jizhong Zhu

71. *Insulators for Icing and Polluted Environments*
    Masoud Farzaneh and William A. Chisholm

72. *PID and Predictive Control of Electric Devices and Power Converters Using MATLAB®/Simulink®*
    Liuping Wang, Shan Chai, Dae Yoo, Lu Gan, and Ki Ng

73. *Power Grid Operation in a Market Environment: Economic Efficiency and Risk Mitigation*
    Hong Chen

74. *Electric Power System Basics for Nonelectrical Professional, Second Edition*
    Steven W. Blume

75. *Energy Production Systems Engineering Thomas*
    Howard Blair

76. *Model Predictive Control of Wind Energy Conversion Systems*
    Venkata Yaramasu and Bin Wu

77. *Understanding Symmetrical Components for Power System Modeling*
    J.C. Das

78. *High-Power Converters and AC Drives, Second Edition*
    Bin Wu and Mehdi Narimani

79. *Current Signature Analysis for Condition Monitoring of Cage Induction Motors: Industrial Application and Case Histories*
    William T. Thomson and Ian Culbert

80. *Introduction to Electric Power and Drive Systems*
    Paul Krause, Oleg Wasynczuk, Timothy O'Connell, and Maher Hasan

81. *Instantaneous Power Theory and Applications to Power Conditioning, Second Edition*
    Hirofumi, Edson Hirokazu Watanabe, and Mauricio Aredes

82. *Practical Lighting Design with LEDs, Second Edition*
    Ron Lenk and Carol Lenk

83. *Introduction to AC Machine Design*
    Thomas A. Lipo

84. *Advances in Electric Power and Energy Systems: Load and Price Forecasting*
Mohamed E. El-Hawary

85. *Electricity Markets: Theories and Applications*
Jeremy Lin and Jernando H. Magnago

86. *Multiphysics Simulation by Design for Electrical Machines, Power Electronics and Drives*
Marius Rosu, Ping Zhou, Dingsheng Lin, Dan M. Ionel, Mircea Popescu, Frede Blaabjerg, Vandana Rallabandi, and David Staton

87. *Modular Multilevel Converters: Analysis, Control, and Applications*
Sixing Du, Apparao Dekka, Bin Wu, and Navid Zargari

88. *Electrical Railway Transportation Systems*
Morris Brenna, Federica Foiadelli, and Dario Zaninelli

89. *Energy Processing and Smart Grid*
James A. Momoh

90. *Handbook of Large Turbo-Generator Operation and Maintenance, 3rd Edition*
Geoff Klempner and Isidor Kerszenbaum

91. *Advanced Control of Doubly Fed Induction Generator for Wind Power Systems*
Dehong Xu, Dr. Frede Blaabjerg, Wenjie Chen, and Nan Zhu

92. *Electric Distribution Systems, 2nd Edition*
Abdelhay A. Sallam and Om P. Malik

93. *Power Electronics in Renewable Energy Systems and Smart Grid: Technology and Applications*
Bimal K. Bose

94. *Distributed Fiber Optic Sensing and Dynamic Rating of Power Cables*
Sudhakar Cherukupalli and George J. Anders

95. *Power System and Control and Stability, Third Edition*
Vijay Vittal, James D. McCalley, Paul M. Anderson, and A.A. Fouad

96. *Electromechanical Motion Devices: Rotating Magnetic Field-Based Analysis and Online Animations, Third Edition*
Paul Krause, Oleg Wasynczuk, Steven D. Pekarek, and Timothy O'Connell

97. *Applications of Modern Heuristic Optimization Methods in Power and Energy Systems*
Kwang Y. Lee and Zita A. Vale

98. *Handbook of Large Hydro Generators: Operation and Maintenance*
Glenn Mottershead, Stefano Bomben, Isidor Kerszenbaum, and Geoff Klempner

99. *Advances in Electric Power and Energy: Static State Estimation*
Mohamed E. El-hawary

100. *Arc Flash Hazard Analysis and Mitigation, Second Edition*
J.C. Das

101. *Maintaining Mission Critical Systems in a 24/7 Environment, Third Edition*
Peter M. Curtis

102. *Real-Time Electromagnetic Transient Simulation of AC-DC Networks*
Venkata Dinavahi and Ning Lin

103. *Probabilistic Power System Expansion Planning with Renewable Energy Resources and Energy Storage Systems*
Jaeseok Choi and Kwang Y. Lee

104. *Power Magnetic Devices: A Multi-Objective Design Approach, Second Edition*
Scott D. Sudhoff

105. *Optimal Coordination of Power Protective Devices with Illustrative Examples*
Ali R. Al-Roomi

106. *Resilient Control Architectures and Power Systems*
Craig Rieger, Ronald Boring, Brian Johnson, and Timothy McJunkin

107. *Alternative Liquid Dielectrics for High Voltage Transformer Insulation Systems: Performance Analysis and Applications*
Edited by U. Mohan Rao, I. Fofana, and R. Sarathi

108. *Introduction to the Analysis of Electromechanical Systems*
Paul C. Krause, Oleg Wasynczuk, and Timothy O'Connell

109. *Power Flow Control Solutions for a Modern Grid using SMART Power Flow Controllers*
Kalyan K. Sen and Mey Ling Sen

110. *Power System Protection: Fundamentals and Applications*
John Ciufo and Aaron Cooperberg

111. *Soft-Switching Technology for Three-phase Power Electronics Converters*
Dehong Xu, Rui Li, Ning He, Jinyi Deng, and Yuying Wu

112. *Power System Protection, Second Edition*
Paul M. Anderson, Charles Henville, Rasheek Rifaat, Brian Johnson, and Sakis Meliopoulos

113. *High Voltage and Electrical Insulation Engineering, Second Edition*
Ravindra Arora and Wolfgang Mosch

114. *Modeling and Control of Modern Electrical Energy Systems*
Masoud Karimi-Ghartemani

115. *Control of Power Electronic Converters with Microgrid Applications*
Armdam Ghosh and Firuz Zare

116. *Coordinated Operation and Planning of Modern Heat and Electricity Incorporated Networks*
Mohammadreza Daneshvar, Behnam Mohammadi-Ivatloo, and Kazem Zare

117. *Smart Energy for Transportation and Health in a Smart City*
Chun Sing Lai, Loi Lei Lai, and Qi Hong Lai

118. *Wireless Power Transfer: Principles and Applications*
Zhen Zhang and Hongliang Pang

119. *Intelligent Data Mining and Analysis in Power and Energy Systems: Models and Applications for Smarter Efficient Power Systems*
Zita Vale, Tiago Pinto, Michael Negnevitsky, and Ganesh Kumar Venayagamoorthy

120. *Introduction to Modern Analysis of Electric Machines and Drives*
Paul C. Krause and Thomas C. Krause

121. *Electromagnetic Analysis and Condition Monitoring of Synchronous Generators*
Hossein Ehya and Jawad Faiz

122. *Transportation Electrification: Breakthroughs in Electrified Vehicles, Aircraft, Rolling Stock, and Watercraft*
Ahmed A. Mohamed, Ahmad Arshan Khan, Ahmed T. Elsayed, and Mohamed A. Elshaer

123. *Modular Multilevel Converters: Control, Fault Detection, and Protection*
Fuji Deng, Chengkai Liu, and Zhe Chen

124. *Stability-Constrained Optimization for Modern Power System Operation and Planning*
Yan Xu, Yuan Chi, and Heling Yuan

125. *Interval Methods for Uncertain Power System Analysis*
Alfredo Vaccaro

126. *Practical Partial Discharge Measurement on Electrical Equipment*
Greg Stone, Andrea Cavallini, Glenn Behrmann, and Claudio Angelo Serafino

127. *Graph Database and Graph Computing for Power System Analysis*
Renchang Dai and Guangyi Liu

128. *The Power of Artificial Intelligence for the Next-Generation Oil and Gas Industry: Envisaging AI-inspired Intelligent Energy Systems and Environments*
Pethuru Raj Chelliah, Venkatraman Jayasankar, Mats Agerstam, B. Sundaravadivazhagan, and Robin Cyriac

129. *Microgrids: Theory and Practice*
Peng Zhang

130. *Electric Machinery and Drives: An Electromagnetics Perspective*
Nabeel A. O. Demerdash, JiangBiao He, and Hao Chen

131. *Smart Cyber-Physical Power Systems, Volume 1: Fundamental Concepts, Challenges, and Solutions*
Ali Parizad, Hamid Reza Baghaee, and Saifur Rahman

132. *Smart Cyber-Physical Power Systems, Volume 2: Solutions from Emerging Technologies*
Ali Parizad, Hamid Reza Baghaee, and Saifur Rahman

133. *Analysis of Electric Machinery and Drive Systems, Fourth Edition*
Paul C. Krause, Oleg Wasynczuk, Scott D. Sudhoff, and Steven D. Pekarek

134. *Understanding Electromagnetic Transients in Power System*
Luiz Cera Zanetta Jr.

135. *Communities for Clean Energy Justice and Equity in Grid Modernization*
Mohammadreza Daneshvar, Behnam Mohammadi-Ivatloo, and Amjad Anvari-Moghaddam

www.ingramcontent.com/pod-product-compliance
Lightning Source LLC
LaVergne TN
LVHW011533200926
848484LV00001B/259
9781394265718